Ueberhuber · Numerical Com

T0250587

Springer
Berlin
Heidelberg
New York
Barcelona
Budapest
Hong Kong
London
Milan
Paris
Santa Clara
Singapore
Tokyo

Christoph W. Ueberhuber

NUMERICAL COMPUTATION 2

Methods, Software, and Analysis

With 73 Figures

 Springer

Christoph W. Ueberhuber

Technical University of Vienna
Wiedner Hauptstrasse 8–10/115
A-1040 Vienna
Austria

e-mail: cwu@uranus.tuwien.ac.at

Title of the German original edition: *Computer-Numerik 2.* Published
by Springer, Berlin Heidelberg 1995.

ROBERT LETTNER created the picture on the cover of this book. He is the person in charge of the central printing and reprography office at the Academy of
Applied Arts in Vienna, Austria. In the past few years Robert Lettner, inspired
by advancements in digital image processing, has developed new forms of
aesthetics.
Sequences of pictures recently created by Robert Lettner include
Das Spiel vom Kommen und Gehen (Coming and Going),
*Dubliner Thesen zur informellen Geometrie (Dublin Theses on Informal
Geometry),*
Bilder zur magischen Geometrie (Magical Geometry).
The illustration on the cover of this book was taken from the sequence
Das Spiel vom Kommen und Gehen.

Mathematics Subject Classification: 65-00, 65-01, 65-04, 65Dxx, 65Fxx,
65Hxx, 65Y10, 65Y15, 65Y20

Library of Congress Cataloging-in-Publication Data

Ueberhuber, Christoph W. 1946-
[Computer Numerik. English] Numerical computation : methods, software, and
analysis /
Christoph W. Ueberhuber.   p.   cm.
Includes bibliographical references and index.
ISBN 3-540-62058-3 (v. 1 : soft : acid-free paper). - -
ISBN 3-540-62057-5 (v. 2 : soft : acid-free paper)
1. Numerical analysis - - Data processing. I. Title.
QA297.U2413 1997   96-46772
519.4' 0285'53 - -dc21   CIP

ISBN 3-540-62057-5 Springer-Verlag Berlin Heidelberg New York

Typesetting: By the author using LaTeX.
SPIN 10543822   41/3143 – 5 4 3 2 1 0 – Printed on acid-free paper

Preface

This book deals with various aspects of scientific numerical computing. No attempt was made to be complete or encyclopedic. The successful solution of a numerical problem has many facets and consequently involves different fields of computer science. Computer numerics—as opposed to computer algebra—is thus based on applied mathematics, numerical analysis and numerical computation as well as on certain areas of computer science such as computer architecture and operating systems.

Applied Mathematics	
Numerical Analysis	Analysis, Algebra
Numerical Computation	Symbolic Computation

Operating Systems
Computer Hardware

Each chapter begins with sample situations taken from specific fields of application. Abstract and general formulations of mathematical problems are then presented. Following this abstract level, a general discussion about principles and methods for the numerical solution of mathematical problems is presented. Relevant algorithms are developed and their efficiency and the accuracy of their results is assessed. It is then explained as to how they can be obtained in the form of numerical software. The reader is presented with various ways of applying the general methods and principles to particular classes of problems and approaches to extracting practically useful solutions with appropriately chosen numerical software are developed. Potential difficulties and obstacles are examined, and ways of avoiding them are discussed.

The volume and diversity of all the available numerical software is tremendous. When confronted with all the available software, it is important for the user to have a broad knowledge of numerical software in general and to know where this software can be optimally applied in order for him to be able to choose the most appropriate software for a specific problem. Assistance in this complicated matter is offered in this book.

This book gives a comprehensive survey of the high quality software products available and the methods and algorithms they are based on. Positive properties and inherent weaknesses of specific software products are pointed out. Special subsections in this book, devoted to software, provide relevant information about commercially available software libraries (IMSL, NAG, etc.) as well as public domain software (such as the NETLIB programs) which can be downloaded from the Internet.

This book addresses people interested in the numerical solution of mathematical problems, who wish to make a good selection from the wealth of available software products, and who wish to utilize the functionality and efficiency of modern numerical software. These people may be students, scientists or engineers. Accordingly, this monograph may be used either as a textbook or as a reference book.

The German version *Computer-Numerik* was published in 1995 by Springer-Verlag, Heidelberg. The English version, however, is not merely a translation; the book has also been revised and updated.

Synopsis

Volume I starts with a short introduction into scientific model building, which is the foundation of all numerical methods which rely on finite models that replace infinite mathematical objects and techniques. This unavoidable finitization (found in floating-point numbers, truncation, discretization, etc.) is introduced in Chapter 2 and implications are discussed briefly.

The peak performance of modern computer hardware has increased remarkably in recent years and continues to increase at a constant rate of nearly 100 % every year. On the other hand, there is a steadily increasing gap between the potential performance and the empirical performance values of contemporary computer systems. Reasons for this development are examined and remedial measures are presented in Chapter 3.

Chapter 4 is dedicated to the objects of all numerical methods—numerical data—and to the operations they are used in. The main emphasis of this chapter is on standardized floating-point systems as found on most computers used for numerical data processing. It is explained as to how portable programs can be written to adapt themselves to the features of a specific number system.

Chapter 5 deals with the foundations of algorithm theory, in so far as this knowledge is important for numerical methods. Floating-point operations and arithmetic algorithms, which are the basic elements of all other numerical algorithms, are dealt with extensively.

Chapter 6 presents the most important quality attributes of numerical software. Particular attention is paid to techniques which provide for the efficient utilization of modern computer systems when solving large-scale numerical problems.

Chapter 7 gives an overview of readily available commercial or public domain software products. Numerical programs (published in TOMS or in other journals or books), program packages (LAPACK, QUADPACK etc.), and software libraries are dealt with. Particular emphasis is placed on software available on the Internet (NETLIB, ELIB etc.).

Chapter 8 deals with modeling by approximation, a technique important in many fields of numerical data processing. The applications of these techniques range from data analysis to the modeling processes that take place in numerical programs (such as the process found in numerical integration programs in which the integrand function is replaced by a piecewise polynomial).

The most effective approach to obtaining model functions which approximate given data is interpolation. Chapter 9 gives the theoretical background necessary to understanding particular interpolation methods and programs. In addition, the algorithmic processing of polynomials and various spline functions is presented.

Volume II begins with Chapter 10 which contains best approximation techniques for linear and nonlinear data fitting applications.

Chapter 11 is dedicated to a very important application of approximation: the Fourier transform. In particular, the fast Fourier transform (FFT) and its implementations are dealt with.

The subject of Chapter 12 is numerical integration. Software products which solve univariate integration problems are systematically presented. Topical algorithms (such as lattice rules) are covered in this chapter to enable the user to produce tailor-made integration software for solving multivariate integration problems (which is very useful since there is a paucity of ready made software for these problems).

The solution of systems of linear equations is undoubtedly the most important field of computer numerics. Accordingly, it is the field in which the greatest number of software products is available. However, this book deals primarily with LAPACK programs. Chapter 13 answers many questions relevant to users: How should algorithms and software products appropriate for a given problem be chosen? What properties of the system matrix influence the choice of programs? How can it be ascertained whether or not a program has produced an adequate solution? What has to be done if a program does not produce a useful result?

Chapter 14 deals with nonlinear algebraic equations. The properties of these equations may differ greatly, which makes it difficult to solve them with black box software. Chapter 15 is devoted to a very special type of nonlinear equations: algebraic eigenproblems. There is a multitude of numerical software available for these problems. Again, this book deals primarily with LAPACK programs.

Chapter 16 takes a closer look at the topics found in previous chapters, especially problems with large and sparse matrices. Problems of this type are not covered using black box software; a solution method has to be selected individually for each of them. Accordingly, this chapter gives hints on how to select appropriate algorithms and programs.

Monte Carlo methods, which are important for solving numerical problems as well as for empirical sensitivity studies, are based on random numbers. The last chapter of the book, therefore, gives a short introduction into the world of random numbers and random number generators.

Acknowledgments

Many people have helped me in the process of writing this book. I want to thank all of them cordially.

Arnold Krommer, with whom I have worked closely for many years, has had a very strong influence on the greater part of this book. Arnold's expertise and knowledge has particularly enhanced the chapter on numerical integration, a field of interest to both of us.

Roman Augustyn and Wilfried Gansterer have contributed significantly to the chapter on computer hardware.

Winfried Auzinger and Hans Stetter reviewed parts of early manuscripts. Their comments have helped to shape the final version.

Many students at the Technical University of Vienna have been of great assistance. Their suggestions and their help with the proofreading contributed in turning my lecture notes on numerical data processing and a subsequent rough draft into a manuscript. In particular the contributions of Bernhard Bodenstorfer, Ernst Haunschmid, Michael Karg and Robert Matzinger have improved and enriched the content of this book. I wish to express my thanks to all of them.

Wolfgang Moser contributed considerably to the process of translating the German version of this book. He organized the technical assistance of many students who produced a rough draft of the English version. He also translated several chapters himself.

Yolanda Goss was a competent and helpful collaborator in linguistic matters. She brought the rough draft of the English manuscript a giant step nearer to the final version. We even managed to squeeze some fun out of the dry business of improving the grammar, style and form of the draft version.

Robert Lynch (Purdue University) has made a great many valuable suggestions for improving the comprehensibility and the formulation of several chapters.

Robert Bursill (University of Sheffield) and Ronnie Sircar (Stanford University) have proofread and revised the manuscript competently.

Jeremy Du Croz (NAG Ltd.) and Richard Hanson (Visual Numerics Inc.) have checked the software sections and made numerous suggestions for improvements.

Christoph Schmid and Thomas Wihan deserve particular appreciation for the appealing text lay out and the design of the diagrams and illustrations. Christoph Schmid created the final LATEX printout with remarkable dedication.

In addition, I would like to acknowledge the financial support of the Austrian Science Foundation.

January 1997   Christoph W. Ueberhuber

Contents

Chapter 10

Best Approximation

> *Far better an approximate answer to the right question,*
> *which is often vague,*
> *than an exact answer to the wrong question,*
> *which can always be made precise.*
>
> JOHN W. TUKEY

In the interpolation problems discussed in Chapter 9, the number N of parameters of the approximating functions is always equal to the number k of data points. In this chapter the case $k > N$ is discussed, where there exist more data points than parameters. Accordingly, the equation $D(\Delta_k g, y) = 0$ cannot generally be satisfied, i.e., the interpolation requirement, that k values of the approximating function coincide with k given data points, cannot be met.

In this chapter the approximating function $g^* : B \subset \mathbb{R}^n \to \mathbb{R}$ is chosen from a class of functions $\mathcal{G}_N$, such that the distance $D(g^*, y)$ is as small as possible (but generally not zero).

Data of Approximation Problems

Depending on the available data, the approximation problem concerns either a discrete approximation (data approximation) or a function approximation. The data usually belong to one of the following three categories:

Mathematically defined functions which are to be approximated may belong, for instance, to the class of elementary functions (*sin, exp, log* etc.), statistical distribution functions, or higher transcendental functions (elliptic integrals, Bessel functions, Legendre functions etc.) which are important in science and engineering.

Usually there is qualitative and quantitative information about the functions of this class (e.g., position of zeros, extrema and inflections; asymptotic behavior etc.) available.

Functions defined by subprograms which specify analytic data, are often supplied as black box programs by the user. Often there is no information available about the function to be approximated except for the function values produced by the subprogram.

Functions defined by discrete data generally occur during the analysis of empirically obtained data. Often the data points are afflicted with errors and useful additional information is hardly ever available.

The applications of the approximation methods discussed in this chapter are varied. They are used, for instance, to

create models for *data analysis* (extraction of information) and *data synthesis* (e. g., for predicting the future behavior of time series). Some examples are:

- Determination of parameters with a special meaning: In an exponential decay model certain parameters have the meaning of half-life periods of the disintegrating substance.

- Determination of parameters without a concrete meaning: For instance, the *learning* of neural networks is nothing other than the parameterization of special nonlinear functions by approximation methods. These parameters are meaningless within the context of the application problem.

- Analysis of deterministic chaos.

- Separation (*smoothing*, filtering) of the systematic and the stochastic part (perturbations) of empirical data. Fields of application are, for instance, signal processing, statistical and econometrical studies etc.

  Example (Variation of Temperature) To obtain a simple model for the variation of the average temperature in a certain place, given temperature data can be approximated e. g., using a trigonometric polynomial S_d. This leads to the trigonometric model function (——) shown in Fig. 10.1, where the average temperature per day in Vienna for the year 1990 is approximated with respect to the Euclidean norm by a trigonometric polynomial

$$S_1(t) = a_0/2 + a_1 \cos \tau + b_1 \sin \tau \qquad \text{with} \quad \tau = \pi \frac{t+1}{365}.$$

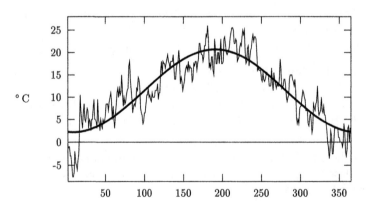

Figure 10.1: The average temperature per day (——) in Vienna in 1990 and the trigonometric approximation function S_1 (——).

compress data to eliminate redundant information.

visualize data by means of smooth curves or surfaces which are defined by a small number of data points; unlike interpolation, the data points are not required to coincide with the curve or surface.

homogenize discrete data to convert discrete data points into analytical data (functions) which allow for further manipulation (integration, differentiation, minimization, etc.).

10.1 Mathematical Foundations

The interpolation principle (see Chapter 9) can be used to determine approximating (model) functions which coincide with the function to be approximated at *given* positions.

If the approximating function g is chosen out of a given class $\mathcal{G}$ of functions whose number of parameters is smaller than the number of data points, the latter requirement cannot be satisfied. In this case the function g is usually chosen as the *best possible* approximating function from the class $\mathcal{G}$.

In a normed space it is obvious to denote functions $g^* \in \mathcal{G}$ as *optimal* or *best approximating* if their approximation error is minimal:

$$\|g^* - f\| \le \|g - f\| \qquad \text{for all} \quad g \in \mathcal{G}.$$

In this section only members from an N-dimensional linear subspace $\mathcal{G}_N$ of a normed space are considered, i.e., only *linear* approximation problems are discussed.

Theorem 10.1.1 (Existence of the Best Linear Approximation) *Let $\mathcal{G}_N$ be an N-dimensional subspace of a normed space $\mathcal{F}$. Then for every $f \in \mathcal{F}$ there exists a best linear approximation $g^* \in \mathcal{G}_N$, i.e., an element g^* with the property*

$$\|g^* - f\| \le \|g - f\| \qquad \text{for all} \quad g \in \mathcal{G}_N.$$

Proof: Davis [40].

Thus, best linear approximation problems necessitate the search for an element g^* of the space $\mathcal{G}_N$ with the smallest distance from f. This can be visualized as follows: Firstly, the sphere

$$S(f,r) := f + r\,S(0,1) \qquad \text{with} \quad S(0,1) := \{\bar{f} \in \mathcal{F} : \|\bar{f}\| \le 1\}$$

is placed around f (various unit spheres are shown in Fig. 10.2), then the radius r is chosen such that the sphere is tangential to the subspace $\mathcal{G}_N$ (see Fig. 10.3). Every point of contact can be taken as a desired function g^*.

This in particular makes it clear that the best approximating functions with respect to different norms generally do not coincide and that best approximation problems related to specific norms (for instance, the L_1- or L_∞-norm) do *not* necessarily have *unique* solutions. Clearly, the unique solvability of the best approximation problem can be guaranteed if the unit sphere $S(0,1)$ with respect to the norm used is strictly convex, i.e., if the line segment connecting two arbitrary non-identical points on the unit sphere is located inside the sphere (except for the endpoints). Generalization of this concept of convexity leads to the following definition:

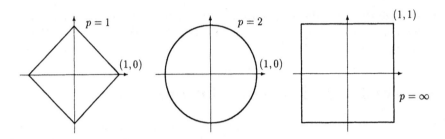

Figure 10.2: Unit spheres $S(0,1)$ in $\mathbb{R}^2$ with respect to $\|\ \|_p$, $p = 1, 2, \infty$ (1-norm $\|\ \|_1$, Euclidean norm $\|\ \|_2$ and maximum norm $\|\ \|_\infty$).

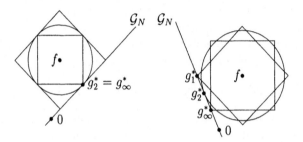

Figure 10.3: Approximation of $f \in \mathcal{F}$ by elements of $\mathcal{G}_N$.

Definition 10.1.1 (Strictly Convex Norm) *A normed space $\mathcal{F}$ has a strictly convex norm if*

$$\|f_1\| = \|f_2\| = 1 \quad and \quad f_1 \neq f_2$$

implies

$$\|f_1 + f_2\| < 2.$$

This definition makes it possible to specify a precise uniqueness condition:

Theorem 10.1.2 *For a strictly convex norm, the solution of the linear approximation problem is always unique.*

Proof: Assume there are two different best approximating elements g_1^* and g_2^* with

$$\|g_1^* - f\| = \|g_2^* - f\| =: E^*(f),$$

then the newly defined element $g_3^* := (g_1^* + g_2^*)/2$ leads to a contradiction:

$$\|g_3^* - f\| = \|(g_1^* + g_2^*)/2 - f\| = \|(g_1^* - f) + (g_2^* - f)\|/2 < E^*(f).$$

$\square$

Condition of Approximation Problems

Even if the existence and the uniqueness of a solution of the best approximation problem are guaranteed, the conditioning of the problem still needs to be investigated: How does the best approximating function change, if a perturbed function

$\tilde{f}$ is approximated instead of f? The next theorem gives a favorable response to this question.

Theorem 10.1.3 *Let $\mathcal{G}_N$ be an N-dimensional subspace of a normed space $\mathcal{F}$. For the elements f, $\tilde{f} \in \mathcal{F}$, the condition estimate*

$$|E_N^*(\tilde{f}) - E_N^*(f)| \leq \|\tilde{f} - f\| \tag{10.1}$$

holds with respect to the best approximation errors

$$E_N^*(f) := \|g^* - f\| \qquad and \qquad E_N^*(\tilde{f}) := \|\tilde{g}^* - \tilde{f}\|.$$

Proof: Since g^* is the best approximating function for f,

$$\|g^* - f\| \leq \|\tilde{g}^* - f\| \leq \|\tilde{g}^* - \tilde{f}\| + \|\tilde{f} - f\|$$

is true, and

$$E_N^*(f) \leq E_N^*(\tilde{f}) + \|\tilde{f} - f\|.$$

Equivalently, the inequality

$$E_N^*(\tilde{f}) \leq E_N^*(f) + \|\tilde{f} - f\|$$

is obtained.   □

As only the triangle inequality is used in the proof, the result is valid for linear as well as *nonlinear* approximation problems.

Inequality (10.1) shows that a perturbation in f increases the error $E_N^*(f)$ in the best approximating function at most by the magnitude $\|\tilde{f} - f\|$ of the perturbation. The approximation problem is therefore well conditioned.

It may not be concluded, however, that the determination of *parameters* (co-efficients) of a best approximating function is also a well-conditioned problem; the approximation by exponential sums—which is extremely ill-conditioned with respect to the determination of parameters—clearly shows this.

10.2 Continuous Least Squares Approximation

In the discussion on distance functions in Chapter 8, it has been pointed out that neither the L_2-norm for function approximation, nor the l_2-norm for data approximation, represents an ideal selection (from the users' point of view). In function approximation it is usually the maximum deviation (of the model function from the original function) which is to be minimized; thus preference is given to the L_∞-norm. In data approximation a distance measure which is insensitive to outliers is desired; for this purpose the l_1-norm, for example, is more suitable than the l_2-norm. In both cases, however, it is the comparatively small effort needed for calculating least squares approximations which makes the Euclidean norm an attractive choice. In function approximation it is even possible to determine a best approximating function with respect to the L_2-norm but to *estimate* the approximation error using, for instance, the L_∞- or the l_∞-norm.

10.2.1 Mathematical Foundations

Inner Products (Scalar Products)

In addition to the assumptions made in Section 10.1 a further assumption regarding the function space is made in this section, which is typical for best approximation in the quadratic mean (i.e., using the L_2-norm): it is assumed that in the linear space $\mathcal{F}$ an *inner product* (a *scalar product*)

$$\langle f, g \rangle \in \mathbb{R} \qquad \text{for all} \quad f, g \in \mathcal{F}$$

can be defined.

Definition 10.2.1 (Inner Product) *A function that uniquely maps two vectors u and v of a real linear space M to a real number, denoted $\langle u, v \rangle$ is called an inner product (scalar product) if the following properties are valid for all $u, u_1, u_2, v \in M$ and $\alpha \in \mathbb{R}$:*

    1. $\langle u_1 + u_2, v \rangle = \langle u_1, v \rangle + \langle u_2, v \rangle$ *(additivity)*,

    2. $\langle \alpha u, v \rangle = \alpha \langle u, v \rangle$ *(homogeneity)*,

    3. $\langle u, v \rangle = \langle v, u \rangle$ *(symmetry)*,

    4. $\langle u, u \rangle \geq 0$
       $\langle u, u \rangle = 0$ *if and only if* $u = 0$ *(positivity)*.

A linear space on which an inner product is defined is said to be an inner product space.

Notation (Scalar Product)   Instead of $\langle\,,\,\rangle$ the alternative notation $(\cdot, \cdot)$ is sometimes used.

Additivity in the second argument

$$\langle u, v_1 + v_2 \rangle = \langle u, v_1 \rangle + \langle u, v_2 \rangle$$

is a consequence of symmetry and additivity in the first argument and thus is not separately required.

In the case of complex linear spaces, the symmetry requirement on

$$\langle\,,\,\rangle : M \times M \to \mathbb{C}$$

has to be replaced by

$$\langle u, v \rangle = \overline{\langle v, u \rangle} \qquad (Hermitian\ symmetry).$$

Example (Scalar Products for n-Vectors)   In $\mathbb{R}^n$ the *Euclidean scalar product* of two vectors, $(\alpha_1, \ldots, \alpha_n)^\top$ and $(\beta_1, \ldots, \beta_n)^\top$, is defined by

$$\langle a, b \rangle := a^\top b = \alpha_1 \beta_1 + \alpha_2 \beta_2 + \cdots + \alpha_n \beta_n. \qquad (10.2)$$

Other inner products are defined in $\mathbb{R}^n$ by

$$\langle a, b \rangle_B := \langle a, Bb \rangle = a^\top B b \qquad (10.3)$$

if $B \in \mathbb{R}^{n \times n}$ is a symmetric and positive definite matrix. It can be proved that every scalar product in $\mathbb{R}^n$ can be represented in this form.

Example (Scalar Products for Continuous Functions) On the linear space $\mathcal{F} = C[a,b]$ of continuous functions on the interval $[a,b] \subset \mathbb{R}$, an inner product is defined by

$$\langle f, g \rangle := \int_a^b f(x)g(x)\,dx \qquad f, g \in C[a,b]. \tag{10.4}$$

Other inner products on $C[a,b]$ can be defined by

$$\langle f, g \rangle_w := \int_a^b w(x)f(x)g(x)\,dx \qquad f, g \in C[a,b]$$

if $w(x) > 0$ holds for all $x \in [a,b]$ or if $w(x) = 0$ only at isolated points .

For every inner product space the (Cauchy-) Schwarz inequality holds.

Theorem 10.2.1 (Schwarz Inequality) *For all u, v in an inner product space, the inequality*

$$|\langle u, v \rangle|^2 \le \langle u, u \rangle \langle v, v \rangle$$

holds. Equality holds if and only if u and v are linearly dependent.

Proof: Davis [40].

The Euclidean Norm

The Schwarz inequality makes it possible to define a norm based on the inner product.

Theorem 10.2.2 (Euclidean Norm, Euclidean Space) *On an inner product space a strictly convex norm is defined by*

$$\|u\|_2 := \sqrt{\langle u, u \rangle}, \tag{10.5}$$

which is referred to as the Euclidean norm. M is then called a Euclidean space.

Proof: Davis [40].

Example (Euclidean Norm for n-Vectors) In $\mathbb{R}^n$ with the Euclidean inner product (10.2), the Euclidean norm (the l_2-norm)

$$\|u\|_2 = \sqrt{\langle u, u \rangle} = \sqrt{u_1^2 + u_2^2 + \cdots + u_n^2}$$

is defined by (10.5) (cf. Section 8.6.2). In a similar way, the general inner product (10.3) induces a norm

$$\|u\|_B := \sqrt{\langle u, u \rangle_B}.$$

Geometrically, the difference between the Euclidean norm $\|\ \|_2$ and the B-norm $\|\ \|_B$ can be illustrated by the shape of the unit sphere

$$S_n := \{u \in \mathbb{R}^n : \|u\|_2^2 = u^\top u \le 1\}$$

which in case of the B-norm turns out to be an ellipsoid

$$E_n(B) := \{u \in \mathbb{R}^n : \|u\|_B^2 = u^\top B u \le 1\}.$$

Example (Euclidean Norm for Continuous Functions)   In the linear space $C[a,b]$ with the inner product (10.4), the definition (10.5) leads to the L_2-norm

$$\|f\|_2 = \sqrt{\langle f, f \rangle} = \left(\int_a^b (f(x))^2 \, dx \right)^{1/2}.$$

Notation (Euclidean Norm)   The neutral symbol $\| \ \|$ is used instead of $\| \ \|_2$ if the context makes it clear that the Euclidean norm is meant.

The Angle between two Vectors

In an inner product space M, the inequality

$$-1 \le \frac{\langle u, v \rangle}{\|u\| \, \|v\|} \le 1 \qquad \text{for all} \quad u, v \in M \setminus \{0\}$$

is valid as a result of the Schwarz inequality. Therefore, there exists a unique number $\varphi \in [0, \pi]$ defined by $\cos \varphi = \langle u, v \rangle / \|u\| \, \|v\|$ which leads to the following definition:

Definition 10.2.2 (The Angle between two Vectors)   *The angle φ between two elements of an inner product space is defined by*

$$\cos \varphi := \frac{\langle u, v \rangle}{\|u\| \, \|v\|}, \qquad \varphi \in [0, \pi]. \tag{10.6}$$

Example (The Angle between two 3-Vectors)   In $\mathbb{R}^3$ the formula (10.6) leads to

$$\cos \varphi = \frac{u_1 v_1 + u_2 v_2 + u_3 v_3}{\sqrt{u_1^2 + u_2^2 + u_3^2} \sqrt{v_1^2 + v_2^2 + v_3^2}},$$

which is a well-known formula in analytical geometry as the angle φ between the two vectors

$$u = (u_1, u_2, u_3)^\top \in \mathbb{R}^3 \qquad \text{and} \qquad v = (v_1, v_2, v_3)^\top \in \mathbb{R}^3.$$

Parallel Vectors and Orthogonality

Two special cases of the angle between two vectors are particularly noteworthy:

Definition 10.2.3 (Parallel Vectors)   *If $\varphi = 0$ then $\cos \varphi = 1$ and*

$$|\langle u, v \rangle| = \|u\| \, \|v\|.$$

This implies that u and v are linearly dependent due to Theorem 10.2.1. In this special case the vectors u and v are said to be parallel.

Definition 10.2.4 (Orthogonal Vectors)   *Two elements $u \neq 0$ and $v \neq 0$ are said to be orthogonal if*

$$\cos \varphi = \frac{\langle u, v \rangle}{\|u\| \, \|v\|} = 0 \qquad \text{or} \qquad \langle u, v \rangle = 0.$$

Example (Orthogonal n-Vectors) Two vectors $u, v \in \mathbb{R}^n \setminus \{0\}$ are orthogonal if their Euclidean scalar product vanishes:

$$\langle u, v \rangle = u^\top v = 0.$$

Example (B-Orthogonal n-Vectors) The two vectors $u, v \in \mathbb{R}^n \setminus \{0\}$ are said to be B-orthogonal (*conjugate*) if the general scalar product (10.3) vanishes:

$$\langle u, v \rangle_B = u^\top B v = 0.$$

This generalized type of orthogonality no longer corresponds to common expectations. Two B-orthogonal vectors are not generally perpendicular to one other.

Example (Orthogonal Functions) The functions $f(x) = \sin x$ and $g(x) = \cos x$ are orthogonal on the interval $[0, 2\pi]$ with respect to the L_2 scalar product:

$$\langle f, g \rangle = \int_0^{2\pi} \sin x \cos x \, dx = \frac{1}{2} \sin^2 x \Big|_0^{2\pi} = 0.$$

From this example it can be seen that orthogonality in function spaces does not correspond to the orthogonality of the curves

$$\{(x, y) : y = f(x)\} \quad \text{and} \quad \{(x, y) : y = g(x)\}.$$

Definition 10.2.5 (w-orthogonal Functions) *Two functions f and g are said to be w-orthogonal (on the interval $[a, b]$ and with respect to the weight function w) if*

$$\langle f, g \rangle_w = 0.$$

In the case $w(x) \equiv 1$ w-orthogonal functions are referred to as orthogonal functions.

Example (w-orthogonal Functions) The functions $f(x) = x$ and $g(x) = 2x^2 - 1$ are orthogonal on the interval $[-1, 1]$ with respect to the weight function $w(x) = (1 - x^2)^{-1/2}$:

$$\langle f, g \rangle_w = \int_{-1}^1 \frac{x(2x^2 - 1)}{\sqrt{1 - x^2}} \, dx = \int_0^\pi (2 \cos^3 t - \cos t) \, dt = 0.$$

The orthogonality of two functions on a certain interval $[a, b]$ and with respect to a special weight function w generally does *not* imply orthogonality of these functions on other intervals and/or with respect to other weight functions.

10.2.2 Best Approximation with Orthogonality

The solution of the best approximation problem with respect to the Euclidean norm can be accomplished by generalizing the fact, easily visualized in $\mathbb{R}^2$, that the shortest connecting line segment between a point and a linear subspace is perpendicular to the subspace (see Fig. 10.4).

Theorem 10.2.3 (Characterization Theorem) *Let $\mathcal{G}_N$ be a finite-dimensional subspace of the Euclidean space $\mathcal{F}$. The element $g^* \in \mathcal{G}_N$ is the best approximation of an element $f \in \mathcal{F}$ if and only if*

$$\langle g^* - f, g \rangle = 0 \qquad \text{for all} \quad g \in \mathcal{G}_N, \tag{10.7}$$

i. e., if the error function $g^* - f$ is orthogonal to all functions in $\mathcal{G}_N$.

Proof: Hämmerlin, Hoffmann [52].

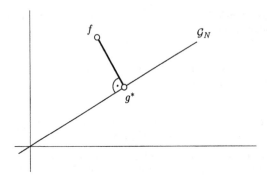

Figure 10.4: The shortest connecting line segment between f and $\mathcal{G}_N$ is perpendicular to $\mathcal{G}_N$.

10.2.3 The Normal Equations

The best approximation g^* to f is a linear combination of the basis functions $g_1, \ldots, g_N$ of the subspace $\mathcal{G}_N$:

$$g^* = \sum_{j=1}^{N} c_j^* g_j. \tag{10.8}$$

The orthogonality relation (10.7) of the characterization theorem is valid for *every* function $g \in \mathcal{G}_N$, in particular for the N basis functions $g_1, \ldots, g_N$:

$$\langle f - g^*, g_i \rangle = \langle f, g_i \rangle - \sum_{j=1}^{N} c_j^* \langle g_j, g_i \rangle = 0, \qquad i = 1, 2, \ldots, N.$$

Thus, the coefficients $c_1^*, \ldots, c_N^*$ of the representation (10.8) of g^* are obtained as the solution of the following system of linear equations:

$$
\begin{aligned}
\langle g_1, g_1 \rangle c_1 + \langle g_2, g_1 \rangle c_2 + \cdots + \langle g_N, g_1 \rangle c_N &= \langle f, g_1 \rangle \\
\langle g_1, g_2 \rangle c_1 + \langle g_2, g_2 \rangle c_2 + \cdots + \langle g_N, g_2 \rangle c_N &= \langle f, g_2 \rangle \\
\vdots \qquad\quad \vdots \qquad\qquad\quad \vdots \qquad\quad \vdots & \\
\langle g_1, g_N \rangle c_1 + \langle g_2, g_N \rangle c_2 + \cdots + \langle g_N, g_N \rangle c_N &= \langle f, g_N \rangle.
\end{aligned}
\tag{10.9}
$$

This system of linear equations is referred to as the (Gaussian) *normal equations*. The coefficient matrix of (10.9) is called the *Gramian* and is regular if and only if the elements $g_1, \ldots, g_N$ are linearly independent. Therefore, if the elements $g_1, \ldots, g_N$ form a basis of $\mathcal{G}_N$, then the system of linear equations (10.9) always possesses a unique solution.

Example (Hilbert Matrices)   In the space $L^2[0,1]$ an inner product is defined by

$$\langle u, v \rangle := \int_0^1 u(t)v(t)\, dt.$$

If the monomials $g_i(x) = x^i$, $i = 0, 1, \ldots, d$ of maximum degree d are chosen as a basis of the $(d+1)$-dimensional space $\mathbb{P}_d$, then

$$\langle g_i, g_j \rangle = \frac{1}{i+j+1}, \qquad i = 0, 1, \ldots, d, \quad j = 0, 1, \ldots, d.$$

This means that the Gramian of the normal equations is the $(d+1)\times(d+1)$ *Hilbert matrix* H_{d+1}, which is regular for all $d \in \mathbb{N}_0$. However, due to the large condition number of these matrices even for small values of d

d	3	4	5	6	7	8
$\mathrm{cond}_2(H_d)$	$5.24 \cdot 10^2$	$1.55 \cdot 10^4$	$4.77 \cdot 10^5$	$1.50 \cdot 10^7$	$4.75 \cdot 10^8$	$1.53 \cdot 10^{10}$

the practical computation of the best approximating polynomial with respect to the L_2-norm *using a monomial basis* $\{1, x, \ldots, x^d\}$ is highly problematic. The reason for this is that—depending on the machine arithmetic—the matrix $(\langle g_i, g_j \rangle) = H_{d+1}$ is already numerically singular for small values of d, i. e., the basis elements $1, x, \ldots, x^d$ are numerically linearly dependent.

The Optimal Selection of the Basis

From a practical point of view, a basis of the subspace $\mathcal{G}_N$ is satisfactory for computing an approximation in the sense of L_2 if the system of normal equations is well conditioned and easy to solve. The optimal condition number (cond $= 1$) can be attained if and only if

$$\langle g_i, g_j \rangle = \delta_{ij} := \begin{cases} 1 & \text{for} \quad i = j \\ 0 & \text{otherwise.} \end{cases}$$

That is, if the basis $\{g_1, \ldots, g_N\}$ is an *orthonormal system*, in which case the Gramian is equal to the identity matrix I_N. In such cases the coefficients c_i^* of the representation (10.8) of g^* are obtained from

$$c_i^* = \langle f, g_i \rangle, \qquad i = 1, 2, \ldots, N.$$

The best approximation function g^* to f is then simply given by

$$g^* = \sum_{i=1}^{N} \langle f, g_i \rangle \, g_i.$$

In a geometrical sense the coefficients $c_i^* = \langle f, g_i \rangle$ of an expansion of f with respect to the orthonormal system $\{g_1, \ldots, g_N\}$ can be interpreted as rectangular coordinates of f in a function space. This reveals a further advantage of using an orthonormal basis. In moving from $\mathcal{G}_N$ to $\mathcal{G}_{N+1} \supset \mathcal{G}_N$, the coefficients $c_1^*, \ldots, c_N^*$ (which have already been calculated in $\mathcal{G}_N$) can be used in $\mathcal{G}_{N+1}$ as well, and it simply suffices to determine c_{N+1}^*.

If the basis $\{g_1, \ldots, g_N\}$ is an orthogonal system which is *not* ortho*normal*, then the coefficients c_i^* of the best approximation are given by

$$c_i^* = \frac{\langle f, g_i \rangle}{\langle g_i, g_i \rangle} = \frac{\langle f, g_i \rangle}{\|g_i\|^2}, \qquad i = 1, 2, \ldots, N. \tag{10.10}$$

Definition 10.2.6 (Fourier Series, Fourier Polynomials)   *Let $g_1, g_2, \ldots$ be a sequence of pairwise orthonormal elements in an inner product space $\mathcal{F}$. Then the series*

$$\sum_{j=1}^{\infty} \langle f, g_j \rangle g_j$$

is called the Fourier series for $f \in \mathcal{F}$ (with respect to the orthonormal system $\{g_j\}$). The coefficients $c_j := \langle f, g_j \rangle$ are the Fourier coefficients of f with respect to $\{g_j\}$. A truncated Fourier series

$$\sum_{j=1}^{N} \langle f, g_j \rangle g_j$$

is referred to as a Fourier polynomial.

Since $\langle f, g_j \rangle g_j$ is the orthogonal projection of f into the subspace spanned by g_j, the Fourier polynomial represents the orthogonal projection of f onto the subspace spanned by $\{g_1, \ldots, g_N\}$ whence the Fourier polynomial is the solution of the best approximation problem with respect to the *Euclidean norm*.

The Fourier Expansion

In the $(2d+1)$-dimensional space of all trigonometric polynomials

$$S_d(x) = \frac{a_0}{2} + \sum_{k=1}^{d} (a_k \cos kx + b_k \sin kx)$$

on $[0, 2\pi]$, the functions

$$g_0(x) := 1, \quad g_1(x) := \sin x, \quad g_3(x) := \sin 2x, \ldots, g_{2d-1}(x) := \sin(dx),$$
$$g_2(x) := \cos x, \quad g_4(x) := \cos 2x, \ldots, g_{2d}(x) := \cos(dx)$$

form an orthogonal basis:

$$\langle g_k, g_l \rangle = \begin{cases} 2\pi & \text{for} \quad k = l = 0 \\ \pi & \text{for} \quad k = l = 1, 2, \ldots, d \\ 0 & \text{for} \quad k \neq l \end{cases}$$

with respect to the scalar product

$$\langle f, g \rangle := \int_0^{2\pi} f(t) g(t) \, dt.$$

Thus, the best approximating trigonometric polynomial S_d^* is given by

$$S_d^*(x) = \sum_{j=0}^{2d} {}' c_j^* g_j(x) = \frac{a_0^*}{2} + \sum_{k=1}^{d} (a_k^* \cos kx + b_k^* \sin kx).$$

The coefficients c_j^* are determined by (10.10):

$$c_{2k}^* = a_k^* = \frac{1}{\pi} \int_0^{2\pi} f(t) \cos kt \, dt, \quad k = 0, 1, \ldots, d,$$

$$c_{2k-1}^* = b_k^* = \frac{1}{\pi} \int_0^{2\pi} f(t) \sin kt \, dt, \quad k = 1, 2, \ldots, d.$$

The best approximating trigonometric polynomial S_d^* with respect to the L_2-norm can also be obtained by truncating the infinite Fourier series

$$S(x) := \frac{a_0}{2} + \sum_{k=1}^{\infty} (a_k \cos kx + b_k \sin kx)$$

whose coefficients are given by

$$a_k = \frac{1}{\pi} \int_0^{2\pi} f(t) \cos kt \, dt, \quad k = 0, 1, 2, \ldots$$

$$b_k = \frac{1}{\pi} \int_0^{2\pi} f(t) \sin kt \, dt, \quad k = 1, 2, 3, \ldots .$$

The Chebyshev Expansion

The Chebyshev polynomials T_k, $k = 0, 1, \ldots, d$ form an orthogonal basis (see (10.11)) of the space $\mathbb{P}_d$ of the polynomials of maximum degree d on $[-1, 1]$ with respect to the weighted inner product

$$\langle f, g \rangle_w := \int_{-1}^{1} \frac{1}{\sqrt{1 - t^2}} f(t) g(t) \, dt.$$

Using the ansatz

$$P_d^*(x) = \sum_{k=0}^{d} {}' c_k^* T_k(x)$$

for the best approximating polynomial P_d^* and the substitution $x := \cos \tau$ together with the relations

$$T_k(\cos x) = \cos kx$$

and

$$\cos x = \cos(2\pi - x),$$

(10.10) leads to the coefficients

$$c_k^* = \frac{\langle f, g_k \rangle}{\langle g_k, g_k \rangle}$$

$$= \frac{2}{\pi} \int_{-1}^{1} \frac{1}{\sqrt{1-t^2}} f(t) T_k(t)\, dt$$

$$= \frac{2}{\pi} \int_{0}^{\pi} f(\cos t) \cos kt\, dt$$

$$= \frac{1}{\pi} \int_{0}^{2\pi} f(\cos t) \cos kt\, dt, \quad k = 0, 1, \ldots, d.$$

In other words, the coefficients of the Chebyshev expansion of the function f are obtained by determining the *Fourier* coefficients a_k^* of the periodic function $f(\cos x)$.

10.2.4 The Approximation Error

The norm of the approximation error for the best approximating g^* to f can be obtained as follows:

$$\begin{aligned}
\|f - g^*\|^2 &= \langle f - g^*, f - g^* \rangle \\
&= \langle f - g^*, f \rangle - \langle f - g^*, g^* \rangle \\
&= \langle f - g^*, f \rangle \\
&= \|f\|^2 - \langle g^*, f \rangle \\
&= \|f\|^2 - \sum_{j=1}^{N} c_j^* \langle f, g_j \rangle.
\end{aligned}$$

In particular, for an orthonormal basis $\{g_j\}$

$$\|f - g^*\|^2 = \|f\|^2 - \sum_{j=1}^{N} \langle f, g_j \rangle^2.$$

Using $\|f - g^*\|^2 \geq 0$ it follows that

$$\sum_{j=1}^{N} (c_j^*)^2 = \sum_{j=1}^{N} \langle f, g_j \rangle^2 \leq \|f\|^2,$$

the *Bessel inequality*, holds. Furthermore, as $N \to \infty$

$$\sum_{j=1}^{\infty} (c_j^*)^2 = \sum_{j=1}^{\infty} \langle f, g_j \rangle^2 \leq \|f\|^2.$$

Therefore,

$$\lim_{j \to \infty} \langle f, g_j \rangle = 0$$

for all $f \in \mathcal{F}$. However, this does *not* necessarily imply that the sequence of Fourier polynomials converges to f (in the sense of the Euclidean norm) (Lanczos [265]).

Example (Failure to Converge) The functions

$$g_j = \frac{1}{\sqrt{\pi}} \sin jx, \quad j = 1, 2, \ldots$$

form an orthonormal sequence on the interval $[-\pi, \pi]$ with respect to the inner product

$$\langle u, v \rangle = \int_{-\pi}^{\pi} u(t)v(t)\, dt.$$

However, since these functions are all odd, the partial sums of the Fourier series—with respect to *this* orthonormal system—obviously cannot converge if f is an even, non-zero function. Thus, the function system $\{g_1, g_2, \ldots\}$ is *not* a basis of a space $\mathcal{F}$ containing even functions in that not every element of the normed space $\mathcal{F}$ can be represented as a Fourier series $\sum_{j=1}^{\infty} c_j g_j$.

Theorem 10.2.4 *If the orthonormal system $\{g_1, g_2, \ldots\}$ is a basis for an inner product space $\mathcal{F}$, then the following holds:*

1) For any $f \in \mathcal{F}$ the truncated Fourier series converges to f:

$$\lim_{N \to \infty} \left\| f - \sum_{j=1}^{N} \langle f, g_j \rangle g_j \right\| = 0.$$

2) The Parseval identity

$$\|f\|^2 = \sum_{j=1}^{\infty} \langle f, g_j \rangle^2$$

and the extended Parseval identity

$$\langle f_1, f_2 \rangle = \sum_{j=1}^{\infty} \langle f_1, g_j \rangle \langle f_2, g_j \rangle$$

hold for arbitrary $f_1, f_2 \in \mathcal{F}$.

3) If $\langle f, g_j \rangle = 0$ for all $j = 1, 2, \ldots$, then f vanishes identically.

4) Each element of $\mathcal{F}$ is uniquely characterized by its Fourier coefficients, i. e.,

$$\langle f_1, g_j \rangle = \langle f_2, g_j \rangle \qquad \text{for all} \quad j = 1, 2, \ldots$$

implies $f_1 = f_2$.

Proof: Davis [40].

The difficulty in practically applying this theorem lies in finding out whether a set $\{g_1, g_2, \ldots\}$ is really a basis for $\mathcal{F}$ or not. Orthogonal polynomials (see the next section) are a basis for the space of continuous functions. Trigonometric polynomials are covered by the results of classical Fourier theory.

10.2.5 Orthogonal Polynomials

If a function $f \in C[a, b]$ is to be approximated by a polynomial of degree d, then an optimally conditioned representation is obtained by using a basis of the space $\mathbb{P}_d$ which consists of w-orthogonal polynomials. Theorem 10.2.3 shows that the best approximating polynomial $P_d^* \in \mathbb{P}_d$ is characterized by an error function $f - P_d^*$ which is w-orthogonal to $\mathbb{P}_d$. As the following theorem shows, P_d^* coincides with f at $d+1$ points of $[a, b]$. It is thus possible to regard the best approximating polynomial P_d^* as a polynomial interpolating the function f.

Theorem 10.2.5   *A function $g \in C[a, b]$ which is w-orthogonal to the space $\mathbb{P}_d$ is either identically zero or has at least $d+1$ zeros in $[a, b]$, where its sign changes.*

Proof: Indirect. Assume that g has only $k \leq d$ zeros $z_1 < z_2 < \ldots < z_k$, where its sign changes. If a polynomial $P \in \mathbb{P}_d$ is defined by

$$P(x) := (x - z_1)(x - z_2) \cdots (x - z_k),$$

then the function $g(x)P(x)$ has the same sign in all intervals

$$(a, z_1), (z_1, z_2), \ldots, (z_k, b),$$

and the scalar product

$$\langle g, P \rangle_w = \int_a^b w(x)g(x)P(x)\, dx$$

cannot vanish, which is a contradiction to the orthogonality assumption.   □

The property of the best approximating polynomial P_d^* of coinciding with f at (at least) $d+1$ points, *cannot* be used for algorithmic purposes because these points are in most cases unknown.

In order to obtain a w-orthogonal sequence $\{P_d \in \mathbb{P}_d, \ d = 0, 1, 2, \ldots\}$ of polynomials with the property

$$P_{d+1} \quad \text{is } w\text{-orthogonal to} \quad \mathbb{P}_d, \qquad d = 0, 1, 2, \ldots,$$

the *Gram-Schmidt orthonormalization process* can, for example, be applied to the monomials $\{1, x, x^2, \ldots\}$ (Isaacson, Keller [61]).

In fact every sequence of w-orthogonal polynomials can be generated using a *three term recursion*, as the following theorem shows:

Theorem 10.2.6   *All polynomials $P_0, P_1, \ldots$, which are pairwise orthogonal with respect to an arbitrary weight function w and do not vanish identically satisfy a three term recursion*

$$P_{d+1} = (a_{d+1}x + b_{d+1})P_d - c_{d+1}P_{d-1}, \qquad d = 1, 2, 3, \ldots,$$

with real coefficients $a_{d+1}, b_{d+1}, c_{d+1}$ which satisfy the relation

$$a_{d+1} = c_{d+1}a_d \frac{\|P_{d-1}\|^2}{\|P_d\|^2}.$$

Proof: Davis [40].

In the following sections some important classes of orthogonal polynomials are discussed.

The Legendre Polynomials

The Legendre polynomials are characterized by the fact that they are orthogonal on $[-1, 1]$ with respect to the weight function $w(x) \equiv 1$. They can be defined in the following manner:

Recursion formula:

$$P_{d+1} := \frac{2d+1}{d+1} x P_d - \frac{d}{d+1} P_{d-1}, \qquad d = 1, 2, 3, \dots$$

$$P_0(x) := 1, \quad P_1(x) := x, \quad P_2(x) = \frac{1}{2}(3x^2 - 1), \ \dots.$$

Orthogonality relations:

$$\langle P_i, P_j \rangle = \int_{-1}^{1} P_i(x) P_j(x) \, dx = \begin{cases} 0 & i \neq j \\ 2/(2i+1) & i = j. \end{cases}$$

Thus, the Legendre polynomials form an orthogonal system, but *not* an orthonormal system.

The Chebyshev Polynomials

This system of polynomials, which is orthogonal on $[-1, 1]$ with respect to the weight function $w(x) = (1 - x^2)^{-1/2}$ is very important in numerical mathematics and data processing. Note that Chebyshev polynomials satisfy the following:

Recursion formula:

$$T_{d+1} := 2x T_d - T_{d-1}, \qquad d = 1, 2, 3, \dots$$

$$T_0(x) := 1, \quad T_1(x) := x, \quad T_2(x) = 2x^2 - 1, \dots$$

Orthogonality relations:

$$\langle T_i, T_j \rangle_w = \int_{-1}^{1} \frac{1}{\sqrt{1 - x^2}} T_i(x) T_j(x) \, dx = \begin{cases} 0 & i \neq j \\ \pi/2 & i = j = 1, 2, 3, \dots \\ \pi & i = j = 0, \end{cases} \qquad (10.11)$$

i. e., the Chebyshev polynomials also form an orthogonal, but not an orthonormal system with respect to $w(x) = (1 - x^2)^{-1/2}$.

Discrete Orthogonality: Chebyshev polynomials are not only continuously orthogonal (10.11); they are also orthogonal with respect to the discrete inner products (9.18) and (9.20), as is dealt with in Section 9.3.2.

The increasing density of the nodes ξ_k used for the discrete inner product $\langle\,,\,\rangle_T$ or the nodes η_k used for $\langle\,,\,\rangle_U$ near the endpoints of the interval $[-1,1]$ has its counterpart in the stronger weighting of the peripheral zones by the weight function

$$w(x) = (1-x^2)^{-1/2}$$

in the case of the continuous inner product $\langle\,,\,\rangle_w$. The special advantages of the Chebyshev polynomials stem from the emphasized weighting of the peripheral zones. These are critical because undesirable oscillations may occur there.

The Chebyshev polynomials are the only class of polynomials which have both properties, continuous and discrete orthogonality. Only the trigonometric polynomials S_d, which are closely related to Chebyshev polynomials, are orthogonal in both the continuous and the discrete case. The discrete orthogonality of trigonometric polynomials holds with respect to N nodes *equidistantly* distributed on the interval $[0, 2\pi)$.

The Chebyshev expansion

$$f(x) = \sum_{j=0}^{\infty}{}' a_j T_j(x), \tag{10.12}$$

i.e., the expansion of f using Chebyshev polynomials, has the advantage of the fastest decreasing coefficients; this property is particularly favorable in comparison with the Taylor expansion. For a function $f : [-1,1] \to \mathbb{R}$, the coefficients of the Chebyshev expansion are given by

$$\begin{aligned} a_j &= \frac{\langle f, T_j \rangle}{\langle T_j, T_j \rangle} \tag{10.13}\\[2mm] &= \frac{2}{\pi} \int_{-1}^{1}{}' f(t) T_j(t) \sqrt{\frac{1}{1-t^2}}\, dt \\[2mm] &= \frac{2}{\pi} \int_{0}^{\pi} f(\cos\vartheta) \cos j\vartheta\, d\vartheta, \qquad j = 0, 1, 2, \dots. \tag{10.14} \end{aligned}$$

After truncating the Chebyshev series (10.12), the polynomial

$$P_d(x) := \sum_{j=0}^{d}{}' a_j T_j(x)$$

is often a very good approximation for f because of the rapidly decreasing coefficients a_j; the approximation error e_d can be estimated by

$$|e_d(x)| = |P_d(x) - f(x)| = \left| \sum_{j=d+1}^{\infty} a_j T_j(x) \right| \leq \sum_{j=d+1}^{\infty} |a_j|.$$

For practical applications the integral (10.14) defining the coefficients a_j generally has to be computed numerically. Using the compound trapezoidal rule,

$$a_j^U = \frac{2}{d} \sum_{k=0}^{d-1} f(\cos \vartheta_k) \cos j\vartheta_k, \qquad \vartheta_k := \frac{k\pi}{d}, \qquad j = 0, 1, \ldots, d$$

is obtained. This is exactly the formula for determining the coefficients of the *interpolation* polynomial with respect to the Chebyshev extrema and the basis $\{T_0, \ldots, T_d\}$. The reason for this inter-connection is the relation $\eta_k = \cos \frac{k\pi}{d}$. Analogously, by using the rectangular rule, the interpolation polynomial (with coefficients a_j^T) with respect to the Chebyshev zeros is obtained. The coefficients a_j, a_j^U and a_j^T, $j = 0, 1, \ldots, d$ are inter-connected by

$$\begin{aligned} a_j^U &= a_j + (a_{2d-j} + a_{2d+j}) + (a_{4d-j} + a_{4d+j}) + \cdots \\ a_j^T &= a_j + (a_{2d+2-j} + a_{2d+2+j}) + (a_{4d+4-j} + a_{4d+4+j}) + \cdots. \end{aligned}$$

If the truncated Chebyshev expansion has an approximation error which can be neglected within a specified tolerance, then the representation of a_j^U and a_j^T implies that the deviations $a_j^U - a_j$ and $a_j^T - a_j$ are also negligible. Accordingly the approximate calculation of the coefficients of a truncated Chebyshev series can be best done by polynomial interpolation using Chebyshev nodes.

All the properties of the Chebyshev polynomials cited in this section, including proofs, can be found in the books by Rivlin [329] and by Fox and Parker [194].

The Laguerre Polynomials

The Laguerre polynomials form a system of orthogonal polynomials on the infinite interval $[0, \infty)$. The *orthonormality* of these polynomials holds with respect to the weight function $w(x) = \exp(-x)$.

Recursion formula:

$$L_{d+1}(x) := \left(\frac{2d+1}{d+1} - \frac{1}{d+1}x \right) L_d(x) - \frac{d}{d+1} L_{d-1}(x),$$

$$L_0(x) := 1, \qquad L_1(x) := 1 - x.$$

The Hermite Polynomials

The Hermite polynomials form an orthogonal system defined on $(-\infty, \infty)$, whose weight function is $w(x) = \exp(-x^2)$.

Recursion formula:

$$H_{d+1}(x) := 2xH_d(x) - 2dH_{d-1}(x), \qquad H_0(x) := 1, \quad H_1(x) := 2x.$$

10.3 Discrete Least Squares Approximation

The discrete least squares method is used especially in the analysis of discrete data, i.e., data which is supplied pointwise.

10.3.1 Linear Least Squares Approximation

If the number m of observations (equations) is larger than the number n of parameters (unknowns) when data are to be fitted by a linear model, then the resulting system of linear equations is overdetermined:

$$Ax = \begin{pmatrix} a_{11}x_1 + a_{12}x_2 + \cdots + a_{1n}x_n \\ a_{21}x_1 + a_{22}x_2 + \cdots + a_{2n}x_n \\ \vdots \quad \vdots \quad \quad \vdots \\ a_{m1}x_1 + a_{m2}x_2 + \cdots + a_{mn}x_n \end{pmatrix} = \begin{pmatrix} b_1 \\ b_2 \\ \vdots \\ b_m \end{pmatrix} = b. \tag{10.15}$$

Best approximation can be applied to solve such a problem by searching in the image space

$$\mathcal{R}(A) := \operatorname{span}\{a_1, a_2, \ldots, a_n\}$$

of the linear mapping (i. e., the subspace spanned by the column vectors $a_1, \ldots, a_n$ of the matrix A) for the vector which has the smallest distance to b.

The Normal Equations

Due to Theorem 10.2.3 (the characterization theorem) the desired optimum vector $x^* \in \mathbb{R}^n$ is the vector for which the residual $Ax^* - b$ is orthogonal to $\mathcal{R}(A)$ (see Fig. 10.5). As in function approximation (Section 10.2) the solution x^* of a

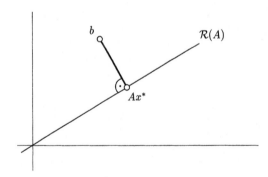

Figure 10.5: Solution of a linear fitting problem in the l_2-norm.

discrete approximation problem is obtained by solving the (Gaussian) normal equations

$$\langle a_1, a_1 \rangle x_1 + \langle a_2, a_1 \rangle x_2 + \cdots + \langle a_n, a_1 \rangle x_n = \langle b, a_1 \rangle$$
$$\langle a_1, a_2 \rangle x_1 + \langle a_2, a_2 \rangle x_2 + \cdots + \langle a_n, a_2 \rangle x_n = \langle b, a_2 \rangle$$
$$\vdots \qquad \qquad \vdots \qquad \qquad \vdots \qquad \qquad \vdots$$
$$\langle a_1, a_n \rangle x_1 + \langle a_2, a_n \rangle x_2 + \cdots + \langle a_n, a_n \rangle x_n = \langle b, a_n \rangle$$

or in matrix notation,

$$A^\top Ax = A^\top b. \tag{10.16}$$

Example (Line Fitting) For the purpose of fitting a line $f(x) = c_0 + c_1 x$ to given data

$$(x_1, y_1), (x_2, y_2), \ldots, (x_m, y_m) \in \mathbb{R}^2$$

such that the distance

$$D_2(\Delta_m f, y) = \left(\sum_{i=1}^m (c_0 + c_1 x_i - y_i)^2 \right)^{1/2}$$

is minimal, the Euclidean norm of the residual vector of the system of equations

$$
\begin{array}{ccc}
c_0 + c_1 x_1 = y_1 \\
c_0 + c_1 x_2 = y_2 \\
\vdots \quad \vdots \quad \vdots \\
c_0 + c_1 x_m = y_m
\end{array}
\quad \text{or} \quad Ax = b \quad \text{with} \quad
A = \begin{pmatrix} 1 & x_1 \\ 1 & x_2 \\ \vdots & \vdots \\ 1 & x_m \end{pmatrix}, \;
b = \begin{pmatrix} y_1 \\ y_2 \\ \vdots \\ y_m \end{pmatrix}, \;
x = \begin{pmatrix} c_0 \\ c_1 \end{pmatrix}
$$

has to be minimized. This leads to the normal equations

$$
\begin{pmatrix} m & \sum_{i=1}^m x_i \\ \sum_{i=1}^m x_i & \sum_{i=1}^m x_i^2 \end{pmatrix}
\begin{pmatrix} c_0 \\ c_1 \end{pmatrix} =
\begin{pmatrix} \sum_{i=1}^m y_i \\ \sum_{i=1}^m x_i y_i \end{pmatrix}
$$

whose solution $(c_0^*, c_1^*)^\top$ is given by

$$
c_0^* = \frac{\sum_{i=1}^m x_i^2 \sum_{i=1}^m y_i - \sum_{i=1}^m x_i y_i \sum_{i=1}^m x_i}{m \sum_{i=1}^m x_i^2 - \left(\sum_{i=1}^m x_i \right)^2}, \qquad
c_1^* = \frac{m \sum_{i=1}^m x_i y_i - \sum_{i=1}^m x_i \sum_{i=1}^m y_i}{m \sum_{i=1}^m x_i^2 - \left(\sum_{i=1}^m x_i \right)^2}.
$$

The matrix coefficients $A^\top A$ of the normal equations is symmetric and, if rank$(A) = n$, then it is also positive definite. These properties make it possible to solve (10.16) using the Cholesky method (see Section 13.12.1) or using a conjugate gradient method (see Sections 16.7.4 and 16.9.1).

Solving a linear fitting problem using the normal equations is very advantageous in so far as the computational effort is concerned: Due to the symmetry of $A^\top A$, it is sufficient to calculate and to store only the subdiagonal elements and the main diagonal of this matrix. The calculation of $A^\top A$ requires about $mn^2/2$ floating-point operations. Solving the system of equations (10.16) using the Cholesky algorithm requires $n^3/6$ additional floating-point operations. In many problems occurring in practice $m \gg n$. In these cases the cost of forming the normal equations is considerably higher than the cost of solving them.

Factorizing the Matrix of a Least Squares Problem

Setting up and solving normal equations is a rather easy and obvious method for linear l_2-approximation, but it has one essential disadvantage: The determination of the parameters of a linear model (the determination of the vector x^*) using this method is much more sensitive to data and computational errors than using methods based on the factorization of A using orthogonal matrices. The

$$QR \text{ decomposition} \qquad A = QR, \quad \text{and the}$$

$$\text{singular value decomposition} \quad A = USV^\top,$$

discussed in Chapter 13, do not deteriorate the conditioning of the original problem (10.15), whereas the condition number of the normal equations is the *square* of the original condition number. This condition number is especially high if the column vectors $a_1, \ldots, a_n$ of A are *nearly linearly dependent*, as is very often the case in modeling.

Example (Socio-Economic Model)   The monthly per capita spending on consumer goods (effected, for example, by the inhabitants of a European town) is to be analyzed by a linear model. For this purpose the following factors could be used in the model: personal net income per month, household income, age, duration of professional training and monthly housing expenditure.

The matrix A, whose coefficients are the concrete values of the factors, come, for example, from a poll, would be very ill-conditioned in this case. The reason for this ill-conditioning is the strong correlation between some of the factors and the resulting *almost linear dependency* of the column vectors of A.

The situation expressed in this example can often be found in the initial stage of a modeling process. When forming a model for the first time, one is often tempted to include in the model all relevant factors for which data are available (cf. Chapter 1).

Software for Linear Least Squares Problems

Programs for overdetermined systems of linear equations which are to be solved by minimizing the l_2-norm of the residual, can be found in LAPACK (see Section 13.16.2).

Software for fitting various functions (which are linear in the parameters) are available in the IMSL and NAG libraries:

polynomials: IMSL/rcurv, NAG/e02adf
polynomials with constraints: NAG/e02agf

cubic splines: NAG/e02baf, NAG/e02bef
splines of general order: IMSL/bslsq, IMSL/bsvls
splines with constraints: IMSL/conft

user-defined basis functions: IMSL/fnlsq

bivariate polynomials: NAG/e02caf

bicubic splines: NAG/e02daf, NAG/e02dcf, NAG/e02ddf
2D tensor product splines of general order: IMSL/bsls2
3D tensor product splines of general order: IMSL/bsls3

10.3.2 Nonlinear Least Squares Approximation

The fitting of approximation functions which are nonlinear in the parameters, require the solution of a nonlinear minimization problem. The parameters of a nonlinear approximation function have to be determined in such a way that the distance between the approximation function and the data is minimized. This minimization problem has to be solved numerically (see Section 14.4) whence undesirable phenomena may be encountered:

1. The distance function to be minimized may have several *local* minima. There are, however, no effective algorithms available to determine the *global* minimum which is the definitive solution of the approximation problem. Most algorithms do not even attempt to find the global minimum; they simply determine one of the *local* minima.

2. Worse still, the iterative scheme may *diverge*, i.e., the minimization algorithm may not even converge to a local minimum.

10.4 Uniform Best Approximation

The calculation of best approximating functions with respect to an L_p-norm with $p \neq 2$ requires a significantly higher computational effort than the solution of least squares problems (whose solutions can be obtained by solving a system of *linear* equations). Furthermore, there are relative few mathematical software products available for approximation problems with respect to $\| \; \|_p$ for $p \neq 2$. Therefore, before taking up such problems in practice, other approaches should be given serious consideration: for instance utilizing the interpolation principle or the least squares principle may lead to a satisfactory solution.

As already mentioned in the context of robust distance functions (Chapter 8), the maximum norm $\| \; \|_\infty$ is greatly disadvantageous for data approximation problems: it depends greatly on outliers i.e., data points in *extreme positions*. Therefore, the maximum norm is inappropriate for data approximation problems where the data is afflicted with errors. The maximum norm is perhaps appropriate for data approximation problems with exact data which are packed very densely with respect to the behavior of the underlying function; but generally the use of a (piecewise) interpolation polynomial is more favorable than uniform best approximation.

The maximum norm is important for *function approximation* since the absolute error quite often represents a natural measure of the approximation quality for this type of problem. Therefore, from a sequence of uniformly best approximating functions with $k = 1, 2, 3, \ldots$ parameters, the optimum function is the one which complies with the accuracy requirement

$$D(g, f) = \|g - f\|_\infty = \max \left\{ |g(x; c) - f(x)| : x \in [a, b] \right\} \leq \tau$$

and has the smallest number of parameters $c = (c_1, \ldots, c_k)$. As every function approximation generally aims at obtaining an easy to manage model function g, the best approximating function with the smallest number of parameters is the most favorable selection from the prescribed class of functions $\mathcal{G}$.

Software (Uniformly Best Approximating Functions) For a user-definable function which is linear in the parameters, the subprogram NAG/e02gcf can be used to determine parameters which minimize the L_∞-norm of the residual. To determine the uniformly best rational approximation functions with given numerator and denominator degrees the program IMSL/ratch is available.

10.4.1 Uniformly Best Approximating Polynomials

Computing the value of a polynomial (using the Horner scheme) requires a computational effort which increases linearly with the degree of the polynomial; the same is true for differentiating or integrating a polynomial. Therefore, the uniformly best approximating polynomials P_d^* represent a kind of quality standard: for a given error tolerance they enable one to choose the lowest degree of all approximating polynomials.

In the light of this advantageous property of uniformly best approximating polynomials, it would seem appropriate to calculate $P_1^*, P_2^*, \ldots$ until an estimate of the error $P_d^* - f$ indicates that the required accuracy has been reached. There are algorithms which iteratively calculate the best approximating polynomial $P_d^* \in \mathbb{P}_d$ of a given degree d. The basis of all these algorithms is the following theorem:

Theorem 10.4.1 (Chebyshev)   *For any function $f \in C[a,b]$ there exists a unique, uniformly best approximating polynomial P_d^* of degree d which is characterized by the existence of $d + 2$ points*

$$a \leq x_0 < \cdots < x_{d+1} \leq b, \qquad (10.17)$$

for which

$$(-1)^i[P_d^*(x_i) - f(x_i)] = \sigma\|P_d^* - f\|_\infty, \qquad i = 0, 1, \ldots, d+1, \qquad (10.18)$$

where $\sigma := \mathrm{sgn}(P_d^*(x_0) - f(x_0))$.

Proof: Davis [40].

Example (Root Function)   The uniformly best approximating polynomial P_1^* is determined for the function $f(x) = \sqrt{x}$ on $[a,b] \subset \mathbb{R}_+$. For

$$P_1^*(x) = c_0 + c_1 x$$

the approximation error

$$e_1(x) = c_0 + c_1 x - \sqrt{x}$$

and its derivative

$$e_1{}'(x) = c_1 - \frac{1}{2\sqrt{x}}$$

are obtained. Setting $e_1{}'$ to zero shows that the function e_1 has its only extreme value in (a, b) at the point

$$x_m = \frac{1}{4c_1^2}.$$

Since according to Theorem 10.4.1 the maximum deviation $E_1 = \pm\|e_1\|_\infty$ occurs at three points in the interval $[a, b]$, two boundary extrema must exist

$$c_0 + c_1 a - \sqrt{a} = E_1$$

$$c_0 + \frac{1}{4c_1} - \frac{1}{2c_1} = -E_1$$

$$c_0 + c_1 b - \sqrt{b} = E_1.$$

This system of nonlinear equations leads to

$$c_0 = \frac{1}{2}\left[\sqrt{a} - \frac{a}{\sqrt{a}+\sqrt{b}} + \frac{\sqrt{a}+\sqrt{b}}{4}\right]$$

$$c_1 = \frac{1}{\sqrt{a}+\sqrt{b}}$$

$$E_1 = c_0 + c_1 a - \sqrt{a}.$$

The Calculation of Uniformly Best Approximating Polynomials

The calculation of P_d^* on the basis of Theorem 10.4.1 is not an easy task for higher polynomial degrees and complicated functions which are to be approximated. Assuming that $f \in C^1[a, b]$, one can determine the $d+1$ coefficients of P_d^*, the d extrema $x_1, \ldots, x_d \in (a, b)$ and the maximum error $\delta := \|P_d^* - f\|_\infty$ under the constraint (10.17) from the system of nonlinear equations

$$
\begin{aligned}
P_d^*(x_i) - f(x_i) &= (-1)^i \delta, & i &= 0, 1, \ldots, d+1 \\
(P_d^*)'(x_i) - f'(x_i) &= 0, & i &= 1, 2, \ldots, d.
\end{aligned}
\tag{10.19}
$$

The *Remez algorithm* applies the Newton method to the system of nonlinear equations (10.19) in a way specially adapted to this approximation problem. The computational effort needed to determine P_d^* (which is only part of the solution of the approximation problem) is substantial. This effort, however, does not pay off when consideration is given to the fact that according to Theorem 9.3.6, the error bound

$$E_d(f) = \|P_d - f\|_\infty \le E_d^*(f)[1 + \Lambda_d(K)]$$

holds for polynomials P_d interpolating at the nodes of the matrix K. By using the Chebyshev zeros or extrema as interpolation nodes, a nearly optimal approximating polynomial P_d is obtained and by utilizing the fast Fourier transform a relatively small computational effort is required.

Example (Exponential Function) The function $f(t) = e^t$ is approximated on $[-1, 1]$ by a cubic polynomial. The uniformly best approximating polynomial P_3^* has an error $E_3^*(f) = 5.53 \cdot 10^{-3}$. Therefore, in the worst case, the polynomial P_3 interpolating f at the Chebyshev zeros has an error $E_3(f)$ which is at most 2.85 times larger than $E_3^*(f)$ as $\Lambda_3(K^T) = 1.85$. Actually, the error $E_3(f) = 6.66 \cdot 10^{-3}$ is only 20 % higher than the optimal error $E_3^*(f)$.

Up to $d = 111$ the polynomial P_d interpolating the Chebyshev zeros is less accurate than the uniformly best approximating polynomial P_d^* by at most half a decimal digit because $\Lambda_{111}(K^T) \le 4\, E_{111}(f) \le 5 E_{111}^*(f)$.

This example is typical as in practice the accuracy lost by using interpolation instead of best approximation is characterized by a factor which is well below $1 + \Lambda_d(K)$. In most cases the insignificant difference between the approximation errors of P_d and P_d^* is easily compensated by increasing the degree of the interpolation polynomial by 1 or 2. Therefore, in practice (piecewise) *interpolation polynomials* are used instead of best approximating polynomials.

Convergence Behavior

The convergence $P_d^* \to f$ as $d \to \infty$ is a direct result of the Weierstrass theorem: Since for every $\varepsilon > 0$ and for every sufficiently high polynomial degree d there exists a polynomial P_d complying with

$$\|P_d - f\|_\infty = \max\left\{|P_d(x) - f(x)| : x \in [a, b]\right\} \leq \varepsilon,$$

this holds particularly for the uniformly best approximating polynomial P_d^*:

$$\|P_d^* - f\|_\infty \leq \|P_d - f\|_\infty \leq \varepsilon.$$

As a result of the oscillation (10.18) of the error function $e = P_d^* - f$, the uniformly best approximating polynomial P_d^* coincides with f at $d + 1$ points in $[a, b]$. P_d^* can thus be seen as an *interpolation polynomial* whose nodes—contrary to the usual situation in polynomial interpolation—are *not* known a priori. Thus, $\{P_d^* : d \in \mathbb{N}\}$ is a convergent sequence of interpolation polynomials for every function f that is approximated. According to Theorem 9.3.5 (Faber) though, there exists *no* common node matrix, which is applicable to any such function.

10.5 Approximation Algorithms

E. Grosse created a database [214] containing many literature references concerning approximation algorithms and software products. Analytic reports which do not contain approximation algorithms or programs only appear in this database if they provide criteria for choosing algorithms or programs. Only high-quality algorithms and software are included.

Expensive commercial software products are not included. The main emphasis has been placed on *public domain* approximation software.

Grosse's database uses a tree-like classification scheme to categorize the large amount of literature. The following properties correspond to the roots of the tree:

Form describes the function space. The category `polyrat`, for instance, comprises approximation algorithms based on polynomial or rational functions, whereas spline functions or finite-element spaces are dealt with in `sfem`.

Norm specifies the underlying distance measure. For example, the selection of `lp` leads to approximation algorithms based on one of the l_p-norms. In the category `interp`, on the other hand, interpolation is the approximation principle. In this case the distance between the approximating function and the given data points is zero.

Variable specifies the position of the data points (*sampling points*) or the position and shape of the approximation domain. Possible coordinate transformations are dealt with as well. Thus, algorithms that approximate a given function on certain one- or higher-dimensional domains (Cartesian product domains, spheres etc.) are found in the category geometry. On the other hand, algorithms dealing with the approximation of functions at grid points or the generation of grids are found in the category mesh.

There are other classification categories covering literature which is only remotely associated with the topic of approximation. These categories, however, cover information within the context of approximation algorithms. The *leaves* of the tree consist of lists with references to literature belonging to the chosen classification features. These literature references are annotated, i.e., contain remarks on the subject matter covered by the publications.

When looking for suitable quotations, the user runs through the tree from the root to the node, which contains only leaves, i.e., the sought references.

The Grosse database is publically accessible in netlib/a/catalog.html, where the extension html indicates that the file format uses the hyper-text markup language (HTML). This file format is used in the WWW system (cf. Section 7.3.5) to link various parts of (possibly different) documents. Here HTML is used to represent the classification tree of the database. The user may thus browse the database using WWW graphical user interfaces (like Mosaic) by selecting the desired classification characteristics.

In the remaining part of this section this literature search process is illustrated by examples. First of all a WWW connection to the Grosse database has to be made. The WWW address is

$$\text{ftp} : //\text{netlib.att.com/netlib/a/catalog.html}.$$

After establishing the connection, the roots of the classification tree are shown in a window (cf. Fig. 10.6). Clicking on a tree node (i.e., a word highlighted by underlining), brings the successors of the selected node on the screen. Clicking on, for instance, form, the window shown in Fig. 10.7 appears. By continued clicking of nodes it is possible to reach the desired documents or software references. Thus, selecting the nodes form, polyrat, poly, Cheby and Cheby-12 produces information concerning least squares polynomial approximation utilizing the Chebyshev basis (cf. Fig. 10.8). By clicking these references detailed bibliographic information is obtained (cf. Fig. 10.9).

Figure 10.6: Roots of the classification tree of the Grosse database [214].

Figure 10.7: The next window after selecting form in Fig. 10.6.

Figure 10.8: References to least squares approximation with Chebyshev polynomials.

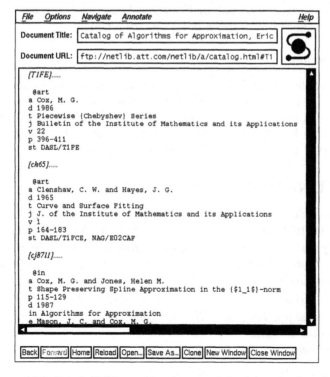

Figure 10.9: Detailed bibliographic references of the Grosse database [214].

10.6 Approximation Software for Special Functions

10.6.1 Elementary Functions

Elementary functions play an important role in applied mathematics. Taking this into account, many elementary functions have been included in scientific programming languages (like Fortran or C) in the form of pre-defined subprograms.

Example (Fortran 90)   In Fortran 90 there are pre-defined function subprograms:

Square root function:	SQRT
Trigonometric functions:	SIN, COS, TAN
Arcus functions:	ASIN, ACOS, ATAN, ATAN2
Exponential function:	EXP
Logarithmic functions:	LOG, LOG10
Hyperbolic functions:	SINH, COSH, TANH

Great care is taken with the implementation of these subprograms. The approximate values produced by these routines are as accurate as, or slightly less accurate than, machine accuracy allows. The run time efficiency of these routines is also optimized as much as possible. Therefore, there is no need to develop such routines oneself.

Should it be necessary for the user to develop standard function subprograms himself, then it is recommended that well-tried and reliable algorithms be used. For instance the reference book by Cody and Waite [8] is a source of excellent algorithms on the implementation of elementary functions.

Standardization

As opposed to the standards for floating-point number systems established many years ago, at the moment there are neither standards for implementing elementary functions nor for their inclusion in scientific programming languages. However, an extension of the ISO/IEC standard [239] covering the mathematical elementary functions—*Part 2: Mathematical Procedures*—is on the way.

10.6.2 FUNPACK

In addition to EISPACK (Cody [138]), FUNPACK (Smith et al. [30]) was also developed in the 1970s as a part of the NATS project (*National Activity to Test software*) by American universities and research institutions (see Cowell [10], Chapter 3). Among other things it contains Fortran subprograms for the calculation of approximate values of the following special functions:

Bessel functions:	$J_0, J_1, I_0, I_1, K_0, K_1, Y_\nu$
Complete elliptic integrals:	$K(m), E(m)$
Exponential integral:	Ei
Psi function:	Ψ
Dawson integral:	$e^{-x^2} \int\limits_0^x e^{t^2} dt$

The FUNPACK can be ordered for a small fee from the *National Energy Software Center, Argonne National Laboratory, 9700 South Cass Avenue, Argonne, Illinois 60439, USA.*

10.6.3 IMSL Libraries

The *special functions* part of the IMSL Fortran library contains approximately 160 subprograms, which are among other things dedicated to the approximation of the following functions:

Complex elementary functions:	sin, cos, tan, ...
Complex hyperbolic functions:	sinh, cosh, tanh, ...
Exponential integrals and the like:	Ei, li, Si, Cin, ...
Gamma function and the like:	Γ, γ, Ψ
Error functions:	erf, erfc
Fresnel integrals:	S, C
Bessel functions:	J_0, J_1, I_0, I_1, K_0, K_1, Y_0, Y_1, ...
Kelvin functions:	ber, bei
Airy functions:	Ai, Bi, Ai', Bi'
Elliptic integrals:	K, E, R_C, R_F, R_D, R_J
Density and distribution functions	

10.6.4 NAG Libraries

The NAG Fortran library contains about 60 subprograms designed to calculate approximate values of the following (and many other) functions:

Hyperbolic functions:	sinh, cosh, tanh, Arsinh, Arcosh, Artanh
Bessel functions:	$J_{\nu+a}$, $Y_{\nu+a}$, $a \geq 0$, $\nu = 0, 1, 2, \ldots$,
	$I_{\nu+a}$, $K_{\nu+a}$, $a \geq 0$, $\nu = 0, 1, 2, \ldots$
Elliptic integrals:	R_C, R_F, R_D, R_J
Airy functions:	Ai, Bi, Ai', Bi'
Hankel functions:	$H_{\nu+a}^{(1)}$, $H_{\nu+a}^{(2)}$, $a \geq 0$, $\nu = 0, 1, 2, \ldots$
Fresnel integrals:	S, C
Kelvin functions:	ber, bei, ker, kei

Chapter 11

The Fourier Transform

For a short time, therefore, allow your thought to leave this world in order to come to see a wholly new one, which I shall cause to be born in the presence of your thought in imaging spaces.

RENE DESCARTES

11.1 Background

Waves, that is regular temporal or spatial variations of state quantities, occur in nature and in all fields of science and engineering. In particular, waves in which state quantities change *periodically* play a central role; for instance, when a variable of state can be described by a univariate periodic function f:

$$f(t + \tau) = f(t) \qquad \text{for all} \quad t \in \mathbb{R}. \tag{11.1}$$

The smallest number $\tau > 0$ which complies with equation (11.1) is the *period* of f; its reciprocal value $1/\tau$ is the *frequency* of the oscillation, and $\omega := 2\pi/\tau$ is its *angular frequency*.

In many technical scientific fields, differential equations are used to model oscillations. One simple example is the harmonic oscillator (used for example to model oscillating mechanical systems or electric circuits):

$$\frac{\mathrm{d}^2 x}{\mathrm{d}t^2} = -\omega_0^2 x$$

with the general sinusoidal solution

$$x(t) = A \cos \omega_0 t + B \sin \omega_0 t = C \sin(\omega_0 t + \varphi).$$

This solution is completely characterized by three parameters: C is the *amplitude*, ω_0 is the *angular frequency*, and φ is the *phase*. Another well-known example is the wave equation

$$\frac{\partial^2 \Psi}{\partial t^2} = \left(\frac{\lambda \omega}{2\pi}\right)^2 \frac{\partial^2 \Psi}{\partial z^2}. \tag{11.2}$$

In this equation the quantity z represents the coordinate of the position, whereas t represents the time. Consequently, the function $\Psi(z, t)$ describes the amplitude of a wave at position z and time t. The wave equation (11.2) is a linear partial differential equation. Solving the equation using a separation ansatz leads to the family of solutions

$$\Psi(z, t) = \cos(\omega t + \varphi)\left[A \sin(2\pi z/\lambda) + B \cos(2\pi z/\lambda)\right] \tag{11.3}$$

for each value of ω.

The first factor in the solution represents the temporal oscillation of the wave, where ω denotes its angular frequency and φ its phase. The second term describes the spatial shape of the wave, where λ denotes the *wavelength*.

In most cases, the wave equation (11.2) contains, instead of $\lambda\omega/2\pi$, a term which depends on the modeled system, and which describes the connection between λ and ω. The behavior of a sinusoidal wave is generally of less interest than the behavior of more waves. If, for example, the form of a wave at time $t=0$ is known, it is of interest to find its temporal development.

On account of the linearity of (11.2), a finite number of solutions of the form (11.3) can be added, and the sum is again a solution of (11.2). Moreover, the collection of terms of the form

$$A\sin(2\pi z/\lambda) + B\cos(2\pi z/\lambda)$$

where $\lambda \in \mathbb{R}$ constitutes a complete system of functions, which can be used to represent any periodic function $f(z)$ (see Section 11.2.1). The method used for achieving this representation is the *Fourier transform* !

The example problems given above and many other problems have one property in common: their solutions are *trigonometric series* or *trigonometric polynomials*

$$S_d(t) = \frac{a_0}{2} + \sum_{k=1}^{d}(a_k \cos kt + b_k \sin kt).$$

Data sequences originating from such problems are therefore often interpolated or approximated using trigonometric polynomials.

Historical Background (Fourier Transform)   Before and after the time of J. de Fourier (1768 – 1830), after whom the whole subject is named, a great number of mathematicians have concerned themselves with the analysis of functions and data using trigonometric functions.

The use of trigonometric series in analysis first occurred in the work of L. Euler (1707 – 1783). He presented the formulas for the coefficients of the Fourier series representation of a function of a real variable. Euler used a trigonometric series to describe sound propagation in an elastic medium.

The stature of Euler in his own time assured that his work was read by his contemporaries, particularly a number of French mathematicians. Among them was A. C. Clairaut (1713 – 1765), who in 1754 published the earliest formula for the discrete Fourier transform. His formula was restricted, however, to a cosine-only finite Fourier series. J. L. Lagrange (1736 – 1813) published a formula for sine-only series in 1762. In 1753 D. Bernoulli (1700 – 1782) expressed the displacement of a vibrating string as a series of sine and cosine terms with both position and time as arguments. This implied that an arbitrary function could be expressed as an infinite sum of sines and cosines.

Clairaut and Lagrange were concerned with orbital mechanics and the problem of determining the details of an orbit from a finite set of observations. C. F. Gauss (1777 – 1855) extended the work of Euler and Lagrange dealing with trigonometric interpolation to include periodic functions which are not necessarily odd or even. This was done while considering the problem of determining the orbit of certain asteroids from sample locations. Gauss developed an algorithm similar to the Cooley-Tukey *Fast Fourier Transform* (see Section 11.6.1), which was based on the reduction of one large Fourier transform to several smaller ones for computing the coefficients of a finite Fourier series. Gauss' treatise describing the algorithm appeared in his collected works as an unpublished manuscript. The presumed year of the creation of his algorithm is 1805. Though not influencing the work of Cooley and Tukey, the roots of the FFT algorithm date from the early *nineteenth* century!

Applications of the Fourier Transform

Problems that can be investigated using the Fourier transform can be found in

Signal Processing: Fourier transforms are used in signal processing to find out what a given signal looks like after it has been sent through a certain filter and how a signal that has been perturbed by a transmission channel can be recovered as precisely as possible.

Image Processing: In digital image processing, a (two-dimensional) image signal can be reduced to its spatial frequencies using the discrete Fourier transform. The spectrum obtained in this way can be modified and, using the inverse Fourier transform, a different image signal can be generated. For example, a damping of the low spatial frequencies allows for the reduction of *slow* signal changes. The reduction of the spatial frequencies allows for an increase in the image contrast, which in turn makes edges and small details easier to see. In this way, image blurredness caused by imperfections in the recording equipment can be reduced.

> **Example (Astronomy)**   In the production of the *Hubble Space Telescope,* a mirror was not constructed correctly. Because of this production error (which has been corrected), the pictures taken by the telescope were blurred. Digital image processing was applied to the blurred pictures in an attempt to compensate for this defect.

Differential Equations: Certain differential equations—such as the wave equation (11.2)—are easier to handle in the frequency domain than in the original domain. Moreover, some methods for the numerical solution of differential equations (especially of boundary value problems) are based on the expansion of the desired solution in terms of its spectral components.

Voice Recognition and Acoustics: Using frequency analysis it is possible to compare the speech characteristics of individuals. It can also be used for voice recognition.

Analysis and Synthesis

Interpolation or approximation of periodic functions requires an appropriate class of functions $\mathcal{G}$, the approximating functions. The choice of $\mathcal{G}$ is dictated by the class of modeling problems at hand.

Synthesis Problem: Approximations $g \in \mathcal{G}$ for functions f are required in terms of as few and as elementary basic functions as possible. The purpose of this kind of approximation is a data reduction effect or the production of a substitute function which is easy to generate using, for example, special electronic devices). For the solution of such problems, not only the trigonometric functions found in this chapter, but also other classes of functions such as *wavelets* (Pittner, Schneid, Ueberhuber [23]) should be considered.

Analysis Problem: The coefficients of an expansion of f with respect to a certain system of functions are to be determined. These coefficients allow for a concrete interpretation of and supply important information about certain properties of the function f which are important for practical applications (especially those properties that can be directly seen from the spectrum; see Section 11.2.2).

Example (Thyristor) A sinusoidal electric current can be controlled using a thyristor (i.e., a special kind of an electronic switch). At each zero-crossing of the voltage the thyristor turns on the current after a variable time delay δ (see Fig. 11.1).

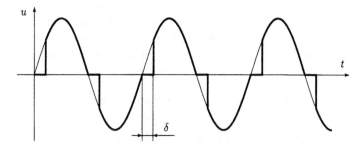

Figure 11.1: Graphical representation of the voltage u influenced by thyristor-controlling.

The application of these electronic switches ranges from household appliances (dimmers, etc.) to power electronics (motor controls, etc). If the voltage behavior of a thyristor is to be simulated, when developing, for example, an anti-interference device for the thyristor, then a simple model-function is required (synthesis problem). The simplest model-function for the given course of voltage is a piecewise defined function with the amplitude $u_{\max}$:

$$\bar{u}(t) := \begin{cases} u_{\max}\sin(\omega t + \varphi) & t \in \left[\dfrac{k\pi + \delta - \varphi}{\omega}, \dfrac{k\pi + \pi - \varphi}{\omega}\right] \\ 0 & t \in \left(\dfrac{k\pi - \varphi}{\omega}, \dfrac{k\pi + \delta - \varphi}{\omega}\right) \end{cases} \quad \text{where} \quad k \in \mathbb{Z}.$$

Since, in real applications, the voltage behavior is not exactly sinusoidal in the intervals

$$[(k\pi + \delta - \varphi)/\omega, (k\pi + \pi - \varphi)/\omega],$$

the function $\bar{u}$ is only a model of reality. Its simplicity has the advantage that for simulation purposes $\bar{u}$ can be generated without difficulty using an electronic function generator. However, if the user is interested in numerically determining those harmonic components interfering with the medium-wave RF-band (approx. 0.5 - 1.5 MHz), the function $\bar{u}$ has to be decomposed into a fundamental wave and harmonic oscillations by solving an analysis problem:

$$\bar{u}(t) = \sum_{k=0}^{\infty} d_k \sin(k\omega t + \varphi_k).$$

The coefficients of the harmonic components whose frequencies belong to the RF-band lead to the desired information.

Signals

In telecommunications terminology, functions of time or of location which convey information are called *signals*.

Terminology (Signal)   In this chapter the expression *signal* is often used for a function which is to be investigated *without* considering the information contained in it.

Signals are transmitted, stored and processed in various ways or evaluated according to their information content (Proakis, Manolakis [319]). In this case as well the two previously mentioned types of problems can be distinguished: For the transmission and storage of a signal, a representation as simple as possible should be sought (synthesis problem), whereas the information content can be recovered more easily from other forms of representation (e. g., from an expansion with respect to an orthogonal system) (analysis problem).

For the expansion of a signal with respect to an orthogonal system, the information conveyed by the signal is contained in the coefficients of the expansion. Among the many orthogonal systems used for representing signals trigonometric functions $\sin kt$, $\cos kt$, $k = 0, 1, 2, \ldots$, play a key role because

- the mapping of the signal into the *frequency domain* defined by such an expansion allows an immediate physical interpretation;

- the trigonometric functions are the eigenfunctions of linear time-invariant systems[1] and, as a result, form a natural basis for the representation of periodic oscillations;

- for the computation of the signal mapping into the frequency domain, the fast Fourier transform (FFT) algorithm (see Section 11.6.1) can be used. With this extremely efficient tool, convolutions and correlations can be calculated with little computational effort since these operations can be carried out as multiplication operations in the frequency domain.

11.2 Mathematical Foundations

The *Fourier transform F* of a function f is defined as follows:

$$F(\omega) := \int\limits_{-\infty}^{\infty} f(t) e^{2\pi i \omega t} dt, \tag{11.4}$$

provided that the integral exists as a Cauchy principal value[2] for every $\omega \in \mathbb{R}$. The Fourier transform maps the function $f(t)$, a signal taken as a function of time,

---

[1]The term *time-invariance* means that for a temporal shift of the input signal by any time increment, the output signal is shifted by the same time increment. For the sake of simplicity, real physical systems are often assumed to be time-invariant.

[2]The Cauchy principal value of an integral from $-\infty$ to $+\infty$ is the limit $\int_{-\infty}^{\infty} x(t)\, dt = \lim_{A \to \infty} \int_{-A}^{A} x(t)\, dt$.

into a complex valued function $F(\omega)$, which represents the signal as a function of the frequency.

The inverse operation to (11.4) is the *inverse Fourier transform*:

$$f(t) = \int\limits_{-\infty}^{\infty} F(\omega)e^{-2\pi i \omega t}\, d\omega. \qquad (11.5)$$

With this transform signals can be (re-)transformed from the frequency domain into the time domain.

The representation of a signal as a function of time is also said to be a representation in the *time domain*; the representation of a signal as a function of frequency is said to be a representation in the *frequency domain*. Both representations portray the same signal. Often the representation in the time domain appears more *natural* (if the signal is recorded or measured in this form), whereas the representation in the frequency domain is more suitable for filtering, i.e., manipulations of the spectrum (see Section 11.2.2 and Section 11.5).

Continuous functions $f(t)$ do not often appear in technical applications. People in practice often have to content themselves with a finite number of individual observations (*samples*) of the function at certain time points. In the following it is assumed that the sampling of f is carried out at regular intervals of size Δ. The quantity $1/\Delta$ is the *sampling-rate*; it gives the number of discrete values of f available per time unit.

Definition 11.2.1 (Nyquist Frequency) *For every sampling interval Δ,*

$$\omega_c := \frac{1}{2\Delta}$$

is called the Nyquist frequency.

The importance of this quantity is made clear in the following section.

Discrete Fourier Transform (DFT)

The following considerations are devoted to the Fourier transform of discrete (sampled) data sequences. Assuming that there are N data points, where, for sake of simplicity, it is presupposed that N is even (although all considerations can, of course, be made with an odd number N):

$$f_k := f(t_k), \qquad t_k := (k_0 N + k)\Delta, \qquad k = 0, 1, \ldots, N-1.$$

The translation constant is written $k_0 N$ for the sake of notational simplicity.

For piecewise continuous functions f which satisfy the condition

$$\int\limits_{-\infty}^{\infty} |f(t)|\, dt < \infty,$$

the Fourier transform F always exists. In this case the infinite integral (11.4) can be approximated by a finite sum (for a suitable choice of N, Δ and k_0) using numerical integration formulas:

$$F(f) = \int_{-\infty}^{\infty} f(t)e^{2\pi i\omega t}dt \approx \sum_{k=0}^{N-1} f_k e^{2\pi i\omega t_k}\Delta.$$

Moreover, f can be interpolated, according to Shannon's sampling theorem (Theorem 11.2.1), by a continuous function g with the properties

$$\lim_{t\to\pm\infty} g(t) = 0$$

and

$$G(\omega) = 0 \qquad \text{for} \quad \omega \notin (-\omega_c, \omega_c)$$

at the points t_k, i.e.,

$$f(t_k) = g(t_k), \qquad k = 0, 1, \ldots, N-1.$$

G denotes the Fourier transform of g and ω_c the Nyquist frequency.

The Fourier transform F of f can now be represented approximately by estimates of its function values at the points

$$\omega_n = \frac{n}{N\Delta}, \qquad n = -\frac{N}{2}, -\frac{N}{2}+1, \ldots, \frac{N}{2}-1:$$

$$F(\omega_n) \approx \Delta \sum_{k=0}^{N-1} f_k e^{2\pi i \frac{n}{N\Delta}(k_0 N+k)\Delta} = \Delta \sum_{k=0}^{N-1} f_k e^{2\pi ikn/N}. \qquad (11.6)$$

Using this formula, N discrete function values f_k lead to N values of F (for the discrete frequencies ω_n). Thus, instead of determining the entire Fourier transform $F(\omega)$ in the range $[-\omega_c, \omega_c]$, F is only determined for the discrete frequencies ω_n. The formula (11.6) is the *discrete Fourier transform* (DFT), and in this book it is always denoted by F_n (which leaves out the factor Δ):

$$F_n := \sum_{k=0}^{N-1} f_k e^{2\pi ikn/N}. \qquad (11.7)$$

This transform is periodic in n with period N, i.e., $F_{N+n} = F_n$ for every $n \in \mathbb{Z}$.

The connection between the continuous Fourier transform F of the function f and the discrete Fourier transform F_n of the data f_k, obtained by sampling f with a sampling interval Δ is represented as follows:

$$F(\omega_n) \approx \Delta F_n. \qquad (11.8)$$

In (11.8) $n = 0$ corresponds to the frequency $\omega = 0$; the positive frequencies $0 < \omega < \omega_c$ correspond to $1 \leq n \leq N/2 - 1$; and the negative frequencies

$-\omega_c < \omega < 0$ correspond to the values $N/2 + 1 \leq n \leq N-1$. For $n = N/2$, n corresponds to both of the frequencies ω_c and $-\omega_c$.

The *inverse discrete Fourier transform* (inverse DFT) is given by:

$$f_k = \frac{1}{N} \sum_{n=0}^{N-1} F_n e^{-2\pi i k n/N}, \tag{11.9}$$

as can be derived from (11.5) analogously to (11.6). The only differences between (11.9) and (11.7) are the negative sign in the exponential and the division of the sum by N. Therefore, to a large extent the calculation of the discrete Fourier transform and its inverse can be done using the same routines.

For a sinusoidal wave the Nyquist frequency ω_c is the highest frequency that is still reconstructible at a fixed length of the sampling interval Δ. This is because, on the one hand, the argument of the expression $\sin(2\pi\omega_0 t)$ with frequency $\omega_0 \leq \frac{1}{2\Delta}$ can move two quadrants, at the most, from one sampling to the next on the unit circle; and if, on the other hand, for the given sampling points, there exists a sinusoidal wave with a frequency $> \frac{1}{2\Delta}$, then there must always be one sinusoidal wave with a frequency $< \frac{1}{2\Delta}$. That is

$$\sin\left(2\pi\left(\frac{1}{2\Delta}+\varepsilon\right)(t_0 + k\Delta)\right) = -\sin\left[2\pi\left(\frac{1}{2\Delta}-\varepsilon\right)\left(\frac{t_0}{\varepsilon\Delta - \frac{1}{2}} + (t_0 + k\Delta)\right)\right]$$

for all values $t_0, \varepsilon \in \mathbb{R}$ and $k \in \mathbb{Z}$. Thus, at the very least, these two sample points per period are required in order to reconstruct a sine wave correctly.

Another restriction on the sampling of a function with frequency ω appears if $N\omega/2\omega_c$ is not an integer. In such cases none of the possible discrete frequencies $\omega_n = 2\omega_c n/N$ appear in the discrete spectrum. The frequency in the discrete spectrum which is nearest to the original frequency ω is clearly the largest; however, all other frequencies also appear more or less strongly in the spectrum. This phenomenon is called *leakage*.

Theorem 11.2.1 (Shannon's Sampling Theorem) *Let f denote a function which satisfies $\int_{-\infty}^{\infty} |f(t)|^2 \, dt < \infty$, i.e., a signal with finite energy. This signal is assumed to be sampled at a rate $1/\Delta$. If f is band limited in its continuous frequency spectrum by the Nyquist frequency $\omega_c = 1/(2\Delta)$, i.e., if for the Fourier transform F the relation $F(\omega) = 0$ holds for all ω with $|\omega| \geq \omega_c$, then the function f can be recovered without any errors from its sample values using the interpolation function*

$$g(t) = \frac{sin(2\pi\omega_c t)}{2\pi\omega_c t}. \tag{11.10}$$

Thus, f may be expressed as

$$f(t) = \sum_{n=-\infty}^{\infty} f_n g(t - n\Delta), \tag{11.11}$$

where $f_n := f(n\Delta)$ are the samples of f.

If the function f is not band limited, then in the Fourier transform of f all components of the frequency spectrum located outside of $[-\omega_c, \omega_c]$ are moved into this frequency range. This spectral overlap is called *aliasing*. In this case the function can not be completely reconstructed using its sample values.

Thus, the length Δ of the sampling interval has to be selected so that the critical frequency ω_c is higher than all frequencies appearing in the spectrum of the data. Whether this requirement is satisfied can be seen if the frequency spectrum $F(\omega)$ of the data approaches 0, as the frequency ω approaches ω_c. If this is not the case, then remedial action can be taken by shortening the sampling interval or by restricting the signal's frequency spectrum (for instance using low-pass filtering) before sampling.

11.2.1 Trigonometric Approximation

For the approximation of periodic functions, the *trigonometric polynomials* of order d,

$$S_d(t) = \frac{a_0}{2} + \sum_{k=1}^{d}(a_k \cos kt + b_k \sin kt), \qquad (11.12)$$

play an important role. The coefficients $a_0, a_1, \ldots, a_d, b_1, b_2, \ldots, b_d$ of S_d are assumed to be real in this chapter. Every trigonometric polynomial of this form is 2π-periodic. Accordingly, if a τ-periodic function f with $\tau \neq 2\pi$ is to be approximated using a trigonometric polynomial of the form (11.12), then scaling is required. That is, the 2π-periodic function

$$\hat{f}(t) := f\left(\frac{\tau}{2\pi}t\right)$$

is approximated by a polynomial S_d of the form (11.12). A τ-periodic, trigonometric polynomial S_d^τ that approximates f is then given by

$$S_d^\tau(t) := S_d\left(\frac{2\pi}{\tau}t\right).$$

A trigonometric polynomial S_d can be represented in various equivalent ways. Utilizing the *Eulerian identities*

$$\sin t = \frac{e^{it} - e^{-it}}{2i}, \qquad \cos t = \frac{e^{it} + e^{-it}}{2},$$

S_d can be given the simple form

$$S_d(t) = \sum_{k=-d}^{d} c_k e^{ikt} \qquad (11.13)$$

with the coefficients

$$c_k = \begin{cases} (a_{-k} + ib_{-k})/2, & k = -d, -d+1, \ldots, -1 \\ a_0/2, & k = 0 \\ (a_k - ib_k)/2, & k = 1, 2, \ldots, d. \end{cases}$$

These relations can be inverted:

$$\begin{aligned} a_k &= c_k + c_{-k}, & k &= 0, 1, \ldots, d \\ b_k &= i(c_k - c_{-k}), & k &= 1, 2, \ldots, d. \end{aligned} \tag{11.14}$$

A complex number z is real if and only if it equals to its complex conjugate $\bar{z}$. This can be applied to the representation (11.13)

$$\overline{\sum_{k=-d}^{d} c_k e^{ikt}} = \sum_{k=-d}^{d} \bar{c}_k e^{-ikt} = \sum_{k=-d}^{d} \bar{c}_{-k} e^{ikt}.$$

S_d is, due to the linear independence of the set $\{e^{ikt} : k \in \mathbb{Z}\}$, real if and only if

$$c_k = \bar{c}_{-k}, \qquad k = 0, 1, \ldots, d.$$

Since S_d has been assumed to be a real function, the relation (11.14) is simplified to

$$\begin{aligned} a_k &= 2\,\mathrm{Re}(c_k), & k &= 0, 1, \ldots, d \\ b_k &= -2\,\mathrm{Im}(c_k), & k &= 1, 2, \ldots, d. \end{aligned}$$

The trigonometric approximation is closely related to the expansion of a periodic function into a *Fourier series*

$$f(t) = \sum_{k=-\infty}^{\infty} c_k e^{ikt}. \tag{11.15}$$

Terminology (Fourier Series) In this section a series of the form (11.15) is called a Fourier series, whereas in Section 10.2 *every* expansion with respect to an orthonormal system is called Fourier series.

The natural definition of an inner product for two piecewise continuous complex valued functions f and g with period 2π is given by

$$\langle f, g \rangle := \frac{1}{2\pi} \int_0^{2\pi} f(t) \bar{g}(t) dt.$$

With respect to this inner product, the complex valued exponentials e^{ikt} form an orthonormal system:

$$\begin{aligned} \langle e^{ikt}, e^{ilt} \rangle &= \frac{1}{2\pi} \int_0^{2\pi} e^{ikt}\, \overline{e^{ilt}}\, dt \\ &= \frac{1}{2\pi} \int_0^{2\pi} e^{i(k-l)t}\, dt \\ &= \begin{cases} \dfrac{1}{2\pi} \displaystyle\int_0^{2\pi} 1\, dt = 1 & \text{for } k = l \\[2ex] \dfrac{1}{2\pi} \dfrac{1}{i(k-l)} e^{i(k-l)t} \Big|_0^{2\pi} = 0 & \text{for } k \neq l. \end{cases} \end{aligned}$$

Accordingly, the results of Section 10.2 lead to the result that the partial sum (the *Fourier polynomial*)

$$\sum_{k=-d}^{d} \langle f(t), e^{ikt} \rangle \, e^{ikt} \tag{11.16}$$

of the Fourier series

$$\sum_{k=-\infty}^{\infty} \langle f(t), e^{ikt} \rangle \, e^{ikt} \tag{11.17}$$

of f is the best-approximating trigonometric polynomial of order d with respect to the L_2-norm

$$\|u\|_2 = \sqrt{\langle u, u \rangle} = \left(\frac{1}{2\pi} \int_0^{2\pi} |u(t)|^2 dt \right)^{\frac{1}{2}}.$$

Generally, the Fourier polynomial (11.16) is *not* the best-approximating trigonometric polynomial with respect to the maximum norm (the L_∞-norm). On the one hand, the Weierstrass theorem also holds for trigonometric polynomials:

Theorem 11.2.2 (Weierstrass)   *For every continuous 2π-periodic function $f : \mathbb{R} \to \mathbb{R}$ and for each $\varepsilon > 0$, there exists a $d = d(\varepsilon) \in \mathbb{N}$ and an S_d such that*

$$|S_d(t) - f(t)| < \varepsilon \qquad \text{for all} \quad t \in \mathbb{R}.$$

On the other hand, there are *continuous* 2π-periodic functions that are *not* represented by their Fourier series. This occurs if and only if the Fourier series diverges at at least one point because, if the Fourier series converges then the limit equals the corresponding function value at the point in question.

In such cases, the partial sums of the Fourier series of f cannot be used to construct the trigonometric polynomial S_d appearing in Theorem 11.2.2; more precisely this partial sum does *not* generally yield the L_∞-best-approximating trigonometric polynomial.

Example (Sawtooth Function)   The 2π-periodic function f, defined on $[-\pi, \pi)$ by $f(t) = t$, is an odd function. All its coefficients a_k therefore vanish. The values of the coefficients b_k determined from (11.20) are

$$b_k = \frac{2(-1)^{k-1}}{k}, \qquad k = 1, 2, 3, \dots .$$

It follows that $|c_k| = k^{-1}$ for all $k \in \mathbb{Z}$ according to (11.28) since f is infinitely differentiable on $(-\pi, \pi)$, but as a periodic function f has jump discontinuities.

In this example the discrepancy between the efficient approximation (a synthesis problem) and the reduction of a signal into a fundamental wave and harmonic oscillations (an analysis problem) can be seen. Due to the slowly decreasing Fourier coefficients for a given tolerance ε, a very large number of terms in the partial sum S_d are required in order to approximate f uniformly with an accuracy ε. Considered as a pure approximation problem, a trigonometric polynomial is in this case an extremely unfavorable approximation function. A *piecewise* defined (periodic) approximation function solves such approximation problems with much greater efficiency.

11.2.2 The Spectrum

The coefficients

$$c_k = \frac{1}{2\pi} \int_0^{2\pi} f(t) e^{-ikt} \, dt, \quad k \in \mathbb{Z}, \tag{11.18}$$

in the complex form of the Fourier series (11.17) and the coefficients

$$a_k = \frac{1}{\pi} \int_0^{2\pi} f(t) \cos(kt) \, dt, \qquad k = 0, 1, 2, \ldots, \tag{11.19}$$

$$b_k = \frac{1}{\pi} \int_0^{2\pi} f(t) \sin(kt) \, dt, \qquad k = 1, 2, 3, \ldots, \tag{11.20}$$

in the real form

$$\frac{a_0}{2} + \sum_{k=1}^{\infty} (a_k \cos kt + b_k \sin kt)$$

convey the information contained in the function f (the signal). For the practical recovery of this information—which should be independent of the selected coordinate-system—both representations have an essential disadvantage: The coefficients depend on the choice of the origin of the coordinate-system: The integrals (11.18), (11.19) and (11.20) are *not* translation invariant. If the substitution

$$t_s = t + s$$

is carried out, then, for example, the coefficient a_k^s for the function

$$f_s(t) = f(t_s) = f(t + s)$$

is given by:

$$a_k^s = \frac{1}{\pi} \int_0^{2\pi} f(t+s) \cos kt \, dt = \frac{1}{\pi} \int_s^{s+2\pi} f(t_s) \cos k(t_s - s) \, dt_s.$$

On account of the 2π-periodicity of the integrands, it follows that

$$a_k^s = \frac{1}{\pi} \int_0^{2\pi} f(t_s) \left(\cos kt_s \cos ks + \sin kt_s \sin ks \right) dt_s =$$

$$= a_k \cos ks + b_k \sin ks.$$

Analogously

$$b_k^s = b_k \cos ks - a_k \sin ks.$$

Thus, $a_k^s \neq a_k$ and $b_k^s \neq b_k$ generally holds, i.e., the coefficients $a_0, a_1, \ldots$ and $b_1, b_2, \ldots$ are *not* translation invariant and therefore *not* suitable for recovering the information content of the signal.

A more suitable representation is achieved by using the exponential form of the complex coefficients

$$c_k = |c_k| e^{i\varphi_k}$$

to represent $f(t)$ with the following Fourier series:

$$f(t) = |c_0| \cos \varphi_0 + 2 \sum_{k=1}^{\infty} |c_k| \cos(kt + \varphi_k). \tag{11.21}$$

The quantities $2|c_k|^2 = 2|c_{-k}|^2 = (a_k^2 + b_k^2)/2$ $(k \in \mathbb{N})$ and $c_0^2 = a_0^2/4$ are translation invariant,

$$\begin{aligned}
(a_k^s)^2 + (b_k^s)^2 &= a_k^2 \cos^2 ks + 2 a_k b_k \cos ks \sin ks + b_k^2 \sin^2 ks + \\
&\quad + a_k^2 \sin^2 ks - 2 a_k b_k \cos ks \sin ks + b_k^2 \cos^2 ks = \\
&= a_k^2 + b_k^2;
\end{aligned}$$

moreover, as can be seen in the next paragraph, they have the physical interpretation as the *power* of the kth component of the series (11.21). Often these quantities are shown graphically in the form of a *discrete power spectrum*.

If a 2π-periodic signal f either represents the electric current through or the voltage across a unit resistor, then the electric power is given by

$$\frac{1}{2\pi} \int_0^{2\pi} f^2(t) dt = \langle f, f \rangle = \|f\|^2.$$

Thus, the power of the kth Fourier component for any $k \in \mathbb{N}$ is

$$\||2|c_k| \cos(kt + \varphi_k)\|^2 = 2|c_k|^2 = \frac{a_k^2 + b_k^2}{2}.$$

From the Parseval equation

$$\|f\|^2 = \sum_{k=-\infty}^{\infty} \left| \langle f, e^{ikt} \rangle \right|^2 = \sum_{k=-\infty}^{\infty} |c_k|^2 = |c_0|^2 + 2 \sum_{k=1}^{\infty} |c_k|^2,$$

it can be seen that the power of a periodic signal is equal to the sum of the powers of its Fourier components. Therefore, the set of the quantities $|c_k|^2$ where $k = 0, 1, 2, \ldots$ is called the *power spectrum* of the function f.

The power spectrum contains less information than the Fourier series as it includes no information about the *phase angles* φ_k of the coefficients c_k. In some applications (e. g., in acoustics) the phase angles are of little practical importance, whereas in other fields (e. g., in optics) they are indispensable.

Using (11.21) a periodic signal f with fundamental frequency F_0 is represented by a series of harmonic components. The kth order harmonic component $2|c_k| \cos(2\pi k F_0 t + \varphi_k)$ has the following characteristic quantities:

1. *amplitude* $2|c_k|$,

2. *phase angle* φ_k,

3. *frequency* kF_0,

4. *angular frequency* $2\pi kF_0$ and

5. *period* (wavelength) $1/kF_0$.

The quantity $|c_k|$ is a measure of how much the kth harmonic component contributes to the signal f. The set $\{|c_0|, |c_1|, \ldots\}$ is called the *spectrum* of f and the set $\{\varphi_0, \varphi_1, \ldots\}$ is the *phase spectrum*.

11.3 Trigonometric Interpolation

If the number of points that have to be interpolated is relatively small, or if the problem requires it, then a trigonometric interpolation polynomial can be calculated directly, using conventional methods.

The discrete Fourier analysis starts from a set of equidistant points in an interval that are extended periodically outside the interval. To simplify the calculations, it is presupposed that this interval is $[-\pi, \pi)$, as every finite set of equidistant points can be mapped into the interval $[-\pi, \pi)$ by a simple linear transformation. This leads to the representation

$$x_j = -\pi + j\frac{\pi}{N} \quad \text{with} \quad j = 0, 1, \ldots, 2N-1$$

for $2N$ given points $\{(x_j, y_j) : j = 0, 1, \ldots, 2N-1\}$. According to the properties of the discrete Fourier transform, the interpolation problem, given by these points, is solved by exactly one polynomial of the form

$$S_N(x) = \frac{a_0 + a_N \cos Nx}{2} + \sum_{k=1}^{N-1}(a_k \cos kx + b_k \sin kx).$$

(The coefficients a_0 and a_N are halved to enable uniformly composed formulas for calculation.) The formulas (11.7) and (11.9) lead to

$$y_j = \frac{1}{2N}\left(\sum_{n=0}^{N-1}(-1)^n Y_n e^{-inx_j} + \sum_{n=1}^{N}(-1)^n \overline{Y_n} e^{inx_j}\right),$$

where

$$(-1)^n Y_n = \sum_{j=0}^{2N-1} y_j e^{inx_j}.$$

As a result the coefficients appear as

$$a_0 = \frac{1}{N}\sum_{j=0}^{2N-1} y_j, \qquad a_N = \frac{1}{N}\sum_{j=0}^{2N-1}(-1)^{j-N} y_j,$$

$$a_k = \frac{1}{N} \sum_{j=0}^{2N-1} y_j \cos k x_j \qquad \text{for} \quad k = 1, 2, \ldots, N-1,$$

$$b_k = \frac{1}{N} \sum_{j=0}^{2N-1} y_j \sin k x_j \qquad \text{for} \quad k = 1, 2, \ldots, N-1.$$

Example (Approximation of a Non-Periodic Function) The function

$$f(x) = 0.2x^2 \exp(\sin x^2) \tag{11.22}$$

is approximated in the interval $[-\pi, \pi]$ by a trigonometric polynomial, using 8 equidistant sampling points $\{(x_j, y_j) : j = 0, 1, \ldots, 7\}$ with $x_j := -\pi + j\frac{\pi}{4}$ and $y_j := f(x_j)$. The coefficients of the trigonometric polynomial are

$$a_k = \frac{1}{4} \sum_{j=0}^{7} \frac{1}{5} \Big((-\pi + j\pi/4)^2 e^{\sin(-\pi + j\pi/4)^2} \Big) \cos k(-\pi + j\pi/4) \qquad \text{for} \quad k = 0, 1, 2, 3, 4$$

$$b_k = \frac{1}{4} \sum_{j=0}^{7} \frac{1}{5} \Big((-\pi + j\pi/4)^2 e^{\sin(-\pi + j\pi/4)^2} \Big) \sin k(-\pi + j\pi/4) \qquad \text{for} \quad k = 1, 2, 3.$$

Thus, the trigonometric interpolation polynomial

$$\begin{aligned} S_4(x) &\approx 0.58810 - 4.4409 \cdot 10^{-17} \sin x - 0.44441 \cos x + 8.8818 \cdot 10^{-17} \sin 2x \\ &\quad - 0.13972 \cos 2x - 1.1102 \cdot 10^{-17} \sin 3x - 0.19742 \cos 3x + 0.19345 \cos 4x \end{aligned}$$

is obtained.

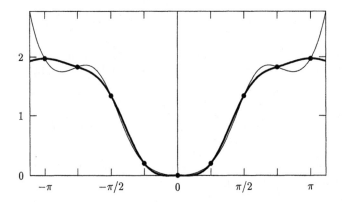

Figure 11.2: Interpolation of the non-periodic function (11.22) (—) using the trigonometric polynomial S_4 (—) at eight equidistant points ($\bullet$) in $[-\pi, \pi)$.

Since f is an even function, the numerical values of the coefficients of the sine terms are negligibly small. In this case, of course, the sine terms could have been omitted beforehand.

In the next example, a function which is not differentiable at certain points is considered. Every trigonometric polynomial is, however, infinitely differentiable. The question thus arises: In such cases, does trigonometric interpolation make satisfactory results at all possible? And if so, is such an interpolation feasible with respect to computational complexity?

Example (Approximation of a Periodic, Non-Differentiable Function) As an extremely simplified model for the speed of the blood flowing from the heart, the following function (with constants a and b) can be used:

$$f(t) = \begin{cases} a \sin bt & \text{for} \quad t \in [0, \pi/b] \\ 0 & \text{for} \quad t \in (\pi/b, 2\pi/b]. \end{cases} \tag{11.23}$$

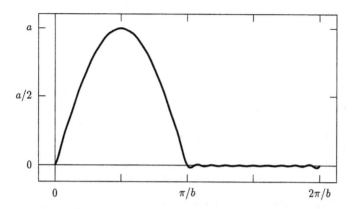

Figure 11.3: Approximation of (11.23) by a trigonometric polynomial S_{20} (—).

The trigonometric interpolation polynomial S_{20} shows undesirable oscillations (see Fig. 11.3) close to the points $t = \pi/b$ and $t = 2\pi/b$, where the function (11.23) is not differentiable. Near these points a good approximation can only be achieved if a very large number of nodes are used in determining the interpolating polynomial.

The Fourier transform is especially useful for the interpolation of data originating from periodic phenomena, as is shown in the following example on the basis of annually recurring air pollution data.

Example (CO Concentration) Having measured the average monthly concentration during the course of a year, the curve of the carbon monoxide (CO) pollution in the air is to be described by a function $CO(x)$. Since domestic fuel and motor traffic are the two main causes of increased concentration of CO in the air, the values are subject to seasonal fluctuations (see Table 11.1).

Table 11.1: Average monthly CO concentrations in the air, Vienna 1993.

month	1	2	3	4	5	6	7	8	9	10	11	12
CO [mg/m³]	1.3	1.4	1.0	0.8	0.7	0.6	0.5	0.6	0.7	1.1	1.2	1.3

At this point a trigonometric polynomial which interpolates this data could be determined, but in general this data is not periodic because, as can be seen from Table 11.2 which lists the average annual concentrations, there is a slowly decreasing long-term trend which inevitably has an effect on the observations during the course of any one year.

Thus, a more sophisticated approach begins with the determination of a long-term trend function which enables the balancing of the short-time sample points. Next, the data modified in

Table 11.2: Average annual CO concentrations in the air, Vienna 1988 – 93.

year	1988	1989	1990	1991	1992	1993
CO [mg/m^3]	1.2	1.3	1.3	1.2	0.9	0.9

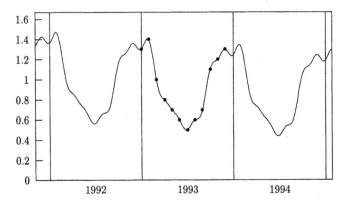

Figure 11.4: Average monthly values (•) of the CO concentration [mg/m^3] in the air of Vienna 1993 interpolated using a trigonometric polynomial and a long-term trend function.

this way are interpolated, and the trigonometric polynomial obtained can again be adapted to the original data using the trend function. In Fig. 11.4 the influence of the decreasing trend function can be recognized. Also the fluctuations during the winter months are conspicuous. The reason for these fluctuations can be found in the fact that the air pollution in winter months is more strongly affected by changes of the weather than in summer months. By using the data observed over several years and by calculating a *statistically significant year*, these fluctuations can be reduced.

Trigonometric interpolation—as opposed to polynomial interpolation—makes it possible to smooth out the corresponding curve by omitting high order harmonic components in the trigonometric polynomial (see Fig. 11.5). The functions obtained are not influenced to this degree by *outliers* in the measurement; and are therefore suitable for prognoses if only a small amount of data is available.

11.4 Convolution

In signal analysis linear operators are often called *linear filters*. If the output b of a filter does not depend on the origin of the coordinates of the signal g, i. e., if

$$b(t + t_0) = \int\limits_{-\infty}^{+\infty} g(\tau + t_0)h(t, \tau)\, d\tau$$

for any $t_0 \in \mathbb{R}$, then the filter is said to be *translation invariant*. Thus, one choice of such a filter is a function h which depends only on the difference of its

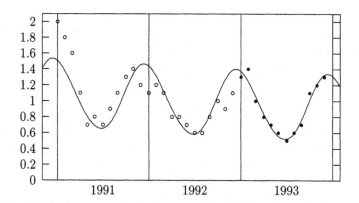

Figure 11.5: Average monthly values (○ and ●) of CO concentration [mg/m³] in the air of Vienna approximated using a trigonometric polynomial S_2 which is determined using data only from the year 1993 (●) and a long-term trend function.

two arguments:

$$b(t) = \int\limits_{-\infty}^{+\infty} g(\tau)h(t - \tau)\, d\tau.$$

This operation is called *convolution*. An important application of the Fourier transform is the (approximate) calculation of this *convolution integral*:

$$(g * h)(t) := \int\limits_{-\infty}^{\infty} g(\tau)h(t - \tau)d\tau. \tag{11.24}$$

The convolution integral $g * h$ is a function which at the point t is the weighted integral of the basic function h. The weight of this integration is the function g whose argument is shifted by τ. In signal processing, h is a signal to be measured, and g is the response function of an imperfect measuring instrument. The convolution $g * h$ then describes the signal *distorted* by the measurement.

Of course the convolution integral could be calculated according to formula (11.24) for two sequences $\{g_j\}$ and $\{h_j\}$ of length N as:

$$(g * h)_l := \sum_{j=0}^{N-1} g_j h_{l-j}, \qquad l = 0, 1, \ldots, N-1. \tag{11.25}$$

The sequence $\{h_j\}$ for the indices $j = -1, -2, \ldots, -(N-1)$ must be continued periodically.

The calculation of the whole sequence $\{(g*h)_l\}$ according to the definition (11.25) would require a computational effort of $O(N^2)$ multiplication operations. For many applications, however, this is far too costly. In this situation the *convolution theorem* which says that the pointwise product of the Fourier transforms G and H is equal to the Fourier transform of the convolution:

$$g * h \leftrightarrow GH, \tag{11.26}$$

is useful. Thus, for the calculation of the convolution, only the two functions (sequences) have to be transformed. The transformed functions have then to be multiplied pointwise (term by term), and the result has to be re-transformed. This leads to a computational complexity of only[3]

$$2O(N \log N) + O(N) + O(N \log N) = O(N \log N)$$

multiplication operations. From the previous discussion, it is clear that, using the convolution, a weighted smoothing is carried out on a function. This property of convolution can be used in order to (approximately) reconstruct functions that have been additively overlaid by stochastic perturbations (*noise*).

In Fig. 11.6 the smoothing effect on a function which is very noisy in the right-hand half of the observation interval can be observed. This smoothing procedure corresponds to a low pass filtering, which is known to reduce high frequencies, but also to leave low frequencies (nearly) unchanged.

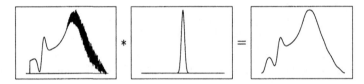

Figure 11.6: The *smoothing* (low pass filtering) of a noisy function using a convolution with a Gaussian curve.

In Fig. 11.7 the input function is *differentiated* by convolving it with a discrete function whose values are zero except the two succeeding values -1 and $+1$. This convolution operation leads to a subtraction of consecutive function values of the input function. This procedure makes the irregular behavior of the basic function even more pronounced, as can be seen in the third frame of the illustration[4].

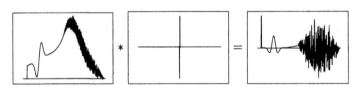

Figure 11.7: *Differentiation* of a function using convolution.

In Fig. 11.8 a similar function is used: $h(t) = -te^{-t^2}$. This function leads to a similar *differentiating* effect, however with some smearing of the derivative. This procedure corresponds to a band pass filtering.

---

[3]For the calculation of the Fourier transform, the FFT algorithm (see Section 11.6.1) which only requires $O(N \log N)$ multiplication operations can be used.

[4]Human hearing works by "measuring" air pressure differences, but not by finding out the level of air pressure. Therefore, some part of the sound processing accomplished by the human nervous system could be modeled by such a convolution operation.

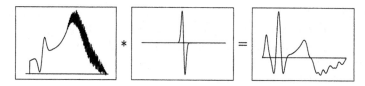

Figure 11.8: *Band pass filtering* of a function using convolution.

Conversely, if m is a (measured) function that has been obtained from an initial function g using a measurement instrument whose distortion can be explained by a convolution with a known function h, then the initial function can be reconstructed using *deconvolution*. In this case the perturbation is nothing other than the convolution $m = g * h$. According to (11.26) in the frequency domain this means

$$M(f) = G(f)H(f) \quad \text{or} \quad G(f) = M(f)/H(f). \quad (11.27)$$

Thus, by dividing the Fourier transform M of the measured function m by the Fourier transform H of the known response function h, the original function g can be reconstructed. The only condition for (11.27) to be applicable is that the Fourier transform $H(f)$ does not vanish in the considered frequency range. This would mean that the measuring instrument *destroys* all information about these frequencies of g.

The improvement of the Hubble telescope images (mentioned on page 34) is an example demonstrating the limits of this method: The representation of dim objects can be improved, but if the brightness of the objects is waning too much, or if too few samples of these objects are observed, then no improvement can be achieved.

11.5 Manipulation of the Signal Spectrum

An important task in signal processing is the treatment of signals in the presence of noise. In particular, the separation of useful and useless information is to be carried out. This can be achieved by utilizing the properties of a signal spectrum. For a smooth function, the spectral components fade away rapidly with an increasing order k, whereas noise has slowly decreasing spectral components. How fast the spectral components of a periodic function f approach zero is closely related to the differentiability of f: It can be shown that

$$|c_k| = O(|k|^{-j}) \quad \text{as} \quad k \to \infty \quad (11.28)$$

if f (taken as a periodic function) has $j-1$ continuous derivatives, and if the jth derivative is piecewise continuous (or of bounded variation).

If a smooth function is perturbed by additive noise, then the signal seems to become smoother and smoother as the noise is diminished. This is what is meant by *smoothing*. A commonly used smoothing technique consists of the calculation of the Fourier coefficients of the perturbed signal, the *filtering*[5] of these coefficients (i.e., the diminution or suppression of certain, often high frequencies) and the subsequent synthesis of the function corresponding to the filtered Fourier coefficients. Such filtering can be specified by simple functions of the spectral components of a signal.

11.5.1 Case Study: The Filtering of a Noisy Signal

In a computer experiment 1024 samples of the function

$$f(t) = 2\sin(2\pi t/500) + \cos(2\pi t/200) - \frac{1}{2}\sin(2\pi t/50)$$

were used as unperturbed signals. The function values and their discrete Fourier transform can be seen in Fig. 11.9 a. These function values were artificially perturbed by random noise whose amplitude was the same size as the original signal. The noisy signal and its Fourier transform are shown in Fig. 11.9 b. In the spectrum of the noisy signal, a nearly uniform distribution of the perturbation over the whole frequency range is seen.

To reconstruct the original function out of noisy data, there are (among others) two filtering techniques:

1. The first technique utilizes the fact that noise in the frequency domain is spread over the whole spectrum, whereas the Fourier components of the useful part of the data are in most cases relatively large compared to this *uniform* spectrum. Consequently, all the components of the noisy spectrum below a certain threshold are removed. The resulting spectrum is shown on the right side in Fig. 11.9 c; the re-transformed signal is shown on the left side. The form of the original signal is clearly visible again, although on closer examination the recovered signal is not exactly equal to the original signal. This results from the fact that some components of the original spectrum (see Fig. 11.9 a), are below the cut-off threshold and are, as a result, irretrievably lost. They cannot be recovered by any filtering technique.

2. The second technique assumes that the frequency spectrum of the original signal is band-limited, i.e., that the spectrum of the original signal has no components above a certain limit frequency. In many technical applications this frequency band is known, or reasonable assumptions about it can be made. As a result, all frequency components outside of this band are removed because they definitely do not belong to the original signal. The modified spectrum is then re-transformed (see Fig. 11.9 d). Again, the original signal is roughly reconstructed, but essential differences are visible as well.

---

[5]The expression *filtering* is used here in the same way as it is used in electronics. In electronics, *filter* denotes an (analog) electric circuit, whose resistance depends on the frequency. In electronics noise filters and rumble filters, for example, are used in hi-fi amplifiers.

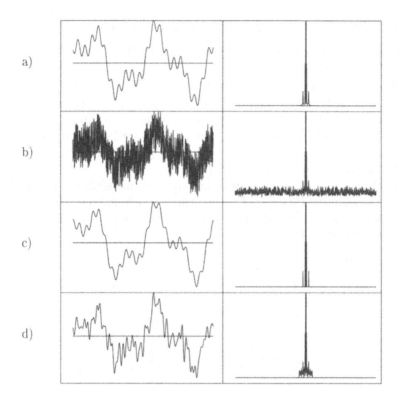

Figure 11.9: Filtering a stochastically perturbed (noisy) signal.

In general, it is clear that both methods have limitations. Of course both methods can be improved (e.g., by sampling a longer time-interval or by using sophisticated filter functions), but the original signal can never be reconstructed perfectly.

11.6 DFT Algorithms

Defining the complex number $W := e^{2\pi i/N}$ (11.7) can be rewritten as

$$F_n = \sum_{k=0}^{N-1} W^{kn} f_k. \qquad (11.29)$$

W is an Nth root of unity: $W^N = 1$. At this point, implementing the discrete Fourier transform could be done simply by N evaluations of the summation formula (11.29). This method would have a computational complexity of $O(N^2)$ multiplication operations, which is rather high for many areas of application. There is, however, a method for the discrete Fourier transform, with a complexity of only $O(N \log N)$ multiplication operations: the *Fast Fourier Transform* (see Elliot, Rao [179], Nussbaumer [302], Van Loan [372]).

11.6.1 The Fast Fourier Transform (FFT)

Unaware of the Gaussian method of 1805, in 1965 J.W. Cooley and J.W. Tukey published a paper [139] about a special DFT algorithm which they called the *fast Fourier transform* (FFT). It was this paper which caused the widespread dissemination of the FFT algorithm. Nowadays it is one of the most important numerical methods. Using FFT programs the calculation of DFTs with as many as $N = 2^{12} = 4\,096$ points became routine. Even DFTs of very long data sequences, e.g., $N = 2^{26} = 67\,108\,864$ (so-called *giant Fourier transforms*), are possible (Bershader, Kraay, Holland [112]).

The description of the FFT algorithm given in this section is based on the following proof of the *Danielson-Lanczos lemma*, which makes it possible to write a discrete Fourier transform of length N (N even) as a sum of two discrete Fourier transforms of the length $N/2$:

$$F_k = \sum_{j=0}^{N-1} e^{2\pi ijk/N} f_j \tag{11.30}$$

$$= \sum_{j=0}^{N/2-1} e^{2\pi ik(2j)/N} f_{2j} + \sum_{j=0}^{N/2-1} e^{2\pi ik(2j+1)/N} f_{2j+1}$$

$$= \sum_{j=0}^{N/2-1} e^{2\pi ikj/(N/2)} f_{2j} + W^k \sum_{j=0}^{N/2-1} e^{2\pi ikj/(N/2)} f_{2j+1} \tag{11.31}$$

$$= F_k^e + W^k F_k^o.$$

F_k^e (F_k^o) denotes the kth component of the Fourier transform of length $N/2$ formed from the even (odd) components of the original f_j's. The transforms F_k^e and F_k^o are periodic in k with period $N/2$ such that, again, the required N components for F_k are obtained.

If $N/2$ is even, (11.31) can be used again on F_k^e and F_k^o. In the next step F_k^e becomes the two Fourier transforms F_k^{ee} and F_k^{eo} of length $N/4$. For $N = 2^r$ (with $r \in \mathbb{N}$) this can be carried out recursively r times, until identities of the form $F_k^{eooe\cdots oee} = f_n$ for any n are achieved. At this point the patterns of the e and o are changed to $e = 0$ and $o = 1$. From this point on, this operation is called a *bit reversal* permutation. If the resulting sequence of bits is interpreted as a binary number, then it is exactly n. It is because of the successive subdivisions of the data into even and odd values that is equivalent to the testing of the least significant bit of the binary representation of n.

Therefore, the first part of the FFT algorithm is to interchange f_n using a bit reversal permutation. The second part has an outer loop which is executed $\log_2 N$ times and calculates, in turn, transforms of the length $2, 4, 8, \ldots, N$. At each stage of this process, two nested inner loops range over the sub-transforms already computed and the elements of each transform implementing the Danielson-Lanczos lemma (11.31). During each stage $O(N)$ arithmetic operations are carried out. Since there are $\log_2 N$ stages, the complexity of the whole FFT algorithm is of order $O(N \log N)$.

The r calculation steps of the FFT algorithm are denoted by $j = 1, 2, \ldots, r$. Each of these steps computes the N quantities

$$
\begin{aligned}
d_n^j &= d_n^{j-1} + W^e d_{n+2^{j-1}}^{j-1} \\
d_{n+2^{j-1}}^j &= d_n^{j-1} - W^e d_{n+2^{j-1}}^{j-1}
\end{aligned}
$$

with $0 \le n \bmod 2^j < 2^{j-1}$ and the exponent $e = n \bmod 2^{j-1}$, where d_k^{j-1} and d_k^j $(k = 0, 1, \ldots, N-1)$ describe the input and output data of the jth step respectively.

The following simple FFT algorithm (in pseudo-code) performs the transform of a data sequence $\{d_i\}$, where the output again appears in $\{d_i\}$:

Interchange using bit reversal
$m_{\max} := 1$
do while $N > m_{\max}$
    $i_{step} := 2m_{\max}$
    $\vartheta := 2\pi/N$
    $W_p := e^{i\vartheta}$
    $W := 1$
    do $m := 0, 1, \ldots, m_{\max} - 1$
        do $i = m, m + i_{step}, \ldots, N-1$
            $j := i + m_{\max}$
            $t := W \cdot d_j$
            $d_j := d_i - t$
            $d_i := d_i + t$
        end do
        $W := W \cdot W_p$
    end do
    $m_{\max} := i_{step}$
end do

The FFT for eight data values $d_0, d_1, \ldots, d_7$ proceeds as follows:

$d_0 := d_0$	$d_0 := d_0 + d_1 W^0$	$d_0 := d_0 + d_2 W^0$	$d_0 := d_0 + d_4 W^0$
$d_1 := d_4$	$d_1 := d_0 - d_1 W^0$	$d_1 := d_1 + d_3 W^1$	$d_1 := d_1 + d_5 W^1$
$d_2 := d_2$	$d_2 := d_2 + d_3 W^0$	$d_2 := d_0 - d_2 W^0$	$d_2 := d_2 + d_6 W^2$
$d_3 := d_6$	$d_3 := d_2 - d_3 W^0$	$d_3 := d_1 - d_3 W^1$	$d_3 := d_3 + d_7 W^3$
$d_4 := d_1$	$d_4 := d_4 + d_5 W^0$	$d_4 := d_4 + d_6 W^0$	$d_4 := d_0 - d_4 W^0$
$d_5 := d_5$	$d_5 := d_4 - d_5 W^0$	$d_5 := d_5 + d_7 W^1$	$d_5 := d_1 - d_5 W^1$
$d_6 := d_3$	$d_6 := d_6 + d_7 W^0$	$d_6 := d_4 - d_6 W^0$	$d_6 := d_2 - d_6 W^2$
$d_7 := d_7$	$d_7 := d_6 - d_7 W^0$	$d_7 := d_5 - d_7 W^1$	$d_7 := d_3 - d_7 W^3$

To start with the array (vector) d contains the input data. In the first column of the table below the bit reversal is carried out. Now in every row there is a sub-transform of length 1. (The lines between the rows show the demarcation between the single sub-transforms.) In the next column these sub-transforms of length 1 are transformed into sub-transforms of length 2. The next two columns describe the transitions from length 2 to length 4 and from 4 to 8. Finally, the array d contains the desired Fourier transform of the original data vector.

The above algorithm requires that the length N of the data vector to be transformed is a power of 2 ($N = 2^n$ with $n \in \mathbb{N}$); but for many practical applications, this represents a major restriction.

The FFT principle applies only if N is *not* a prime, i. e., N has prime factorization $p_1^{n_1} p_2^{n_2} \cdots p_i^{n_i}$. High efficiency is only achieved, however, for values of N which can be reduced to many small prime factors, as can be seen in the example above with $N = 2^n$. FFT algorithms of this type are also called *Cooley-Tukey radix-2 FFT algorithms*.

The radix-2 algorithm was the basis for an intensive research effort to develop higher radix, mixed radix, prime factor, Winograd and most recently the split radix FFT algorithms.

These algorithms make different assumptions about the length N of the data vector. For example, the *mixed radix FFT algorithms* do not require the restriction $N = 2^n$. For this reason they are used for most of the FFT routines included in scientific software libraries. Thus, for example, the FFT programs included in the IBM ESSL (Engineering and Scientific Software Library) allow lengths

$$N = 2^h 3^i 5^j 7^k 11^m \quad \text{with} \quad h \in \{1, 2, \ldots, 25\}, \ i \in \{0, 1, 2\}, \ j, k, m \in \{0, 1\},$$

as long as $N \leq 37\,748\,736$.

Example (FFT Algorithms and Programs)   To compare the floating-point complexity of three FFT algorithms—written by D. Fulker, P. N. Swarztrauber and C. Temperton—the number $a(N)$ of floating-point operations was empirically determined and normalized by the asymptotic complexity $N \log_2 N$ of the FFT algorithms (see Fig. 11.10). As can be seen from the nearly constant behavior of the normalized complexities in the range under investigation— data vectors with the lengths $N = 32, 64, \ldots, 1\,048\,576$—the empirical complexity $a(N)$ is nearly proportional to the asymptotic complexity. Swarztrauber's program requires the smallest number of floating-point operations for the investigated lengths of the vectors (especially for short data vectors).

When comparing the computing times $T(N)$ (see Fig. 11.11) it becomes clear that (especially for long vectors) the higher complexity of the Temperton algorithm does not result in proportionally prolonged computing times. The reason for this phenomenon can be seen in Fig. 11.12. The Temperton program attains the best floating-point performance of all three programs for long data vectors, and for shorter data vectors, it is nearly as good as Swarztrauber's program. Thus, the *implementation* of the Temperton algorithm is the reason for the favorable computing times (cf. Chapter 6).

The increase of the normalized computing time for data vectors longer than $N = 2^{15} = 32\,768$ is a symptom of a performance collapse caused by the exhaustion of the cache-memory of the workstation used.

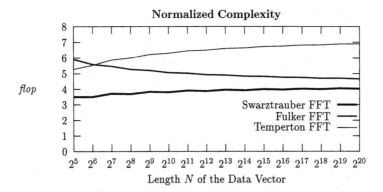

Figure 11.10: The normalized number $a(N)/(N \log_2 N)$ of floating-point operations [*flop*] of three FFT programs: `FFTPACK/cfftf` by P.N. Swarztrauber, one program by D. Fulker and one by C. Temperton.

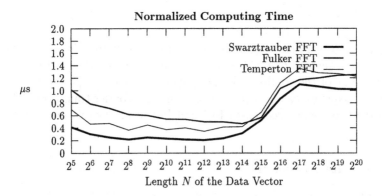

Figure 11.11: Normalized computing time $T(N)/(N \log_2 N)$ in microseconds.

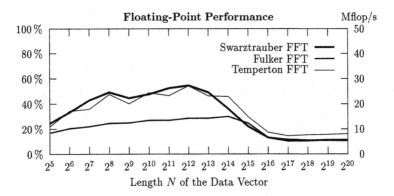

Figure 11.12: Floating-point performance (Mflop/s) and efficiency (%).

11.6.2 FFT of Real Functions

In most applications data vectors appear as real-valued sequences. If this is the case, it is too costly to compute the FFT of a complex-valued sequence. However, due to the properties of the Fourier transform of real functions, simplifications can be made.

FFT of two Real Functions

For the continuous Fourier transform, some symmetry properties hold: For a real-valued function h the Fourier transform H satisfies $H(-f) = \overline{H(f)}$. The corresponding relation for the discrete Fourier transform H_n is

$$H_{N-n} = \overline{H_n} \qquad \text{for all} \quad n \in \mathbb{Z} . \tag{11.32}$$

The Fourier transform G of a purely imaginary function g satisfies $G(-f) = -\overline{G(f)}$. The corresponding relation for the discrete Fourier transform G_n is

$$G_{N-n} = -\overline{G_n} \qquad \text{for all} \quad n \in \mathbb{Z} . \tag{11.33}$$

If there are two real data sequences to transform, they are joined together into one complex sequence: one real sequence forms the real part and the other the imaginary part of the resulting complex sequence. The complex Fourier transform of this sequence can be separated into the two Fourier transforms $\{A_i\}$ and $\{B_i\}$ of the two original sequences $\{a_i\}$ and $\{b_i\}$ using the relations (11.32) and (11.33).

FFT of one Real Function

The Fourier transforms of two data sequences are needed for some applications (especially for convolutions). Normally, however, only the Fourier transform of one data sequence is sought.

In order to compute this transform efficiently, the original real-valued data sequence $\{f_i\}$ is taken, and a complex sequence $\{h_j\}$ of length $N/2$

$$h_j = f_{2j} + i f_{2j+1}, \qquad j = 0, 1, \ldots, N/2 - 1$$

is built out of it. A complex Fourier transform is then applied. The resulting complex transform H_n can be reduced to two complex sequences $\{F_n^e\}$ and $\{F_n^o\}$:

$$
\begin{aligned}
H_n &= F_n^e + i F_n^o, \qquad n = 0, 1, \ldots, N/2 - 1, \\
F_n^e &= \sum_{k=0}^{N/2-1} f_{2k} e^{2\pi i k n/(N/2)}, \\
F_n^o &= \sum_{k=0}^{N/2-1} f_{2k+1} e^{2\pi i k n/(N/2)}.
\end{aligned}
$$

The two sequences $\{F_n^e\}$ and $\{F_n^o\}$ can be separated as described above in the transform of two real functions. The Fourier transform F_n of the original function f_i can be obtained using the relation

$$F_n = F_n^e + e^{2\pi i n/N} F_n^o, \qquad n = 0, 1, \ldots, N-1.$$

Expressing F_n directly in terms of H_n, F_n results in

$$F_n = \frac{1}{2}(H_n + \overline{H_{N/2-n}}) - \frac{i}{2}(H_n - \overline{H_{N/2-n}})e^{2\pi i n/N}.$$

Fast Sine and Cosine Transforms

For some applications of the representation (11.13), either the real or the imaginary parts of the complex exponentials is required. For example, when solving a differential equation with the boundary conditions $y(0) = 0$ and $y(2\pi) = 0$, a sum of expressions of the form $a_n \sin nx$ $(n \in \mathbb{N})$ would be a possible ansatz. In the same way, for the boundary conditions $\dot{y}(0) = 0$ and $\dot{y}(2\pi) = 0$, a sum of expressions of the form $b_n \cos nx$ $(n \in \mathbb{N}_0)$ could be a suitable ansatz. When the boundary conditions stated above are used, no complete complex Fourier transform has to be carried out; an expansion with respect to terms of the form $\sin nx$ or $\cos nx$ suffices.

Discrete Sine Transform

The *discrete sine transform* (DST) of a sequence $\{f_j\}$ with $j = 0, 1, \ldots, N-1$ and $f_0 := 0$ is defined by

$$F_k := \sum_{j=1}^{N-1} f_j \sin(\pi jk/N) \tag{11.34}$$

and looks, except for a factor of 2 in the sine, like the imaginary part of (11.7).

Therefore, to transform the sequence $\{f_j\}$ using the discrete Fourier transform, the sequence has to be expanded into an odd discrete sequence of double the length: Let $f_N := 0$ and $f_{2N-j} := -f_j$ for $j = 0, 1, \ldots, N-1$. By expanding the sequence this way, the upper half of the sum in the Fourier transform (indices $j = N, N+1, \ldots, 2N-1$) can be paraphrased using the substitution $j' := 2N-j$ in the following way:

$$\sum_{j=N}^{2N-1} f_j e^{2\pi i jk/(2N)} = \sum_{j'=1}^{N} f_{2N-j'} e^{2\pi i(2N-j')k/(2N)} = -\sum_{j'=0}^{N-1} f_{j'} e^{-2\pi i j'k/(2N)}$$

such that the Fourier transform of the modified function can be written as

$$F_k = \sum_{j=0}^{N-1} f_j \left[e^{2\pi i jk/(2N)} - e^{-2\pi i jk/(2N)} \right] = 2i \sum_{j=0}^{N-1} f_j \sin(\pi jk/N).$$

This means that the sine transform (except for the factor of $2i$) can be obtained from the discrete Fourier transform of the augmented sequence $\{f_j\}$. The only disadvantage of using this method is that the transformed sequence is twice as long as the original data sequence, and the computing time is nearly doubled. To cope with this increase in computing time, the Fourier transform of a purely real function can be used since usually the data sequence to transform is purely real-valued.

The sine transform (11.34) is virtually its own inverse. Therefore, applying the sine transform (11.34) of length $2N$ two times, gives the original data multiplied by $N/2$.

Discrete Cosine Transform

For the *discrete cosine transform* (DCT), the matter is not quite so clear, because the transform can be defined in various ways. Two methods are introduced in the following.

The first method is obtained by expanding a data sequence of $N+1$ data points into a data sequence of length $2N$, such that the result is an even sequence with a central point $j = N$: $f_{2N-j} := f_j$ for $j = 0, 1, \ldots, N-1$. The cosine transform obtained in this way is

$$F_k = \tfrac{1}{2} \left(f_0 + (-1)^k f_N \right) + \sum_{j=1}^{N-1} f_j \cos(\pi j k / N). \tag{11.35}$$

The cosine transform is, like the sine transform, its own inverse. Applying the cosine transform twice gives the original data multiplied by $N/2$.

The second method is to expand a sequence of N data points into a sequence of length $2N$, such that the resulting sequence is even with a central point $N - \tfrac{1}{2}$. This symmetry property is called *quarter wave even*. For this reason, the resulting transform is called a *quarter wave cosine transform*. The formula obtained in this way is

$$F_k = \sum_{j=0}^{N-1} f_j \cos \left(\pi k (j + \tfrac{1}{2}) / N \right).$$

The inverse transform produces

$$f_j = \frac{2}{N} \sum_{k=0}^{N-1} {}' F_k \cos \left(\pi k (j + \tfrac{1}{2}) / N \right).$$

The accent ($'$) above the summation symbol means that for $k = 0$ the coefficient has to be halved.

For both cosine transforms presented in this section, it is possible to reduce the computational complexity by an additional factor of two.

Example (Image Compression) The increasing use of *digital image processing* in multimedia systems, in digital television (*high definition television*, HDTV), etc. has made the standardization of methods for compressing digital images necessary. All three standards

JPEG	used to store single images,
MPEG	used to store sequences of video images and
CCITT H.261 (Px64)	used for television phones

use the discrete cosine transform (DCT) as a tool for compressing digital image data (Watson [375]).

11.6.3 FFT in Two or More Dimensions

In many applications (especially in image processing) higher-dimensional data arrays have to be used.

For a two-dimensional function $h(k_1, k_2)$, defined on a grid with coordinates

$$0 \leq k_1 \leq N_1 - 1, \quad 0 \leq k_2 \leq N_2 - 1,$$

the Fourier transform $H(n_1, n_2)$ is defined by

$$H(n_1, n_2) := \sum_{k_2=0}^{N_2-1} \sum_{k_1=0}^{N_1-1} e^{2\pi i k_2 n_2 / N_2} e^{2\pi i k_1 n_1 / N_1} h(k_1, k_2) \tag{11.36}$$

on a grid of the same size. Since in (11.36) the order of the summations can be interchanged, the order of the Fourier transforms can also be changed. Thus, the two-dimensional Fourier transform can be reduced to one-dimensional Fourier transforms, for which the FFT algorithm can be used.

For an L-dimensional Fourier transform

$$H(n_1, n_2, \ldots, n_L)$$
$$:= \sum_{k_L=0}^{N_L-1} \sum_{k_{L-1}=0}^{N_{L-1}-1} \cdots \sum_{k_1=0}^{N_1-1} e^{2\pi i k_L n_L / N_L} e^{2\pi i k_{L-1} n_{L-1} / N_{L-1}} \cdots e^{2\pi i k_1 n_1 / N_1} h(k_1, \ldots, k_L)$$

this procedure can be generalized accordingly.

In many applications (e. g., digital image processing) large higher-dimensional FFTs are to be performed as fast as possible. Therefore, parallel computers are clearly a good choice for this task. For parallelization, higher-dimensional FFTs are especially suitable since many one-dimensional FFTs are to be performed independently of each other, as previously mentioned, and can easily be assigned to different processors.

11.7 FFT Software Packages

11.7.1 FFTPACK

FFTPACK is a software package developed by P. N. Swarztrauber (*National Center for Atmospheric Research*, USA) for the calculation of various forms of the fast Fourier transform as well as for the corresponding inverse transforms.

The subprogram FFTPACK/rfftf serves to compute the FFT of a real data sequence, FFTPACK/sint or FFTPACK/cost perform the sine or cosine transform and FFTPACK/cfftf can be used for the FFT of a complex data sequence.

Also, FFTPACK/sinqf and FFTPACK/cosqf compute the quarter wave sine and quarter wave cosine transform of a real data sequence with only odd wave numbers. Corresponding inverse transforms are available.

In all FFTPACK subprograms the number N of data points can be chosen to be any natural number (not only a power of two) although the algorithms are most efficient if N is a product of small primes. A power of two is optimal and a large prime is most unfavorable.

The FFTPACK is available in NETLIB, in the directory fftpack. It is also included in CMLIB, IMSL, SLATEC and other libraries.

11.7.2 VFFTPK

VFFTPK was designed in exactly the same way as FFTPACK and differs from it only in the fact that only transforms for real-valued data sequences are included and in that FFTs of M different data sequences can be computed simultaneously (by *one* subprogram call). The names of the VFFTPK subprograms are obtained by placing the letter v before the names of the FFTPACK routines.

If a nest of loops appears in a VFFTPK program, then the loop $m = 1, 2, \ldots, M$ over the various sequences is always at the *innermost* position. Since the calculations for the various sequences are completely independent of each other, if the number M of the sequences is high enough, then well vectorizable code (cf. Chapter 6) is obtained.

The software package VFFTPK is available in NETLIB, in the directory vfftpk.

11.8 FFT Routines in Software Libraries

11.8.1 IMSL Software Libraries

In the IMSL Fortran library, FFTPACK subprograms were included, but their names were changed. Details can be found in the IMSL documentation [31].

The programs IMSL/MATH LIBRARY/fft2d and IMSL/MATH LIBRARY/fft2b compute two-dimensional complex FFTs or their inverses. For three-dimensional complex FFTs, IMSL/MATH LIBRARY/fft3d and IMSL/MATH LIBRARY/fft3b are available. Any number of data points can be chosen in any dimension. The products of small primes are especially favorable for program efficiency.

In the IMSL Fortran library, there are also subprograms for the fast calculation of convolutions and correlations: IMSL/MATH LIBRARY/rconv calculates the convolution of two real data sequences, and IMSL/MATH LIBRARY/cconv calculates the convolution of two complex data sequences. The correlation of data sequences can be computed using IMSL/MATH LIBRARY/rcorl or IMSL/MATH LIBRARY/ccorl.

11.8.2 NAG Software Libraries

Two subprograms, `NAG/c06eaf` and `NAG/c06ebf`, which compute the FFT of a real data sequence or the corresponding inverse transform, are included in the NAG Fortran library. `NAG/c06ecf` computes the FFT of a complex data sequence. For the inverse transform of a complex data sequence, no special subprogram is available. The inverse transform can easily be implemented though by using the FFT and by forming the conjugate complex data sequence (subprogram `NAG/c06gcf`). The corresponding subprograms `NAG/c06faf`, `NAG/c06fbf` and `NAG/c06fcf` implement the same function, but they are optimized with respect to runtime, for which additional working storage is required. For all these NAG programs the selection of the number N of data points in one dimension is restricted: In the prime factorization of N, no more than 20 factors are allowed to appear (every prime factor is counted according to its multiplicity), and the highest prime factor may not be greater than 19.

In the NAG Fortran library, a set of subprograms—`NAG/c06fpf`, `NAG/c06fqf` and `NAG/c06frf`—which calculate the respective transforms for M different data sequences simultaneously is available. This leads to better vectorizability. These routines place no restrictions on the number N of data points.

The sine and cosine transforms (for one or more data vectors) can be calculated using the subprograms `NAG/c06haf` and `NAG/c06hbf`. For quarter wave transforms with odd wave numbers, `NAG/c06hcf` and `NAG/c06hdf` are available.

For the calculation of higher-dimensional FFTs of complex data sequences, the subprogram `NAG/c06fjf`, with which a general L-dimensional FFT can be computed is available. For two-dimensional FFTs the routine `NAG/c06fuf`, which calls `NAG/c06frf` internally is available. `NAG/c06fuf` attains an essentially higher performance than `NAG/c06fjf` using vectorization. Similarly `NAG/c06fxf` is available for three-dimensional FFTs. The subprogram `NAG/c06fff` calculates all one-dimensional FFTs for a given dimension of an L-dimensional array.

11.9 Other FFT Programs

Many additional FFT programs can be obtained via anonymous-FTP. For instance, there is the FFT package developed by D. Fulker (*National Center for Atmospheric Research*, USA). It contains FFT programs for real and complex data sequences as well as programs for the inverse transform.

Another FFT package available via anonymous-FTP is the program CFFT99 developed by C. Temperton (*European Center for Medium-Range Weather Forecasts*, England), with which the FFT or its inverse for several complex data sequences can be calculated simultaneously.

11.9.1 TOMS Collection

`NETLIB/TOMS/545` contains subprograms for the calculation of real and complex FFTs. The distinct advantage of this FFT implementation is the minimization of read/write operations to secondary storage media if the data sequence does not fit into main memory.

Moreover, the user can further improve the program performance by implementing his own read/write operations, which then provide the *swap* operations.

11.9.2 Various NETLIB Software

The NETLIB directory go contains the files fft.f and realtr.f. Both the complex and the real FFT and their inverse transforms can be calculated using these subprograms. They are also suitable for the simultaneous calculation of several FFTs.

In the directory misc the file fft.f is available. This subprogram is a simple implementation of the complex FFT. It is suitable only if the data vector length is a power of two.

The files fft.f and ffc.f are available in the directory napack. These subprograms can be used to calculate the complex FFT and its inverse, where the length N of the sequence is not restricted to particular values.

Chapter 12

Numerical Integration

*As I have received your paper about the approximate integration,
I no longer can withstand to thank you for the great pleasure
you have given to me.*

From a letter written by BESSEL to GAUSS

The calculation of areas and volumes is one of the oldest mathematical problems. Even in ancient Greece, mathematicians were interested in *quadrature*, the conversion of areas into quadrats (squares) of the same size.

In modern scientific computing, many methods require the calculation of integrals, for example finite element methods, integral transforms, and many statistical methods.

Numerical integration is covered in thousands of articles in mathematical journals, a number of books (Brass [123], Davis, Rabinowitz [41], Engels [46], Evans [47], Krommer, Ueberhuber [64], [258], Krylov [260], Stroud [362]), dozens of published algorithms and programs (e.g., in ACM Transactions on Mathematical Software), a special software package (QUADPACK [22]), and extensive sections of numerical program libraries (IMSL [31], [32]; NAG [21], etc.). Because of these wide-ranging and detailed discussions of the theoretical and practical aspects of a seemingly simple topic, one might conclude that almost every integral of practical relevance can be approximated with sufficient precision and efficiency using existing software. On closer examination it becomes clear, however, that this is *not* the case.

For some types of integration problems—especially for univariate integrands (one-dimensional problems)—high quality software is available. Yet, if one wants to use these algorithms correctly, it is necessary to understand their principles, for example when trying to interpret unexpected incorrect results. It is also very difficult to find good software for high dimensional problems. In this case, the user is often forced to find and implement algorithms on his own.

This chapter focuses on *numerical* integration. Of course non-numerical methods (e.g., symbolic methods) can also be used to calculate integrals.

For calculating certain kinds of integrals, numerical methods that are *not* related to numerical integration can be used. Such integrals occur, for example, when defining special mathematical functions like the Gamma function or the Bessel functions.

Software (Multivariate Normal Distribution)   A typical example of an integral which is related to a special mathematical function is the cumulative n-dimensional normal (Gaussian) distribution function

$$F(b_1, \ldots, b_n) := \frac{1}{\sqrt{(2\pi)^n \det V}} \int\limits_{-\infty}^{b_1} \cdots \int\limits_{-\infty}^{b_n} \exp\left(-\frac{1}{2} x^\top V^{-1} x\right) dx_n \cdots dx_1, \qquad (12.1)$$

where $V \in \mathbb{R}^{n \times n}$ denotes the symmetric, positive definite variance-covariance matrix. The IMSL/STAT-LIBRARY contains subroutines for computing (12.1) when $n = 1$ and $n = 2$. For $n \geq 3$, no reliable and efficient software for evaluating F is currently available. Recently published methods for solving this problem (Genz [210], Drezner [172]) are based on general numerical integration schemes.

Software (Special Integrals) Most of the special functions defined using integrals belong to one of the following classes: exponential or logarithmic integrals, cosine or sine integrals, the Gamma function, the error function, Bessel functions, elliptic integrals, and distribution functions.

Both the IMSL and the NAG libraries contain subroutines for computing values of the most important special functions defined by integrals. Another important source of software for evaluating such functions is the ACM *Transactions on Mathematical Software* (TOMS).

12.1 Fundamentals of Integration

In this chapter, the *mathematical problem* of calculating the definite integral $\mathrm{I}f \in \mathbb{R}$

$$\mathrm{I}f := \int\limits_B f(x)\, dx \qquad (12.2)$$

for a given integrand

$$f : B \subseteq \mathbb{R}^n \to \mathbb{R}$$

and a given region B is treated. The integral (12.2) is usually assumed to exist in the Riemann sense either as a proper integral or an improper integral.[1]

In many applications, an additional *weight function* $w : B \subseteq \mathbb{R}^n \to \mathbb{R}$ is included in the integral:

$$\mathrm{I}f = \int\limits_B w(x) f(x)\, dx.$$

These weight functions are usually non-negative ($w(x) \geq 0$ for all $x \in B$) or are oscillatory.

Some applications require the computation of m different integrals with the same weight function w and the same region B, but m different integrands $f_1, f_2, \ldots, f_m : B \to \mathbb{R}$:

$$\mathrm{I}f_1 = \int\limits_B w(x) f_1(x)\, dx,$$

$$\vdots$$

$$\mathrm{I}f_m = \int\limits_B w(x) f_m(x)\, dx.$$

---

[1] For some applications it is necessary to compute integrals which exist only in a generalized sense, for example as *Cauchy principal value integrals*. In this book, such cases are only dealt with if suitable software is available.

12.1.1 Integration Regions

Most numerical methods for the approximation of integrals use certain *standardized* integration regions. If the integral is to be computed over a region B and B is not standard, the corresponding integration problem has to undergo appropriate transformations in order to make these numerical schemes applicable (see Section 12.2).

The following table contains standard regions used by most of the existing methods for numerical integration (Stroud [362]):

Region	Notation	Definition
entire space	E_n	$\mathbb{R}^n$
unit cube	C_n	$\{ x \in \mathbb{R}^n \ : \ \|x\|_\infty \leq 1 \}$
positive unit cube	W_n	$C_n \cap \left(\mathbb{R}_0^+ \right)^n$
cubical shell	$C_n^{\text{shell}}(K_1, K_2)$	$\{ x \in \mathbb{R}^n \ : \ K_1 \leq \|x\|_\infty \leq K_2 \}$
unit sphere	S_n	$\{ x \in \mathbb{R}^n \ : \ \|x\|_2 \leq 1 \}$
spherical shell	$S_n^{\text{shell}}(K_1, K_2)$	$\{ x \in \mathbb{R}^n \ : \ K_1 \leq \|x\|_2 \leq K_2 \}$
spherical surface	U_n	$\{ x \in \mathbb{R}^n \ : \ \|x\|_2 = 1 \}$
unit octahedron	G_n	$\{ x \in \mathbb{R}^n \ : \ \|x\|_1 \leq 1 \}$
unit simplex	T_n	$G_n \cap \left(\mathbb{R}_0^+ \right)^n$

12.1.2 Weight Functions

The following list contains the most frequently used weight functions together with the corresponding *one-dimensional* integration region (Piessens [314]):

Interval	Weight Function $w(x)$	Name
$[-1, 1]$	1	Legendre
$[-1, 1]$	$(1 - x^2)^{-1/2}$	Chebyshev, first kind
$[-1, 1]$	$(1 - x^2)^{1/2}$	Chebyshev, second kind
$[-1, 1]$	$(1 - x)^\alpha (1 + x)^\beta, \ \alpha, \beta > -1$	Jacobi
$[0, \infty)$	$\exp(-x)$	Laguerre
$[0, \infty)$	$x^\alpha \exp(-x), \ \alpha > -1$	generalized Laguerre
$(-\infty, \infty)$	$\exp(-x^2)$	Hermite
$(-\infty, \infty)$	$1/\cosh(x)$	hyperbolic cosine
$(-\infty, \infty)$	$\cos \omega x, \ \sin \omega x, \ \omega \in \mathbb{R}$	trigonometric
$[0, 2\pi]$	$\cos kx, \ \sin kx, \ k \in \mathbb{N}$	finite trigonometric

Standard weight functions for multi-dimensional integration problems can often be obtained by parameterizing the integration region B and by forming a product of n one-dimensional standard weight functions.

Example (Spherical Coordinates) The two-dimensional space $E_2 = \mathbb{R}^2$ can be parameterized as follows:

$$x_1 = r\cos\varphi, \qquad x_2 = r\sin\varphi \qquad \text{where} \quad (r,\varphi) \in B := [0,\infty) \times [0,2\pi).$$

Assigning the Laguerre weight function $w_1(r) := \exp(-r)$ to the interval $[0,\infty)$, and the Legendre weight function $w_2(\varphi) \equiv 1$ to the interval $[0,2\pi)$ yields the bivariate weight function

$$w(x_1,x_2) = \exp\left(-\sqrt{x_1^2 + x_2^2}\right)$$

for E_2.

12.1.3 Integration Methods

In order to simplify matters, this section only deals with the *one-dimensional* integration (quadrature) problem:

$$\mathrm{I}f := \int_a^b f(x)\,dx \qquad \text{where} \quad f : [a,b] \subset \mathbb{R} \to \mathbb{R}.$$

Manual Analytical Integration

If it is possible to find an *indefinite integral* (*antiderivative*) F of f that can be represented by elementary functions, then the value of the definite integral can be obtained immediately from the fundamental theorem of integral calculus:

$$\mathrm{I}f = F(b) - F(a).$$

Example (Partial Fractions) The integral

$$\int_0^1 \frac{1}{1+x^4}\,dx \tag{12.3}$$

can be calculated analytically. By decomposing the integrand into partial fractions, F can be composed of *elementary* indefinite integrals:

$$P\int \frac{1}{1+x^4}\,dx = \frac{1}{4\sqrt{2}}\log\frac{x^2 + x\sqrt{2} + 1}{x^2 - x\sqrt{2} + 1} + \frac{1}{2\sqrt{2}}\arctan\frac{x\sqrt{2}}{1 - x^2} + \text{const.} \tag{12.4}$$

Unfortunately, this formula cannot be used directly to calculate a numerical value for (12.3) because at $x = 1$ *division by zero* occurs. Suitable transformations lead to the alternative formula

$$\int_0^1 \frac{1}{1+x^4}\,dx = \frac{1}{4\sqrt{2}}\log\frac{2+\sqrt{2}}{2-\sqrt{2}} + \frac{1}{2\sqrt{2}}\left[\arctan\frac{1}{\sqrt{2}-1} + \arctan\frac{1}{\sqrt{2}+1}\right],$$

which can be evaluated numerically.

It has to be noted that the elementary functions $\sqrt{\ }$, *log* and *arctan* can only be evaluated to a certain degree of accuracy, although an *exact* formula has been obtained.

In order to avoid certain disadvantages of using manual integration (such as the large amount of time required and the likelihood of mistakes), tables often have been used (see, for example, Brytschkow et al. [6] or Zwillinger [33]). The probability of making errors, however, is not negligible, even when using relevant literature. In a survey of eight widespread tables of integrals, the frequency of errors was found to be larger than 5 % (Klerer, Grossman [253]).

Moreover, it is often not possible to find *explicit* formulas for many functions of practical relevance (even for some elementary functions).

Example (Non-Integrable Functions)   The integration of $f(x) = \exp(-x^2)$ leads to a function that *cannot* be represented by a finite number of algebraic, logarithmic, and exponential expressions (Rosenlicht [331]).

Also, integrands are often not known *exactly* (as formulas) but are given only in tabular form or defined as the solution of differential equations which cannot be solved directly. This may also be the case for the integration region $B \subset \mathbb{R}^n$, which, for instance, may be given in an *implicit* form that cannot be transformed into an explicit one. Such problems cannot be solved analytically.

Software (Indefinite Integrals)   The subroutine NAG/e02ajf computes the *indefinite* integral of a polynomial given as a linear combination of Chebyshev polynomials. It is thus useful in situations where integrands are first approximated by polynomials in Chebyshev representation.

Symbolic Integration

If the integrand is given analytically in closed form (as a formula), an attempt can be made to compute the integral symbolically using computer algebra software.

Example (Mathematica)   The software package MATHEMATICA can be used to compute the indefinite integral

$$\int \frac{1}{1+x^4}\, dx.$$

The command

```
Integrate[(1+x^4)^(-1),x]
```

yields

$$\int \frac{1}{1+x^4}\, dx = \frac{\arctan\left(\frac{-\sqrt{2}+2x}{\sqrt{2}}\right)}{2^{3/2}} + \frac{\arctan\left(\frac{\sqrt{2}+2x}{\sqrt{2}}\right)}{2^{3/2}} - \frac{\log(1-\sqrt{2}x+x^2)}{2^{5/2}} + \frac{\log(1+\sqrt{2}x+x^2)}{2^{5/2}}, \tag{12.5}$$

which is equivalent to (12.4). This indefinite integral can easily be evaluated at the boundaries of the interval $[0, 1]$:

$$\int_0^1 \frac{1}{1+x^4}\, dx = \frac{\arctan\left(\frac{2-\sqrt{2}}{\sqrt{2}}\right)}{2^{3/2}} + \frac{\arctan\left(\frac{2+\sqrt{2}}{\sqrt{2}}\right)}{2^{3/2}} - \frac{\log(2-\sqrt{2})}{2^{5/2}} + \frac{\log(2+\sqrt{2})}{2^{5/2}}.$$

The restrictions for the symbolic solution of integrals are similar to those for analytical integration done by hand:

1. For many functions, no elementary representation of an antiderivative exists. These cases can however be recognized automatically. There are algorithms that decide whether or not an indefinite integral can be represented by elementary functions and which yield this representation as a result if such a representation exists (Bronstein [128]).

Example (Mathematica, Maple) The indefinite integral

$$\int \sin(\sin x)\, dx$$

is not an elementary function. The symbolic software packages MATHEMATICA and MAPLE stop after no success at simplification.

2. Multi-dimensional integrals can only be dealt with if they can be expressed as *iterated integrals* (see Section 12.2.3) in which both the integrand and the integration boundaries are expressed explicitly in an analytical form. If, for instance, the integration region B is given in an implicit representation which cannot be transformed into an explicit one, then integrals over B *cannot* be calculated analytically or symbolically.

3. Algorithms for symbolic integration sometimes compute antiderivatives that are discontinuous in the integration region, although they could be represented in a continuous form.

Example (Maple) When computing

$$\int\limits_{\pi/3}^{3\pi/2} (\cot x + \operatorname{cosec} x)\, dx = \ln 2,$$

MAPLE produces the following antiderivative:

$$F(x) = \ln \sin x + \ln(\operatorname{cosec} x - \cot x).$$

As F is discontinuous at $x = \pi$, the integral evaluates to $\ln 2 + 2\pi i$.

Example (Maple, Mathematica, Macsyma, Axiom) For

$$I(f; [a, b]) = \int\limits_a^b \frac{3}{5 - 4\cos x}\, dx$$

the antiderivative of the integrand can be determined. The computation results in the following representations of the indefinite integral:

MAPLE, MATHEMATICA: $F(x) = 2\arctan(3\tan(x/2))$,
MACSYMA: $F(x) = 2\arctan(3\sin x/(\cos x + 1))$,
AXIOM: $F(x) = -\arctan(-3\sin x/(5\cos x - 4))$.

None of these functions are continuous for all $x \in \mathbb{R}$.

4. Even if a program for symbolic manipulation is able to determine the indefinite integral, the evaluation of the antiderivative F can lead to numerical difficulties such as cancellation, division by zero, etc.

5. The computation of the antiderivative F and the subsequent evaluation of F at the end-points of the integration interval take often much more time than the computation of the respective definite integral using numerical methods.

Example (Computing Time) The turnaround time for the computation of the indefinite integral (12.5) using MATHEMATICA on an HP workstation took about 6 seconds.

The same definite integral can be evaluated in double precision arithmetic over the region $B = [0, 1]$ with a relative error of $6.2 \cdot 10^{-15}$ by using the 15-point Gauss-Kronrod formula implemented in the subroutine QUADPACK/dqk15. The computing time required to evaluate this Gauss-Kronrod formula is several orders of magnitude smaller than the time needed for the corresponding symbolic manipulations. In fact, it was significantly smaller than 10 milliseconds, the resolution of the available clock routines.

Despite their deficiencies, symbolic integration methods are highly valuable if *indefinite* rather than definite integrals—i. e., formulas rather than numerical values—are to be computed. In this case, *numerical* integration methods are *not* applicable.

The principles of symbolic integration belong to computer algebra and are not part of computer numerics. They are not, therefore, dealt with in this book. More information about symbolic integration can be found, for example, in Davenport [146], [147], Davenport et al. [148], Geddes et al. [207], and Krommer, Ueberhuber [64].

Numerical Integration

For the reasons stated in previous sections, it is often advisable to calculate a solution Qf of the *numerical problem*

$$\begin{aligned} \text{Input-Data:} \quad & f, B, \varepsilon \\ \text{Output-Data:} \quad & Qf \quad \text{where} \quad |Qf - If| \leq \varepsilon \end{aligned} \qquad (12.6)$$

instead of looking for a solution of the mathematical problem (12.2) using analytical and/or symbolic techniques or using integral tables.

Terminology (Quadrature, Cubature) In connection with the numerical problem (12.6), the term *numerical quadrature* is often used for one-dimensional (univariate) problems. For multi-dimensional (multivariate) integration problems, the term *cubature* is commonly used. This terminology makes it possible to distinguish between the calculation of integrals and the integration of differential equations.

The choice of an appropriate method for solving a numerical integration problem largely depends on the form and the content of available information about the integrand:

Data points: Values $f(x_1), \ldots, f(x_N)$ of the integrand $f : B \subseteq \mathbb{R}^n \to \mathbb{R}$ are available only for a given fixed set of points $\{x_i \in B, \ i = 1, 2, \ldots, N\}$. These points may be distributed systematically or irregularly over the region B.

In this setting, the values $\{f(x_i)\}$ are often results of measurements (describing the result of experiments, for instance). A substantial amount of stochastic noise is often superimposed on such data.

Subroutine: The function f is defined and can be evaluated at any point $x \in B$. This assumption is taken as the basis for most integration programs solving problem (12.6). These programs are usually structured so that the integrand is usually specified as a function subprogram which supplies the value $f(x)$ of the integrand for any $x \in B$ on request.

Example (Fortran) The integrand $f(x) = 1/(1 + x^4)$ can be specified in the following Fortran 90 function procedure:

```
FUNCTION f(x) RESULT (f_integrand)
    REAL, INTENT (IN) :: x
    REAL              :: f_integrand

    f_integrand = 1./(1. + x**4)

END FUNCTION f
```

The name f of this function is passed to an integration subroutine which chooses suitable sample points $x_1, x_2, \ldots, x_N$ and evaluates $f(x_1), f(x_2), \ldots, f(x_N)$ by calling f.

Symbolic representation: The integrand has an explicit form suitable for symbolic manipulation. In this case, the integration problem can often be solved using computer algebra software. However, for the reasons discussed earlier, numerical integration methods are often preferred to symbolic techniques for the solution of the numerical problem (12.6).

Integration Using Given Data Points

In some cases, only a finite set of data points

$$\{(x_1, f(x_1)), \ldots, (x_N, f(x_N)) : x_i \in B, \ i = 1, 2, \ldots, N\}$$

is known. Usually a function g is chosen from a suitable linear space $\mathcal{M}$ of model functions in order to interpolate or to approximate these points. Ig then is used as an approximation of If.

The quality of this type of integral approximation (see Section 12.3.1) depends on the quality of the approximation of f by g and thus on the set of given data points. The calculation of Ig of a given function g is relatively simple compared to the selection of suitable spaces $\mathcal{M}$ of model functions and the computation of an interpolation function g. Integration problems of this type are, therefore, dominated by the problem of solving interpolation or approximation problems (see Chapters 9 and 10).

In conjunction with solving numerical integration problems, model spaces containing piecewise polynomials—in particular, polynomial spline functions (see Section 9.5)—are often preferred.

Software (Univariate Integration Using Given Data Points) The equivalent subroutines IMSL/MATH-LIBRARY/bsitg $\approx$ CMLIB/bsqad $\approx$ SLATEC/bsqad, as well as the procedures IMSL/MATH-LIBRARY/csitg and IMSL/MATH-LIBRARY/ppitg compute definite integrals of B-splines, cubic spline functions and other piecewise polynomials over bounded intervals. The programs CMLIB/ppqad $\approx$ SLATEC/ppqad also compute definite integrals of B-splines. They only differ from routines like IMSL/MATH-LIBRARY/bsitg in the way they represent B-splines internally.

The subroutines NAG/e01bhf and NAG/e02bdf calculate definite integrals of piecewise cubic Hermite interpolation polynomials and cubic splines.

The subroutine NAG/d01gaf computes the approximation of the integral of f in the region $[x_1, x_N]$ for given data points

$$\{(x_i, f(x_i)), \; i = 1, 2, \ldots, N\} \qquad \text{where} \quad x_1 < x_2 < \cdots < x_N.$$

In each subregion $[x_i, x_{i+1}]$, $i = 2, 3, \ldots, N - 2$, the integrand f is approximated by a cubic polynomial that interpolates the points

$$(x_{i-1}, f(x_{i-1})), \quad (x_i, f(x_i)), \quad (x_{i+1}, f(x_{i+1})), \quad (x_{i+2}, f(x_{i+2})).$$

In order to interpolate f in the intervals $[x_1, x_2]$ and $[x_{N-1}, x_N]$, the points

$$\{(x_j, f(x_j)), \; j = 1, 2, 3, 4\} \qquad \text{and} \qquad \{(x_j, f(x_j)), \; j = N - 3, N - 2, N - 1, N\}$$

are used. The subroutine SLATEC/avint uses a similar method.

Software (Multivariate Integration Using Given Data Points) The subroutines IMSL/MATH-LIBRARY/bs2ig and IMSL/MATH-LIBRARY/bs3ig calculate definite integrals of tensor-product splines over two- and three-dimensional cubes.

Integration by Sampling

In the remaining sections of this chapter, only the solution of numerical problems for which the integrand function is given as a user-supplied subprogram is dealt with. Only numerical integration methods that use a finite set of values of f (the *sampling information*),

$$S(f) := (f(x_1), f(x_2), \ldots, f(x_N)),$$

to calculate an approximation Qf for If satisfying the inequality

$$|Qf - If| \leq \varepsilon$$

for a given tolerance $\varepsilon > 0$ are discussed. As in the case of data points, a finite number of values of the integrand is used. However, the abscissa values $x_1, \ldots, x_N$ used to calculate the function values $f(x_1), \ldots, f(x_N)$ can be chosen arbitrarily by the integration algorithm. When computing Qf, *linear formulas* or *algorithms*

$$Qf := \sum_{i=1}^{N} c_i f(x_i) \tag{12.7}$$

are most often employed. Other methods (except for those based on nonlinear extrapolation) are, in fact, of minor importance.

12.1.4 Sensitivity of Integration Problems

Changes in the Integrand

The sensitivity of the integral $\mathrm{I}f$ with respect to changes in the integrand f is characterized by

$$|\mathrm{I}\tilde{f} - \mathrm{I}f| \leq l \cdot \|\tilde{f} - f\|_\infty. \tag{12.8}$$

The (absolute) *condition number* K_f of the integration problem is the smallest number satisfying (12.8). It is easily shown that K_f equals $\mathrm{vol}(B)$, the volume of the n-dimensional integration region:

$$\begin{aligned} |\mathrm{I}\tilde{f} - \mathrm{I}f| &= \left| \int_B \tilde{f}(x)\,dx - \int_B f(x)\,dx \right| \leq \int_B |\tilde{f}(x) - f(x)|\,dx \\ &\leq \mathrm{vol}(B) \cdot \|\tilde{f} - f\|_\infty. \end{aligned} \tag{12.9}$$

To prove that $\mathrm{vol}(B)$ is the smallest number satisfying (12.8), it is sufficient to choose $f \equiv 0$ and $\tilde{f} \equiv c$.

The condition estimate (12.9) indicates that the *mathematical* problem of integration is *very well* conditioned with respect to variations in the integrand. This is why in numerical integration the integrand f is often replaced by a neighboring function g (a polynomial, for instance) that can be integrated more easily than the original function f. To do this, g has to be close to f with respect to the maximum norm $\| \ \|_\infty$.

It may happen, however, that an integral $\mathrm{I}f$ has a very small absolute value despite the fact that the associated integrand is significantly different from zero. Such problems have a large *relative* condition number, i.e., the quantity K_f^{rel} in

$$\frac{|\mathrm{I}\tilde{f} - \mathrm{I}f|}{|\mathrm{I}f|} \leq K_f^{\mathrm{rel}} \cdot \frac{\|\tilde{f} - f\|_\infty}{\|f\|_\infty} \tag{12.10}$$

is a large number. Small changes in the integrand (caused, for instance, by rounding errors) may produce large relative errors in the result $\mathrm{I}\tilde{f}$.

Example (Ill-Conditioned Quadrature Problem)   For $m = 2, 4, 6, \ldots$, the integral

$$\mathrm{I}f := \int_0^1 \cos(m\pi x)\,dx$$

evaluates exactly to 0. Using (12.10) to calculate the relative condition number for m with respect to additive variations of the integrand $\tilde{f} := f + \delta$, $\delta \in \mathbb{R}$, yields

$$K_f^{\mathrm{rel}} = \left| \frac{m\pi}{\sin(m\pi)} \right|.$$

For $m \approx 2k$, $k \in \mathbb{N}$, the asymptotic approximation

$$K_f^{\mathrm{rel}} \approx \left| \frac{2k}{2k - m} \right|$$

holds. For $m \approx 20$, the following values of the condition number are obtained:

m	19.90	19.92	19.94	19.96	19.98	20
K_f^{rel}	202.3	251.6	333.3	500.3	999.7	∞

The approximating functional Qf with respect to modifications of the integrand

$$|Q\tilde{f} - Qf| \le \overline{K}_f \cdot \|\tilde{f} - f\|_\infty$$

is characterized by the (absolute) condition number $\overline{K}_f := \sum |c_i|$ viz.

$$
\begin{aligned}
|Q\tilde{f} - Qf| &= |Q(\tilde{f} - f)| \le \sum_{i=1}^{N} |c_i| \cdot |\tilde{f}(x_i) - f(x_i)| \\
&\le \left(\sum_{i=1}^{N} |c_i| \right) \cdot \|\tilde{f} - f\|_\infty .
\end{aligned}
$$

If Q yields the exact value If for all constant functions $f(x) \equiv c$, i.e., $Qf = If = \mathrm{vol}(B)\,c$, then

$$\sum_{i=1}^{N} c_i = \mathrm{vol}(B).$$

In this case, the condition number $\overline{K}_f$ of the approximating functional Q is identical to that of the original functional I if and only if

$$\overline{K}_f = \sum_{i=1}^{N} |c_i| = \sum_{i=1}^{N} c_i = \mathrm{vol}(B) = K_f,$$

i.e., if all weights c_i of Q are *non-negative*. In these instances the numerical integration problem is also *very well* conditioned with respect to changes in the integrand.

All integration formulas in practical use have either positive weights, or their condition numbers $\overline{K}_f = \sum |c_i|$ are only slightly larger than the optimal condition numbers $K_f = \mathrm{vol}(B)$.

Change in Region

In addition to the integrand f, the integration region B also belongs to the input data $\mathcal{D}$ of a numerical integration problem. The condition with respect to changes in the integration region can be considerably worse. In order to demonstrate this phenomenon, finite one-dimensional integration regions, i.e., intervals $[a, b] \subset \mathbb{R}$, are considered in this section. Changes in the integration region are measured by some distance function $\mathrm{dist} : \mathbb{R}^2 \times \mathbb{R}^2 \to \mathbb{R}_+$ that can, for instance, be defined as follows:

$$\mathrm{dist}\left((\tilde{a}, \tilde{b}), (a, b) \right) := |\tilde{a} - a| + |\tilde{b} - b|.$$

The fundamental theorem of integral calculus shows that the (absolute) condition number $K_{a,b}$ of the integral $I(f; a, b)$ with respect to changes in the boundaries a and b,

$$|I(f; [\tilde{a}, \tilde{b}]) - I(f; [a, b])| \le K_{a,b} \,\, \mathrm{dist}\left((\tilde{a}, \tilde{b}), (a, b) \right),$$

is determined by $f(a)$ and $f(b)$.

Theorem 12.1.1 (Fundamental Theorem of Integral Calculus) *For $f \in$ $C[a, b]$ the derivative of*

$$F(t) := \int_a^t f(x) \, dx, \qquad t \in [a, b],$$

exists and is given for $t \in (a, b)$ by

$$F'(t) = \frac{d}{dt} \int_a^t f(x) \, dx = f(t).$$

Not surprisingly, the condition number $K_{a,b}$ is large if f is singular at or near the boundary of B.

Example (Ill-Conditioned Quadrature Problem) The integral

$$I(f; [0, b]) := \int_0^b (1 - x)^{-0.9} \, dx = -10(1 - b)^{0.1} + 10 \tag{12.11}$$

is extremely sensitive to variations in the upper boundary b if b is close to 1.

b	$1 - 10^{-5}$	$1 - 10^{-10}$	$1 - 10^{-15}$	$1 - 10^{-20}$
$I(f; [0, b])$	6.8377	9.0000	9.6838	9.9000
$K_{0,b} = f(b)$	$3.16 \cdot 10^4$	$1.00 \cdot 10^9$	$3.16 \cdot 10^{13}$	$1.00 \cdot 10^{18}$

On a computer with IEC/IEEE floating-point arithmetic, the distance between the machine numbers slightly smaller than 1 is

$$\Delta x = 2^{-24} \approx 5.96 \cdot 10^{-8} \quad (single\ precision)\ \text{or}$$
$$\Delta x = 2^{-53} \approx 1.11 \cdot 10^{-16} \quad (double\ precision).$$

The optimal rounding of b to the nearest machine number may, in the worst case, produce rounding errors of $2.98 \cdot 10^{-8}$ and $5.55 \cdot 10^{-17}$ respectively. This rounding error has an influence on $I(f; [0, b])$, which is magnified by the condition number. Thus, the result of an algorithm of the form (12.11) may have *no* correct decimal digits at all.

12.1.5 Inherent Uncertainty of Numerical Integration

The condition estimate (12.9) shows that *small* variations in the integrand have little impact on the value of the integral. However, the available *discrete* information—the sample $S(f)$—does *not* allow an assessment of the actual distance $\|\tilde{f} - f\|_\infty$. The subset

$$V(f; \mathcal{R}, S) := \{\tilde{f} \in \mathcal{R} \; : \; S(\tilde{f}) = S(f)\}$$

of those Riemann-integrable functions which cannot be distinguished from f on the basis of the sample information S is an *infinite* set.

The integrals of the functions in $V(f; \mathcal{R}, S)$ are real numbers; in fact, the set of all these values is the set of *all* real numbers $\mathbb{R}$. This can easily be understood by considering, for example, the following set of functions:

$$g_c(x) := f(x) + c \prod_{i=1}^{N} \|x - x_i\|_2^2, \quad c \in \mathbb{R}.$$

These functions have the property $S(g_c) = S(f)$, i.e., they cannot be distinguished from f at the sample points $x_1, x_2, \ldots, x_N$,

$$g_c(x_i) = f(x_i), \quad i = 1, 2, \ldots, N.$$

The respective integrals

$$\mathrm{I}\, g_c = \mathrm{I}f + c \int_B \prod_{i=1}^{N} \|x - x_i\|_2^2 \, dx,$$

however, can assume *any* real value, i.e., the distance $\|\tilde{f} - f\|_\infty$ can be *arbitrarily large* for $\tilde{f} \in V(f; \mathcal{R}, S)$.

The good condition number of the integration problem with respect to the perturbation $\Delta f = \tilde{f} - f$ cannot prevent or even reduce this total uncertainty of the integration problem. The set

$$\{\mathrm{I}\tilde{f} \; : \; \tilde{f} \in V(f; \mathcal{F}, S)\}$$

can only be reduced to a finite interval $[\mathrm{I}_{\min}, \mathrm{I}_{\max}] \subset \mathbb{R}$ by restricting the set of admissible functions $\mathcal{F}$ appropriately. For integrals over finite intervals $B = [a, b]$ this can be achieved, for instance, by imposing an upper bound on one of the derivatives:

$$\mathcal{F} = \left\{ f \in C^k[a, b] \; : \; |f^{(k)}(x)| \le M_k, \; \forall x \in [a, b] \right\}, \quad k \in \mathbb{N}. \tag{12.12}$$

Clearly, the best accuracy bound which can be obtained in such a restricted setting is given by the *radius of information* (Wozniakowski [383]):

$$r(S) := \frac{\mathrm{I}_{\max} - \mathrm{I}_{\min}}{2}.$$

In most practical cases, however, a restriction like (12.12) cannot be made.

12.2 Preprocessing of Integration Problems

Quite often, numerical schemes for approximately calculating integrals cannot—or at least cannot *effectively*—be used for integration problems occurring in practice. For instance, the integration region may differ from the standard integration regions in Section 12.1.1. Algorithms which adaptively subdivide the integration region (see Section 12.5.2) are normally applicable only if the underlying integration region is a bounded n-dimensional cube. In order to apply an adaptive subdivision algorithm to integrals over unbounded and/or non-cubic regions, the corresponding integral has to undergo an appropriate (manual or automatic) *pre-processing phase*.

In this section, the most important preprocessing methods are discussed. These methods often also play a central role when developing new integration formulas (see Sections 12.3.1, 12.3.3 and 12.4.1).

12.2.1 Transformation of Integrals

Let $\psi : \mathbb{R}^n \to \mathbb{R}^n$ denote a continuously differentiable one-to-one transformation from $\overline{B}$ to B, and let J be its *Jacobian matrix*

$$J(\overline{x}) = \begin{pmatrix} \dfrac{\partial \psi_1}{\partial \overline{x}_1}(\overline{x}) & \dfrac{\partial \psi_1}{\partial \overline{x}_2}(\overline{x}) & \cdots & \dfrac{\partial \psi_1}{\partial \overline{x}_n}(\overline{x}) \\[2ex] \dfrac{\partial \psi_2}{\partial \overline{x}_1}(\overline{x}) & \dfrac{\partial \psi_2}{\partial \overline{x}_2}(\overline{x}) & \cdots & \dfrac{\partial \psi_2}{\partial \overline{x}_n}(\overline{x}) \\[2ex] \vdots & \vdots & & \vdots \\[2ex] \dfrac{\partial \psi_n}{\partial \overline{x}_1}(\overline{x}) & \dfrac{\partial \psi_n}{\partial \overline{x}_2}(\overline{x}) & \cdots & \dfrac{\partial \psi_n}{\partial \overline{x}_n}(\overline{x}) \end{pmatrix}.$$

Theorem 12.2.1 (Multidimensional Transformation Rule)   *Let J be non-singular on $\overline{B}$, i.e.,*

$$\det J(\overline{x}) \neq 0 \quad \text{for all} \quad \overline{x} \in \overline{B}.$$

Then,

$$\mathrm{I}f = \int_B f(x)\,dx = \int_{\overline{B}} f(\psi(\overline{x}))\,|\det J(\overline{x})|\,d\overline{x} = \int_{\overline{B}} \overline{f}(\overline{x})\,d\overline{x}, \tag{12.13}$$

where

$$\overline{f}(\overline{x}) = f(\psi(\overline{x}))\,|\det J(\overline{x})|, \quad \overline{x} \in \overline{B}.$$

This rule becomes particularly simple if ψ is an *affine transformation*:

$$\psi(\overline{x}) = A\overline{x} + b,$$

where $A \in \mathbb{R}^{n \times n}$ and $b \in \mathbb{R}^n$. In this case, the Jacobian matrix is independent of $\overline{x}$,

$$\det J(\overline{x}) := \det A,$$

and thus the transformation rule is reduced to

$$\mathrm{I}f = \int_B f(x)\,dx = |\det A| \int_{\overline{B}} f(\psi(\overline{x}))\,d\overline{x}.$$

The transformation rule (12.13) also becomes significantly simpler for univariate functions. Because $\det J = \psi'$, the substitution

$$\int_B f(x)\, dx = \int_{\overline{B}} f(\psi(\overline{x}))\, |\psi'(\overline{x})|\, d\overline{x}$$

holds. If $\overline{B} \subset \mathbb{R}$ is an interval, then ψ' is either strictly positive or strictly negative on $\overline{B}$ (because of its continuity), resulting in the univariate transformation formulas

$$\int_B f(x)\, dx = \int_{\overline{B}} f(\psi(\overline{x}))\, \psi'(\overline{x})\, d\overline{x} \qquad \text{for} \quad \psi' > 0,$$

$$\int_B f(x)\, dx = -\int_{\overline{B}} f(\psi(\overline{x}))\, \psi'(\overline{x})\, d\overline{x} \qquad \text{for} \quad \psi' < 0.$$

Transformations on Standard Regions

Preprocessing integrals using appropriate transformations is often done when constructing integration formulas for regions B different from the standard regions $\overline{B}$ described in Section 12.1.1. Provided that

$$\overline{Q}_N \overline{f} = \sum_{i=1}^{N} \overline{c}_i \overline{f}(\overline{x}_i)$$

is an integration formula for the approximate integration of functions $\overline{f} \colon \overline{B} \to \mathbb{R}$, the transformed integration formula Q_N for $f \colon B \to \mathbb{R}$ is given by

$$Q_N f = \sum_{i=1}^{N} c_i f(x_i),$$

with the following abscissas and weights (see Section 12.4.1):

$$x_i = \psi(\overline{x}_i), \quad c_i = \overline{c}_i\, |\det J(\overline{x}_i)|, \quad i = 1, 2, \ldots, N.$$

In other words, the abscissas of Q_N are simply the images of the abscissas of $\overline{Q}_N$ with respect to the transformation ψ. The weights of Q_N are obtained by multiplying the corresponding weights of $\overline{Q}_N$ by the absolute values of the determinant of the Jacobian matrix at the corresponding abscissas.

The application of the transformed integration formula Q_N to a function $f \colon B \to \mathbb{R}$ can also be interpreted as an application of the original integration formula $\overline{Q}_N$ to the transformed integrand $\overline{f} \colon \overline{B} \to \mathbb{R}$ (see Section 12.4.1):

$$\overline{f} = |\det J(\overline{x})|\, f(\psi(\overline{x})).$$

This representation is more conducive to analysis of the error in $Q_N f = \overline{Q}_N \overline{f}$.

The error $Q_N f - I f$ can be expected to be small if the transformed integrand $\overline{f}$ belongs to the class of admissible integrands $\overline{\mathcal{F}}$ of the original integration formula

$\overline{Q}_N$. Whether or not this is the case depends on both the original integrand f and the transformation ψ (and its Jacobian J). For instance, $\mathcal{F}$ is often equal to $C^k(\overline{B})$, $k \in \mathbb{N}$, one of the spaces of continuously differentiable functions. Transformations ψ which map bounded regions $\overline{B}$ onto unbounded regions B have singularities at the boundary of B. In this case it generally cannot be expected that $\overline{f} \in \mathcal{F}$.

A thorough knowledge and understanding of the integrand of a specific integration problem may enable the development of *tailor-made* transformations. Such transformations for integration problems occurring in statistical computation are described, for instance, by Genz [210]. In this section, however, two types of *general purpose* transformations are described.

If the integration problem is given in analytic form, the transformation of integrals can sometimes be facilitated using symbolic computation software.

Example (General Purpose and Tailor-Made Transformations) The integral

$$I f = \int_{-\infty}^{\infty} \exp(-x^2) f(x) \, dx \tag{12.14}$$

is considered, i.e., the integration of f with respect to the Hermite weight function $w(x) = \exp(-x^2)$, in this example. The infinite interval of integration $(-\infty, \infty)$ can be transformed into a finite interval by using the general purpose transformation

$$\psi(\overline{x}) = \frac{\overline{x}}{1 - |\overline{x}|}, \quad -1 < \overline{x} < 1.$$

Because

$$\psi'(\overline{x}) = \frac{1}{(1 - |\overline{x}|)^2},$$

the integral (12.14) is transformed into

$$I f = \int_{-1}^{1} \frac{1}{(1 - |\overline{x}|)^2} \exp\left(-\left(\frac{\overline{x}}{1 - |\overline{x}|}\right)^2\right) f\left(\frac{\overline{x}}{1 - |\overline{x}|}\right) d\overline{x}. \tag{12.15}$$

On the other hand, by substituting $x = t/\sqrt{2}$, (12.14) can be written as

$$I f = \frac{1}{\sqrt{2}} \int_{-\infty}^{\infty} \exp(-t^2/2) f(t/\sqrt{2}) \, dt.$$

The infinite interval of integration $(-\infty, \infty)$ can now be transformed into a finite interval by using the tailor-made transformation

$$t = \psi(\overline{x}) = \Phi^{-1}(\overline{x})$$

where

$$\Phi(t) = \frac{1}{\sqrt{2\pi}} \int_{-\infty}^{t} \exp(-u^2/2) \, du$$

is the distribution function of the standard normal distribution. Because

$$\psi'(\overline{x}) = \frac{1}{\Phi'(t)} = \sqrt{2\pi} \exp(t^2/2),$$

the integral (12.14) is transformed into

$$If = \sqrt{\pi} \int_0^1 f\left(\Phi^{-1}(\overline{x})/\sqrt{2}\right) d\overline{x}. \tag{12.16}$$

A comparison of (12.15) and (12.16) shows that the integrand in (12.16) has a much simpler structure. On the other hand, the integrand in (12.15) contains only elementary functions, whereas the integrand in (12.16) includes Φ^{-1}, which is not an elementary function. Nevertheless, efficient software for the evaluation of Φ^{-1} is available (for instance, the subroutine IMSL/STAT-LIBRARY/anorin). As a result, it is not clear which of the integrals (12.15) and (12.16) is easier to compute numerically.

Software (Transformation into Bounded Intervals) The subroutines CMLIB/qk15i $\approx$ QUADPACK/qk15i $\approx$ SLATEC/qk15i transform unbounded intervals

$$B = [a, \infty) \quad \text{and} \quad B = (-\infty, \infty)$$

into the unit interval $\overline{B} = [0, 1]$. For the first unbounded interval, the transformation

$$x = \psi(\overline{x}) = a + \overline{x}/(1 - \overline{x}) \tag{12.17}$$

is used, and for the second interval the integration region is divided at $a = 0$ into the intervals $(-\infty, 0]$ and $[0, \infty)$, which are then transformed by (12.17). The transformed integrand is then integrated using a 15-point Gauss-Kronrod formula.

Periodizing Transformations of Univariate Integrands

Certain classes of integration formulas prove to be much less effective when applied to non-periodic integrands than when applied to periodic integrands (or integrands which can be extended periodically without changes in their differentiability). In order to accelerate the order of convergence in such cases, non-periodic integrands have to be subjected to *periodizing transformations*.

For one-dimensional integration problems, a number of such transformations based on the transformation rule for integrals have been proposed. In these transformations the independent variable x is transformed by

$$x = \psi(\overline{x}),$$

where the function ψ is a monotonic mapping of the interval $[0, 1]$ onto itself. This gives rise to the following relationship:

$$\int_0^1 f(x)\, dx = \int_0^1 f(\psi(\overline{x}))\psi'(\overline{x})\, d\overline{x}.$$

Definition 12.2.1 (Polynomial Transformations) *For polynomial transformations (Hua, Wang [232])*

$$\psi(\overline{x}) = D_\alpha \int_0^{\overline{x}} (t(1 - t))^{\alpha - 1}\, dt, \quad \alpha \in \{2, 3, 4, \ldots\} \tag{12.18}$$

is chosen, where the scaling factor D_α, is given by

$$D_\alpha = \left(\int_0^1 (t(1-t))^{\alpha-1} dt \right)^{-1}.$$

The function ψ has the property

$$\psi^{(k)}(0) = \psi^{(k)}(1) = 0 \qquad \text{for} \quad k = 1, 2, \ldots, \alpha - 2.$$

As a result, if the integrand f is in $C^{\alpha-2}[0,1]$, the transformed integrand $(f \circ \psi)\psi'$ is also in $C^{\alpha-2}[0,1]$, but all its derivatives up to the maximum order $\alpha - 2$ vanish at the integral boundaries. Therefore, the transformed integrand $(f \circ \psi)\psi'$ has a periodic extension that does not change its differentiability. The parameter α can be used to control the smoothness of the transformed integrand at the boundaries of the integration interval.

Iri, Moriguti and Takasawa [237] propose the so-called periodizing *IMT-transformations*:

Definition 12.2.2 (IMT-Transformations) *IMT-transformations are defined by*

$$\psi(\bar{x}) = D_c \int_0^{\bar{x}} \exp\left(-\frac{c}{t(1-t)} \right) dt. \tag{12.19}$$

c is positive and D_c is given by

$$D_c = \left(\int_0^1 \exp\left(-\frac{c}{t(1-t)} \right) dt \right)^{-1}.$$

It can be verified that *all* derivatives of ψ vanish at the boundaries 0 and 1, i.e.,

$$\psi^{(k)}(0) = \psi^{(k)}(1) = 0, \quad k = 1, 2, 3, \ldots.$$

As a result, if the integrand f is in $C^\infty[0,1]$, the transformed integrand $(f \circ \psi)\psi'$ is also in $C^\infty[0,1]$, and all its derivatives vanish at the interval boundaries. The transformed integrand thus has an optimally smooth, periodic extension. The parameter c determines how fast the derivatives of ψ converge to zero at the boundaries 0 and 1.

Definition 12.2.3 (tanh-Transformations) *tanh-transformations are defined by the following formula (Sag, Szekeres [338]):*

$$\psi(\bar{x}) = \frac{1}{2}\left(1 + \tanh\left(\frac{c}{2}\left(\frac{1}{1-\bar{x}} - \frac{1}{\bar{x}} \right) \right) \right). \tag{12.20}$$

Definition 12.2.4 (Double Exponential Transformations) *Double exponential transforms (Beckers, Haegemans [109]) are defined by*

$$\psi(\bar{x}) = \frac{1}{2}\left(1 + \tanh\left(c \sinh\left(d \left(\frac{1}{1-\bar{x}} - \frac{1}{\bar{x}} \right) \right) \right) \right).$$

Both the *tanh*-transformation and the double exponential transformation have the same properties as the IMT-transformation: *All* derivatives of ψ vanish at the interval boundaries 0 and 1. The parameters c and d determine how fast the derivatives of ψ converge to zero at the boundaries 0 and 1. These transformations are therefore called "IMT-type transformations".

Software (Double Exponential Transformations)   Double exponential transformations are used by the integration programs JACM/defint and JACM/dehint.

IMT-type transformations have a particular advantage: Even if the original integrand f has a boundary singularity, the transformed integrand $(f \circ \psi)\psi'$ is often continuously differentiable at this point due to the rapid decline of ψ' near the boundaries. This property usually significantly improves the accuracy of the integration rules mentioned earlier. However, the use of periodizing transformations in connection with integrands with end-point singularities may give rise to numerical instabilities, as the following example demonstrates:

Example (Numerical Instability due to Periodizing Transformations)   Consider the computation of the integral

$$If = \int\limits_0^1 (1-x)^{-0.9}\, dx$$

using a 26-point compound trapezoidal rule (Section 12.3.3):

$$T_{25}f = \frac{1}{25}\left(\frac{1}{2}f(0) + \sum_{i=1}^{24} f\left(\frac{i}{25}\right) + \frac{1}{2}f(1)\right).$$

The original integrand $f(x) = (1-x)^{-0.9}$ has a singularity at the right-hand boundary $x = 1$. It can easily be seen that the *tanh*-transformation removes this singularity.

For the choice $c = 4$ in (12.20), the value of ψ at $\overline{x}_1 = 0.04$ is $\psi(\overline{x}_1) = 1.0 \cdot 10^{-21}$. For reasons of symmetry, $\psi(\overline{x}_{24}) = 1 - \psi(\overline{x}_1) = 1 - 1.0 \cdot 10^{-21}$. However, $\psi(\overline{x}_{24})$ has to be represented by a machine number < 1. Assuming a *double precision* IEC/IEEE arithmetic, $\psi(\overline{x}_{24})$ is represented by a machine number $\Box\psi(\overline{x}_{24}) \approx 1 - 1.1 \cdot 10^{-16}$. Thus, instead of $f(\psi(\overline{x}_{24})) = 7.9 \cdot 10^{18}$, $f(\Box\psi(\overline{x}_{24})) \approx 2.3 \cdot 10^{14}$ is obtained. Since the weight $c_{24} = \psi'(\overline{x}_{24})/25$ is $c_{24} = 7.8 \cdot 10^{-20}$, there is an inaccuracy of

$$c_{24}\left[f(\psi(\overline{x}_{24})) - f(\Box\psi(\overline{x}_{24}))\right] \approx 4.2 \cdot 10^{-1}$$

As the exact integral value is $If = 10$, the quite substantial relative error of $4.2 \cdot 10^{-2}$ is encountered.

Periodizing Transformations for Multivariate Integrands

Multi-dimensional integrals have to be transformed such that the integration region B becomes the positive unit cube $W_n = [0,1]^n$. The one-dimensional transformations mentioned above can then be applied to each independent variable:

$$x_1 = \psi_1(\overline{x}_1),\quad x_2 = \psi_2(\overline{x}_2),\quad \ldots\ ,\quad x_n = \psi_n(\overline{x}_n).$$

The transformation rule (12.13) for multi-dimensional integrals leads to

$$\int\limits_{[0,1]^n} f(x_1, x_2, \ldots, x_n)\, dx_1 dx_2 \cdots dx_n = \tag{12.21}$$

$$= \int\limits_{[0,1]^n} f(\psi_1(\overline{x}_1), \psi_2(\overline{x}_2), \ldots, \psi_n(\overline{x}_n))\, \psi_1'(\overline{x}_1)\psi_2'(\overline{x}_2) \cdots \psi_n'(\overline{x}_n)\, d\overline{x}_1 d\overline{x}_2 \cdots d\overline{x}_n.$$

Example (Embedded Lattice Rules)   Joe and Sloan [244] discuss the construction of sequences of embedded lattice rules (see Section 12.4.5):

$$Q_{N_0} \subset Q_{N_1} \subset \cdots \subset Q_{N_n},$$

where each rule Q_{N_j}, $j = 0, 1, \ldots, n$, $N_j = 2^j r$ has distinct abscissas. These lattice rules are used for the numerical calculation of the integral

$$\int\limits_{[0,1]^6} \cos(-5.5 + 2\|x\|_1)\, dx = \cos(0.5)\sin^6(1) \approx 0.312. \tag{12.22}$$

It has to be noted that the integrand cannot be extended periodically beyond $[0, 1]^n$ without any loss of differentiability.

The following table shows the absolute values of relative integration errors for a sequence of lattice rules, proposed by Joe and Sloan [244], together with the effect of two different periodizing transformations:

			transformation		
j	N	—	polynomial	tanh	
0	619	$4.3 \cdot 10^{-2}$	$2.4 \cdot 10^{-3}$	$1.6 \cdot 10^{-1}$	
1	1238	$3.3 \cdot 10^{-2}$	$1.8 \cdot 10^{-3}$	$1.1 \cdot 10^{-1}$	
2	2476	$1.9 \cdot 10^{-2}$	$1.5 \cdot 10^{-3}$	$5.3 \cdot 10^{-2}$	
3	4952	$1.8 \cdot 10^{-2}$	$5.6 \cdot 10^{-4}$	$1.4 \cdot 10^{-2}$	
4	9904	$2.6 \cdot 10^{-2}$	$7.2 \cdot 10^{-4}$	$1.4 \cdot 10^{-2}$	
5	19808	$1.4 \cdot 10^{-2}$	$2.8 \cdot 10^{-4}$	$1.0 \cdot 10^{-3}$	
6	39616	$9.9 \cdot 10^{-3}$	$2.2 \cdot 10^{-5}$	$3.1 \cdot 10^{-4}$	

It is useful to note that the errors decrease quite slowly when there is no integrand transformation at all. An increase in the number of abscissas by a factor of 64 ($= 2^6$) results in a reduction of the error by a factor of 4.3 only.

In order to obtain a periodic integrand, the polynomial transformation

$$x_k = 3\overline{x}_k^2 - 2\overline{x}_k^3, \quad \overline{x}_k \in [0, 1], \quad k = 1, 2, \ldots, 6,$$

was applied. As shown in the table, the increased smoothness of the integrand improves the accuracy of all formulas. More importantly, the error is reduced by a factor of 109 between the least accurate and the most accurate results.

Finally, the particular *tanh*-transformation (12.20) with $c = 1$,

$$x_k = \frac{1}{2}\left(1 + \tanh\left(\frac{1}{2}\left(\frac{1}{1 - \overline{x}_k} - \frac{1}{\overline{x}_k}\right)\right)\right), \quad \overline{x}_k \in [0, 1], \quad k = 1, 2, \ldots, 6, \tag{12.23}$$

was applied. This transformation results in a periodic integrand which is infinitely many times differentiable. The positive effect of this transformation is a significantly improved error reduction—by a factor of 516 from the least accurate to the most accurate result. However, the absolute level of the integration errors of all rules is *larger* than those obtained after the polynomial transformation.

For a large number of abscissas ($N \gg 39\,616$), the tanh-transformation (12.23) proves to be superior. For small values of N, however, the integration errors resulting from the tanh-transformation are larger than those of untransformed integrands.

Transformations of Unbounded Regions

One way of dealing with integrals over unbounded regions is to transform them into integrals over bounded regions using the transformation rule. The following list contains a number of frequently used transformations for the semi-infinite interval $[0, \infty)$. The integration interval resulting from the transformation is in all cases $[0, 1)$.

$\psi(\overline{x})$	$\psi'(\overline{x})$
$-\alpha \log(1 - \overline{x})$, $\alpha > 0$	$\alpha/(1 - \overline{x})$
$\overline{x}/(1 - \overline{x})$	$1/(1 - \overline{x})^2$
$(\overline{x}/(1 - \overline{x}))^2$	$2\overline{x}/(1 - \overline{x})^3$

The infinite interval $(-\infty, \infty)$ may be transformed into the interval $(-1, 1)$ using one of the following transformations:

$\psi(\overline{x})$	$\psi'(\overline{x})$
$\overline{x}/(1 - |\overline{x}|)$	$1/(1 - |\overline{x}|)^2$
$\tan(\pi \overline{x}/2)$	$(\pi/2)(1 + \tan^2(\pi \overline{x}/2))$

More transformations for unbounded regions can be found, for instance, in Genz [210]. It has to be noted that ψ' is generally singular at those points $\overline{x}$ for which $\psi(\overline{x}) = \pm\infty$. This may entail singularities of the transformed integrand $(f \circ \psi)\psi'$.

Example (Singularities due to Transformation) If the convergent integral

$$\int_{-\infty}^{\infty} \frac{1}{1 + |x|^\alpha}\, dx, \quad 1 < \alpha < 2,$$

is transformed by

$$x = \psi(\overline{x}) = \overline{x}/(1 - |\overline{x}|), \quad \text{where} \quad \psi'(\overline{x}) = 1/(1 - |\overline{x}|)^2,$$

then the integral

$$\int_{-1}^{1} \frac{1}{(1 - |\overline{x}|)^2 + (1 - |\overline{x}|)^{2-\alpha}|\overline{x}|^\alpha}\, d\overline{x}$$

is obtained. Because $\alpha < 2$, its integrand is singular at both endpoints of the interval $(-1, 1)$.

Multi-dimensional unbounded regions first have to be transformed into a Cartesian product of (infinite) intervals. The transformations mentioned above can then be applied to each coordinate.

12.2.2 Decomposition of Integration Regions

Integrals are *additive* with respect to a decomposition of B (the region of integration) into pairwise disjoint subregions $B_1, B_2, \ldots, B_L$:

$$\int_B f(\mathbf{x})\, d\mathbf{x} = \int_{B_1} f(\mathbf{x})\, d\mathbf{x} + \int_{B_2} f(\mathbf{x})\, d\mathbf{x} + \cdots + \int_{B_L} f(\mathbf{x})\, d\mathbf{x}.$$

Such a decomposition is usually used when the shape of the region B is too complicated to allow for the straightforward application of numerical schemes. Subregions $B_1, \ldots, B_L$ which have a simpler shape than B are chosen. For instance, an integral over a polyhedron B is usually computed by partitioning B into a finite number of simplexes and/or cubes $B_1, \ldots, B_L$. A number of numerical integration schemes exist for integration regions in the form of n-dimensional rectangles or simplexes.

The problem of determining what error tolerances to impose on the subregions $B_1, \ldots, B_L$ in order to meet a given accuracy requirement for the overall integral over B is addressed within the context of adaptive numerical integration algorithms (see Section 12.5.2).

12.2.3 Iteration of Integrals

The decomposition of an integration region results in subregions $B_1, B_2, \ldots, B_L$ which are geometrically simpler than B but have the same dimension. By contrast, the *iteration* of integrals aims at decomposing an integration problem into integrals of reduced dimension.

Let $n = n_1 + n_2$ and thus $\mathbb{R}^n = \mathbb{R}^{n_1} \times \mathbb{R}^{n_2}$. In addition, let B_1 and B_2 be defined by

$$B_1 := \{ x_1 \in \mathbb{R}^{n_1} : (\{x_1\} \times \mathbb{R}^{n_2}) \cap B \neq \emptyset \},$$
$$B_2(x_1) := \{ x_2 \in \mathbb{R}^{n_2} : (x_1, x_2) \in B \}, \quad x_1 \in B_1.$$

Then the integral $\mathrm{I}f$ can be *iterated* as follows:

$$\mathrm{I}f = \int_B f(x)\, dx = \int_{B_1} \int_{B_2(x_1)} f(x_1, x_2)\, dx_2\, dx_1. \qquad (12.24)$$

The iteration process can be repeated for both the outer and the inner integral in (12.24) until $\mathrm{I}f$ is reduced to a sequence of *one*-dimensional integrals:

$$\mathrm{I}f = \int_B f(x)\, dx$$
$$= \int_{B_1} \int_{B_2(x_1)} \int_{B_3(x_1,x_2)} \cdots \int_{B_n(x_1,x_2,\ldots,x_{n-1})} f(x_1, x_2, \ldots, x_n)\, dx_n dx_{n-1} \cdots dx_1.$$

The *order* of iteration may have a profound impact on the computational complexity of the subsequent numerical integration procedure.

Example (Iteration Order) Consider the two-dimensional integral

$$I f := \int_B 1 \, dx$$

where B is defined by

$$B := \{ (x_1, x_2) \in [0,1]^2 \ : \ 0 \le x_1 \le \sqrt{1 - x_2} \}.$$

Iterating If yields

$$If = \int_0^1 \int_0^{\sqrt{1-x_2}} dx_1 dx_2 = \int_0^1 \sqrt{1 - x_2} \, dx_2 \qquad (12.25)$$

and

$$If = \int_0^1 \int_0^{1-x_1^2} dx_2 dx_1 = \int_0^1 (1 - x_1^2) \, dx_1. \qquad (12.26)$$

The integrand $(1 - x_2)^{1/2}$ in (12.25) has a cusp at $x_2 = 1$, while $1 - x_1^2$ is a polynomial of second degree that can easily be integrated. The numerical calculation of (12.25) is thus significantly more difficult than that of (12.26).

The problem of deciding what error tolerances have to be imposed on the inner integrals of an iterated integral in order to meet a given accuracy requirement for the overall integral is non-trivial. This issue is addressed, for instance, in Fritsch, Kahaner and Lyness [204].

Software (Iterated Integrals) NAG/d01daf and IMSL/MATH-LIBRARY/twodq are subroutines for the numerical computation of two-dimensional iterated integrals.

12.3 Univariate Integration Formulas

Definition 12.3.1 (Integration Formula) *A linear combination of function values*

$$Q_N f = \sum_{i=1}^N c_i f(x_i) \qquad (12.27)$$

is called a numerical N-point integration formula if it can be used as an approximation of If.

The N different sampling points $x_1, x_2, \ldots, x_N$ are called integration abscissas (integration nodes); the values $c_1, c_2, \ldots, c_N$ are called integration weights (or integration coefficients).

In order to obtain highly efficient integration methods, the abscissas and weights should be chosen such that a given level of accuracy is reached with the smallest possible value of N.

This problem is addressed in several stages. First, the questions as to (i) which principles can be used for the construction of formulas (12.27), and (ii) whether such formulas are applicable for the solution of numerical integration problems (12.6), are answered. In other words, for given ε and f, is it possible to find an

N which solves (12.6)? This section thus deals with the construction and the convergence of integration formulas.

Section 12.5 discusses the task of finding abscissas $x_1, x_2, \ldots, x_N$ and weights $c_1, c_2, \ldots, c_N$ for a given N, and how to construct computer programs which can find them automatically.

12.3.1 Construction of Integration Formulas

A sequence of integration formulas $\{Q_N\}$ can only be used to construct integration algorithms for a class of functions $\mathcal{F}$ if for all $f \in \mathcal{F}$ and $\varepsilon > 0$ there exists $N \in \mathbb{N}$ such that

$$|Q_N f - If| \leq \varepsilon.$$

Riemann Sums

Riemann integrals of univariate functions $f : [a, b] \subset \mathbb{R} \to \mathbb{R}$ are defined by Riemann sums.

Definition 12.3.2 (Univariate Riemann Sums) *For any subdivision of the integration interval $[a, b]$ into N subintervals*

$$a = x_1 < x_2 < \ldots < x_N < x_{N+1} = b,$$

and any choice of N points

$$\xi_i \in [x_i, x_{i+1}], \quad i = 1, 2, \ldots, N,$$

a Riemann sum is defined by

$$R_N := \sum_{i=1}^{N} (x_{i+1} - x_i) f(\xi_i). \tag{12.28}$$

Definition 12.3.3 (Riemann Integral) *If all sequences $\{R_N\}$ of Riemann sums for which*

$$\Delta_N = \max\{x_2 - x_1, x_3 - x_2, \ldots, x_{N+1} - x_N\} \to 0 \qquad as \quad N \to \infty$$

approach the same limit R, then f is said to be Riemann integrable on $[a, b]$ and

$$\int_a^b f(x) \, dx := R.$$

One obvious solution of the numerical problem (12.6) would be to use a Riemann sum (12.28) as integration formula. Thus, in principle, it is possible to compute *any* Riemann integral with *any* given accuracy requirement (without taking into account the effects of data and rounding errors).

Simple Riemann Sums

The easiest way to obtain a sequence of Riemann sums is to divide the interval $[a, b]$ into N subintervals of the *same* length

$$x_i = a + ih, \quad i = 0, 1, \dots, N \quad \text{where} \quad h := (b - a)/N.$$

An obvious choice of ξ_i is either the right or the left endpoint of the respective subinterval $[x_{i-1}, x_i]$. In the first case, with $\xi_i := x_i$,

$$R_N^r f := \sum_{i=1}^N (x_i - x_{i-1}) f(x_i) = h \sum_{i=1}^N f(x_i)$$

is obtained, while the latter choice of $\xi_i := x_{i-1}$ yields

$$R_N^l f := \sum_{i=1}^N (x_i - x_{i-1}) f(x_{i-1}) = h \sum_{i=1}^N f(x_{i-1}).$$

Hence, two constant-weight quadrature formulas with $c_1 = c_2 = \cdots = c_N = h$ have been derived. As both are Riemann sums, their convergence

$$R_N^r f \to I f, \quad N \to \infty, \quad \text{and} \quad R_N^l f \to I f, \quad N \to \infty,$$

is guaranteed for every Riemann integrable function f. Thus, in principle, both R_N^r and R_N^l are able to solve the numerical integration problem (12.6). Their respective efficiencies, measured by the number of required f-values, can only be assessed on the basis of bounds on the absolute error $Q_N f - I f$. These bounds show the relationship between N and certain properties of f.

For *continuous* integrands f the absolute discretization error can be estimated in terms of the *modulus of continuity*

$$\omega(f; \delta) := \max\{|f(x_1) - f(x_2)| \ : \ x_1, x_2 \in [a, b], \ |x_1 - x_2| \le \delta\}.$$

Theorem 12.3.1 *If $f \in C[a, b]$, then*

$$|R_N^r f - I f| \le (b - a) \, \omega \left(f; \frac{b - a}{N} \right). \tag{12.29}$$

Proof: Davis, Rabinowitz [41].

For the left-hand Riemann sums R_N^l, the same bound can be used to estimate the discretization error $R_N^l f - I f$. Of course, this bound characterizes the convergence of $\{R_N^r f\}$ and $\{R_N^l f\}$ to $I f$ only in the *worst case* sense. In general, if additional restrictions such as smoothness properties are imposed on the class $\mathcal{F}$, the speed of convergence can be better characterized.

Example (Simple Riemann Sums) The Riemann sums $R_N^l f_i$ were calculated for three specific integrands $f_1(x) = x$, $f_2(x) = \sqrt{x}$ and $f_3(x) = \sin \pi x$, each over the interval $[0, 1]$ (see Table 12.1 and Figure 12.1).

Table 12.1: Values of Riemann Sums R_N^l.

N	$R_N^l f_1$	$R_N^l f_2$	$R_N^l f_3$
4	0.3750 000	0.5182 831	0.6035 534
16	0.4687 500	0.6323 312	0.6345 732
64	0.4921 875	0.6584 584	0.6364 920
256	0.4980 469	0.6646 632	0.6366 118
1024	0.4995 117	0.6661 723	0.6366 193
4096	0.4998 779	0.6665 431	0.6366 200
exact	*0.5000 000*	*0.6666 666*	*0.6366 198*

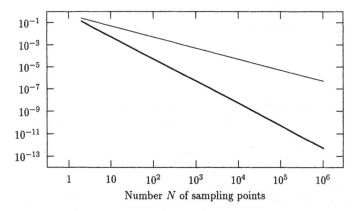

Figure 12.1: (Simple Riemann Sums) Absolute values of the errors of the Riemann sums R_N^l for the integrands $f_2(x) = \sqrt{x}$ (—) and $f_3(x) = \sin \pi x$ (—) over $[0,1]$.

The exact error of f_1 is $R_N^l f_1 - I f_1 = -(2N)^{-1}$. With $N = 4096$, the observed error $-1.221 \cdot 10^{-4}$ equals the expected error $-1/8192$.

The modulus of continuity for the integrand f_2 in $[0,1]$ is $\omega(f_2; \delta) = \sqrt{\delta}$ and the error bound for $|R_N^l f_2 - I f_2|$ is thus $1/\sqrt{N}$. With $N = 4$ the error *bound* is 0.5, whereas the actual error is only -0.1484. For $N = 4096$, the error bound $1.563 \cdot 10^{-2}$ is considerably larger than the actual error $-1.221 \cdot 10^{-4}$. Hence, convergence is in practice considerably faster than that which could be expected from the error bound. This is because the determination of the error bound (12.29) is based on the modulus of continuity with respect to the *whole* interval $[a, b]$. However, the low degree of smoothness of the integrand f near $x = 0$ is only of *local* influence near this endpoint and *not* of global influence on $[0,1]$.

The Riemann sums $R_N^l f_3$ show the most accurate numerical results. An error of $-5 \cdot 10^{-7}$ is reached for as few as 1024 sampling points. At this point R_N^l has converged to within machine precision.

The convergence of the simple Riemann sums R_N^r and R_N^l is, in most practical cases, unacceptably slow. In fact there are families of quadrature formulas $\{Q_N f\}$ which converge to $I f$ much faster than the primitive rules $R_N^r f$ and $R_N^l f$ and which can be proven to be Riemann sums, for instance Gauss formulas (see Section 12.3.2). However, the mere definition of Riemann sums does not give any hint as to how to construct such formulas.

Formula Construction Using Approximation

One of the most effective *constructive* methods for obtaining efficient quadrature formulas is approximation. The basic idea of this approach is to substitute the integrand f with a model function g whose indefinite integral can be expressed as an explicit formula which is easy to evaluate. The definite integral $\mathrm{I} g$ of g is calculated and used as an approximation of $\mathrm{I} f$.

If the approximation function g satisfies

$$\|g - f\|_\infty \le \frac{\varepsilon}{b - a},$$

then the error bound

$$|\mathrm{I} g - \mathrm{I} f| = \left| \int_a^b g(x)\, dx - \int_a^b f(x)\, dx \right| \le \int_a^b |g(x) - f(x)|\, dx \le$$
$$\le (b - a)\|g - f\|_\infty \le \varepsilon$$

can be derived. Approximation functions which are sufficiently close to f with respect to the L_∞ norm are suitable candidates for solving the numerical integration problem (12.6).

Example (Integration of a Model Function) The integral

$$I = \int_0^1 \sqrt{x}\, \Gamma(x + 1)\, dx$$

is to be calculated numerically with the maximal error $\varepsilon = 10^{-3}$. The following approximation can be obtained from Hart et al. [17]:

$$\Gamma(x + 1) = a_0 + a_1 x + a_2 x^2 + a_3 x^3 + e(x),$$

where $a_0 = 0.9991\,0836$, $a_1 = 0.4497\,361$, $a_2 = 0.2855\,737$, $a_3 = 0.2646\,888$, and

$$|e(x)| \le 9 \cdot 10^{-4} \quad \text{for all} \quad x \in [0, 1].$$

Thus

$$I = \frac{2}{3} a_0 + \frac{2}{5} a_1 + \frac{2}{7} a_2 + \frac{2}{9} a_3 + \int_0^1 \sqrt{x} \cdot e(x)\, dx = 0.9863\,789 + \eta$$

is obtained, with the error estimate

$$|\eta| \le \frac{2}{3} 9 \cdot 10^{-4} < \varepsilon,$$

i.e., $I \in [0.9857, 0.9870]$.

Polynomials

$$g(x) = P_d(x) = \sum_{i=0}^d \alpha_i x^i$$

are especially attractive model functions because they can be integrated very easily:

$$\int\limits_a^b P_d(x)\,dx = \sum_{i=0}^d \frac{\alpha_i}{i+1}(b^{i+1} - a^{i+1}).$$

In order to construct an algorithm which is able to meet any given accuracy requirement—without taking into account rounding errors—a *sequence* of model functions $\{P_d\}$ is needed for which convergence to the integrand f can be guaranteed:

$$P_d \to f \qquad \text{as} \qquad d \to \infty.$$

Example (Bernstein Polynomials) Bernstein polynomials $b_{d,i} \in \mathbb{P}_d$ (ref. Chapter 9) play an important role in approximation theory because asymptotically they approximate continuous functions with arbitrary accuracy on compact intervals. For $f \in C[0,1]$ and $d \to \infty$, the sequence $\{B_d(f)\}$ defined by

$$B_d(f)(x) := \sum_{i=0}^d f\left(\frac{i}{d}\right) b_{d,i}(x) = \sum_{i=0}^d f\left(\frac{i}{d}\right)\binom{d}{i} x^i (1-x)^{d-i}$$

converges on $[0,1]$ *uniformly* to f. By integrating

$$\int\limits_0^1 B_d(f)(x)\,dx = \sum_{i=0}^d f\left(\frac{i}{d}\right)\binom{d}{i}\int\limits_0^1 x^i(1-x)^{d-i}\,dx = \frac{1}{d+1}\sum_{i=0}^d f\left(\frac{i}{d}\right) \qquad (12.30)$$

the following quadrature formula is obtained:

$$x_i = \frac{i-1}{d}, \quad c_i = \frac{1}{d+1}, \qquad i = 1, 2, \ldots, d+1,$$

which corresponds to the mean value of the integrand for an equidistant partition of the interval $[0,1]$. Due to the very slow convergence $B_d(f) \to f$, the family of formulas obtained by this type of approximation does *not* show any substantial improvement compared to the simple rules R_N^r and R_N^l: The integration formula (12.30) differs only slightly from the simple Riemann sums on the interval $[0,1]$:

$$R_N^r f = \frac{1}{N}\sum_{i=1}^N f\left(\frac{i}{N}\right), \qquad R_N^l f = \frac{1}{N}\sum_{i=0}^{N-1} f\left(\frac{i}{N}\right).$$

Interpolatory Quadrature Formulas

Approximation based on *polynomial interpolation* plays an important role in the construction of quadrature formulas. An approximating polynomial $P_{N-1} \in \mathbb{P}_{N-1}$, which is identical to the integrand f at the distinct abscissas $x_1, x_2, \ldots, x_N$ (ref. Chapter 9),

$$P_{N-1}(x_i) = f(x_i), \quad i = 1, 2, \ldots, N,$$

can be represented as a linear combination of Lagrange polynomials $\varphi_{N-1,i}$:

$$P_{N-1}(x) = \sum_{i=1}^N f(x_i)\varphi_{N-1,i}(x), \qquad \varphi_{N-1,i}(x) := \prod_{\substack{j=1 \\ j \neq i}}^N \frac{x - x_j}{x_i - x_j}.$$

By calculating the integral of P_{N-1},

$$IP_{N-1} = \int_a^b P_{N-1}(x)\,dx = \sum_{i=1}^N f(x_i) \int_a^b \varphi_{N-1,i}(x)\,dx,$$

and by defining the weights

$$c_i := \int_a^b \varphi_{N-1,i}(x)\,dx,$$

a quadrature formula

$$Q_N f = IP_{N-1} = \sum_{i=1}^N c_i f(x_i)$$

is obtained. This formula can also be derived using a different approach: For given abscissas $x_1, x_2, \ldots, x_N$, the weights $c_1, c_2, \ldots, c_N$ are chosen so that $Q_N f$ *exactly* integrates the elements of a basis $\{b_0, b_1, \ldots, b_{N-1}\}$ of the space of polynomials $\mathbb{P}_{N-1}$, for example $\{1, x, x^2, \ldots, x^{N-1}\}$. The quadrature weights can be obtained by solving the following system of linear equations:

$$c_1 + c_2 + \cdots + c_N \;=\; \int_a^b 1\,dx = b - a$$

$$c_1 x_1 + c_2 x_2 + \cdots + c_N x_N \;=\; \int_a^b x\,dx = \frac{1}{2}(b^2 - a^2)$$

$$\vdots \qquad\qquad (12.31)$$

$$c_1 x_1^{N-1} + c_2 x_2^{N-1} + \cdots + c_N x_N^{N-1} \;=\; \int_a^b x^{N-1}\,dx = \frac{1}{N}(b^N - a^N).$$

The coefficient matrix of this system of linear equations is the $N{\times}N$ *Vandermonde matrix* $V(x_1, x_2, \ldots, x_N)$, whose determinant is known (Davis [40]):

$$\det V(x_1, \ldots, x_N) = \prod_{i=2}^N \prod_{j=1}^{i-1} (x_i - x_j). \qquad (12.32)$$

From (12.32) it follows that if all the abscissas $x_1, \ldots, x_N$ are distinct—which is part of the presupposition—then $V(x_1, \ldots, x_N)$ is nonsingular. Hence, there is a uniquely defined set of integration weights $c_1, c_2, \ldots, c_N$ that can be obtained by solving (12.31).

Definition 12.3.4 (Interpolatory Quadrature Formulas)   *Quadrature formulas are said to be interpolatory if they are obtained using one of the (equivalent) methods described above. They can be interpreted as an integral of a polynomial which interpolates f at the quadrature abscissas.*

The absolute *discretization error* of an interpolatory quadrature formula can be estimated by using the approximation error $P_{N-1} - f$:

$$
\begin{aligned}
|Q_N f - I f| &= |I P_{N-1} - I f| = |I(P_{N-1} - f)| \\
&\leq |b - a| \, \|P_{N-1} - f\|_\infty \\
&= |b - a| \, e_{N-1}(f).
\end{aligned}
$$

The error maximum $e_{N-1}(f) := \|P_{N-1} - f\|_\infty$, therefore characterizes the rate of convergence of $Q_N f \to I f$.

There are two ways of reducing the error maximum $e_{N-1}(f)$ of polynomial interpolation below a given bound:

1. increasing the degree of the approximating polynomial and

2. using an increasing number of polynomials of *constant* degree.

Both methods are used in practice to obtain quadrature formulas. Those obtained using the first procedure are called *simple* quadrature formulas, whereas formulas of the second type are called *compound* quadrature formulas.

There is some freedom in choosing the quadrature abscissas, which leads to a number of different numerical quadrature formulas. In order to show which formulas should be used for a given problem, Sections 12.3.2 and 12.3.3 discuss properties of the most important simple and compound formulas.

Formula Construction Using Acceleration Algorithms

If the sequence of the Riemann sums $\{R_N^l\}$ in Table 12.1 (page 90) with $N = 2^2, 2^4, 2^6, \ldots$ quadrature abscissas is examined closely, it becomes clear that the error is approximately divided by 4 every time the number of sampling points is quadrupled, i.e., the relation

$$
R_N^l f - I f \approx \frac{A}{N}, \tag{12.33}
$$

$$
R_{4N}^l f - I f \approx \frac{A}{4N} \tag{12.34}
$$

holds where $A = const.$ Multiplying (12.34) by 4 and subtracting (12.33) leads to

$$
(4R_{4N}^l f - R_N^l f)/3 - I f \approx 0.
$$

It is to be expected that $(4R_{4N}^l f - R_N^l f)/3$ is a better approximation of $I f$ than $R_{4N}^l f$. This is indeed the case. The value

$$
(4R_{64}^l f_2 - R_{16}^l f_2)/3 = 0.6671\,674
$$

has an error of $5.01 \cdot 10^{-4}$, which is comparable to that of $R_{1024}^l f_2$. Thus, by using the knowledge of the error structure as given by (12.33) and (12.34), it is possible to construct a new quadrature formula

$$
Q_{4N} f := (4R_{4N}^l f - R_N^l f)/3
$$

which is significantly more accurate than $R_{4N}^l f$.

Definition 12.3.5 (Acceleration Algorithm) *Algorithms which transform a convergent sequence $\{Q_N\}$ into a sequence $\{\hat{Q}_N\}$ with a higher rate of convergence are called acceleration algorithms.*

Eliminating the first term of an error expansion

$$Q_N - Q = AN^{-\alpha} + BN^{-\beta} + CN^{-\gamma} + \cdots \tag{12.35}$$

with known exponents $0 < \alpha < \beta < \gamma < \cdots$ and generally unknown coefficients $A, B, C, \ldots$ is a special form of a more general acceleration algorithm, the so-called *Richardson extrapolation*.

If, for two different values N_1 and N_2,

$$\begin{aligned}
Q &= Q_{N_1} - AN_1^{-\alpha} + O(N_1^{-\beta}) \\
Q &= Q_{N_2} - AN_2^{-\alpha} + O(N_2^{-\beta})
\end{aligned}$$

then it follows that

$$\overline{Q} = Q_{N_1} - AN_1^{-\alpha} \tag{12.36}$$

$$\overline{Q} = Q_{N_2} - AN_2^{-\alpha} = Q_{N_2} - A\left(\frac{N_2}{N_1}\right)^{-\alpha} N_1^{-\alpha} \tag{12.37}$$

after omitting higher order terms. Multiplying (12.36) by $(N_2/N_1)^{-\alpha}$ and subtracting the result from (12.37) leads to

$$\overline{Q} - \overline{Q}\left(\frac{N_2}{N_1}\right)^{-\alpha} = Q_{N_2} - \left(\frac{N_2}{N_1}\right)^{-\alpha} Q_{N_1},$$

whence

$$\overline{Q} = \frac{Q_{N_2} - (N_2/N_1)^{-\alpha}Q_{N_1}}{1 - (N_2/N_1)^{-\alpha}} = \frac{(N_2/N_1)^{\alpha}Q_{N_2} - Q_{N_1}}{(N_2/N_1)^{\alpha} - 1}. \tag{12.38}$$

Equation (12.38) can thus be used to define a new sequence of approximate values in which the term $AN^{-\alpha}$ of the expansion (12.35) is eliminated. By applying (12.38) to the newly defined sequence—this time with the exponent β—the term $BN^{-\beta}$ can be eliminated and so on.

Since the Richardson extrapolation corresponds to a linear combination of quadrature formulas, the result itself can be interpreted as a quadrature formula:

$$\hat{Q}_N := \sum_{j=1}^{J} a_j Q_{N_j} = \sum_{j=1}^{J} a_j \sum_{i=1}^{N_j} c_i^j f(x_i^j) =: \sum_{i=1}^{\hat{N}} \hat{c}_i f(\hat{x}_i).$$

The Richardson extrapolation is a method of constructing new quadrature formulas even if this *explicit* representation is not used in practice.

For the compound trapezoidal rule the result of the Richardson extrapolation is the family of *Romberg formulas*, which is discussed in more detail in Section 12.3.4.

For *compound* quadrature formulas $k \times Q$ (see Section 12.3.3) asymptotic expansions of the form

$$(k \times Q)f - If = Ak^{-\alpha} + Bk^{-\beta} + Ck^{-\gamma} + \cdots \qquad (12.39)$$

can be obtained for specific integrand classes. For instance, if the integrand is sufficiently smooth, the Euler-Maclaurin sum formula (Theorem 12.3.10) shows that the exponents for the compound trapezoidal rule are $\alpha = 2$, $\beta = 4$, $\gamma = 6, \ldots$. For integrands with a low order of continuity or differentiability (with algebraic or logarithmic singularities etc.), the exponents depend on certain analytic properties of the integrand (Lyness, Ninham [275]).

These exponents $\alpha, \beta, \gamma, \ldots$ have to be known *explicitly* if Richardson's extrapolation is to be applied. In practice, the user is generally not able to supply this information. In order to overcome this difficulty the exponents in equation (12.39) can be regarded as unknown quantities which have to be determined. This process is, however totally different from the elimination of the coefficients $A, B, C, \ldots$, since the right-hand side of (12.39) is a *nonlinear* function of the exponents. Thus, *nonlinear* extrapolation techniques like the ε-algorithm (Kahaner [247]) have to be applied.

Since nonlinear extrapolation corresponds to a nonlinear combination of quadrature formulas, the result itself cannot be interpreted as a quadrature formula in the original sense. The discussion of quadrature schemes based on nonlinear extrapolation methods is therefore deferred to Section 12.3.5.

Software (Linear Extrapolation) The subroutine NAG/d01paf computes a sequence of trapezoidal rules for n-dimensional simplices and uses Richardson's extrapolation for convergence acceleration.

Linear Transformation of Quadrature Formulas

Many quadrature formulas are specified for particular integration intervals. Gauss formulas, for instance, are specified for the interval $[-1, 1]$. In order to apply such formulas to other intervals, appropriate transformations have to be applied to the formulas. Since any two finite intervals $[a, b]$ and $[c, d]$ can be mapped onto each other using affine transformations, only the transformation of quadrature formulas using such affine transformations is considered in this section.

Nonlinear transformations are only used in conjunction with IMT formulas (see Section 12.3.3).

$\bar{x} \in [\bar{a}, \bar{b}]$ is mapped onto $x \in [a, b]$ using the following linear transformation:

$$x = \gamma \bar{x} + \beta, \qquad \bar{x} = \frac{1}{\gamma}\left(x - \beta\right), \qquad (12.40)$$

with the constants

$$\gamma := \frac{b - a}{\bar{b} - \bar{a}}, \qquad \beta := \frac{a\bar{b} - \bar{a}b}{\bar{b} - \bar{a}}.$$

Hence,

$$\int_a^b w(x)f(x)dx = \gamma \int_{\bar{a}}^{\bar{b}} \overline{w}(\overline{x})\overline{f}(\overline{x})d\overline{x}$$

where

$$\overline{w}(\overline{x}) := w(\gamma\overline{x} + \beta), \qquad \overline{f}(\overline{x}) := f(\gamma\overline{x} + \beta).$$

The affine transformation (12.40) converts the quadrature formula

$$\int_{\bar{a}}^{\bar{b}} \overline{w}(\overline{x})\overline{f}(\overline{x})d\overline{x} = \sum_{i=1}^N \overline{c}_i\overline{f}(\overline{x}_i) + \overline{E}(\overline{f})$$

into the formula

$$\int_a^b w(x)f(x)dx = \sum_{i=1}^N c_i f(x_i) + E(f)$$

with abscissas and weights given by

$$x_i := \gamma\overline{x}_i + \beta, \quad c_i := \gamma\overline{c}_i, \quad i = 1, 2, \ldots, N,$$

and $E(f) = \gamma\overline{E}(\overline{f})$ (see, for instance, Stroud [77]).

12.3.2 Simple Interpolatory Quadrature Formulas

All interpolatory quadrature formulas are uniquely characterized by the choice of their abscissas $x_1, x_2, \ldots, x_N$. The most important sets of quadrature abscissas (which are discussed later in more detail) are

Abscissas	**Class of Formulas**
equidistant subdivision	*Newton-Cotes* formulas
including endpoints	*closed* Newton-Cotes formulas
without endpoints	*open* Newton-Cotes formulas
zeros of	*Gauss* formulas
Legendre polynomials P_N	Gauss-(*Legendre*) formulas
Laguerre polynomials L_N	Gauss-*Laguerre* formulas
Hermite polynomials H_N	Gauss-*Hermite* formulas
Jacobi polynomials $P_N^{\alpha,\beta}$	Gauss-*Jacobi* formulas
zeros of	
$[P_{N-1}(x) + P_N(x)]$	*Radau* formulas
$(x^2 - 1)\,P_{N-1}'$	*Lobatto* formulas
Chebyshev extrema	(practical) *Clenshaw-Curtis* formulas
Chebyshev zeros	(classical) *Clenshaw-Curtis* formulas

All extrema and zeros of orthogonal polynomials which are used as quadrature abscissas must be mapped from their usual domain of definition $\overline{B} = [-1, 1]$ onto the general interval of integration $B = [a, b]$:

$$x_i := \overline{x}_i \frac{b-a}{2} + \frac{a+b}{2}. \tag{12.41}$$

The corresponding transformation of the quadrature weights is given by

$$c_i := \overline{c}_i \, (b-a)/2.$$

Software (Interpolatory Quadrature Formulas)   TOMS/655 contains subroutines which compute the weights $c_1, c_2, \ldots, c_N$ of the corresponding quadrature formula Q_N for most of the positive weight functions mentioned in Section 12.1.2 and for arbitrary abscissas $x_1, x_2, \ldots, x_N$. The given abscissas do not have to be distinct. If a sampling point is given k_i times ($k_i \in \{2, 3, 4, \ldots\}$), then the corresponding interpolatory polynomial interpolates not only f but also its derivatives $f', f'', \ldots, f^{(k_i-1)}$ at x_i (*Hermite-Birkhoff interpolation*).

TOMS/655 also contains subroutines for using the computed quadrature formula in order to calculate the definite integral of integrands f supplied by the user. The functions mentioned above can be applied to any given positive weight function w if the value of

$$I(w; [a, b]) = \int\limits_a^b w(x) \, dx$$

and the sequence of orthogonal polynomials with respect to w is specified by using the recurrence formula discussed in Section 10.2.5.

Degree of Accuracy

An important characteristic of every quadrature formula is its *degree of accuracy*.

Definition 12.3.6 (Degree of Accuracy)   *The degree of accuracy of a quadrature formula Q_N is D if*

$$Q_N x^k \;=\; Ix^k, \qquad k = 0, 1, \ldots, D,$$
$$Q_N x^{D+1} \;\neq\; Ix^{D+1}$$

i. e., if Q_N is the exact integral operator for all polynomials of degree $d \leq D$ and if there is at least one polynomial of degree $d = D + 1$ for which Q_N is not exact.

Clearly, the degree of accuracy of an interpolatory quadrature formula Q_N is $D \geq N-1$. The following theorem shows that the converse is also true.

Theorem 12.3.2   *Any N-point quadrature formula Q_N with a degree of accuracy $D \geq N-1$ is an interpolatory formula.*

Proof: Engels [46].

The relevance of the degree of accuracy is due to the following theorem, which relates the error of an interpolatory quadrature formula to the error of the best approximating polynomial P_D^* with respect to the L_∞-norm:

Theorem 12.3.3 *Any interpolatory quadrature formula Q_N with a degree of accuracy D satisfies the following inequality:*

$$|Q_N f - If| \leq \left(|b - a| + \sum_{i=1}^{N} |c_i|\right) e_D^*(f), \qquad (12.42)$$

where

$$e_D^*(f) := \inf\{\|P_D - f\|_\infty : P_D \in \mathbb{P}_D\}.$$

If all quadrature weights of Q_N are positive, then

$$|Q_N f - If| \leq 2|b - a| e_D^*(f). \qquad (12.43)$$

Proof: Davis, Rabinowitz [41].

Convergence Behavior

If the coefficients of a sequence of interpolatory quadrature formulas $\{Q_N\}$ are *positive* (for all $N \in \mathbb{N}$), then (12.43) implies convergence for all $f \in C[a, b]$:

$$Q_N f \to If \qquad \text{as} \quad N \to \infty,$$

since in this case (according to the Weierstrass theorem; Davis [40])

$$e_D^*(f) \to 0 \qquad \text{as} \quad D \to \infty.$$

In particular,

$$e_{D+1}^*(f) \leq e_D^*(f)$$

along with (12.43) implies that formulas with a higher degree of accuracy result in a smaller error *estimate* (when using the same number of integration abscissas).

In summary, among all N-point interpolatory quadrature formulas, those with positive weights and maximum degrees of accuracy are of particular importance. These are the *Gauss*-(Legendre) *formulas*, i.e., formulas in which the quadrature abscissas are the zeros of Legendre polynomials (see the section on Gauss formulas on page 103).

Without additional information about f—such as information about bounds on certain derivatives or on the continuity modulus of f—no statement about the *rate* of convergence of the sequence $\{Q_N f\}$ can be made. The following theorem shows that the rate of convergence can indeed be arbitrarily slow.

Theorem 12.3.4 *If $\{Q_N f\}$ converges to If for all $f \in C[a, b]$,*

$$Q_N f \to If \qquad \text{for} \quad N \to \infty,$$

then, for every convergent sequence $\{\lambda_N\}$ (with an arbitrarily slow rate of convergence),

$$\lambda_N \in \mathbb{R}, \qquad \lambda_N \geq 0, \qquad \lim_{N \to \infty} \lambda_N = 0,$$

there exists a function $f \in C[a, b]$ for which the discretization error of $\{Q_N f\}$ converges even slower:

$$|Q_N f - If| \geq \lambda_N \qquad \text{for all} \quad N \in \mathbb{N}.$$

Proof: Lipow, Stenger [269].

Closed Newton-Cotes Formulas

Closed Newton-Cotes formulas Q_N, $N = 2, 3, 4, \ldots$, are interpolatory quadrature formulas based on the equidistant quadrature abscissas

$$x_i = a + (i - 1)h, \quad i = 1, 2, \ldots, N \qquad \text{where} \qquad h := \frac{b - a}{N - 1}.$$

They are denoted "closed" because the endpoints a and b of the integration interval are the outermost abscissas of the quadrature formula. The weights c_i can be obtained by integrating the respective Lagrange polynomials.

Example (Trapezoidal Rule)   Perhaps the most important closed Newton-Cotes formula is the *trapezoidal rule*

$$Q_2(f; a, b) = \frac{b - a}{2}[f(a) + f(b)] \approx \int_a^b f(x)\, dx.$$

Linear interpolation using a polynomial $P_1(x) = \alpha_0 + \alpha_1 x$ leads to a trapezoid which is bounded by the lines $x = a$, $x = b$, the x-axis and the chord between $(a, f(a))$ and $(b, f(b))$. The area of this trapezoid is used as an estimate of the required integral value.

The first three closed Newton-Cotes formulas are ($f_i := f(x_i)$):

$$Q_2 f = \frac{h}{2}(f_1 + f_2) \qquad \text{(trapezoidal rule)},$$

$$Q_3 f = \frac{2h}{6}(f_1 + 4f_2 + f_3) \qquad \text{(Simpson's rule)} \quad \text{and}$$

$$Q_4 f = \frac{3h}{8}(f_1 + 3f_2 + 3f_3 + f_4).$$

The closed Newton-Cotes formula of degree $D = N - 1 = 8$,

$$Q_9 f = \frac{8h}{28350}(989 f_1 + 5888 f_2 - 928 f_3 + 10496 f_4 - 4540 f_5 + \cdots),$$

and *all* formulas of degree $D \geq 10$ contain some *negative* coefficients. While interpolatory formulas which contain only *positive* coefficients comply with the equation

$$\sum_{i=1}^N |c_i| = \sum_{i=1}^N c_i = |b - a|$$

(because they are exact for $f \equiv 1$), this is *not* the case for Newton-Cotes formulas of higher degree; for Q_9, for example,

$$\sum_{i=1}^9 |c_i| \approx 1.45\, |b - a|.$$

Moreover, the following theorem holds:

Theorem 12.3.5 (Kusmin) *Let $c_1^N, c_2^N, \ldots, c_N^N$ be the coefficients of the closed Newton-Cotes formulas Q_N. Then*

$$\sum_{i=1}^{N} |c_i^N| \to \infty \qquad as \qquad N \to \infty.$$

Proof: Werner, Schaback [79].

Because the absolute values of the coefficients increase with N, the impact of errors (perturbations) in $f(x_i)$ on the result $Q_N f$ are amplified by a factor c_i^N. Hence, for large values of N, Newton-Cotes formulas are *numerically unstable*.

The *convergence*, $Q_N f \to I f$ as $N \to \infty$, must be guaranteed in order to make the error $Q_N f - I f$ smaller than any desired tolerance. The convergence of Newton-Cotes formulas *cannot* be derived using (12.43) because of the presence of negative weights. As a result of the increase in the absolute values of the weights (see Theorem 12.3.5), the inequality (12.42) cannot be used either.

The convergence $Q_N f \to I f$ as $N \to \infty$ *cannot* even be assured for analytic integrands. Convergence of the closed Newton-Cotes formulas can only be guaranteed for those analytic functions which have no poles in a certain "oval" region of the complex plane around the real interval $[a, b]$ (Krylov [260]). This is not surprising though, considering the stringent conditions that have to be complied with to guarantee the convergence of a sequence of interpolation polynomials $P_1, P_2, P_3, \ldots$ with equidistant abscissas to f (see Chapter 9).

Newton-Cotes formulas are important because of their use in compound quadrature formulas. For this purpose it is important to have a more precise characterization of the error $Q_N f - I f$. Bounds for this error are easy to derive since the *quadrature* error can be related to the *approximation* error (9.35).

Example (Error Bounds for the Trapezoidal Rule) The trapezoidal rule is obtained by integrating the interpolatory polynomial $P_1 \in \mathbb{P}_1$ with the abscissas $x_1 = a$ and $x_2 = b$. From

$$e_1(f) = \|P_1 - f\|_\infty \leq \frac{(b-a)^2}{8} M_2 \qquad \text{where} \qquad M_2 := \max\{|f''(x)| : x \in [a, b]\},$$

the error estimate

$$|Q_2 f - I f| \leq \frac{(b-a)^3}{8} M_2 = \frac{h^3}{8} M_2$$

follows. More detailed investigations make it possible to improve this result:

$$Q_2 f - I f = \frac{h^3}{12} f''(\xi), \quad \xi \in [a, b], \qquad \text{and} \qquad |Q_2 f - I f| \leq \frac{h^3}{12} M_2.$$

Such error bounds have been derived for all Newton-Cotes formulas (Isaacson, Keller [61]).

Open Newton-Cotes Formulas

The *open* Newton-Cotes formulas Q_N, $N = 1, 2, 3, \ldots$, are interpolatory quadrature formulas based on the abscissas

$$x_i = a + ih, \quad i = 1, 2, \ldots, N \qquad \text{where} \qquad h := \frac{b-a}{N+1}.$$

The term "open" indicates that the endpoints a and b of the integration interval are *not* used as quadrature abscissas. The first three open Newton-Cotes formulas are

$$Q_1 f \;=\; 2h f_1 \qquad\qquad \textit{(rectangular rule)},$$

$$Q_2 f \;=\; \frac{3h}{2}(f_1 + f_2) \qquad\qquad \text{and}$$

$$Q_3 f \;=\; \frac{4h}{3}(2f_1 - f_2 + 2f_3).$$

Example (Error Bound for the Rectangular Rule)   The error of the rectangular rule can be expressed as follows:

$$Q_1 f - If = -\frac{h^3}{3} f''(\xi), \;\; \xi \in (a, b) \qquad \text{and} \qquad |Q_2 f - If| \le \frac{h^3}{3} M_2.$$

Software (Newton-Cotes Formulas)   HARWELL/qa01 contains quadrature subroutines based on the most important Newton-Cotes formulas.

The IMSL and NAG libraries contain *no* subroutines using Newton-Cotes formulas because Newton-Cotes formulas are inefficient compared to other quadrature formulas, like the Gauss-Kronrod formulas.

Clenshaw-Curtis Formulas

Chebyshev abscissas (zeros and extrema of Chebyshev polynomials) are significantly superior to equidistant abscissas for the approximation of a function using interpolation polynomials. The same can be expected when using them as quadrature abscissas. In fact, both the Chebyshev zeros

$$x_i = \cos \frac{(2i - 1)\pi}{2N}, \;\; i = 1, 2, \ldots, N,$$

and the abscissas of the Chebyshev extreme values

$$x_i = \cos \frac{(i - 1)\pi}{N - 1}, \;\; i = 1, 2, \ldots, N,$$

result in interpolatory quadrature formulas with *positive* weights (Fejér [187], Imhof [236]). Thus, in both cases the convergence

$$Q_N f \to If \qquad as \qquad N \to \infty,$$

is guaranteed for all $f \in C[a, b]$. However, the degree of accuracy is only $D = N$–1.

In practice, Chebyshev extrema are preferred to Chebyshev zeros as the transition from Q_N to Q_{2N-1} requires the computation of only $N - 1$ new values of f (as opposed to $2N$ new values if zeros are used). This is why Chebyshev extrema are called *practical* Clenshaw-Curtis abscissas.[2]

---

[2]Classical Clenshaw-Curtis formulas are also called *Fejér formulas*.

Computing the weights of practical Clenshaw-Curtis rules Q_N requires $O(N^2)$ arithmetic operations (Davis, Rabinowitz [41]). More efficient implementations of Clenshaw-Curtis formulas, however, do not compute the weights explicitly. Instead, the basis of Chebyshev polynomials is used to represent the interpolation polynomial P_d, $d = N - 1$,

$$P_d(x) = \sum_{i=0}^{d}{}' c_i T_i(x) := \frac{c_0}{2} + c_1 T_1(x) + \cdots + c_d T_d(x).$$

Because

$$Q_N f = \mathrm{I} P_d = \sum_{i=0}^{d}{}' c_i \, \mathrm{I} T_i,$$

the computation of $Q_N f$ requires the calculation of the coefficients $c_0, c_1, \ldots, c_d$ as well as the calculation of the so-called *moments* $\mathrm{I} T_0, \mathrm{I} T_1, \ldots, \mathrm{I} T_d$.

When using Chebyshev extrema and Chebyshev zeros as interpolation abscissas (i.e., when applying practical and classical Clenshaw-Curtis formulas), the coefficients $c_0, c_1, \ldots, c_d$ can be obtained from (9.25) and (9.26) respectively. In both cases a variant of the *Fast Fourier Transform* enables the calculation of all coefficients requiring only $O(N \log N)$ arithmetic operations (Gentleman [209]).

The moments $\mathrm{I} T_0, \mathrm{I} T_1, \ldots, \mathrm{I} T_d$ can be obtained by using recurrences available for a number of weight functions[3] w of practical importance (Piessens et al. [22]).

Software (Clenshaw-Curtis Formulas)   For a given integrand f, the integration subroutines `CMLIB/qc25s` $\approx$ `QUADPACK/qc25s` $\approx$ `SLATEC/qc25s` compute the modified 25-point-Clenshaw-Curtis formula with respect to weight functions with algebraic or algebraic-logarithmic endpoint singularities. In addition, a 13-point-Clenshaw-Curtis formula is used to obtain an error estimate.

The subroutines `CMLIB/qc25f` $\approx$ `QUADPACK/qc25f` $\approx$ `SLATEC/qc25f` are similar to the routines described above, except that they use the trigonometric weight functions $\cos(\omega x)$ and $\sin(\omega x)$.

The subroutines `CMLIB/qc25c` $\approx$ `QUADPACK/qc25c` $\approx$ `SLATEC/qc25c` use the Cauchy principle value weight function $w(x; c) := 1/(x - c)$.

The subroutine `IMSL/MATH-LIBRARY/fqrul` calculates abscissas and weights of the Fejér formula (the classical Clenshaw-Curtis formula) for a user defined number N of abscissas. Moreover, the abscissas and weights of the *modified* Fejér formula can be calculated for weight functions with algebraic or algebraic-logarithmic endpoint singularities or for the Cauchy-principle value weight function.

Gauss Formulas

All interpolatory quadrature formulas Q_N have degrees of accuracy $D \geq N - 1$ for N arbitrarily located abscissas. In general, the degree of accuracy cannot be expected to be larger than $N - 1$ since for *given* abscissas the remaining N parameters (the quadrature coefficients) suffice only to integrate polynomials with N parameters (i.e., the coefficients of the polynomials of degree $N - 1$) exactly.

---

[3]If $w \not\equiv 1$ the moments are said to be *modified*.

The maximal degree of accuracy can only be attained if *all* $2N$ parameters of an N-point formula—N coefficients *and* N abscissas—are chosen appropriately. In this way it is possible to derive exact formulas for polynomials with $2N$ parameters (i.e., polynomials of degree $2N - 1$).

Theorem 12.3.6　*The maximal degree of accuracy of an N-point quadrature formula*

$$Q_N f = \sum_{i=1}^{N} c_i f(x_i)$$

is $D = 2N - 1$. This degree of accuracy can be obtained by using the zeros of the Nth orthogonal polynomial (of degree $d = N$) with respect to the weight function w in $[a, b]$ as the abscissas $x_1, x_2, \ldots, x_N$ of the interpolatory formula.

Proof: Davis, Rabinowitz [41].

For $[a, b] = [-1, 1]$ and $w(x) \equiv 1$ the *optimal* abscissas $x_1, x_2, \ldots, x_N$ are the zeros of the Legendre polynomial of degree $d = N$. When applying such a formula to an arbitrary interval $[a, b]$, the abscissas and the weights have to be transformed according to (12.41). The quadrature formulas G_N obtained in this way are called *Gauss formulas* or *Gauss-Legendre formulas*.

Terminology (Gauss-Legendre Formulas)　The term Gauss-*Legendre* formulas is used to prevent any confusion with Gauss-*Laguerre*, Gauss-*Hermite* and Gauss-*Jacobi* formulas which are obtained by using Laguerre, Hermite or Jacobi weight functions respectively (see Section 12.1.2).

Theorem 12.3.7　*The coefficients $c_1^N, c_2^N, \ldots, c_N^N$ of Gauss formulas G_N are all positive:*

$$c_i^N > 0, \quad i = 1, 2, \ldots, N, \quad N = 1, 2, 3, \ldots .$$

Proof: Engels [46].

Since all Gauss formulas have positive weights, the convergence

$$G_N f \to I f \quad \text{as} \quad N \to \infty,$$

for all $f \in C[a, b]$ can be derived from Theorem 12.3.3. Moreover, the convergence of Gauss formulas is assured for all functions that are Riemann integrable in $[a, b]$:

Theorem 12.3.8　*All Gauss formulas G_N, $N = 1, 2, 3, \ldots$ are Riemann sums.*

Proof: Stroud [77].

Software (Gauss Formulas)　The subroutine IMSL/MATH-LIBRARY/gqrul calculates the abscissas and weights of Gauss formulas for all positive weight functions that are mentioned in Section 12.1.2 and for any given number of abscissas N.

　　IMSL/MATH-LIBRARY/gqrcf even calculates abscissas and weights for arbitrary weight functions. The weight function w itself *cannot* be specified explicitly. Instead, the user has to supply the sequence of orthogonal polynomials with respect to the weight function w by specifying the coefficients of recurrence relationship (10.2.6).

The subroutine IMSL/MATH-LIBRARY/recqr works inversely: For N given abscissas and weights of a quadrature formula, it computes the sequence of the first N corresponding recurrence coefficients. For the positive weight functions mentioned in Section 12.1.2, the sequence of recurrence coefficients can be obtained by using IMSL/MATH-LIBRARY/reccf.

For a given integrand f, the subroutine NAG/d01baf evaluates Gauss formulas. The user can choose between different types of formulas (Gauss-Legendre, Gauss-Laguerre etc.) as well as specify the number N of abscissas. The subroutine NAG/d01bbf calculates the weights and abscissas of these formulas. NAG/d01bcf can be used to compute the Gauss formulas for an *arbitrary* number N of abscissas.

TOMS/655 contains subroutines that calculate weights and abscissas for most of the positive weight functions which have been mentioned in Section 12.1.2 and for an arbitrary number N of abscissas. In addition, the computed Gauss formula can be used to integrate a user-specified integrand f. All modes of operation can be executed for any positive weight function w as long as the user specifies the associated recurrence coefficients.

Radau and Lobatto Formulas

While Gauss formulas use all the $2N$ available parameters (N weights and N abscissas) of an N-point quadrature formula to obtain the maximal degree of accuracy $D = 2N - 1$, in Radau and Lobatto formulas some sampling points are fixed in advance. The remaining abscissas and all weights are determined such that a maximal degree of accuracy D is obtained.

Radau formulas have only *one* fixed abscissa: the left endpoint $x_1 = a$ or the right endpoint $x_N = b$. Using the remaining $2N - 1$ parameters, it is possible to construct formulas of a degree of accuracy $D = 2N - 2$. *Lobatto* formulas have *two* fixed abscissas: $x_1 = a$ and $x_N = b$. The resulting formulas have a degree of accuracy of $D = 2N - 3$.

Due to their high degree of accuracy, both Radau and Lobatto formulas are interpolatory. As a result, all such formulas are uniquely determined by specifying their abscissas.

Radau and Lobatto formulas are not as important for numerical quadrature as for the integration of differential equations or the solution of integral equations, where they form the basis of many high order methods.

Software (Radau and Lobatto Formulas) For any given number N of abscissas, the subroutine IMSL/MATH-LIBRARY/gqrul calculates the weights and abscissas of quadrature formulas with maximal degrees of accuracy for all the positive weight functions that are mentioned in Section 12.1.2 (ref. to the Gauss formulas software section). It is possible to specify, at most, two abscissas of the required quadrature formula. If, for example, one of the interval endpoints $[-1, 1]$ is chosen, the resulting quadrature formula with respect to the Legendre weight function is the corresponding Radau formula. If both endpoints of $[-1, 1]$ are chosen as abscissas the Lobatto formula is obtained. By using IMSL/MATH-LIBRARY/gqrcf, this construction can be carried out for arbitrary weight functions w. Here, w has to be specified implicitly by the sequence of polynomials orthogonal with respect to w.

Gauss-Kronrod Formulas

In practice, Gauss formulas have one severe disadvantage: two arbitrary Gauss formulas G_{N_2} and G_{N_1} with $N_2 > N_1$ do *not* have any identical abscissas (except, perhaps, for the interval center). There is, therefore, no economical method for obtaining a practical error estimate. The usual procedure for obtaining a practical error estimate—evaluating formulas with different numbers of nodes and using the difference as an error estimate, requires too many integrand values and therefore too much computational effort.

This disadvantage was overcome using a seemingly obvious method, published as recently as 1965 by A. S. Kronrod [259]: The N abscissas of G_N are fixed (which makes the construction of Gauss-Kronrod formulas similar to the construction of Radau and Lobatto formulas), and a $(2N + 1)$-point formula is constructed such that its degree of accuracy is the highest possible, i. e., $D = 3N + 1$ (N even) or $D = 3N + 2$ (N odd). The new abscissas are contained in the intervals

$$(a, x_1), \ (x_1, x_2), \ (x_2, x_3), \ldots, (x_N, b),$$

where $x_1, x_2, \ldots, x_N$ are the abscissas of G_N.

Software (Gauss-Kronrod Formulas)   The subroutines   CMLIB/qk$N \approx$ QUADPACK/qk$N$ $\approx$ SLATEC/qk$N$, $N = 15, 21, 31, 41, 51, 61$ apply the N-point Gauss-Kronrod formula to a given integrand f. In addition, another Gauss formula is used to calculate an error estimate.

The subroutines CMLIB/qk15w $\approx$ QUADPACK/qk15w $\approx$ SLATEC/qk15w enable the specification of an additional weight function.

Piessens and Branders [315] published a Fortran program that calculates N Gauss abscissas and weights as well as Kronrod abscissas and weights for a given number of abscissas.

Patterson Formulas

Patterson used Kronrod's idea of extending Gauss formulas from G_N to K_{2N+1} by adding another $2N+2$ abscissas in order to obtain a formula with the degree of accuracy $6N+4$. In this way a sequence of formulas using 3, 7, 15, 31, 63, 127 and 255 nodes was constructed (Patterson [312]).

Software (Patterson Formulas)   TOMS/672 contains a subroutine that computes the weights $c_1, c_2, \ldots c_N, c_{N+1}, \ldots, c_{N+M}$ and the additional abscissas $x_{N+1}, x_{N+2}, \ldots, x_{N+M}$ for a given positive weight function w, for arbitrary abscissas $x_1, x_2, \ldots, x_N$, and for a natural number M such that the resulting quadrature formula Q_{N+M} has a maximal degree of accuracy. The weight function must be specified using the recurrence formula of its corresponding orthogonal polynomials.

The non-adaptive integration program QUADPACK/qng uses a sequence of 10-, 21-, 43-, and 87-point Patterson formulas, while TOMS/699 uses the original sequence of 3-, 7-, 15-, 31-, 63-, 127-, and 255-point quadrature formulas.

12.3.3 Compound Quadrature Formulas

In Section 12.3.2 several *simple* quadrature formulas are discussed, i. e., formulas obtained by integrating *one* interpolation polynomial on the corresponding integration interval. This section introduces formulas which are obtained by integrating a *piecewise* polynomial. An equivalent approach results from the decomposition of the interval $[a, b]$ into subintervals and the application of a simple interpolatory formula to each subinterval.

The fact that a definite integral over an interval $[a, b]$ can be decomposed into integrals over disjoint subintervals

$$\int\limits_a^b f(x)\,dx = \int\limits_a^{\bar{x}_1} f(x)\,dx + \int\limits_{\bar{x}_1}^{\bar{x}_2} f(x)\,dx + \cdots + \int\limits_{\bar{x}_{k-1}}^{b} f(x)\,dx,$$

is of great importance for numerical integration. Throughout this section the assumption is made that $[a, b]$ is subdivided into k intervals by $k + 1$ *equidistant* partition points[4]

$$a = \bar{x}_0 < \bar{x}_1 < \cdots < \bar{x}_k = b.$$

An N-point quadrature formula Q_N is applied to each of the k subintervals. The resulting compound quadrature formula is denoted by $k \times Q_N$.

If Q_N is a *closed* formula—*both* endpoints are quadrature abscissas—then the number of abscissas $k(N-1)+1$ of the compound formula $k \times Q_N$ is smaller than $k\,N$ since the abscissas at the interval endpoints $\bar{x}_1, \bar{x}_2, \ldots, \bar{x}_{k-1}$ are counted only once (only *one* f-value is needed).

Convergence Behavior

In contrast to simple quadrature formulas, the convergence of compound quadrature formulas can be guaranteed for *all* Riemann integrable functions.

Theorem 12.3.9 *Let Q_N be an N- point quadrature formula, and let its degree of accuracy be $D \geq 0$, i. e., $f \equiv 1$ is integrated exactly by Q_N. Then for all Riemann integrable functions f bounded in $[a, b]$, the compound quadrature formula converges:*

$$(k \times Q_N)f \to If \qquad as \qquad k \to \infty.$$

Proof: Davis, Rabinowitz [41].

This theorem provides a strong argument in favor of compound formulas. Another advantage of compound formulas is that they make it possible to use an *irregular* (adaptively derived) refinement of the grid. This subject is discussed in Section 12.5.2. Practically all computer programs for numerical quadrature are, in one way or another, based on compound formulas.

Compound Trapezoidal Rule

The compound trapezoidal rule $T_l := l \times T$,

$$T_l f = h \left[\frac{1}{2} f(a) + f(a + h) + \ldots + f(a + (l-1)h) + \frac{1}{2} f(b) \right],$$

is of special importance. The error of the $(l+1)$-point formula T_l can be characterized accurately (provided f is sufficiently smooth) by the following theorem:

---

[4]Non-equidistant subdivisions of $[a, b]$ in connection with integration algorithms are treated in Section 12.5.

Theorem 12.3.10 (Euler-Maclaurin Sum Formula) *If $f \in C^{2k+1}[a, b]$ then*

$$
\begin{aligned}
T_l f - I f \;=\;& \frac{B_2}{2!} h^2 [f'(b) - f'(a)] + \frac{B_4}{4!} h^4 [f^{(3)}(b) - f^{(3)}(a)] + \cdots \\
& \cdots + \frac{B_{2k}}{(2k)!} h^{2k} [f^{(2k-1)}(b) - f^{(2k-1)}(a)] \qquad\qquad (12.44) \\
& + h^{2k+1} \int_a^b \overline{P}_{2k+1} \left(l \frac{x - a}{b - a} \right) f^{(2k+1)}(x)\, dx,
\end{aligned}
$$

where the constants are the Bernoulli numbers

$$B_2 = 1/6, \quad B_4 = -1/30, \quad B_6 = 1/42, \quad B_8 = -1/30, \quad B_{10} = 5/66, \quad \ldots$$

and $\overline{P}_{2k+1}$ is characterized as follows:

$$\overline{P}_{2k+1}(x) := (-1)^{k-1} \sum_{i=1}^{\infty} 2(2i\pi)^{-2k-1} \sin(2\pi i x).$$

Proof: Davis, Rabinowitz [41].

Formula (12.44) shows, among other things, that the compound trapezoidal rule T_l gives outstandingly accurate results for integrands whose odd derivatives assume equal values at the endpoints of the integration interval. For example, this is the case for all $(b - a)$-periodic integrands.

By applying T_l to an integrand $f \in C^{2k+1}[a, b]$, where

$$f'(a) = f'(b), \quad f^{(3)}(a) = f^{(3)}(b), \quad \ldots, f^{(2k-1)}(a) = f^{(2k-1)}(b)$$

and

$$|f^{(2k+1)}(x)| \le M_{2k+1} \qquad \text{for all } x \in [a, b],$$

then

$$|T_l f - I f| \le C h^{2k+1}, \qquad h = (b - a)/l,$$

denotes the order of convergence of $T_l f \to I f$. Theorems characterizing the rate of convergence of $\{T_l f\}$ for analytic functions can be found, for example, in Davis, Rabinowitz [41].

Example (Trapezoidal Rule for Periodic Integrands)   The function

$$f(x) = \frac{1}{1 + r\sin(2j\pi x)}, \qquad |r| < 1, \quad j \in \mathbb{Z}, \qquad\qquad (12.45)$$

has a period of length 1 and is differentiable infinitely many times. Thus, the rate of convergence of the compound trapezoidal rule T_l is better than l^{-k} for *any* $k \in \mathbb{N}$.

Using $r = 0.5$, $j = 5$, $a = 0$, and $b = 1$ results in the values of the absolute error $|T_l f - I f|$ shown in Figure 12.2.

   The four curves represent—from left to right—the following errors: l is an odd number, but not a multiple of 5; l is an even number, but not a multiple of 5; l is an odd multiple of 5; l is an even multiple of 5. The special impact of the factor 5 on the behavior of the error is related to the zeros of $\sin(10\pi x)$.

   If this function is integrated using a Gauss formula with the same number of abscissas then the obtained errors are far larger (—).

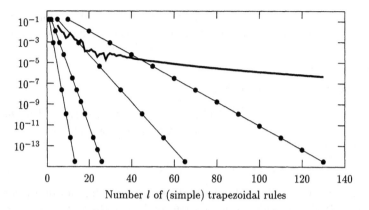

Number l of (simple) trapezoidal rules

Figure 12.2: (Trapezoidal Rule) Absolute errors of the compound trapezoidal rules T_l (—) and of the Gauss formulas G_{l+1} (━) for $l = 1, 2, \ldots, 130$, for the integrand function (12.45).

IMT Formulas

The rapid convergence of the compound trapezoidal rule can also be achieved for *non*-periodic integrands if a suitable periodizing transformation is used in advance (see Section 12.2.1).

For instance, the IMT-transformation (12.19)

$$\psi(\bar{x}) := a + \frac{b-a}{\gamma} \int\limits_{-1}^{\bar{x}} \exp\left(\frac{-c}{1-t^2}\right) dt \qquad \text{where} \quad \gamma := \int\limits_{-1}^{1} \exp\left(\frac{-c}{1-t^2}\right) dt \quad (12.46)$$

provides the transformed function g with the advantageous property

$$g^{(j)}(-1) = g^{(j)}(1) = 0, \qquad j = 0, 1, 2, \ldots,$$

provided the transformation is scaled properly. The parameter c can be chosen from $\mathbb{R}_+$ (in practice, $c = 4$ has turned out to be a good choice). The transformation (12.46) leads to

$$\int\limits_{a}^{b} f(x)dx = \int\limits_{-1}^{1} f(\psi(\bar{x}))\psi'(\bar{x})\, d\bar{x} = \int\limits_{-1}^{1} g(\bar{x})\, d\bar{x}.$$

Expressing $T_l g$ as a quadrature formula for f results in a formula Q_N with $N = l + 1$ abscissas, which is, despite its derivation from the l-times compound trapezoidal rule, *simple* and not compound. Because of the use of IMT-transformation, such formulas are called *IMT-formulas* and have the following *advantages*:

1. Due to the properties of the transformation ψ, and the fact that no boundary abscissas occur in Q_N, they are especially suitable for integrands with (integrable) singularities at the endpoints a and b of the integration interval. This is also true for integrals over infinite intervals ($a = -\infty$ and/or $b = \infty$), which are transformed into finite intervals and which in general have endpoint singularities (see Section 12.2.1).

2. For a sequence $\{Q_N\}$ of formulas, where $N = 1, 3, 5, \ldots$, all f-values computed for preceding formulas of the sequence can be reused in the next IMT formula Q_N.

12.3.4 Romberg Formulas

The Euler-Maclaurin expansion (12.44) shows that the error $T_l f - If$ of the compound trapezoidal rule T_l has the following structure, provided that the integrand is sufficiently smooth:

$$T_l f - If = C_2 h^2 + C_4 h^4 + C_6 h^6 + \cdots . \tag{12.47}$$

The constants $C_2, C_4, \ldots$ depend only on the integrand f but not on the step size $h = (b-a)/l$. If T_{2l} is calculated in addition to T_l, then the first term of the error expansion can be eliminated and the new formula

$$T_l^1 f := \frac{4 T_{2l} f - T_l f}{3}$$

with the error expansion

$$T_l^1 f - If = C_4^1 h^4 + C_6^1 h^6 + \cdots$$

is obtained. This procedure can be continued recursively,

$$T_l^k := \frac{4^k T_{2l}^{k-1} - T_l^{k-1}}{4^k - 1},$$

where $T_l^0 := T_l$ denotes the original trapezoidal sums (see (12.38)). In this way, the array

$$
\begin{aligned}
T_1 &= T_1^0 \\
T_2 &= T_2^0 \quad T_1^1 \\
T_4 &= T_4^0 \quad T_2^1 \quad T_1^2 \\
T_8 &= T_8^0 \quad T_4^1 \quad T_2^2 \quad T_1^3 \\
&\;\;\vdots
\end{aligned}
$$

where T_l^k is a linear combination of the left and upper left values, is obtained. $T_l^1, T_l^2, T_l^3, \ldots$ are called *Romberg formulas*. Since the formula T_l^k is a linear combination of quadrature formulas

$$T_l f = \sum_{i=1}^{l+1} c_i f(x_i)$$

T_l^k can be written as

$$T_l^k f = \sum_{i=1}^{N+1} c_i^k f(x_i^k), \qquad N = l\,2^k.$$

The degree of accuracy of Romberg formulas $T_1^k, T_2^k, T_4^k, \ldots$ is $D = 2k+1$ (Bauer, Rutishauser, Stiefel [106]). Because $D = 2k + 1 < 2^k$ for $k \geq 3$, Romberg formulas $T_l^3, T_l^4, \ldots$ are *not* interpolatory. Nevertheless, they are Riemann sums (Baker [103]).

Efficiency of Romberg Formulas

When applied to smooth functions, the efficiency of Romberg formulas is very high. However, this efficiency decreases rapidly if they are applied to less smooth functions.

Example (Romberg Formulas)   Figure 12.3 depicts the absolute errors of Romberg formulas $T_1^0, T_1^1, \ldots, T_1^9$ when approximating the integrals

$$\mathrm{I}f_k := \int_0^1 (k + 3/2)x^{k+1/2}\, dx = 1, \qquad k = 0, 1, \ldots, 4. \tag{12.48}$$

The error curves show the relation between the rate of convergence and the smoothness of the integrand. The error of the four times differentiable integrand f_4 is considerably smaller than that of the integrand f_0, which is only continuous.

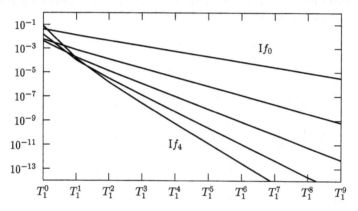

Figure 12.3: (**Romberg Formulas**) Absolute errors of Romberg formulas for the five integrals (12.48).

Software (Romberg Formulas)   RECIPES/qromb and RECIPES/qromo are quadrature programs based on Romberg formulas.

12.3.5 Nonlinear Extrapolation

The Euler-Maclaurin expansion can be applied to a more general class of integrands f. For instance, if f has an *algebraic end-point singularity*,

$$f(x) = x^\beta h(x), \quad -1 < \beta \le 0, \quad h \in C^{p+1}[a, b],$$

then (Lyness, Ninham [275])

$$(k \times \mathrm{Q}_N)f - \mathrm{I}f = \sum_{q=1}^p a_q k^{-\beta - q} + \sum_{q=1}^p b_q k^{-q} + O(k^{-p-1}). \tag{12.49}$$

Similar—though more complicated—expansions can be obtained for integrands with algebraic-logarithmic end-point singularities and for integrands with interior algebraic singularities (Lyness, Ninham [275]).

It has to be noted that the (linear) Richardson extrapolation can only be used to accelerate the rate of convergence if the exponent β is known *explicitly*. If this is not the case, *nonlinear* extrapolation has to be applied in order to accelerate the rate of convergence of $(k \times Q_N)f$.

In order to accelerate the convergence

$$s_k := (k \times Q_N)f \to \mathrm{I}f \qquad \text{as} \qquad k \to \infty,$$

the *Shanks' transformation* (Shanks [346]) has been proven to be useful provided that the error expansion (12.49) holds.

Epsilon Algorithm

Shanks' transformation is generally implemented using the *epsilon algorithm* (ε-*algorithm*) (Wynn [384], [385]):

$$
\begin{aligned}
\varepsilon_{-1}^{(m)} &:= 0, & m = 1, 2, 3, \ldots \\
\varepsilon_{0}^{(m)} &:= s_m, & m = 0, 1, 2 \ldots \\
\varepsilon_{l+1}^{(m)} &:= \varepsilon_{l-1}^{(m+1)} + \frac{1}{\varepsilon_{l}^{(m+1)} - \varepsilon_{l}^{(m)}}, & m, l \geq 0.
\end{aligned}
$$

The data flow in the ε-algorithm is depicted in Figure 12.4.

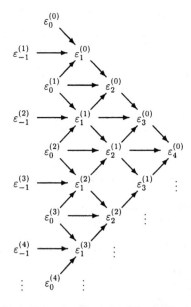

Figure 12.4: Dependency graph of the epsilon algorithm.

Besides the Richardson extrapolation, the ε-algorithm is the most important extrapolation algorithm for numerical integration. For example, all convergence acceleration schemes used in QUADPACK [22] are based on this algorithm.

Example (Epsilon Algorithm) Figure 12.5 depicts the absolute errors obtained when numerically computing the integrals

$$\int_0^1 (3/2)\sqrt{x}\, dx = 1 \tag{12.50}$$

using the Romberg formulas T_2^k, $k = 0, 1, \ldots, 15$.

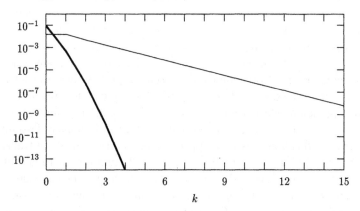

Figure 12.5: (Epsilon Algorithm) Absolute errors of the Romberg formulas (—) and of the ε-algorithm (—) for the calculation of the integral (12.50).

In addition, by applying the ε-algorithm, the sequence of trapezoidal rules in these Romberg formulas was transformed into a sequence $\varepsilon_k^{(0)}$, $k = 0, 1, \ldots, 15$. The order of magnitude of $\varepsilon_5^{(0)}$ is the same as the accumulated rounding error (using double precision IEC/IEEE arithmetic). More accurate results cannot be obtained by increasing the number of sample points unless the program is run at higher precision.

Software (Epsilon Algorithm) An implementation of the ε-algorithm can be found, for instance, in CMLIB/ea.

12.3.6 Special Methods

Beside the integration methods mentioned so far, other methods for numerical quadrature exist which cannot be treated within this monograph. *Sinc methods*, for instance, are particularly useful for integrands which are analytic in a complex domain that contains the integration interval (Lund, Bowers [271]).

Software (Sinc Methods) Integration formulas based on sinc methods can be found in TOMS/614.

The convergence of an alternating series can often be accelerated by using the *Euler transformation* (Davis, Rabinowitz [41]). Integrals of periodically oscillating functions can be calculated efficiently by dividing the integration interval into a sequence of subintervals where the integrand has alternating signs. As the sequence of the resulting integrals is alternating, the sequence of integral approximations obtained using the Euler transformation often converges rapidly. This procedure is particularly useful when applied to integrals over unbounded regions.

Software (Euler Transformation) The subroutine `TOMS/639/oscint` calculates the integrals of certain oscillating functions by applying the Euler transformation.

12.4 Multivariate Integration Formulas

Most one-dimensional integration formulas used in practice are based on (piecewise) polynomial interpolation (see Section 12.3). Abscissas and weights of such quadrature formulas are usually specified for particular integration intervals. Transformations are necessary to make such formulas applicable to other intervals. To be more precise, *affine* transformations[5] have to be used since they map general polynomials onto polynomials of the *same degree*. Other transformations would *not* preserve the degree of accuracy of interpolatory integration formulas.

   The construction of efficient multi-dimensional integration formulas is far more demanding than the construction of efficient quadrature formulas: Two intervals $[a, b] \subset \mathbb{R}$ and $[c, d]$ can always be mapped onto each other by affine transformations (see Section 12.3.1). For any dimension $n \geq 2$, however, there are infinitely many measurable and connected regions in $\mathbb{R}^n$ which are *not* equivalent under affine transformations, i.e., which cannot be mapped onto each other using affine transformations (Stroud [362]). Cubature formulas with polynomial degrees of accuracy devised for any of these regions are fundamentally different for any other region.

Example (Regions without Affine Equivalence Relation) The unit sphere

$$S_2 := \left\{ (x_1, x_2) \ : \ x_1^2 + x_2^2 \leq 1 \right\}$$

can be mapped onto the unit cube

$$C_2 := \left\{ (r, \varphi) \ : \ 0 \leq r \leq 1, 0 \leq \varphi \leq 1 \right\}$$

by using the *non*-affine transformation

$$x_1 \ = \ r \cos(2\pi\varphi), \qquad x_2 \ = \ r \sin(2\pi\varphi)$$

Each affine transformation maps C_2 onto a parallelogram. Thus, there is no affine transformation that maps C_2 onto S_2.

In addition to this, the theory of integration formulas with polynomial degrees of accuracy is closely connected to theories of polynomial interpolation and orthogonal polynomials, fields that are considerably more complex for a number of variables than for a single one (Cools [140]).

   For these reasons, methods which are *not* based on polynomial interpolation are often used when constructing multivariate integration formulas. Such methods, like *Monte Carlo* and *number theoretic* methods, are of practical importance only in multi-dimensional integration problems.

---

[5] A transformation $H : \mathbb{R}^n \to \mathbb{R}^n$ is called *affine* if $H(x) = Ax + b$ and if $A \in \mathbb{R}^{n \times n}$ is regular.

12.4.1 Construction Principles

Riemann Sums

Riemann integrals of multivariate functions $f \colon B \subset \mathbb{R}^n \to \mathbb{R},\ n \geq 2$

$$\mathrm{I}f = \int\limits_B f(x)\,dx$$

are defined for n-dimensional cubes

$$B = [a_1, b_1] \times [a_2, b_2] \times \cdots \times [a_n, b_n] \subset \mathbb{R}^n$$

as follows: The decomposition

$$a_k = x_1^{(k)} < x_2^{(k)} < \cdots < x_{N_k+1}^{(k)} = b_k, \qquad k = 1, 2, \ldots, n,$$

of the intervals $[a_k, b_k],\ k = 1, 2, \ldots, n$, into N_k subintervals is used in order to define subcubes of B:

$$[x_{l_1}^{(1)}, x_{l_1+1}^{(1)}] \times [x_{l_2}^{(2)}, x_{l_2+1}^{(2)}] \times \cdots \times [x_{l_n}^{(n)}, x_{l_n+1}^{(n)}], \quad l_k = 1, 2, \ldots, N_k, \quad k = 1, 2, \ldots, n.$$

The set $\{C_{l_1, l_2, \ldots, l_n}\}$ of the resulting $N = N_1 \cdot N_2 \cdots N_n$ cubes defines a partition $\mathcal{P}$ of the given cube B.

Definition 12.4.1 (Multivariate Riemann Sums) *For any choice of N points*

$$x_{l_1, l_2, \ldots, l_n} \in C_{l_1, l_2, \ldots, l_n}, \qquad C_{l_1, l_2, \ldots, l_n} \in \mathcal{P},$$

a multivariate Riemann sum is defined by

$$R_N := \sum_{C_{l_1, l_2, \ldots, l_n} \in \mathcal{P}} \mathrm{vol}(C_{l_1, l_2, \ldots, l_n})\, f(x_{l_1, l_2, \ldots, l_n}). \tag{12.51}$$

Definition 12.4.2 (Multivariate Riemann Integral) *If all sequences $\{R_N\}$ of multivariate Riemann sums with*

$$\Delta_N := \max\{\,\mathrm{vol}(C_{l_1, l_2, \ldots, l_n}) \; : \; C_{l_1, l_2, \ldots, l_n} \in \mathcal{P}\,\} \to 0 \qquad as \quad N \to \infty$$

approach the same limit R, then the function f is said to be Riemann integrable over B. The multivariate Riemann integral is then defined by

$$\int\limits_B f(x)\,dx := R. \tag{12.52}$$

In general, if the integration region B is bounded, there is an n-dimensional cube $[a_1, b_1] \times [a_2, b_2] \times \cdots \times [a_n, b_n]$ such that

$$B \subseteq [a_1, b_1] \times [a_2, b_2] \times \cdots \times [a_n, b_n].$$

The Riemann integral of f over B is then defined by the Riemann integral of the function $c_B f$ over the cube $[a_1, b_1] \times [a_2, b_2] \times \cdots \times [a_n, b_n]$, where c_B denotes the *characteristic function* of the region B:

$$c_B(x) := \begin{cases} 1 & \text{for} \quad x \in B \\ 0 & \text{otherwise.} \end{cases}$$

If the Riemann integral of $c_B f$ exists, then f is Riemann integrable over B.

As with one-dimensional integrals (see Section 12.3.1), multi-dimensional Riemann sums (12.51) can be used—though very inefficiently—to approximate *all* Riemann integrals (12.52) with arbitrary accuracy.

For *continuous* integrands f and unions of cubes B parallel to the axes, the discretization error can be characterized by the *multi-dimensional modulus of continuity*. For $n \geq 2$ integration regions occurring in practice generally *cannot* be represented by a union of cubes parallel to the axes. In such cases the transition from f to $c_B f$ (which is necessary in the definition of the Riemann integral over non-cubical regions B) does *not* maintain the property of continuity; $c_B f$ is generally discontinuous at the boundary of B. Therefore, the characterization of the discretization error using the modulus of continuity is not useful.

The procedure of transforming integrals over non-elementary integration regions B into integrals over elementary regions B' by replacing the integrand f over B with the integrand $c_B f$ over B' is only of minor practical importance. Integration schemes which are based on polynomial or trigonometric approximation (see Sections 12.4.2 and 12.4.5) work efficiently only for integrands that are sufficiently smooth. Since in general the transition from f to $c_B f$ reduces the degree of continuity and the smoothness of the integrand, such a procedure would require an outstandingly large number of evaluations of the integrand $c_B f$, even for low accuracy requirements. The preprocessing methods described in Section 12.2 (transformation, iteration, and decomposition) are thus preferable to the transition from f to $c_B f$.

The situation is different for pseudo-random or number theoretic methods (see Sections 12.4.4 and 12.4.3). When using these methods, the discretization error is fairly insensitive to the analytical smoothness of the integrand. The procedure described for non-elementary regions can therefore be useful in this case.

As with one-dimensional integrals, the rate of convergence for multivariate integration formulas based on Riemann sums is far too low to enable efficient computation. Simple Riemann sums are thus only of minor practical importance for the construction of multivariate integration formulas and algorithms.

Construction Using Approximation

The construction of integration formulas using approximation is based on the error estimate (12.9)

$$|Ig - If| \leq \text{vol}(B) \, \|g - f\|_\infty,$$

which does not depend on the dimension of the integration problem. Thus, the approximation approach can be used to construct multivariate numerical integration formulas. Every approximating function g which satisfies

$$\|g - f\|_\infty \le \frac{\varepsilon}{\text{vol}(B)}$$

is suitable for solving the numerical problem (12.6).

In practice, mainly (piecewise) polynomials are used as approximating functions. The resulting integration formulas are discussed in Section 12.4.2.

Construction Using Transformation

In Section 12.2.1 it is shown that the numerical computation of integrals can sometimes be facilitated by transforming the integral using the multi-dimensional transformation rule (12.13). In this section it is demonstrated that the transformation rule can also be used for the construction of cubature formulas.

In general, cubature formulas $\overline{Q}_N$ are specific to particular regions $\overline{B}$, usually for one of the standard regions described in Section 12.1.1. In order to make such a formula applicable to another integration region B, the formula has to be transformed appropriately.

Let $\psi = (\psi_1, \psi_2, \ldots, \psi_n)$ be a continuously differentiable one-to-one transformation from $\overline{B}$ to B with a nonsingular Jacobian matrix, i.e.,

$$\det J(\overline{x}) \ne 0 \quad \text{for all} \quad \overline{x} \in \overline{B}.$$

The multi-dimensional transformation rule (12.13) yields

$$\int_{\overline{B}} |\det J(\overline{x})| \, \overline{w}(\overline{x}) \, \overline{f}(\overline{x}) \, d\overline{x} = \int_B w(x) f(x) \, dx$$

where $\overline{w}(\overline{x}) := w(\psi(\overline{x}))$ and $\overline{f}(\overline{x}) := f(\psi(\overline{x}))$.

A cubature formula

$$\int_{\overline{B}} |\det J(\overline{x})| \, \overline{w}(\overline{x}) \, \overline{f}(\overline{x}) \, d\overline{x} = \sum_{i=1}^{N} \overline{c}_i \, |\det J(\overline{x}_i)| \, \overline{f}(\overline{x}_i) + \overline{E}(|\det J| \overline{f})$$

is transformed by ψ into the modified rule

$$\int_B w(x) f(x) \, dx = \sum_{i=1}^{N} c_i f(x_i) + E(f),$$

with abscissas and weights given by

$$x_i = \psi(\overline{x}_i), \quad c_i = \overline{c}_i \, |\det J(\overline{x}_i)|, \quad i = 1, 2, \ldots, N,$$

and error term $E(f) := \overline{E}\left(|\det J| \overline{f}\right)$.

For affine transformations

$$\psi(\overline{x}) = A\overline{x} + b, \quad A \in \mathbb{R}^{n \times n}, \quad b \in \mathbb{R}^n,$$

(where $\det J(\overline{x}) = \det A$ for all $\overline{x} \in \overline{B}$), these relations are particularly simple:

$$x_i = A\overline{x}_i + b, \quad c_i = |\det A| \, \overline{c}_i, \quad i = 1, 2, \ldots, N; \qquad E(f) := |\det A| \, \overline{E}(\overline{f}).$$

Construction Using Iteration

In Section 12.2.3 it is shown that the iteration of integrals can be used to prepro-
cess multivariate integrals. The iteration of integrals, however, can also be used
for the construction of multi-dimensional integration formulas.

Definition 12.4.3 (Product Rule) *Let the integration region $B \subseteq \mathbb{R}^n$ be the
Cartesian product $B = B_1 \times B_2 \times \cdots \times B_K$ of K regions*

$$B_1 \subseteq \mathbb{R}^{n_1}, \quad B_2 \subseteq \mathbb{R}^{n_2}, \ldots, \quad B_K \subseteq \mathbb{R}^{n_K}.$$

and let

$$Q^k_{N_k} f_k = \sum_{i_k=1}^{N_k} w^k_{i_k} f_k(x^k_{i_k}), \quad k = 1, 2, \ldots, K,$$

be K multivariate integration formulas for the integrals

$$I_k f_k := \int_{B_k} f_k(x^k) \, dx^k, \quad k = 1, 2, \ldots, K.$$

Then the product rule $(Q^1_{N_1} \times Q^2_{N_2} \times \cdots \times Q^K_{N_K}) f$ for the integral If is defined by:

$$(Q^1_{N_1} \times Q^2_{N_2} \times \cdots \times Q^K_{N_K}) f := \sum_{i_1=1}^{N_1} \sum_{i_2=1}^{N_2} \cdots \sum_{i_K=1}^{N_K} w^1_{i_1} w^2_{i_2} \cdots w^K_{i_K} f(x^1_{i_1}, x^2_{i_2}, \ldots, x^K_{i_K}).$$

Theorem 12.4.1 *If for $k = 1, 2, \ldots, K$ the exact integral of f_k over B_k is
calculated by $Q^k_{N_k}$, then the exact integral of their product*

$$f(x^1, x^2, \ldots, x^K) := f_1(x^1) \, f_2(x^2) \cdots f_K(x^K), \quad x^k \in B_k, \ k = 1, 2, \ldots, K.$$

over $B := B_1 \times \cdots \times B_K$ is calculated using the product rule $Q^1_{N_1} \times \cdots \times Q^K_{N_K}$.

Proof: Davis, Rabinowitz [41].

Example (Two-dimensional Product Rule) The bivariate integral

$$If = \int_{C_2} f(x_1, x_2) \, dx_1 dx_2 \qquad (12.53)$$

is computed over the unit cube $C_2 = [-1, 1] \times [-1, 1]$. $Q^1_{N_1}$ and $Q^2_{N_2}$ are two one-dimensional
formulas for the unit interval $[-1, 1]$. Thus, their product rule $Q^1_{N_1} \times Q^2_{N_2}$ is a multivariate
formula for (12.53).

If, for instance, $Q^1_{N_1}$ and $Q^2_{N_2}$ are chosen to be the Gauss-Legendre formulas (see Section 12.3.2) G_{N_1} and G_{N_2}, it follows that the product rule $G_{N_1} \times G_{N_2}$ is exact for all polynomials[6]

$$P \in \text{span}\{x_1^{k_1} x_2^{k_2} : k_1 = 0, 1, \ldots, 2N_1 - 1, \ k_2 = 0, 1, \ldots, 2N_2 - 1\}$$

since G_{N_1} and G_{N_2} compute the exact integral of all polynomials with maximal degrees $d = 2N_1 - 1$ and $d = 2N_2 - 1$ respectively. Thus, $G_{N_1} \times G_{N_2}$ is exact for all polynomials $P \in \mathbb{P}^2$
with the maximal degree $d \leq 2 \min\{N_1, N_2\} - 1$ (see Section 9.10).

---

[6]span$\{a_1, a_2, \ldots, a_k\}$ denotes the set of all linear combinations of $a_1, a_2, \ldots, a_k$.

It is a characteristic property of product rules $Q^1_{N_1} \times Q^2_{N_2} \times \cdots \times Q^K_{N_K}$ that the number of abscissas N is the product $N = N_1 N_2 \cdots N_K$ of the numbers of abscissas in the constituent formulas.

If the constituent formulas are chosen to be univariate integration formulas, then the number of abscissas of the product rule increases rapidly with the dimension $n = K$ of the integration problem. This unproportional increase in the computational effort usually makes product rules impracticable for numerical integration problems with dimensions $n > 4$.

The notation of product rules that are composed of *identical* one-dimensional factor rules

$$Q^1_{N_1} = Q^2_{N_2} = \cdots = Q^n_{N_n} = Q_N$$

is $(Q_N)^n$.

The formula $(Q_N)^n$ has N^n abscissas and so if the dimension n is increased to $n + 1$, then the number of abscissas is increased by a factor of N. Even for the moderate value $N = 10$ the number of abscissas is increased by one order of magnitude for each additional dimension. Here the disadvantage of this method becomes painfully apparent.

Software (Product Rules) The subroutine NAG/d01fbf enables the evaluation of product rules $Q^1_{N_1} \times Q^2_{N_2} \times \cdots \times Q^K_{N_K}$ with *univariate* factor formulas $Q^k_{N_k}$, $k = 1, 2, \ldots, K$. The user must supply the weights and abscissas of the formulas $Q^k_{N_k}$. The number K of factor formulas (= dimension n of the integral) can be chosen between 1 and 20.

Product rules for one-dimensional Gauss formulas are used in IMSL/MATH-LIBRARY/qand for the calculation of integrals over n-dimensional ($n \leq 20$) cubes parallel to the axes.

In JCAM/dtria product rules for one-dimensional Gauss formulas are used to calculate integrals over triangular regions.

Construction Using Decomposition

In Section 12.2.2 the preprocessing of integration problems based on a partition of the integration region B into pairwise disjoint subregions $B_1, B_2, \ldots, B_L$

$$If = \int_B f(x)\, dx = \int_{B_1} f(x)\, dx + \int_{B_2} f(x)\, dx + \cdots + \int_{B_L} f(x)\, dx,$$

is discussed. This kind of preprocessing is used when the integration region B is too complicated to allow for the straightforward application of numerical integration methods. In this section it is shown that the decomposition approach can also be used for the construction of cubature formulas.

Definition 12.4.4 (Compound Multivariate Integration Formulas) *Let*

$$Q^l_{N_l} f = \sum_{i_l=1}^{N_l} c^l_{i_l} f(x^l_{i_l}), \quad l = 1, 2, \ldots, L,$$

be integration formulas for the integrals

$$\int\limits_{B_l} f(x)\,dx = Q^l_{N_l} f + E^l(f), \quad l = 1, 2, \ldots, L.$$

Then, the compound integration formula for $\mathrm{I}f$ *is defined by:*

$$(Q^1_{N_1} + Q^2_{N_2} + \cdots + Q^L_{N_L})f := \sum_{l=1}^{L} \sum_{i_l=1}^{N_l} c^l_{i_l} f(x^l_{i_l}). \tag{12.54}$$

The absolute discretization error of the compound integration formula is given by

$$(Q^1_{N_1} + Q^2_{N_2} + \cdots + Q^L_{N_L})f - \mathrm{I}f = E^1(f) + E^2(f) + \cdots + E^L(f).$$

Without additional assumptions about the integration region B, the subregions $B_1, B_2, \ldots, B_L$, and the integration formulas $Q^1_{N_1}, Q^2_{N_2}, \ldots, Q^L_{N_L}$, it is not possible to say anything about the error of the compound integration formula (12.54).

The error of compound *univariate* formulas, however, can be characterized much more accurately (see Sections 12.3.3, 12.3.4 and 12.3.5). Above all, the Euler-Maclaurin sum formula (12.44) provides an asymptotic error expansion for the compound trapezoidal rule T_l. This makes it possible to apply extrapolation methods for convergence acceleration.

Asymptotic error expansions for compound univariate formulas can be obtained only if the integration interval $[a, b]$ is divided into subintervals of equal length and the same quadrature formula is applied to each of them.

For multivariate compound formulas, asymptotic error expansions can also only be obtained under similar conditions. Above all, the subdivision of the integration region B into disjoint subregions $B_1, \ldots, B_L$ must be highly regular, and the same cubature formula has to be applied to each of the subregions after a suitable transformation.

In the following discussion, the integration region is assumed to be the unit cube C_n. Most of the results obtained for C_n also hold when the integration region is chosen as the unit simplex T_n. The interval $[-1, 1]$ is assumed to be subdivided into k intervals by $k + 1$ equidistant partition points

$$-1 = \overline{x}_0 < \overline{x}_1 < \ldots < \overline{x}_{k-1} < \overline{x}_k = 1.$$

The set of k^n subcubes

$$[\overline{x}_{l_1}, \overline{x}_{l_1+1}] \times [\overline{x}_{l_2}, \overline{x}_{l_2+1}] \times \cdots \times [\overline{x}_{l_n}, \overline{x}_{l_n+1}], \quad l_j \in \{0, 1, \ldots, k-1\}, \quad j = 1, 2, \ldots, n,$$

is then a partition $\mathcal{P}$ of C_n. If an N-point cubature formula Q_N for $[-1, 1]^n$ is applied to each of the subcubes in its properly transformed form, then the resulting compound formula $k^n \times Q_N$ is called the k^n-*copy-rule* of Q_N.

Assuming that the basic formula Q_N is exact for all constant polynomials P_0, the convergence

$$(k^n \times Q_N)f \to \mathrm{I}f \quad \text{as} \quad k \to \infty$$

can be guaranteed for all bounded Riemann integrable functions f (see Theorem 12.3.9; Davis,Rabinowitz [41]).

Similarly, asymptotic expansions of the Euler-Maclaurin type can be derived for the compound formulas $k^n \times Q_N$, provided that the basic formula Q_N is exact for all polynomials P_d of degree $d \leq D$ (de Doncker [153]).

The computational effort for the evaluation of integration formulas is characterized by the number of evaluations of f, i.e., by the number of different integration abscissas. In general, a compound formula $k^n \times Q_N$ has $k^n N$ abscissas. However, if some abscissas of the basis formula Q_N are located on the boundary of the integration region C_n, the number of abscissas can be decreased significantly. Since an abscissa of Q_N cannot appear in more than 2^n of the transformed formulas (for the subcubes in $\mathcal{P}$), the total number of abscissas cannot be reduced by more than a factor of 2^n. Hence, the number of abscissas in n-dimensional compound formulas $k^n \times Q_N$ is at least $(k/2)^n N$.

For constant k the number of abscissas in $\{k^n \times Q_N\}$ increases *exponentially* with the dimension n. This rapid growth restricts the number k of subdivisions which can be performed with an acceptable computational effort. As a result, the compound formulas $k^n \times Q_N$ are hardly ever considered for numerical integration problems in dimensions $n > 3$.

12.4.2 Polynomial Integration Formulas

Polynomials of N variables are particularly well-suited model functions when the approximation principle is employed for the construction of integration formulas. A polynomial can be integrated algebraically over an integration region in the shape of a polyhedron. For instance, the integration of a *monomial* $x_1^{d_1} x_2^{d_2} \cdots x_n^{d_n}$ over a cube $C := [a_1, b_1] \times [a_2, b_2] \times \cdots \times [a_n, b_n]$ is performed as follows:

$$
\int_C x_1^{d_1} x_2^{d_2} \cdots x_n^{d_n} \, dx_1 dx_2 \cdots dx_n = \prod_{k=1}^{n} \int_{a_k}^{b_k} x_k^{d_k} \, dx_k
$$

$$
= \prod_{k=1}^{n} \frac{1}{d_k + 1} (b_k^{d_k+1} - a_k^{d_k+1}).
$$

Integrals of polynomials over n-dimensional spheres

$$
\{x \in \mathbb{R}^n \ : \ \|x - m\|_2 \leq r\}, \quad m \in \mathbb{R}^n
$$

and their surfaces

$$
\{x \in \mathbb{R}^n \ : \ \|x - m\|_2 = r\}, \quad m \in \mathbb{R}^n
$$

can be expressed in closed form in terms of the Gamma function (see e.g., Davis, Rabinowitz [41]). Integrals of polynomials over other important integration regions can be found, for example, in Stroud [362] and Engels [46].

Interpolatory Integration Formulas

One-dimensional interpolatory (polynomial) integration formulas are usually obtained by interpolating discrete sample points of the integrand using univariate polynomials and by subsequently integrating these interpolation polynomials. Unfortunately, a straightforward generalization of this approach for multidimensional integrals is *not* possible, as shown in this section.

An important property of interpolatory integration formulas is their degree of accuracy (see Section 12.3.2).

Definition 12.4.5 (Degree of Accuracy) *The degree of accuracy of an n-dimensional integration formula Q_N is D if Q_N is exact for all polynomials of n variables of degree $d \leq D$ and not exact for at least one polynomial of degree $d = D + 1$, in other words, if the following relations hold:*

$$Q_N x^d = I x^d \quad \text{for all monomials } x^d \text{ with} \quad \deg x^d \leq D,$$
$$Q_N x^d \neq I x^d \quad \text{for at least one monomial } x^d \text{ with} \quad \deg x^d = D + 1.$$

The interpolatory univariate integration formulas Q_N with N distinct abscissas is uniquely characterized by the fact that its degree of accuracy satisfies $D \geq N - 1$ (see section "Interpolatory Quadrature Formulas", page 92).

It would seem reasonable to construct multivariate integration formulas Q_N by generalizing this univariate approach, i.e., by requiring Q_N to integrate all polynomials of maximum degree $d \leq D$ exactly. For N distinct abscissas $x_1, x_2, \ldots, x_N \in \mathbb{R}^n$, the weights $c_1, c_2, \ldots, c_N$ of the desired cubature formula

$$Q_N f = \sum_{i=1}^{N} c_i f(x_i)$$

are chosen such that

$$Q_N P_d = I P_d \qquad \text{for all} \quad P_d \in \mathbb{P}_d^n. \tag{12.55}$$

Since $\mathbb{P}_d^n$ is a vector space of dimension $\dim(d, n)$ (see Section 9.10) and both the operators I and Q_N are linear functionals over $\mathbb{P}_d^n$, it is sufficient if the correspondence (12.55) holds for one particular basis $\{b_1, b_2, \ldots, b_{\dim(d,n)}\}$ of $\mathbb{P}_d^n$:

$$Q_N b_j = I b_j, \quad j = 1, 2, \ldots, \dim(d, n). \tag{12.56}$$

The resulting equations are referred to as *moment equations*. For a given choice of the basis $\{b_1, b_2, \ldots, b_{\dim(d,n)}\}$ and of the abscissas $x_1, x_2, \ldots, x_N$, the requirement (12.56) defines a system of algebraic equations:

$$\sum_{i=1}^{N} c_i b_j(x_i) = I b_j, \quad j = 1, 2, \ldots, \dim(d, n). \tag{12.57}$$

Definition 12.4.6 (Interpolatory Cubature Formulas) *Q_N is an interpolatory cubature formula if the system of equations (12.57) has a unique solution $c_1, c_2, \ldots, c_N$, i. e., if the weights of the cubature formula Q_N which satisfy the moment equations (12.56) are uniquely determined by its abscissas.*

An interpolatory cubature formula must satisfy

$$N \le \dim(d, n)$$

so that (12.57) contains at least as many equations as variables. For $n = 1$, Q_N is identical to the quadrature formula obtained by integrating interpolation polynomials. However, this does *not* hold for dimensions $n \ge 2$: For given distinct points $x_1, x_2, \ldots, x_N$ the moment equations (12.57) generally do *not* have a unique solution.

The integration abscissas $x_1, x_2, \ldots, x_N$ cannot be supplied as the data of the problem. Instead, they are the unknowns of the system of equations (12.57). Thus, when solving (12.57) the problem is not only to find the weights $c_1, c_2, \ldots, c_N$ for a given set of abscissas but also to determine the abscissas $x_1, x_2, \ldots, x_N$ themselves. It has to be noted that (12.57) is a *nonlinear* (polynomial) system of equations in the unknown abscissas $x_1, x_2, \ldots, x_N$, whereas for given abscissas the system is *linear* in the weights $c_1, c_2, \ldots, c_N$.

Each abscissa $x_i \in \mathbb{R}^n$ introduces $n + 1$ scalar unknowns into the system: the weight c_i and the n coordinates of x_i. The desired cubature formula Q_N thus has to satisfy a system of $\dim(d, n)$ nonlinear equations in $N(n + 1)$ unknowns. For non-trivial values of n and N, these nonlinear equations are too complex to be solved directly. Practical methods for obtaining multivariate interpolatory integration formulas are described in Cools [140].

Example (Complexity of Moment Equations) The number of moment equations for the degree of accuracy $D = 5$ and the relatively small dimension $n = 5$ is

$$\dim(5, 5) = \binom{10}{5} = 252.$$

In $n = 10$ dimensions and for the same degree of accuracy, the number of moment equations increases to

$$\dim(5, 10) = \binom{15}{5} = 3003.$$

Neither the exact (analytical) solution nor a numerical solution of this system of 3003 nonlinear algebraic equations can be computed with a reasonable amount of effort.

The construction of interpolatory cubature formulas is still the issue of active research. Formulas with a *minimal number of points*—i. e., formulas whose number of abscissas is equal to the corresponding lower bound on the number of abscissas—are only known for low dimensions n and low degrees of accuracy D.

A list of all multivariate interpolatory integration formulas known in 1971 was published by Stroud [362]. Additional formulas derived between 1971 and 1991 can be found in Cools and Rabinowitz [141].

12.4.3 Number-Theoretic Integration Formulas

Number-theoretic integration formulas are based on equidistributed sequences of integration abscissas. The term *equidistribution* is defined as follows:

Definition 12.4.7 (Equidistribution of an Infinite Sequence)   *A sequence of vectors $x_1, x_2, \ldots$, where $x_i \in \mathbb{R}^n$, is said to be equidistributed (or uniformly distributed) in the cube $W_n = [0,1]^N$ if for all Riemann integrable functions $f : W_n \to \mathbb{R}$ the sequence $\{Q_N f\}$ of integration formulas*

$$Q_N := [f(x_1) + f(x_2) + \cdots + f(x_N)]/N$$

converges to $\mathrm{I}(f; W_n)$:

$$\lim_{N \to \infty} \frac{1}{N} \sum_{i=1}^{N} f(x_i) = \int_{[0,1]^n} f(x)\,dx = \mathrm{I}(f; W_n).$$

The characteristic function c_E of a subcube $E \subseteq W_n$ is obviously Riemann integrable. Thus, for any equidistributed sequence $x_1, x_2, \ldots$,

$$\lim_{N \to \infty} \frac{1}{N} \sum_{i=1}^{N} c_E(x_i) = \lim_{N \to \infty} \frac{A(E; N)}{N} = \int_{[0,1]^n} c_E(x)\,dx = \mathrm{vol}(E),$$

where

$$A(E; N) := \sum_{i=1}^{N} c_E(x_i)$$

counts the number of points $x_i \in E$. That is, the fraction $A(E; N)/N$ of points of an equidistributed sequence which lie in any subcube of $W_n = [0,1]^n$ is asymptotically equal to the volume of the subcube.

Example (Equidistributed Sequences)   Van der Corput sequences (see Definition 12.4.12) and their multi-dimensional counterparts, the Halton sequences (see Definition 12.4.13), are often used in practice.

If $x_1, x_2, \ldots$ is an equidistributed sequence in $W_n = [0,1]^n$, then there is a corresponding family of integration formulas $Q_1, Q_2, \ldots$ defined by

$$Q_N f := \frac{1}{N} \sum_{i=1}^{N} f(x_i), \quad N = 1, 2, \ldots ,$$

which is convergent for all Riemann integrable functions $f : W_n \to \mathbb{R}$ according to Definition 12.4.7:

$$\lim_{N \to \infty} Q_N f = \int_{W_n} f(x)\,dx.$$

There is a striking similarity between integration formulas $Q_1, Q_2, \ldots$ based on equidistributed sequences and integration formulas based on Monte Carlo methods (see Section 12.4.4) as far as their definitions and convergence properties are concerned. This is why computational techniques based on equidistributed sequences are sometimes called *Quasi-Monte Carlo methods*.

The only difference between integration formulas based on equidistributed sequences and integration formulas based on Monte Carlo methods is the provenance of the abscissas $\{x_i\}$. In Monte Carlo methods the abscissas are supposed to be samples of independent random variables, whereas in equidistributed sequences the abscissas are generated by deterministic mechanisms.

The convergence result $Q_1 f, Q_2 f, \ldots \to I(f; W_n)$ does not provide any information concerning the error $Q_N f - If$. A bound of $Q_n f - If$ in terms of the *discrepancy* of the finite sequence $x_1, \ldots, x_N$ and the *variation* of the integrand f is given by the *Koksma-Hlawka inequality* (Theorem 12.4.2).

Discrepancy of a Sequence

The value $A(E; N)/N$ derived from points of a sequence $x_1, x_2, \ldots, x_N$, which is uniformly distributed in the cube $W_n = [0,1]^n$, could be expected to approximate the volume of a subcube $E \subseteq W_n$ effectively.

The *discrepancy* of a finite sequence $x_1, x_2, \ldots, x_N$ is an index of the deviation of the sequence's distribution from a hypothetical distribution, for which, ideally, the equality

$$\frac{A(E; N)}{N} = \int_{W_n} c_E(x) \, dx \qquad \text{for all} \quad E \in \mathcal{M}$$

holds for a whole family $\mathcal{M}$ of subsets of W_n.

Definition 12.4.8 (Discrepancy of a Finite Sequence) *For a non-empty set $\mathcal{M}$ of subsets of $W_n = [0,1]^n$*

$$D_N^{\mathcal{M}}(x_1, x_2, \ldots, x_N) := \sup_{E \in \mathcal{M}} \left| \frac{A(E; N)}{N} - \int_{W_n} c_E(x) \, dx \right|$$

denotes the discrepancy of the finite sequence $x_1, x_2, \ldots, x_N \in W_n$.

Note that $\mathcal{M}_1 \subseteq \mathcal{M}_2$ implies

$$D_N^{\mathcal{M}_1}(x_1, x_2, \ldots, x_N) \le D_N^{\mathcal{M}_2}(x_1, x_2, \ldots, x_N).$$

In the context of number theoretic integration rules, the following two choices of $\mathcal{M}$ are the most important:

$$\mathcal{M} = \mathcal{C} := \{[a_1, b_1) \times [a_2, b_2) \times \cdots \times [a_n, b_n) \subseteq W_n\} \qquad \text{and}$$
$$\mathcal{M} = \mathcal{C}_0 := \{[0, b_1) \times [0, b_2) \times \cdots \times [0, b_n) \subseteq W_n\}.$$

The corresponding discrepancies are denoted

$$D_N := D_N^{\mathcal{C}} \qquad \text{and} \qquad D_N^* := D_N^{\mathcal{C}_0} \quad \textit{(star discrepancy)}.$$

Equidistributed sequences can be characterized by *asymptotically optimal discrepancies*: An infinite sequence $x_1, x_2, \ldots$ is equidistributed if and only if the discrepancy of its finite subsequences approaches zero:

$$\lim_{N \to \infty} D_N(x_1, x_2, \ldots, x_N) = 0.$$

Variation of a Function

The *variation* of a univariate function $f : [a, b] \to \mathbb{R}$ characterizes its regularity on the interval $[a, b]$ (Niederreiter [298]). For a partition $\mathcal{P}$ of the interval $[a, b]$ into N subintervals,

$$\mathcal{P} := \{x_i \; : \; a = x_0 < x_1 < \cdots < x_{N-1} < x_N = b\},$$

the sum

$$\sum_{i=1}^{N} |f(x_i) - f(x_{i-1})|$$

is a measure of the variation of f with respect to the specific partition $\mathcal{P}$.

Definition 12.4.9 (Variation of a Univariate Function)   *The variation of a real, univariate function $f : [a, b] \to \mathbb{R}$ is the supremum over all discrete partitions $\mathcal{P}$:*

$$V(f) := \sup_{\mathcal{P}} \left\{ \sum_{i=1}^{N} |f(x_i) - f(x_{i-1})| \right\}.$$

Variation in the Sense of Vitali

The univariate concept of the variation of a function can be generalized to multivariate functions $f : [0, 1]^n \to \mathbb{R}$. Using n partitions of $[0, 1]$

$$0 = x_0^{(k)} < x_1^{(k)} < \cdots < x_{m_k}^{(k)} = 1, \qquad k = 1, 2, \ldots, n,$$

a partition $\mathcal{P}$ of the cube $W_n = [0, 1]^n$ into subcubes can be constructed:

$$W_n' := [x_{l_1}^{(1)}, x_{l_1+1}^{(1)}] \times [x_{l_2}^{(2)}, x_{l_2+1}^{(2)}] \times \cdots \times [x_{l_n}^{(n)}, x_{l_n+1}^{(n)}],$$
$$l_k = 0, 1, \ldots, m_k - 1, \quad k = 1, 2, \ldots, n.$$

For any subcube

$$W_n' := [a_1, b_1] \times [a_2, b_2] \times \cdots \times [a_n, b_n] \quad \text{where} \quad W_n' \subseteq W_n \subset \mathbb{R}^n,$$

the n-dimensional *difference operator* Δ is defined by

$$\Delta(f; W_n') := \sum_{j_1=0}^{1} \sum_{j_2=0}^{1} \cdots \sum_{j_n=0}^{1} (-1)^{\sum_{k=1}^{n} j_k} f(j_1 a_1 + (1 - j_1)b_1, \ldots, j_n a_n + (1 - j_n)b_n).$$

Definition 12.4.10 (Variation in the Sense of Vitali)   *The variation $V^{(n)}(f)$ of a function $f : [0, 1]^n \to \mathbb{R}$ in the sense of Vitali is defined by*

$$V^{(n)}(f) := \sup_{\mathcal{P}} \sum_{W_n' \in \mathcal{P}} |\Delta(f; W_n')|,$$

where the supremum is extended over all partitions $\mathcal{P}$ of W_n. If $V^{(n)}(f)$ is finite then f is said to have bounded variation on W_n in the sense of Vitali.

The variation in the sense of Vitali has a serious disadvantage. If f is constant in one of its n variables, then $\Delta(f; W_n') = 0$ and $V^{(n)}(f) = 0$. In this case the variation $V^{(n)}$ is independent of the function behavior; so $V^{(n)}$ is not a suitable measure of the variability of functions.

Variation in the Sense of Hardy and Krause

A more suitable notion of variation than that of Vitali can be obtained by taking into account the behavior of f when its domain is restricted to the faces of W_n: For $1 \leq k \leq n$ and $1 \leq i_1 < \cdots < i_k \leq n$ let $V^{(k)}(f; i_1, i_2, \ldots, i_k)$ be the k-dimensional variation in the sense of Vitali of the restriction of f to the face

$$W_n^{i_1, i_2, \ldots, i_k} := \{(x_1, x_2, \ldots, x_n) \in W_n \; : \; x_j = 0 \text{ for all } j \neq i_1, i_2, \ldots, i_k\}.$$

Definition 12.4.11 (Variation in the Sense of Hardy and Krause)   *The variation $V(f)$ of f on $W_n = [0,1]^n$ in the sense of Hardy and Krause is defined by*

$$V(f) := \sum_{1 \leq i_1 \leq n} V^{(1)}(f; i_1) \; + \sum_{1 \leq i_1 \leq i_2 \leq n} V^{(2)}(f; i_1, i_2) + \cdots + V^{(n)}(f; 1, 2, \ldots, n).$$

If all variations of the form $V^{(k)}(f; i_1, i_2, \ldots, i_k)$ are finite, then f is said to have bounded variation in the sense of Hardy and Krause.

If $f : W_n \to \mathbb{R}$ is constant in one of its n variables—for instance in x_n—then f is *not* automatically of bounded variation in the sense of Hardy and Krause.

Koksma-Hlawka Inequality

The relevance of the mathematical concepts of discrepancy and variation to numerical integration is established by the Koksma-Hlawka inequality.

Theorem 12.4.2 (Koksma-Hlawka Inequality)   *For every integration formula*

$$Q_N f = \frac{1}{N} \sum_{i=1}^{N} f(x_i) \qquad for \qquad \mathrm{I}f = \int_{[0,1]^n} f(x) \, dx,$$

the error bound

$$|Q_N f - \mathrm{I}f| \; \leq \; V(f) \, D_N^*(x_1, x_2, \ldots, x_N) \tag{12.58}$$

holds, where $V(f)$ is the variation of f in the sense of Hardy and Krause and $D_N^*(x_1, x_2, \ldots, x_N)$ is the star discrepancy of the sequence $x_1, x_2, \ldots, x_N$.

Proof: Hlawka [227].

The Koksma-Hlawka inequality (12.58) characterizes the discretization error of an integration formula Q_N derived from number theory by using the regularity and variability of the integrand f and the regularity of the abscissa distribution.

   The Koksma-Hlawka inequality (12.58) can be generalized to integrals over arbitrary regions $B \subseteq W_n$ (see Niederreiter [299]). In these *generalized Koksma-Hlawka inequalities*, it is the discrepancy D_N—not the star discrepancy D_N^* as in (12.58)—which characterizes the regularity of the abscissas.

   Small error bounds (12.58), i.e., efficient integration formulas Q_N, for all integrands f of bounded variation can be obtained if the abscissas $x_1, x_2, \ldots, x_N$

are chosen so that their discrepancy becomes small. Low-discrepancy sequences are provided and discussed in the following sections.

In practice, it is usually not possible to determine the number of abscissas N necessary to comply with a given error requirement in advance. It is therefore important that the number of abscissas N used in integration formulas Q_N can be increased without disposing of previously calculated function values. This is the reason *infinite* sequences of low discrepancy are also studied in this section. For the construction of a family of integration formulas $\{Q_N\}$, the first N elements of such infinite sequences are used, enabling all data from earlier computations to be reused.

One-Dimensional Low Discrepancy Sequences

One-dimensional low discrepancy sequences are discussed in this section merely to form a basis for the description of multi-dimensional low discrepancy sequences. Univariate integration formulas with low discrepancy are very inefficient in comparison to formulas based on polynomial approximation.

The discrepancies D_N^* and D_N of a general finite sequence $x_1, \ldots, x_N$ in $[0, 1]$ are given by the following formulas:

Theorem 12.4.3 *Let* $0 \leq x_1 \leq x_2 \leq \cdots \leq x_N \leq 1$. *Then*

$$D_N^*(x_1, x_2, \ldots, x_N) \;=\; \frac{1}{2N} + \max\left\{ \left| x_i - \tfrac{2i-1}{2N} \right| \; : \; i = 1, 2, \ldots, N \right\} \quad (12.59)$$

and

$$D_N(x_1, x_2, \ldots, x_N) \;=\; \frac{1}{N} + \max\left\{ \tfrac{i}{N} - x_i \; : \; i = 1, 2, \ldots, N \right\} \quad (12.60)$$

$$- \min\left\{ \tfrac{i}{n} - x_i \; : \; i = 1, 2, \ldots, N \right\}.$$

Proof: Niederreiter [299].

The formulas (12.59) and (12.60) directly imply the following lower bounds for $D_N^*(x_1, x_2, \ldots, x_N)$ and $D_N(x_1, x_2, \ldots, x_N)$:

$$D_N^*(x_1, x_2, \ldots, x_N) \geq \frac{1}{2N} \quad \text{and} \quad D_N(x_1, x_2, \ldots, x_N) \geq \frac{1}{N}. \quad (12.61)$$

Both bounds are attained if the abscissas $x_1, x_2, \ldots, x_N$ are defined by

$$x_i := \frac{2i - 1}{2N}, \quad i = 1, 2, \ldots, N.$$

The corresponding quadrature formula with optimal (lowest) error bounds is the compound midpoint Gauss-Legendre formula, i.e., the compound midpoint formula.

Van der Corput Sequences

Each integer $i \in \mathbb{N}_0$ has a unique *representation* with respect to a *base* (radix) $b \in \{2, 3, \ldots\}$

$$i = \sum_{j=0}^{J(i)} d_j(i)\, b^j$$

with the *digits*

$$d_j(i) \in \{0, 1, \ldots, b-1\}, \quad j = 0, 1, \ldots, J(i).$$

Without using the base explicitly, i can be written as $d_{J(i)} \cdots d_2 d_1 d_0$.

The *radical inverse function* $\varphi_b : \mathbb{N}_0 \to [0, 1)$ *with respect to the base* b defines a reflection of the digit sequence of an integer i:

$$\varphi_b : \quad d_{J(i)} \cdots d_2 d_1 d_0 \mapsto 0 \,.\, d_0 d_1 d_2 \cdots d_{J(i)}.$$

The function φ_b can be used for the definition of equidistributed sequences.

Definition 12.4.12 (Van der Corput Sequence) *The van der Corput sequence with respect to the base b is an infinite sequence $x_1, x_2, \ldots$ defined by*

$$x_i := \varphi_b(i - 1), \quad i = 1, 2, 3, \ldots \tag{12.62}$$

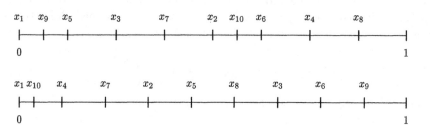

Figure 12.6: Van der Corput sequences with respect to the bases $b = 2$ and $b = 3$.

Van der Corput sequences (see Figure 12.6) have an optimal asymptotic discrepancy (Niederreiter [299]):

$$D_N^* = O(N^{-1} \log N) \quad \text{and} \quad D_N = O(N^{-1} \log N) \quad \text{as} \quad N \to \infty.$$

Example (Van der Corput Sequences) The integral

$$If = \int_0^1 |x - 1/3|^{1/2}\, dx = 2/3 \left[(2/3)^{3/2} + (1/3)^{3/2} \right] \approx 0.49 \tag{12.63}$$

was approximated using quadrature formulas $Q_N^{(b)}$ based on the first N elements of the van der Corput sequences with respect to the bases $b = 2$, $b = 3$, and $b = 5$. Figure 12.7 depicts the integration errors for different choices of the number N of abscissas.

The integral (12.63) was also computed using QUADPACK/qag, a globally adaptive integration program based on pairs of Gauss-Kronrod formulas (see Section 12.3.2). For $N = 225$ function evaluations If was approximated with the error $|Q_N f - If| = 2.2 \cdot 10^{-6}$. This result was more accurate than the one obtained using the quadrature formula $Q_N^{(b)}$ for $N = 100\,000$ abscissas. Due to this dramatic inefficiency, integration formulas based on equidistributed sequences are only used for dimensions $n \geq 2$.

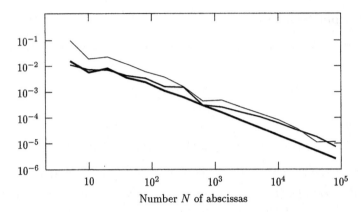

Number N of abscissas

Figure 12.7: (**Van der Corput Sequences**) Absolute errors of the quadrature formulas based on van der Corput sequences in bases $b=2$ (——), $b=3$ (—) and $b=5$ (—).

Multi-Dimensional Low Discrepancy Sequences

(12.61) provides the optimal lower bound for the discrepancy of a *univariate* finite sequence $x_1, \ldots, x_N$; this lower bound is attained for the compound midpoint rule. In contrast, optimal bounds for *multivariate* finite sequences and corresponding optimal finite sequences $x_1, \ldots, x_N$ have *not* yet been established. Nevertheless, it is conjectured that for any finite n-dimensional sequence $x_1, x_2, \ldots, x_N$ the estimate

$$D_N(x_1, x_2, \ldots, x_N) \geq c(n) \frac{(\log N)^{n-1}}{N} \qquad (12.64)$$

holds, and for any infinite sequence $x_1, x_2, \ldots$ the estimate

$$D_N(x_1, x_2, \ldots, x_N) \geq c'(n) \frac{(\log N)^n}{N}$$

is valid for infinitely many N (Niederreiter [299]). The constants $c(n)$ and $c'(n)$ of the two lower bounds only depend on the dimension n, i.e., they are independent of the sequence.

For $n = 1$ (12.64) is identical to inequality (12.61). For $n = 2$, inequality (12.64) was proven by Schmidt [341] in 1972. For $n \geq 3$ it remains an unproven conjecture. Nevertheless, there are particular infinite sequences with the asymptotic discrepancy

$$D_N(x_1, x_2, \ldots, x_N) = O\left(N^{-1}(\log N)^n\right) \qquad \text{as} \quad N \to \infty$$

for which the above inequality holds, as discussed in the following sections.

Halton Sequences

Halton sequences are multi-dimensional extensions of univariate van der Corput sequences (12.62).

Definition 12.4.13 (Halton Sequence)   *The Halton sequence* $x_1, x_2, x_3, \ldots \in$ $\mathbb{R}^n$ *with respect to the bases* $b_1, b_2, \ldots, b_n$ *is defined by*

$$x_i := (\varphi_{b_1}(i-1), \varphi_{b_2}(i-1), \ldots, \varphi_{b_n}(i-1))^\top, \quad i = 1, 2, 3, \ldots, \qquad (12.65)$$

where $\varphi_{b_1}, \varphi_{b_2}, \ldots, \varphi_{b_n}$ *are the radical inverse functions with respect to the bases* $b_1, b_2, \ldots, b_n$.

Theorem 12.4.4   *If the bases* $b_1, b_2, \ldots, b_n$ *are pairwise relatively prime, then the discrepancy of the corresponding Halton sequence* $x_1, x_2, x_3, \ldots$ *satisfies the relation*

$$D_N^*(x_1, x_2, \ldots, x_N) \leq c(b_1, b_2, \ldots, b_n) \frac{(\log N)^n}{N} + O(N^{-1}(\log N)^{n-1}). \quad (12.66)$$

The coefficient c *of the leading term is given by*

$$c(b_1, b_2, \ldots, b_n) = \prod_{k=1}^n \frac{b_k - 1}{2 \log b_k}. \qquad (12.67)$$

Proof: Niederreiter [299].

Since the function $x \mapsto (x - 1)/(2 \log x)$ in (12.67) increases monotonically for $x \geq 2$, the coefficient $c(b_1, b_2, \ldots, b_n)$ attains its minimum when the first n prime numbers $p_1, p_2, \ldots, p_n$ are chosen as the bases $b_1, b_2, \ldots, b_n$ in (12.65).

Example (Halton Sequences)   The n-variate integral

$$\begin{aligned} I_n f_n &:= c_n \int\limits_{[0,1]^n} \prod_{k=1}^n |x_k - 1/3|^{1/2} \, dx_k \\ &= c_n (2/3)^n ((2/3)^{3/2} + (1/3)^{3/2})^n \approx c_n 0.49^n \end{aligned} \qquad (12.68)$$

is an n-dimensional generalization of the integral (12.63). In the following experiments the parameter c_n was chosen such that $I_n f_n = 1$.

For $n = 5, 10$, and 15, the integral (12.68) was computed using cubature formulas Q_N^n based on the first N elements of Halton sequences with respect to the bases $p_1, p_2, \ldots, p_n$. Figure 12.8 depicts the error $|Q_N^n f_n - I_n f_n|$ for different numbers of abscissas N.

Software (Equidistributed Sequences)   Halton sequences for dimensions $n \leq 40$ are generated by a subroutine contained in TOMS/647. Another method for generating equidistributed sequences is proposed by Faure [186]. Faure's method is implemented in TOMS/659. In addition, TOMS/659 contains a subroutine for the computation of uniformly distributed pseudo-random numbers based on a method proposed by Sobol [355].

Hammersley Sequences

In contrast to the infinite Halton sequences, *Hammersley sequences* are finite.

Definition 12.4.14 (Hammersley Sequence)   *The N-point Hammersley sequence with respect to the bases* $b_1, b_2, \ldots, b_{n-1}$ *is the sequence* $x_1, x_2, \ldots, x_N \in \mathbb{R}^n$ *defined by*

$$x_i := \left(\frac{i-1}{N}, \varphi_{b_1}(i-1), \varphi_{b_2}(i-1), \ldots, \varphi_{b_{n-1}}(i-1) \right), \quad i = 1, 2, \ldots, N.$$

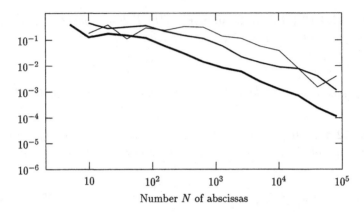

Figure 12.8: (Halton Sequences) Absolute errors of integration formulas based on N elements of Halton sequences. The integral (12.68) was computed for dimensions $n = 5$ (—), $n = 10$ (—), and $n = 15$ (—).

Hammersley and Halton sequences differ only in the definition of the first component of the sequence vectors $x_i \in \mathbb{R}^n$. It can be shown that the bound (12.66) for the discrepancy of Halton sequences also holds for Hammersley sequences when the number of bases n is replaced by $n - 1$:

$$D_N^*(x_1, x_2, \ldots, x_N) \leq c(b_1, b_2, \ldots, b_{n-1}) \frac{(\log N)^{n-1}}{N} + O\left(N^{-1}(\log N)^{n-2}\right).$$

For the optimal choice of the bases in Halton and Hammersley sequences as the first n prime numbers, (12.67) implies

$$\frac{c(p_1, p_2, \ldots, p_{n-1})}{c(p_1, p_2, \ldots, p_n)} = \frac{2 \log p_n}{p_n - 1}.$$

This demonstrates that the bound for the discrepancy of the N-point Hammersley sequence is smaller than the bound for the discrepancy of the corresponding Halton sequence by the factor $(2 \log p_n)^{-1}(p_n - 1) \log N$.

Example (Hammersley Sequences)   The integral (12.68) was computed for $n = 5$ using cubature formulas Q_N with abscissas equal to the elements of an N-point Hammersley sequence with respect to the bases $p_1, p_2, \ldots, p_{n-1}$.

Figure 12.9 shows the error $|Q_N f_5 - I_5 f_5|$ for different numbers N of abscissas. The errors in formulas based on Hammersley sequences are significantly smaller than the errors in the corresponding formulas based on Halton sequences (see Section 12.8).

12.4.4 Monte Carlo Techniques

So far it has been assumed that the sample points $x_i \in B$ are chosen *deterministically*. In *randomized formulas* the choice of these points is controlled by random quantities characterized by an appropriate probability distribution. In numerical integration a special type of randomized algorithm is used: *Monte Carlo methods*

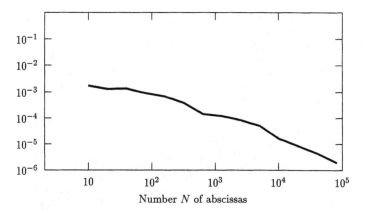

Figure 12.9: (Hammersley Sequence) Absolute errors of integration formulas based on N elements of the Hammersley sequence. The integral (12.68) was calculated for $n=5$.

(see Section 5.2.6). In Monte Carlo methods the integral If is interpreted as the *stochastic mean value*:

$$If = \int_B f(x)\,dx = \text{vol}(B) \int_{\mathbb{R}^n} \text{vol}(B)^{-1} c_B(x) f(x)\,dx = \text{vol}(B) \cdot \mu(f). \qquad (12.69)$$

$\mu(f)$ is the mean value of the function $f(X)$, where X is a continuous random variable uniformly distributed in the region $B \subseteq \mathbb{R}^n$ with the density function $\text{vol}(B)^{-1} c_B$ (c_B denotes the characteristic function of the region B).

A basic statistical technique is to estimate a mean value using a sample mean. The *Monte Carlo estimate* of the mean value $\mu(f)$ is obtained by taking N independent samples $x_1, x_2, \ldots, x_N \in \mathbb{R}^n$ with density $\text{vol}(B)^{-1} c_B$ and estimating

$$\bar{f} := \frac{1}{N} \sum_{i=1}^{N} f(x_i) = Q_N f.$$

From a stochastic point of view, this procedure can be justified as follows: The samples $x_1, \ldots, x_N$ are regarded as realizations of a finite sequence $X_1, X_2, \ldots, X_N$ of independent, identically distributed random variables with common density $vol(B)^{-1} c_B$. According to the *strong law of large numbers*, the convergence of

$$Q_N f = \frac{1}{N} \sum_{i=1}^{N} f(X_i) \quad \to \quad \mu(f) \qquad \text{as} \quad N \to \infty$$

is almost certain, i. e., it happens with a probability of 1.

In order to develop useful Monte Carlo methods for numerical integration, a sequence of pseudo-random numbers $x_1, x_2, \ldots, x_N$ calculated deterministically using a random-number generator (see Section 17) must be used instead of a real sample. Such formulas are thus referred to as *pseudo-random formulas*.

Example (Monte Carlo Integration) The n-variate integral $I_n f_n$ (12.68) was computed using pseudo-random formulas Q_N^n. The abscissas $x_1, x_2, \ldots, x_N$ of Q_N^n were generated by the subroutine `NAG/g05faf`. Figure 12.10 depicts the discretization errors $|Q_N^n f_n - I_n f_n|$ for different numbers N of abscissas and for the dimensions $n = 5, 10$, and 15. It can be seen that the behavior of the errors $|Q_N^n f_n - I_n f_n|$ is indeed irregular. For a fixed dimension n, for example, an increase in the number N of abscissas does not necessarily result in a decrease of the discretization error. Compared with the corresponding results of formulas based on Halton sequences (page 131), the results do not indicate that there is any advantage in using low-discrepancy formulas, except for dimension $n = 5$.

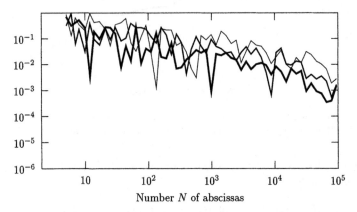

Number N of abscissas

Figure 12.10: (Monte Carlo Integration) Absolute errors of pseudo-random integration formulas for the integral (12.68) with dimensions $n=5$ (——), $n=10$ (——), and $n=15$ (——).

Error Estimates for Monte Carlo Methods

While the strong law of large numbers essentially guarantees that $\overline{f} = Q_N f$ converges to the desired quantity $\mu(f)$ as $N \to \infty$, it does not provide any information about the magnitude of the error $Q_N f - \mu(f)$ for a specific sample of size N.

A *probabilistic estimate* can be obtained by calculating the *mean deviation* $Q_N f - \mu(f)$. A measure for this deviation is, for instance, the *standard deviation* $\sigma(Q_N f)$ of the sample mean $Q_N f$:

$$\sigma(Q_N f) := \sqrt{\mu\left(Q_N f - \mu(f)\right)^2}.$$

If f—again interpreted as a random variable—has the standard deviation $\sigma(f)$, then the standard deviation of the sample mean $Q_N f$ is

$$\sigma(Q_N f) = \frac{\sigma(f)}{\sqrt{N}}.$$

Since the standard deviation is the quadratic mean of the deviation of a random variable and its mean value, the quadratic mean of the error $Q_N f - \mu(f)$ of a Monte Carlo method is given by $\sigma(f)N^{-1/2}$, resulting in the order of convergence $O(N^{-1/2})$ in the statistical error estimate $\sigma(Q_N f)$ as $N \to \infty$. It is an important property of Monte Carlo integration that the order of convergence does *not* depend on the dimension n of the integration problem.

In order to obtain useful error estimates, the generally unknown standard deviation $\sigma(f)$ must be replaced by the *empirical standard deviation* (see Section 12.5.1)

$$s(f) = \sqrt{\frac{1}{N-1} \sum_{i=1}^{N} (f(x_i) - Q_N f)^2}.$$

When using Monte Carlo methods (12.69), the computation of If not only requires the determination of $\mu(f)$, but also that of $\mathrm{vol}(B)$. If $\mathrm{vol}(B)$ cannot be calculated easily, the Monte Carlo method can be modified in the following way:

$$\begin{aligned}
If &= \int_B f(x)\,dx = \int_{B'} c_B(x) f(x)\,dx \\
&= \mathrm{vol}(B') \int_{\mathbb{R}^n} \mathrm{vol}(B')^{-1} c_{B'}(x)[c_B(x)f(x)]\,dx = \mathrm{vol}(B')\,\mu(c_B f).
\end{aligned}$$

Region B' (of which B is a subset) can be chosen so that $\mathrm{vol}(B')$ can be computed easily. In this modified Monte Carlo integration method, c_B is not considered to be a part of the density function but of the integrand $c_B f$: The expected value of $c_B f$ is computed with respect to the density function $\mathrm{vol}(B')^{-1} c_{B'}$.

The application of Monte Carlo methods to the solution of integration problems[7] involves the following difficulties:

1. The error bounds for $Q_N f - If$ are probabilistic. There is no guarantee that the discretization error $Q_N f - If$ is lower than the *expected* accuracy $\sigma(Q_N f)$ in specific situations.

2. The order of convergence $O(N^{-1/2})$ does not reflect the smoothness of the integrand. In contrast to polynomial integration formulas, smooth functions are *not* integrated more accurately or more efficiently than non-smooth functions.

3. The abscissas $\{x_i\}$ must be samples of independent random variables. There is the question as to how such "random sequences" should be generated. In practice, so-called *pseudo-random* sequences are used, i. e., *deterministic* sequences which—in a sense—simulate samples of random variables (see Chapter 17).

Variance Reduction Using Importance Sampling

The stochastic nature of the error estimate $\sigma(f)N^{-1/2}$ in Monte Carlo methods implies two alternative approaches to the reduction of the discretization error $Q_N f - If$:

(i) increasing the number N of function evaluations and

(ii) reducing the variance of the integrand f.

---

[7]For the sake of simplicity, it is assumed in the remainder of this section that the integration problem is scaled such that $\mathrm{vol}(B) = 1$. As a result, $\mu(f) = If$.

The first method is not efficient (see Section 12.10): Due to the $1/\sqrt{N}$-law it is necessary to increase the number N of function evaluations by a factor of 100 in order to improve the accuracy by a factor of 10. Alternative methods have therefore been developed to decrease the error by reducing the variance $\sigma(f)$.

The most important method for variance reduction and, therefore, for improving the efficiency of Monte Carlo methods, is the use of *importance sampling*. The basic idea is to introduce a standardized weight function

$$\mathrm{I}f = \int_B f(x)\, dx = \int_B w(x)\frac{f(x)}{w(x)}\, dx \quad \text{and} \quad \int_B w(x)\, dx = 1 \qquad (12.70)$$

which satisfies $w(x) > 0$ on B and is therefore a non-vanishing probability density on B. Equation (12.70) can thus be written as

$$\mathrm{I}f = \mu_w\left(\frac{f}{w}\right),$$

where μ_w denotes the mean value with respect to the probability density w on B.

The *Monte Carlo estimate* of the expected value $\mu_w(f/w)$ is obtained by generating N independent samples $x_1, \ldots, x_N$ with density w and estimating

$$\mathrm{I}f = \mu_w(f/w) \approx Q_N(f/w) := \frac{1}{N}\sum_{i=1}^{N}\frac{f(x_i)}{w(x_i)}. \qquad (12.71)$$

As in the basic Monte Carlo method, the standard deviation $\sigma_w(Q_N(f/w))$ with respect to the density w is a probabilistic error estimate for $|Q_N(f/w) - \mu_w(f/w)|$. Moreover, if f/w has standard deviation $\sigma_w(f/w)$ with respect to the density w then the standard deviation of the sample mean $Q_N(f/w)$ is

$$\sigma_w(Q_N(f/w)) = \sigma_w(f/w)/\sqrt{N}.$$

This modification of the Monte Carlo method leads to a reduction of the variance and thus to a decreased error whenever

$$\sigma_w^2(f/w) < \sigma^2(f).$$

For a positive function f, the particular choice

$$w(x) := \frac{f(x)}{\mathrm{I}f}$$

results in the smallest possible variance

$$\sigma_w^2(f/w) = \sigma_w^2(\mathrm{I}f) = 0.$$

This choice of w presupposes that the desired integral value $\mathrm{I}f$ is already known in advance, which is of course not the case. However, these considerations suggest that $\sigma_w^2(f/w)$ can be expected to be small if w is chosen so that f/w is nearly constant.

If f has both negative and positive values and a finite lower bound,

$$f(x) > M \qquad \text{for all} \quad x \in B,$$

then this variance reduction scheme can be applied to the function $f - M$.

It has to be noted that the abscissas $x_1, \ldots, x_N$ in (12.71) are distributed according to the density $w(x)$. The generation of such *non-uniformly* distributed random numbers is usually a non-trivial task (see Section 17.3).

In general, sampling according to a *non-constant* density function is computationally more expensive than sampling according to a uniform density function. Moreover, in addition to f, the density function w has to be evaluated at the abscissas $x_1, x_2, \ldots, x_N$. Hence, more computing time is required for the evaluation of N-point integration formulas for importance sampling than for uniform sampling.

Example (Variance Reduction)   In order to derive a hydrological model of extreme flooding situations using Bayesian statistics, independent samples of the water level $x_1, x_2, \ldots, x_M$ are gathered. The distribution[8] of these samples depends on the unknown state of the system $(\vartheta_1, \vartheta_2)^\top$ according to

$$v_i(x_i | \vartheta_1, \vartheta_2) = d \exp(-g(x_i) - \exp(-g(x_i))), \qquad i = 1, 2, \ldots, M,$$

where

$$g(x) := d \left(x - \vartheta_1 + \frac{\gamma}{d} \right), \qquad d = \frac{\pi}{\sqrt{6} \vartheta_1 \vartheta_2}, \qquad \gamma \approx 0.5772157.$$

The a priori information about the state of the system $(\vartheta_1, \vartheta_2)^\top$ is modeled by the distribution

$$\pi(\vartheta_1, \vartheta_2 | \mu_1, \sigma_1, \mu_2, \sigma_2) = \left(\begin{array}{c} \frac{1}{\sqrt{2\pi} \vartheta_1 \sigma_1} \exp\left(-\frac{1}{2\sigma_1^2} (\ln \vartheta_1 - \ln \mu_1)^2 \right) \\ \frac{1}{\sqrt{2\pi} \vartheta_2 \sigma_2} \exp\left(-\frac{1}{2\sigma_2^2} (\ln \vartheta_2 - \ln \mu_2)^2 \right) \end{array} \right)$$

of the system state $(\vartheta_1, \vartheta_2)^\top$.[9] The parameters μ_1, σ_1, and μ_2, σ_2 of the distributions of ϑ_1 and ϑ_2 are derived from experience. In order to calculate the a posteriori density, the integral

$$\int_0^\infty \int_0^\infty f(\vartheta_1, \vartheta_2)\, d\vartheta_1 d\vartheta_2, \qquad f(\vartheta_1, \vartheta_2) = \pi(\vartheta_1, \vartheta_2 | \mu_1, \sigma_1, \mu_2, \sigma_2) v(x_1, \ldots, x_M | \vartheta_1, \vartheta_2), \quad (12.72)$$

in which v is the common density

$$v(x_1, \ldots, x_M | \vartheta_1, \vartheta_2) = \prod_{i=1}^M v_i(x_i | \vartheta_1, \vartheta_2)$$

of $x_1, x_2, \ldots, x_M$, must be determined.

First of all, the direct application of Monte Carlo methods for the approximation of the integral (12.72) is not possible because the integration region $B = [0, \infty)^2$ is unbounded. Since the integrand f decreases quickly as $\vartheta_1 \to \infty$ and $\vartheta_2 \to \infty$, an obvious approach is to restrict

---

[8]This distribution is referred to as *Gumbel distribution*.
[9]Both ϑ_1 and ϑ_2 comply with a *log-normal* distribution.

the original region of integration to a bounded region B'. For example, if (12.72) is to be determined with an accuracy of three digits, it is sufficient to integrate f over $B' := [0,1]^2$. In order to approximate the modified integral, Monte Carlo integration formulas Q_N with uniformly distributed abscissas in B' can be used.

Another method is based on the fact that the integrand f in (12.72) contains the a priori density π whence $w := \pi$ is an obvious choice for the density for variance reduction. The corresponding Monte Carlo integration formulas Q_N^π for approximating (12.72) thus have abscissas that are distributed according to the a priori density π.

Figure 12.11 shows the discretization errors from the methods Q_N and Q_N^π for different numbers N of abscissas. The required random numbers were generated by the subroutines NAG/g05faf and NAG/g05def. The average error in Q_N^π is smaller than that in Q_N by a factor of 10. In order to recover the same accuracy, Q_N must therefore use 100 times more abscissas than Q_N^π. The (measured) computational effort per abscissa is 1.3 times higher for Q_N^π than for Q_N. So the overall computational effort is 77 times higher for Q_N than for Q_N^π.

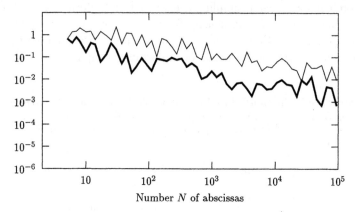

Figure 12.11: (Monte Carlo Method with Variance Reduction) Absolute errors of the Monte Carlo methods Q_N (—) and Q_N^π (—) for the integral (12.72).

Software (Monte Carlo Methods)   The subroutine NAG/d01gbf uses an adaptive compound Monte Carlo integration method to compute integrals over n-dimensional cubes. The region of integration is divided into subcubes, and a Monte Carlo method is applied to each subcube. If the required accuracy is not attained then the subcubes can be divided further.

12.4.5 Lattice Rules

The univariate compound trapezoidal rule T_N is defined as the compound 2-point closed Newton-Cotes formula (Section 12.3.3). Since multivariate Newton-Cotes formulas do not exist, this kind of derivation cannot be generalized to multidimensional integration problems. Another approach is described in this section.

Formula Construction Using Harmonic Analysis

The strategy for generalizing the univariate trapezoidal rule discussed in this section makes use of the particular structure of the error in T_N, which can be characterized by the Euler-Maclaurin sum formula (12.44). This formula shows—among other things—that the compound trapezoidal rule gives outstandingly accurate results for smooth, $(b-a)$-*periodic* integrands (or integrands which can be extended outside (a, b) as smooth, $(b-a)$-periodic integrands). This property suggests that the error behavior of compound trapezoidal rules can be studied more thoroughly using *harmonic analysis*, i.e., by expanding the integrand f in a Fourier series. In this way a multi-dimensional generalization of the univariate trapezoidal rule can be developed.

Univariate Trapezoidal Rule for Periodic Integrands

In the following, without loss of generality, a periodic function $f: \mathbb{R} \to \mathbb{R}$ with period 1 is considered, i.e., f is assumed to satisfy the functional equation

$$f(x + 1) = f(x).$$

If f is sufficiently smooth, then it may be expanded as a (complex) Fourier series:

$$f(x) = \sum_{m \in \mathbb{Z}} c_m e^{2\pi i m x}. \tag{12.73}$$

The Fourier coefficients $c_m \in \mathbb{C}$ are given by

$$c_m = \int_0^1 f(x) e^{-2\pi i m x} \, dx, \qquad m \in \mathbb{Z}.$$

Absolute convergence of the Fourier series (12.73) is assured if f, for example, is *Hölder continuous*

$$|f(x) - f(y)| \leq K|x - y|^\alpha, \quad K > 0,$$

with exponent $\alpha > 1/2$ (Rees et al. [69]). If f has bounded variation (see Definition 12.4.9), then it is even sufficient that the Hölder exponent satisfies $\alpha > 0$.

By substituting (12.73) into a quadrature rule Q_N and changing the order of summation (which can be justified by the absolute convergence of the Fourier series), the formula

$$Q_N f = \sum_{m \in \mathbb{Z}} c_m Q_N(e^{2\pi i m x}) \tag{12.74}$$

can be derived. Because

$$c_0 = \int_0^1 f(x) e^{-2\pi i 0 x} \, dx = \int_0^1 f(x) \, dx = If,$$

(12.74) yields the error representation

$$Q_N f - If = \sum_{m \in \mathbb{Z} \setminus \{0\}} c_m Q_N(e^{2\pi i m x})$$

for Q_N (Lyness [273]). In order to develop an efficient rule Q_N, i.e., an integration formula for which the error $Q_N f - If$ is as small as possible for a large class of functions, it is necessary to construct Q_N so that basis functions $e^{2\pi i m x}$ with large Fourier coefficients c_m are damped or even annihilated by Q_N in as many cases as possible.

Since for smooth functions the most significant terms c_m are usually the ones where $|m|$ is small, quadrature rules Q_N that annihilate as many *low-frequency* functions $e^{2\pi i x}, e^{4\pi i x}, e^{6\pi i x}, \dots$ as possible are the most appropriate.[10] This optimal result is attained for the *displaced trapezoidal rules* $(v \in [0, 1))$

$$T_N(v)f = \frac{1}{N} \sum_{i=0}^{N-1} f\left(\text{fraction}\left(\frac{i}{N} + v\right)\right), \qquad (12.75)$$

where the function

$$\text{fraction} : \mathbb{R} \to [0, 1)$$

returns the fractional part of a real number. These integration formulas annihilate all *trigonometric polynomials* $e^{2\pi i m x}$, except for those where $m = \pm kN$ for $k = 0, 1, 2, \dots$. As a result, the discretization error of the displaced trapezoidal rules (12.75) can be represented as

$$T_N(v)f - \mathrm{I}f = \sum_{k \in \mathbb{Z} \setminus \{0\}} c_{kN} T_N(v)(e^{2\pi i k N x}).$$

This shows that formulas belonging to the family (12.75) are well suited to integrands f whose Fourier coefficients c_m decay rapidly as $m \to \infty$. This decay rate itself is strongly correlated to the smoothness of f. For instance, if f is infinitely many times differentiable, then the Fourier coefficients decay particularly fast (Lyness [273]):

$$|c_m| = o(|m|^{-k}) \quad \text{for all} \quad k \in \mathbb{N}.$$

The coefficient sequences $c_0, c_1, c_2, \dots$ and $c_0, c_{-1}, c_{-2}, \dots$ converge to zero faster than m^{-k} for any $k \in \mathbb{N}$. The compound trapezoidal rule proves to be highly efficient for such integrands (see the example on page 108).

Multivariate Integration Formulas for Periodic Integrands

For the sake of simplicity, it is again assumed that the function $f : \mathbb{R}^n \to \mathbb{R}$ is periodic with period 1, i. e.,

$$f(x) = f(x + z) \qquad \text{for all} \quad x \in \mathbb{R}^n \text{ and } z \in \mathbb{Z}^n$$

and that f can be represented as an absolutely convergent Fourier series

$$f(x) = \sum_{m \in \mathbb{Z}^n} c_m\, e^{2\pi i m \cdot x} \qquad (12.76)$$

with

$$c_m = \int_C f(x)\, e^{-2\pi i m \cdot x}\, dx, \qquad c_m \in \mathbb{C}, \quad m \in \mathbb{Z}^n \qquad (12.77)$$

---

[10]Because $c_{-m} = \overline{c_m}$ for real functions $f : \mathbb{R} \to \mathbb{R}$, all corresponding low-frequency functions with negative indices ($e^{-2\pi i x}, e^{-4\pi i x}, e^{-6\pi i x}, \dots$) are annihilated.

where $m \cdot x$ denotes the inner product

$$m \cdot x := m_1 x_1 + m_2 x_2 + \cdots + m_n x_n$$

of the vectors m and x.

Substituting the Fourier series (12.76) of f into the multivariate integration rule Q_N and changing the order of summation (which can again be justified by the absolute convergence of the Fourier series) yields

$$Q_N f = \sum_{m \in \mathbf{Z}^n} c_m Q_N(e^{2\pi i m \cdot x})$$

as well as the error representation

$$Q_N f - I f = \sum_{m \in \mathbf{Z}^n \setminus \{0\}} c_m Q_N(e^{2\pi i m \cdot x}). \tag{12.78}$$

When trying to construct efficient multivariate integration formulas Q_N similar to one-dimensional trapezoidal rules, i.e., formulas whose discretization error (12.78) is *as small as possible* for a large set of integrands, the following problems arise:

- In order to obtain an efficient integration rule, Q_N should be constructed so that it annihilates those functions $e^{2\pi i m \cdot x}$ for which the Fourier coefficient c_m is "large". The problem is that it is difficult to characterize the behavior of multi-dimensional Fourier coefficients as this behavior depends heavily on the exact degree of smoothness of f. This results in various ordering schemes of Fourier coefficients $c_m \in \mathbf{C}$, each representing the decay behavior of a certain set of functions. There is *no* such order for *all* multi-dimensional smooth functions.

- The construction of multi-dimensional rules that annihilate as many relevant (with respect to a given order) functions $e^{2\pi i m \cdot x}$ as possible is generally an intractable problem (Lyness [273]), in contrast to the corresponding one-dimensional problem, which has a well-defined solution.

Lattice Rules

Lattices are sets of specially located points in $\mathbf{R}^n$. They form the basis for the construction of a large class of multivariate integration formulas which can be regarded as generalizations of the univariate trapezoidal rule: *lattice rules*.

Definition 12.4.15 (Lattice)   *A lattice L is a subset of $\mathbf{R}^n$ with the following properties:*

1. *If x_1 and x_2 belong to L, then so do $x_1 + x_2$ and $x_1 - x_2$;*

2. *L contains n linearly independent points; and*

3. 0 *is an isolated point of* L, *i.e., there exists a neighborhood of* 0 *whose inter-section with* L *contains only* 0.[11]

It follows that every lattice L is a discrete subset of $\mathbb{R}^n$. It can be derived from the third property that all lattice points are *isolated*.

Terminology (Lattice) Care has to be taken not to confuse a *lattice* (which is used in number theory) with a *mesh* (or *grid*), which is mainly used in numerical mathematics.

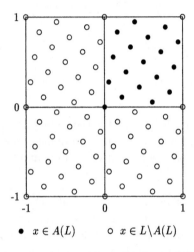

● $x \in A(L)$ ○ $x \in L \backslash A(L)$

Figure 12.12: An integration lattice in $\mathbb{R}^2$.

Definition 12.4.16 (Integration Lattice) *An* integration lattice L *in* $\mathbb{R}^n$ *is a lattice that contains* $\mathbb{Z}^n$ *as a sublattice, i.e.,* $\mathbb{Z}^n \subseteq L$.

Lattice rules, which are special multivariate integration formulas, can be defined using integration lattices (Sloan [350], Sloan, Joe [351]).

Definition 12.4.17 (Lattice Rule) *A* lattice rule *for the numerical integration of multivariate functions* $f \colon [0,1)^n \to \mathbb{R}$ *is a multivariate integration formula*[12] *with equal weights* $w_0 = w_1 = \cdots = w_{N-1} = 1/N$,

$$Q_N f := \frac{1}{N} \sum_{i=0}^{N-1} f(x_i),$$

whose abscissas $x_0, \ldots, x_{N-1}$ *are those points of an integration lattice* L *which are located in the cube* $W_n = [0,1)^n$.

---

[11]It can be derived from the first property that 0 always is an element of L.

[12]As the construction of lattice rules is related to concept of remainder classes, it is convenient to let the summation index run from $i = 0$ to $i = N - 1$.

Definition 12.4.18 (Abscissa Set)   *The abscissa set $A(L)$ of a lattice rule is given by*

$$A(L) := \{x_0, \ldots, x_{N-1}\} = L \cap W_n.$$

The set $A(L)$ is non-empty because $0 \in L$ and finite because L contains only isolated points.

Example (Product-Trapezoidal Rule)   The most obvious integration lattice is the *cubical lattice*

$$L = \left\{ \left(\frac{i_1}{m}, \ldots, \frac{i_n}{m} \right) \ : \ i_1, \ldots, i_n \in \mathbb{Z} \right\}, \tag{12.79}$$

where $m \in \mathbb{N}$. The corresponding lattice rule is the *product-trapezoidal rule*

$$T_N f = \frac{1}{N} \sum_{i_1=0}^{m-1} \cdots \sum_{i_n=0}^{m-1} f \left(\frac{i_1}{m}, \ldots, \frac{i_n}{m} \right),$$

whose number of abscissas is given by $N = m^n$.

Example (Method of Good Lattice Points)   The so-called *method of good lattice points* is the multi-dimensional generalization of the displaced trapezoidal rule (12.75),

$$Q_N f = \frac{1}{N} \sum_{i=0}^{N-1} f \left(\text{fraction} \left(\frac{i}{N} p \right) \right), \tag{12.80}$$

where $p = (p_1, p_2, \ldots, p_n)^{\mathsf{T}} \in \mathbb{Z}^n$ is a vector of integers. The function

$$\text{fraction} \colon \mathbb{R}^n \to [0,1)^n$$

returns the decimal part of the components of a real vector. It is assumed that the greatest common divisor of $p_1, p_2, \ldots, p_n$ and N is 1 such that all N abscissas are distinct.

Clearly, (12.80) is a lattice rule corresponding to the lattice

$$L = \left\{ \text{fraction} \left(\frac{i}{N} p \right) + z \ : \ i \in \mathbb{Z}, \ z \in \mathbb{Z}^n \right\}. \tag{12.81}$$

Figure 12.12 depicts an integration lattice in $\mathbb{R}^2$ described by Beckers and Cools [108], which pertains to the corresponding method of good lattice points Q_{18} for the parameter vector $p = (1,5)^{\mathsf{T}}$.

A *displaced lattice rule* for an integration lattice L and a vector $v \in \mathbb{R}^n$ is given by

$$Q_N(v)f = \frac{1}{N} \sum_{i=0}^{N-1} f(\text{fraction}(x_i + v)),$$

where $\{x_0, \ldots, x_{N-1}\} = A(L)$. Clearly, the choice $v \in L$ leads to a conventional lattice rule:

$$Q_N(v) = Q_N \qquad \text{for} \quad v \in L.$$

For example, v can be chosen so that all cubature abscissas are in the interior of W_n (in order to avoid e. g., the evaluation of f at boundary singularities).

Representation of Lattice Rules in k-Cycle Form

A useful representation of lattice rule abscissas can be derived by generalizing the representations (12.79) for product-trapezoidal rules and (12.81) for the method of good lattice points as follows: For any combination of k integer vectors $z_1, \ldots, z_k \in \mathbb{Z}^n$ and k numbers $N_1, \ldots, N_k \in \mathbb{N}$, the set

$$L = \left\{ \tfrac{i_1}{N_1} z_1 + \cdots + \tfrac{i_k}{N_k} z_k + z \ : \quad i_j = 0, 1, \ldots, N_j - 1, \atop j = 1, 2, \ldots, k, \ z \in \mathbb{Z}^n \right\}$$

constitutes an integration lattice. The abscissa set for the corresponding lattice rule is given by

$$A(L) = \left\{ \text{fraction} \left(\tfrac{i_1}{N_1} z_1 + \cdots + \tfrac{i_k}{N_k} z_k \right) \ : \quad i_j = 0, 1, \ldots, N_j - 1, \atop j = 1, 2, \ldots, k \right\}. \tag{12.82}$$

This form is called the *k-cycle* form of lattice rules (Lyness, Sloan [276]). In a numerical integration program, this lattice rule can be represented in a compact form by simply storing the k integer vectors $z_1, \ldots, z_k \in \mathbb{Z}^n$ and the k numbers $N_1, \ldots, N_k \in \mathbb{N}$. The abscissas of the corresponding lattice rule can easily be computed by forming all linear combinations (12.82).

Error Analysis of Lattice Rules

The usefulness of a class of integration formulas depends crucially on the behavior of the discretization error $Q_N f - If$ for a relevant class of integrands $\mathcal{F}$. In this section $f : \mathbb{R}^n \to \mathbb{R}$ is assumed to be a continuous function which is periodic with period 1 in every coordinate. In such cases, the speed of convergence of $Q_N f \to If$, which depends on the smoothness of f, is generally higher than that of non-periodic integrands.

Integrands occurring in practice are generally *not* periodic. Such non-periodic integrands have to be submitted to periodizing transformations (see Section 12.2.1) in order to accelerate the convergence of a sequence $\{Q_N\}$ of lattice rules.

If f can be expanded into an absolutely convergent Fourier series then the error $Q_N f - If$ can be represented by (12.78):

$$Q_N f - If = \sum_{m \in \mathbb{Z}^n \setminus \{0\}} c_m \frac{1}{N} \sum_{i=0}^{N-1} e^{2\pi i m \cdot x_i}. \tag{12.83}$$

This implies (Sloan, Kachoyan [352])

$$\frac{1}{N} \sum_{i=0}^{N-1} e^{2\pi i m \cdot x_i} = \begin{cases} 1 & \text{for } m \cdot x_i \in \mathbb{Z}, \quad i = 0, 1, \ldots, N-1, \\ 0 & \text{otherwise.} \end{cases} \tag{12.84}$$

The set

$$L^\perp := \{ m \in \mathbb{R}^n : m \cdot x \in \mathbb{Z} \text{ for all } x \in L \}$$

itself is a lattice in $\mathbb{Z}^n$ and is referred to as the *dual lattice* of L.

Example (Dual Lattice) The dual lattice $L^{\perp}$ of L, described in Figure 12.12, is given by

$$L^{\perp} = \{a_1 z_1 + a_2 z_2 \ : \ a_1, a_2 \in \mathbb{Z}\},$$

where

$$z_1 = (18, 0)^{\top} \quad \text{and} \quad z_2 = (-5, 1)^{\top}.$$

It is depicted in Figure 12.13.

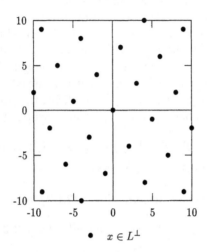

$$\bullet \quad x \in L^{\perp}$$

Figure 12.13: Dual lattice $L^{\perp}$ of L described in Figure 12.12.

Combining (12.83) and (12.84) yields the following fundamental theorem:

Theorem 12.4.5 *For an integration lattice L the absolute error of the corresponding lattice rule Q_N is given by*

$$Q_N f - I f = \sum_{m \in L^{\perp} \setminus \{0\}} c_m.$$

The absolute error of the displaced lattice rule $Q_N(v)$ is given by

$$Q_N(v) f - I f = \sum_{m \in L^{\perp} \setminus \{0\}} e^{2\pi i m \cdot v} c_m, \tag{12.85}$$

where $L^{\perp}$ denotes the dual lattice of L.

Proof: Sloan, Kachoyan [352].

Korobov Spaces

According to Theorem 12.4.5 the construction of a lattice rule with a small discretization error requires the determination of a lattice L which minimizes the sum of the Fourier coefficients

$$\sum_{m \in L^{\perp} \setminus \{0\}} c_m \tag{12.86}$$

over the dual lattice $L^{\perp}$. Thus, only few vectors $m \in \mathbf{Z}^n$ with large Fourier coefficients c_m should be contained in the dual lattice $L^{\perp}$.

In order to obtain lattices L that are suitable for a whole class $\mathcal{F}$ of integrands, it is necessary that the decay behavior of the Fourier coefficients c_m of all functions $f \in \mathcal{F}$ is the same. Without this property it is impossible to find a lattice L for which (12.86) is uniformly small for all $f \in \mathcal{F}$. The question is which class of functions is relevant to a lattice rule and whether general statements about the behavior of their Fourier coefficients c_m can be made.

In the determination of such classes it is often assumed that the functions f are infinitely many times differentiable in the cube $W_n = [0,1]^n$, but are *not* 1-periodic. This assumption is reasonable for most integrands that appear in multi-dimensional integration problems. Moreover it is assumed that, in transforming the integral according to (12.21), the same periodizing transformation ψ is applied to each independent variable $x_1, x_2, \ldots, x_n$. The transformation ψ either satisfies

$$
\begin{aligned}
\psi^{(k)}(0) &= \psi^{(k)}(1) = 0, \quad k = 1, 2 \ldots, \alpha, \\
\psi^{(\alpha+1)}(0) &= -\psi^{(\alpha+1)}(1) \neq 0
\end{aligned}
\tag{12.87}
$$

for $\alpha \in \mathbf{N}$ (as in polynomial transformations (12.18)), or

$$
\psi^{(k)}(0) = \psi^{(k)}(1) = 0 \qquad \text{for } \textit{all} \quad k = 1, 2, 3, \ldots
\tag{12.88}
$$

as in the IMT transformation (12.19). If (12.88) holds, then all partial derivatives of the transformed integrand

$$
f(\psi(\overline{x}_1), \psi(\overline{x}_2), \ldots, \psi_n(\overline{x}_n)) \, \psi'(\overline{x}_1) \, \psi'(\overline{x}_2) \cdots \psi'(\overline{x}_n)
\tag{12.89}
$$

vanish at the boundary of W_n. If (12.87) holds, then as can be shown by differentiating (12.89), this is true only for the following partial differential operators:

$$
\frac{\partial^{\alpha_1 + \cdots + \alpha_n}}{\partial \overline{x}_1^{\alpha_1} \partial \overline{x}_2^{\alpha_1} \cdots \partial \overline{x}_n^{\alpha_n}}, \qquad \alpha_j = 0, 1, \ldots, \alpha - 1, \quad j = 1, 2, \ldots, n.
\tag{12.90}
$$

Since the partial derivatives (12.90) vanish on the interval boundaries, partial integration of the Fourier coefficients (12.77) yields the following bound for the Fourier coefficients c_m (Zygmund [387]):

$$
|c_m| \leq \frac{c}{(\overline{m}_1 \cdot \overline{m}_2 \cdots \overline{m}_n)^{\alpha}},
\tag{12.91}
$$

where

$$
\overline{m}_k := \begin{cases} |m_k| & \text{for } |m_k| \geq 1, \\ 1 & \text{for } |m_k| < 1 \end{cases}
$$

and $c > 0$ is a constant. Thus, if $f \in C^{\infty}([0,1]^n)$ then the decay rate of the Fourier coefficients of integrands transformed by (12.89) can be characterized by (12.91) with the exponent α given by (12.87). If the periodizing transformation ψ satisfies (12.88) then (12.91) holds for *any* $\alpha \in \mathbf{N}$.

Definition 12.4.19 (Korobov Spaces) $E_n^\alpha(c)$ *denotes the set of functions f whose Fourier coefficients c_m satisfy the inequality (12.91). E_n^α denotes the union of the sets $\{E_n^\alpha(c) \ : \ c \in \mathbb{R}_+\}$,*

$$E_n^\alpha := \bigcup_{c \in \mathbb{R}_+} E_n^\alpha(c).$$

The function spaces $E_n^\alpha(c)$ and E_n^α are called Korobov spaces.

When constructing lattice rules, Korobov spaces $E_n^\alpha(c)$ and E_n^α are often used as integrand classes $\mathcal{F}$.

Convergence of Lattice Rules

If the integrand f is a member of one of the Korobov spaces $E_n^\alpha(c)$, the convergence rate $Q_N f \to If$ as $N \to \infty$ of a sequence of lattice rules $\{Q_N\}$ can be characterized as follows. For $f \in E_n^\alpha(c)$, Theorem 12.4.5, (12.91), and

$$P_\alpha(Q_N) := \sum_{m \in L^\perp \setminus \{0\}} \frac{1}{(\overline{m}_1 \cdots \overline{m}_n)^\alpha}$$

yield the error estimate

$$|Q_N f - If| \leq \sum_{m \in L^\perp \setminus \{0\}} \frac{c}{(\overline{m}_1 \cdots \overline{m}_n)^\alpha} = c\, P_\alpha(Q_N). \tag{12.92}$$

This error bound is tight since there exists an $f \in E_n^\alpha(c)$ for which equality is attained in (12.92) (Niederreiter [299]).

The following fundamental theorem explains why lattice rules are important for multi-dimensional integration:

Theorem 12.4.6 *For $\alpha > 1$ and $n \geq 2$ and any $N \in \mathbb{N}$ there exists an N-point lattice rule Q_N such that*

$$P_\alpha(Q_N) \leq d(n, \alpha) \frac{(\log N)^{c(n,\alpha)}}{N^\alpha},$$

where c and d do not depend on N.

Proof: Niederreiter [299].

$c\, P_\alpha(Q_N)$ is the worst-case integration error for integrands $f \in E_n^\alpha(c)$. For $f \in E_n^\alpha(c)$, Theorem 12.4.6 shows that for any dimension n there exists a sequence of lattice rules $\{Q_N\}$ whose asymptotic integration error can be characterized by

$$|Q_N f - If| = O\left(\frac{(\log N)^c}{N^\alpha}\right) \quad \text{as} \quad N \to \infty. \tag{12.93}$$

Because $(\log N)^c = o(N^\alpha)$ as $N \to \infty$, the order of convergence (at least theoretically) does not deteriorate significantly as the dimension is increased. Hence, the convergence behavior depends primarily on the decay rate of the Fourier coefficients, i.e., on the smoothness of the integrand f. In contrast, the convergence rate of pseudo-random formulas (see Section 12.4.4) and number-theoretic formulas (see Section 12.4.3) are independent of the smoothness of the integrand. Hence, lattice rules are more suitable for smooth integrands.

Practical Results

The proof of Theorem 12.4.6 is *non-constructive*. It is therefore *not* possible to find efficient lattice rules using Theorem 12.4.6. At present such lattice rules have to be determined by computer search. It turns out that the enormous number of lattice rules for most practically relevant values of N and n prohibits an *exhaustive* computer search for efficient lattice rules. Computer searches thus have to be restricted to certain classes of lattice rules. The main problem is finding appropriate classes of lattice rules, i. e., *small* classes which contain efficient lattice rules with *high* probability. A thorough discussion of search methods for efficient lattice rules can be found in Krommer and Ueberhuber [257].

Efficient lattice rules can be found in a number of publications. Maisonneuve [280] contains rank-1 lattice rules for dimensions $n = 3, 4, \ldots, 10$. More rank-1 lattice rules can be found in Kedem and Zaremba [250], Bourdeau and Pitre [122] and Keng and Yuan [251]. Optimal lattice rules for dimension $n = 3$ are described in Lyness and Soerevik [277], and for $n = 4$ in Lyness and Soerevik [278].

Software (Lattice Rules)   NAG/d01gyf and NAG/d01gzf are subroutines for finding good lattice point methods (12.80). In NAG/d01gyf the search is restricted to those lattice rules (12.80) for which the number of abscissas N is prime and the vector $p \in \{0, 1, \ldots, N-1\}^n$ is of the *Korobov form*

$$p = (1, a, a^2, \ldots, a^{n-1})^\top.$$

The lattice rules obtained have a minimal P_2 value and are, therefore, optimal for the integrand set E_2. NAG/d01gzf computes lattice rules that are optimal in the same sense. However, their numbers of abscissas N are products of two distinct primes.

The subroutine NAG/d01gcf computes iterated integrals of dimension $n \leq 20$ by applying lattice rules belonging to the class of methods of good lattice points (12.80). The integrand is periodized by a polynomial transformation (12.18) with $\alpha = 2$. The user can select the number of abscissas N from a list of 6 distinct primes. NAG/d01gcf then chooses optimal Korobov vectors p. Alternatively, the user can specify the number of abscissas N and a *good* vector $p \in \{0, 1, \ldots, N-1\}^n$ himself, for which he can use the subroutines mentioned above. In the first case, NAG/d01gcf itself chooses optimal Korobov vectors p. The randomizing method described in Section 12.5.1 is used for the error estimation.

The subroutine NAG/d01gdf is similar to NAG/d01gcf, yet the integrand f must be specified in a different way. The user-specified function subroutine can be evaluated at several abscissas concurrently. If the integrand evaluation can be vectorized then the computational speed can be accelerated significantly by using vector computers.

12.4.6 Special Methods

In this section various special numerical cubature methods are outlined. Further methods, like e. g., those based on mean values, can be found in Davis and Rabinowitz [41].

In the method of Sag and Szekeres [338], the integration region B is transformed into the unit sphere S_n such that the transformed integrand and all its derivatives vanish at the surface of S_n. The resulting integral over S_n is computed using a product-trapezoidal rule.

Software (Method of Sag and Szekeres)   The subroutine NAG/d01jaf computes definite integrals for dimensions $n = 2, 3$ and 4 over n-dimensional spheres by using the method of Sag and Szekeres. The user can choose between different transformations. The subroutine NAG/d01fdf uses the method of Sag and Szekeres to compute definite integrals over *arbitrary* product regions for dimensions $n \leq 40$. However, the user cannot choose between different transformations using NAG/d01fdf.

Haber's method of *stratified sampling* (Haber [216], [217]) uses compound formulas $k^n \times Q_N$, where Q_N is a pseudo-random formula. In this sense, Haber's method of stratified sampling can be referred to as a *compound Monte Carlo method*. If the integrand f is continuously differentiable on W_n then the probabilistic error estimate for $k^n \times Q_N$ is smaller than the probabilistic error estimate for a pseudo-random formula $Q'_{N'}$ on W_n with the same number of abscissas $N' = k^n N$. A further reduction of the probabilistic error estimate can be obtained by using *antithetic variables*. In this method, the basic pseudo-random formula Q_N is extended symmetrically to a formula $\overline{Q}_{2N}$ by appending the abscissa $-x_i$ for each abscissa x_i used in Q_N. Recent work on stratified Monte Carlo methods can be found in Masry and Cambanis [286], [134].

12.5 Integration Algorithms

Integration formulas, as discussed in Chapters 12.3 and 12.4, are used to approximate definite integrals. When applied to specific integration problems, bounds for the error of integration formulas depend on analytic properties of the integrand which are generally not known. Consequently, calculating an approximate value Qf for the desired result If which satisfies a given error tolerance

$$|Qf - If| \leq \varepsilon \tag{12.94}$$

usually requires more than just one formula evaluation. This makes it necessary to apply an integration *algorithm*.

Terminology (Integration Formula)   The evaluation of one integration formula is, strictly speaking, already an integration algorithm. However, as this is normally *inadequate* for solving a given integration problem (12.6), the term integration *algorithm* is *not* used in the context of a single formula evaluation.

In Sections 12.3 and 12.4 the *convergence*

$$Q_N f \to If \qquad \text{as} \quad N \to \infty; \quad f \in \mathcal{F} \tag{12.95}$$

is demonstrated for different families $\{Q_N\}$ of one-dimensional and multi-dimensional integration formulas and for large classes $\mathcal{F}$ of integrands. Convergence theorems of type (12.95) assert the *existence* of an $N \in \mathbb{N}$ for which $Qf = Q_N f$ satisfies the error bound (12.94).

However, specific numerical problems *cannot* be solved constructively by simply providing a convergent sequence of formulas. Even precise statements about the convergence behavior such as

$$|Q_N f - If| = O(n^{-p}) \qquad \text{or} \qquad Q_N f - If = C\, f^{(p)}(\xi)\, n^{-p},$$

which can be made if certain restrictions are imposed on the derivatives of f, *cannot* be used for the *a priori* determination of an appropriate value of N for a given numerical integration problem. In practice, there is normally not enough information available about f and its derivatives.

In order to enable the development of integration algorithms, the discretization error $Q_N f - I f$ of a suitable formula Q_N has to be known at least approximately in order to decide whether or not $Q_N f$ is a satisfactory solution of a specific integration (sub) problem. Therefore, appropriate error estimates are basic elements of all integration algorithms.

12.5.1 Error Estimation

An approximate value for the error of a numerically calculated quantity is called an *error estimate*.[13] In most cases, two or more approximate values for the required quantity—which are calculated in different ways—are compared in order to obtain an error estimate. In numerical integration the results of either

1. two or more *different* integration formulas (with different accuracies),

2. *one* (simple or compound) integration formula, applied first to the whole region and then to subregions of the integration region, or

3. several *randomized* rules (with similar accuracies)

are compared. In each case, the differences in the results are used to obtain error estimates.

For the sake of efficiency, the set of sampling points used to calculate the final result (the approximation with the highest accuracy) usually contains all abscissas previously used in the computational process of calculating lower accuracy results so as to avoid additional integrand evaluations. However, different linear combinations of the integrand function values have to be computed in order to obtain results with different degrees of accuracy. These additional linear combinations constitute the main effort necessary to calculate error estimates in numerical integration programs.

Fundamental Uncertainty

Since all numerical integration algorithms are subject to the inherent uncertainty of any numerical integration problem (see Section 12.1.5), their error estimates are subject to this *total uncertainty* as well.

Whatever discrete information

$$S_e(f) := (f(x_1^e), f(x_2^e), \ldots, f(x_{N_e}^e))$$

is used to obtain information about the error of the approximate result $Q_N f$, it does *not* enable reliable estimates.

---

[13]The term *error estimate* usually refers to the truncation (discretization) error only.

For any bound $\lambda > 0$ it is possible to find a function $f \in C^\infty(B)$ for which the estimated error $\mathrm{E}(f)$ differs from the actual error $E(f)$ by more than λ:

$$|\mathrm{E}(f) - E(f)| > \lambda.$$

Only additional global information about f, which covers the whole region B, like e.g., global bounds of a derivative of the integrand

$$f \in C^k[a, b] \ : \ |f^{(k)}(x)| \leq M_k \quad \text{for all} \quad x \in [a, b],$$

enables a reliable error estimation in numerical integration schemes. However, global information of this kind is rarely available in practice.

Use of Integration Formulas with Different Accuracies

An obvious way to obtain an error estimate is to compare a pair of integration formulas Q_{n_1} and Q_{n_2}. In this technique it is *assumed* that Q_{n_2} generally gives more accurate results than Q_{n_1}.

Example (Polynomial Rules) If formula Q_{n_1} has a degree of accuracy D_1 (see Sections 12.3.2 and 12.4.2) then it integrates all polynomials $P \in \mathbb{P}_{D_1}$ exactly. A formula Q_{n_2} with a degree of accuracy D_2 with $D_2 > D_1$ is exact for a *larger* set of functions $\mathbb{P}_{D_2} \supset \mathbb{P}_{d_1}$, i.e., Q_{n_2} gives better results for polynomials $P \in \mathbb{P}_{D_2} \setminus \mathbb{P}_{D_1}$ than Q_{n_1}.

The assumption that Q_{n_2} returns significantly better results than Q_{n_1}, i.e.,

$$|Q_{n_2}f - If| \ll |Q_{n_1}f - If|,$$

implies that the estimate

$$Q_{n_1}f - If \ \approx \ Q_{n_1}f - Q_{n_2}f \tag{12.96}$$

should be true for *many* $f \in \mathcal{F}$. Q_{n_2} is therefore heuristically regarded as the *exact value* when compared with the less accurate value Q_{n_1}.

The estimate (12.96) is disappointing in the sense that $Q_{n_1} - Q_{n_2}$ only provides information about the error in the *less* accurate value Q_{n_1}. The more accurate value Q_{n_2}, which generally requires more computational effort, is used only to obtain an error estimate.

Moreover, in practice there is no guarantee that Q_{n_2} is indeed more accurate than Q_{n_1} for a specific integrand. If this is the case, the whole integration algorithm may fail, i.e., it may produce a total error which is considerably larger than the maximum admissible error.

Tests for Asymptotic Behavior

If Q_{n_1} and Q_{n_2} are polynomial integration formulas with degrees of accuracy D_1 and D_2 respectively, a test for *asymptotic error behavior* is often used to increase the reliability and accuracy of the error estimation procedure based on $Q_{n_1} - Q_{n_2}$. Such tests are carried out in order to determine whether the application of a formula with a higher degree of accuracy D_2 can actually be expected to result in a more precise integral estimate, i. e., whether or not f is sufficiently smooth and B is sufficiently small. If an asymptotically predictable error behavior is observed, then specific knowledge of the convergence behavior such as

$$\left|Q_{n_1}(f; B_l) - I(f; B_l)\right| \leq C_1 \,(\text{diam } B_l)^{D_1+1}$$
$$\left|Q_{n_2}(f; B_l) - I(f; B_l)\right| \leq C_2 \,(\text{diam } B_l)^{D_2+1},$$

can be utilized in order to obtain more precise and/or more reliable error estimates.

Tests for asymptotic error behavior are often highly heuristic and usually depend on the underlying integration formulas. Thus, critical algorithmic parameters are often based on experimental data rather than on mathematical reasoning.

Example (Pairs of Gauss-Kronrod Formulas) The 6 subroutines QUADPACK/qk15, QUADPACK/qk21, QUADPACK/qk31, QUADPACK/qk41, QUADPACK/qk51, and QUADPACK/qk61, for instance, use the Gauss formulas G_7, G_{10}, G_{15}, G_{20}, G_{25} and G_{30} with the corresponding Kronrod formulas K_{15}, K_{21}, K_{31}, K_{41}, K_{51} and K_{61} (see Section 12.3.2). The approximate values of the Kronrod formulas are used as the numerical integration result, and the difference between the Gauss and the Kronrod results serves as an error estimate.

The error estimation process for pairs of Gauss-Kronrod formulas in QUADPACK includes the following test for asymptotic error behavior: The *mean value* of f on $[x_{l-1}, x_l]$ is computed numerically,

$$M(f; x_{l-1}, x_l) := \frac{1}{x_l - x_{l-1}} K_{2N+1}(f; x_{l-1}, x_l);$$

the absolute deviation $K_{2N+1}(|f - M|; x_{l-1}, x_l)$ of f from $M(f; x_{l-1}, x_l)$ on $[x_{l-1}, x_l]$ provides a measure for the smoothness of f on $[x_{l-1}, x_l]$. The test for asymptotic behavior compares $|G_N - K_{2N+1}|$ with $K_{2N+1}(|f - M|; x_{l-1}, x_l)$. If this ratio is small then the difference between the two quadrature formulas is small when compared to the variation in f on $[x_{l-1}, x_l]$, i.e., the discretization in the quadrature formulas is sufficiently fine in relation to the smoothness of f. In this case K_{2N+1} can be expected to yield a better approximation than G_N.

The reliability of error estimation procedures may be jeopardized by *phase factor* effects (Lyness, Kaganove [274]). While *bounds* for the error $|Q_N f - If|$ often decrease monotonically as $N \to \infty$, the actual error does not necessarily behave in this way. This can eventually lead to values of $|Q_{n_1} f - Q_{n_2} f|$ which are too small when compared to $|Q_{n_2} f - If|$, resulting in a significant *underestimation* of the true error and hence making the algorithm in fact unreliable.

Example (Phase Factor Effects) All members of the parameterized family of functions

$$f_{\lambda,\mu}(x) := \frac{10^{-\mu}}{(x - \lambda)^2 + 10^{-2\mu}}$$

have a peak at $x = \lambda$. Integrating $f_{\lambda,\mu}$ over the interval $[-1, 1]$ gives

$$\int_{-1}^{1} f_{\lambda,\mu}(x)\,dx = \arctan(10^\mu(1 - \lambda)) - \arctan(10^\mu(-1 - \lambda)).$$

This integral was approximated numerically using the two Clenshaw-Curtis formulas Q_7 and Q_9. Figure 12.14 shows the absolute errors $Q_7 f_{\lambda,\mu} - I f_{\lambda,\mu}$ and $Q_9 f_{\lambda,\mu} - I f_{\lambda,\mu}$ for $\mu = 0.35$

as a function of the position of the peak $\lambda \in [-1, 1]$. In general, it can be seen that $Q_9 f_\lambda$ approximates If_λ significantly better than $Q_7 f_\lambda$ does.

However, $|Q_7 f_\lambda - Q_9 f_\lambda|$ is smaller than $|Q_9 f_\lambda - If_\lambda|$ in about 6% of all cases. Both error curves oscillate around zero, but the respective frequencies are different. As a result, the two error curves intersect at points different from the zeros of $Q_9 f_\lambda - If_\lambda$. Phase factor effects occur at these points.

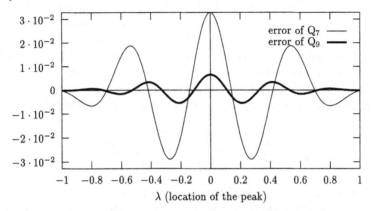

Figure 12.14: Error curves for Clenshaw-Curtis formulas Q_7 and Q_9.

Application of the Same Formula to Different Subdivisions

A simple way to obtain two different estimates Q_{n_1} and Q_{n_2} for If is to use compound rules. This section is limited to one-dimensional formulas. Since the following results are based on the existence of asymptotic error expansions of the Euler-Maclaurin type, they can all be generalized easily to multi-dimensional cubature rules $k^n \times Q_N$, $n \geq 2$.

Q_N is applied first to $[a, b]$ and then to the two subintervals $[a, (a+b)/2]$ and $[(a+b)/2, b]$. Adding up the results of both subintervals leads to a formula Q_{2N} (or Q_{2N-1}) which is presumably more accurate than Q_N. If, for instance, Q_2 is the trapezoidal rule, then

$$Q_3 = \frac{b-a}{2}\left[\frac{1}{2}f(a) + f\left(\frac{a+b}{2}\right) + \frac{1}{2}f(b)\right].$$

In this case, detailed knowledge of the error expansion of the trapezoidal rule (see Theorem 12.3.10) makes it possible to apply Richardson's extrapolation to eliminate one term of the error expansion: If Q_N has the order of convergence p, then[14]

$$Q_{2N} - I \approx \frac{1}{2^p}(Q_N - I). \tag{12.97}$$

Treating (12.97) as an equation leads to

$$I \approx Q_{2N} + \frac{Q_{2N} - Q_N}{2^p - 1} \tag{12.98}$$

---

[14]The symbol $\approx$ is to mean equality except for a remainder term which becomes smaller as the interval $[a, b]$ becomes smaller and f becomes smoother.

and to the following error estimate for the *more accurate* approximate value Q_{2N}:

$$Q_{2N} - I \approx -\frac{Q_{2N} - Q_N}{2^p - 1}.$$

This error estimate can still be arbitrarily large if the assumptions required for (12.97) are *not* satisfied, in particular if f is not a sufficiently smooth function.

The extrapolation formula (12.98) can also be used to construct quadrature formulas with an improved accuracy by interpreting its right-hand side as a new quadrature formula. However, no error estimate for this new formula is available.

If *three* approximate values T_N, T_{2N}, and T_{4N} are calculated according to the Romberg scheme, then the following relation holds, provided (12.47) gives a good characterization of the error structure of the compound trapezoidal rule:

$$\frac{T_{2N} - T_n}{T_{4N} - T_{2N}} \approx \frac{C_2 \left(\frac{h}{2}\right)^2 - C_2 h^2}{C_2 \left(\frac{h}{4}\right)^2 - C_2 \left(\frac{h}{2}\right)^2} = 2^2 = 4. \tag{12.99}$$

The validity of (12.99) can be easily examined in a computer program. Similarly, the values of the kth column of the Romberg scheme can be examined: the respective relations must be of order 2^{k+2}. The reliability of any implementation of the Romberg algorithm can only be improved distinctly by continuing the calculations if the quotients determined in this examination do not deviate too much from the expected value 2^{2k+2}.

Software (Cautious Romberg Integration) A type of Romberg integration which checks the asymptotic error behavior, called the *cautious Romberg integration*, can be found e. g., in cadre (Davis, Rabinowitz [41]), a quadrature program written by de Boor [150]. Another implementation is HARWELL/qa05.

Error Estimation Using Randomized Formulas

The use of *randomized* integration formulas for error estimation purposes is based on the Monte Carlo principle (see Section 5.2.6). The (deterministic) result of an integration formula is regarded as a stochastic entity whose expected value is the desired integral.

If a sequence of stochastically independent integral approximations is generated then the mean value of this sequence converges to the integral value with probability 1 in accordance with the *strong law of large numbers*. Standard statistical methods can be used to establish *confidence intervals* for the integral value. This section deals with the use of randomized rules in connection with lattice rules (see Section 12.4.5). Nevertheless, the obtained results can be generalized to *pseudo-random* formulas (see Section 12.4.4).

It is assumed that the displacement V of a lattice rule $Q_N(V)$ is a random variable with multivariate uniform distribution on $W_N = [0, 1]^N$. The error of the displaced rule is given by (12.85),

$$Q_N(V)f - If = \sum_{m \in L^\perp \setminus \{0\}} c_m e^{2\pi i m \cdot V}.$$

From the expected value

$$\mu(e^{2\pi i m \cdot V}) = \int_{W_N} e^{2\pi i m \cdot v} dv = \prod_{j=1}^{N} \int_0^1 e^{2\pi i m_j v_j} dv_j = 0 \qquad \text{for} \quad m \neq 0$$

together with the linearity of the operator μ (which calculates the expected value) it follows that

$$\mu(Q_N(V)f - If) = 0.$$

Suppose $V_1, V_2, \ldots, V_q$ is a sequence of independent random variables identically distributed with a uniform distribution on W_N. Then the mean value $\overline{Q}f$ of the sequence of the corresponding displaced lattice rules,

$$\overline{Q}f := \frac{1}{q} \sum_{j=1}^{q} Q_N(V_j)f,$$

is an unbiased estimate for If, i.e.,

$$\mu(\overline{Q}f) = If.$$

The standard deviation of $\overline{Q}f$ from the mean value If (i.e., the quadratic mean of the error) is given by

$$\sigma(\overline{Q}f) = \frac{\sigma'}{\sqrt{q}},$$

where $\sigma' := \sigma(Q_N(V_j)f)$ is the standard deviation of $Q_N(V_j)f$, $j = 1, \ldots, q$. This expected size $|\overline{Q}f - If|$ can be approximately calculated using the unbiased estimate of the sample variance,

$$\frac{1}{q(q-1)} \sum_{j=1}^{q} \left(Q_N(V_j)f - \overline{Q}f\right)^2 \approx \sigma^2(\overline{Q}f)$$

By applying the Chebyshev inequality (Brunk [36])

$$P\left(|\overline{Q}f - If| < \varepsilon\right) \geq 1 - \frac{\sigma^2(\overline{Q}f)}{\varepsilon^2},$$

confidence intervals for the approximate value $\overline{Q}f$ can be determined.

Due to the stochastic nature of this error estimation procedure, there is an additional degree of freedom in specifying the error requirement. This additional degree of freedom is the *confidence level* s:

$$P\left(|\overline{Q}f - If| < \varepsilon\right) \geq s.$$

The confidence level s is usually not specified by the user; instead, s is chosen by the integration program itself.

Software (Randomized Error Estimate) The subroutine NAG/d01gbf automatically sets the confidence level $s = 0.9$.

There is a trade-off in the choice of an appropriate value of s: For small values of s, there is a tendency to underestimate the actual integration error, which decreases the reliability of the integration program. For large values of s, the actual integration error tends to be overestimated, which decreases the efficiency of the integration program.

Example (Randomization of Lattice Rules) The method of randomization was applied to the lattice rules described in the example on page 84. The experiments were carried out using the built-in random number generator of a workstation. Prior to numerical integration the integrand was transformed using the *tanh*-transformation (12.23). The following table shows the confidence levels required in order to prevent an underestimation of the integration error for the integration problem (12.22):

j	n	$q = 2$	$q = 3$	$q = 4$	$q = 5$	$q = 6$
0	619	0.89	0.96	—	—	—
1	1238	—	—	—	—	—
2	2476	0.53	—	—	—	—
3	4952	0.94	—	—	—	—
4	9904	0.99	0.96	0.70	0.16	—
5	19808	0.92	—	0.15	0.66	0.18
6	39616	0.86	—	—	—	0.07

A confidence level "—" indicates that the estimate for $\sigma(\overline{Q}f - If)$ is larger than the actual integration error e. In this case, because

$$P(|\overline{Q}f - If| < e) \geq 1 - \frac{\sigma^2(\overline{Q}f - If)}{e^2} < 0,$$

all confidence intervals with a positive probability contain the exact integral value If, i.e., an arbitrary, positive confidence level s is sufficient to prevent an underestimation of the integration error. In most cases, the values of s decrease with increasing values of q. The accuracy of the rule in use, however, has no apparent influence on the size of s.

The maximal value of s is $s = 0.99$. This maximal value occurs when the actual integration error is ten times as large as the estimate for $\sigma(\overline{Q}f - If)$.

12.5.2 Sampling Strategy

An important decision in the process of developing numerical integration algorithms concerns the choice of *sample points*. If all points $\{x_i\} \in B$ are chosen independently of any information about f, the sample

$$S^{na}(f) = (f(x_1), f(x_2), \ldots, f(x_N))$$

conveys *non-adaptive information* about f. The required function values $f(x_1), f(x_2), \ldots, f(x_N)$ can be computed in any order, even *concurrent* (parallel) evaluation is possible. Thus, non-adaptive information is also denoted *parallel information*.

Non-adaptive information is generally used by those parts of integration algorithms which implement integration formulas

$$Q_n(f; \overline{B}) = \sum_{i=1}^{n} \overline{c}_i f(\overline{x}_i), \quad \overline{B} \subseteq B.$$

In order to keep the number of required function values as small as possible and thus enhance efficiency, sample points are chosen according to information about the integrand f: At first, f is evaluated at a sample point x_1 which is determined in advance. The choice of the next point x_2 is based on knowledge of x_1 *and* $f(x_1)$. After the evaluation of $f(x_2)$, x_3 is chosen according to what is known about x_1, $f(x_1)$, x_2 and $f(x_2)$. Generally, the choice of x_i is based on what is currently known, i.e., x_1, $f(x_1)$, x_2, $f(x_2)$, ..., x_{i-1}, $f(x_{i-1})$.

$$S^a(f) = (f(x_1), f(x_2), \ldots, f(x_N))$$

constitutes *adaptive* sample information, which is inherently *sequential*.

In most software products for uniprocessor machines (like QUADPACK [22]), non-adaptive integration modules (implementing integration formulas and error estimates) are applied adaptively to an appropriately chosen sequence $\{B_s\}$ of sub-regions $B_s \subseteq B$ (see Section 12.5.3). Parallel integration algorithms suitable for multiprocessor machines have to take into account inefficiencies due to the sequential nature of adaptive schemes (Krommer, Ueberhuber [258]).

Non-Adaptive Sampling

As with the approximation problems in Chapter 8, the meta-algorithm in Table 12.2 can be used as a starting point for the development of *non-adaptive* integration algorithms based on a given family $\{Q_n\}$ of integration formulas.

   Integration algorithms whose control strategy is determined in advance—as is the case in the meta-algorithm in Table 12.2—need an unnecessarily large computational effort (in terms of the number of function evaluations) for certain types of integrands, like integrands f with boundary singularities (Wasilkowski, Gao [374]).

Software (Non-Adaptive Algorithms) The five (similar) subroutines CMLIB/qng $\approx$ IMSL/MATH-LIBRARY/qdng $\approx$ NAG/d01bdf $\approx$ QUADPACK/qng $\approx$ SLATEC/qng use the non-adaptive Patterson sequence (see Section 12.3.2) of 10-, 21-, 43- and 87-point formulas for the calculation of univariate integrals over bounded intervals. NAG/d01arf uses the Patterson sequence of 1-, 3-, 7-, 15-, 31-, 63-, 127- and 255-point formulas. NAG/d01daf computes two-dimensional iterated integrals. In order to approximate both the outer and inner integrals, the Patterson sequence of 1-, 3-, 7-, 15-, 31-, 63-, 127-, and 255-point formulas is used.

Adaptive Discretization

The aim of constructing integration algorithms is to approximate If with a given error tolerance ε and as few function evaluations as possible. Adaptive algorithms decide *dynamically* how many function evaluations are needed. The information

Table 12.2: Meta Algorithm for *non*-adaptive Integration.

$N := 0$;

$e_0 := \text{HUGE}(e_0)$;

do while $e_N > \varepsilon$

    $N := N +$ increment;

    calculate the approximate value $q_N := Q_N(f; B)$;

    calculate the error estimate $e_N := E_N(f; B)$ *for* q_N

end do

for such decisions is derived from numerical experiments based on the integrand. In general, no a priori information about this decision process is available. The efficiency and reliability of such algorithms therefore depends on the applied subdivision strategy.

The decision as to whether or not a subregion has to be further subdivided is based on either *local* or *global* knowledge. This leads to local and global subdivision strategies respectively. Local knowledge is based only on the considered subregion, whereas global knowledge is based on knowledge about *all* subregions of the integration region.

In any case, the *depth* of the subdivision process is determined dynamically.

12.5.3 Adaptive Integration Algorithms and Programs

Adaptive integration algorithms and programs usually contain the following components:

1. A *basic integration module* receives an *input signal* (the values of the integrand sampled at certain points $x_i \in B_s \subseteq B$ of the integration region) and calculates an approximate value for the integral

$$Q(f; B_s) \approx I(f; B_s)$$

as well as a suitable error estimate $E(f; B_s)$, which ought to satisfy the inequality

$$|Q(f; B_s) - I(f; B_s)| \leq E(f; B_s)$$

in as many cases as possible.

2. An *identification module* is responsible for the steady collection and processing of data (approximate values of the integral and error estimates in subregions) received from the basic module. It provides an a posteriori data supplement (e. g., the calculation of appropriate values for the estimation of the asymptotic error behavior).

3. Information obtained in this way is made available to a *decision module*, which implements a control strategy for an adaptive modification of the input data received by the basic integration module. The intervention required for implementing this control strategy is carried out in a *modification module* by adjusting the corresponding control parameters. These parameters mainly decide *where* the function should be evaluated next; moreover, they may contain additional information for deciding which formula is to be used next.

Structure of an Adaptive Program

The overall structure of an adaptive integration program has a striking logical similarity to the basic structure of a von Neumann computer: it has

1. a *processor* (*arithmetic logic unit*) which executes calculation and comparison operations; in an integration program the data consists of samples of the integrand and information deduced from these function values (approximate values for the integral, error estimates, etc.);

2. *memory* which is responsible for data storage (integration programs usually administer their data autonomously in heaps, lists, etc.);

3. an *input processor* which forms an interface between the integration program and its calling program as well as the module which defines the integrand; and

4. an *output processor* which is responsible for the summary of all collected and processed information and its transmission to the calling program.

The meta-algorithm in Table 12.3 is an abstract description of the mechanisms involved in adaptive integration. It can be used, for instance, as a starting point for the development of adaptive integration algorithms based on a given formula Q_N with an error estimator E.

Data Collection

During an adaptive integration process, numerical calculations are carried out on subregions[15] $B_s \subseteq B$. Various types of data are associated with these subregions:

1. parameters determining B_s, like the vertices of an n-dimensional simplex;

2. approximations of integrals $Q(f; B_s) \approx I(f; B_s)$;

3. error estimates $E(f; B_s) \approx |Q(f; B_s) - I(f; B_s)|$; and

4. additional information, such as

---

[15]The procedure of numerical integration can be terminated after just *one* execution of the integration module; in that case no subregions are processed, rather the overall integration region is used as a whole.

Table 12.3: Meta Algorithm for *Adaptive* Integration.

$q := Q(f; B);\quad e := E(f; B);$

insert (B, q, e) into the data structure;

do while $e > \varepsilon$

    choose an element of the data structure (with index s);

    subdivide the chosen region B_s into subregions

        $B_l, \quad l = 1, 2, \ldots, L;$

    calculate approximate values for integrals over $B_1, \ldots, B_L$

        $q_l := Q_N(f; B_l), \quad l = 1, 2, \ldots, L;$

    calculate corresponding error estimates

        $e_l := E(f; B_l), \quad l = 1, 2, \ldots, L;$

    remove the old data (B_s, q_s, e_s) from the data structure;

    insert $(B_1, q_1, e_1), \ldots, (B_L, q_L, e_L)$ into the data structure;

    $q := \sum_i q_i; \quad e := \sum_i e_i$

end do

(i)   the number of subdivisions which have led to the subregion,

(ii) the eligibility of B_s for further subdivision (depending on whether the status of B_s is *active* or *inactive*),

(iii) function values which could be used in later calculations.

Convergence Testing

Testing for convergence in the meta-algorithm—expressed by the statement **do while** $e > \varepsilon$—corresponds to testing the *absolute* error provided that the accuracy requirement ε is a *constant*. However, most integration programs use one of the following alternative definitions, where ε depends on the current integral value:

1. $\varepsilon := \varepsilon_{abs} + \varepsilon_{rel} |q|$

2. $\varepsilon := \max(\varepsilon_{abs}, \varepsilon_{rel} |q|)$

3. $\varepsilon := \max(\varepsilon_{abs}, \varepsilon_{rel} q_{abs}), \quad$ where $\quad q_{abs} := Q(|f|; B).$

The first two definitions are *combinations* of a pure test for the convergence of the *absolute* error

$$\varepsilon_{rel} = 0 \quad \Rightarrow \quad |E(f; B)| < \varepsilon_{abs}$$

and a pure test for the convergence of the *relative* error

$$\varepsilon_{abs} = 0 \quad \Rightarrow \quad \frac{|E(f; B)|}{|Q(f; B)|} < \varepsilon_{rel}. \tag{12.100}$$

Convergence tests of the strictly relative form (12.100) often lead to extremely large numbers of function evaluations if the *relative* condition number of the integration problem is large, i.e., if $I(f; B) \approx 0$. Unless all the values of f are *small*, the approximate value $Q(f; B) \approx 0$ is due to cancellation. This makes the convergence test (12.100) extremely unreliable.

The third test in the list uses a relative L_1 convergence test instead of (12.100):

$$\varepsilon_{\text{abs}} := 0 \quad \Rightarrow \quad \frac{|E(f; B)|}{Q(|f|; B)} < \varepsilon_{rel}, \tag{12.101}$$

where $Q(f; B)$ is replaced by $Q(|f|; B)$ in order to avoid cancellation. The additional calculation of $Q(|f|; B)$ causes only a minor increase in the overall computational effort as no additional function evaluations are needed.

Although (12.101) is more reliable, this definition of the relative error is *unfamiliar* to most users. The estimates of the relative error in (12.100) and (12.101) coincide only if the integrand is strictly positive or strictly negative:

$$f(x) > 0 \quad \text{for all} \quad x \in B \quad \text{or} \quad f(x) < 0 \quad \text{for all} \quad x \in B.$$

Subdivision of the Domain of Integration

The type of spatial subdivision schemes used in modern integration programs depends on the form of the integration region and on the dimension n of the integral.

One-dimensional integration (quadrature) programs are usually based on *bisecting* the integration intervals. Some programs (like DQAINT [184]) use an interval *trisection*.

For multi-dimensional integration the domain of integration B is often divided into more than two parts. For instance, a two-dimensional triangular domain is often subdivided into *four* congruent subtriangles by bisecting the sides of the triangle (Berntsen, Espelid [110]).

Local Subdivision Strategy

The local subdivision strategy is characterized by two properties:

1. there are *active* and *inactive* subregions, and
2. the active-inactive classification of regions is based on local information.

Every subregion $B_s \subseteq B$ is assigned an individual tolerance

$$\varepsilon_{\text{abs}}(B_s) := \frac{\text{vol}(B_s)}{\text{vol}(B)} \varepsilon_{\text{abs}}, \tag{12.102}$$

which is proportional to its volume $\text{vol}(B_s)$. By splitting the tolerance in this way, it is possible to process subregions of B separately. As soon as the inequality

$$|E(f; B_s)| \leq \varepsilon_{\text{abs}}(B_s)$$

is satisfied, the region B_s is not subdivided any further; it is rendered *inactive*.
If the individual error estimates

$$|Q(f; B_s) - I(f; B_s)| \leq |E(f; B_s)|$$

hold in *all* subregions B_s obtained by an adaptive subdivision of B then for the
overall error

$$
\left| \sum_{B_s} Q(f; B_s) - \int_B f(x)\, dx \right| = \left| \sum_{B_s} \left(Q(f; B_s) - \int_{B_s} f(x)\, dx \right) \right|
$$

$$
\leq \sum_{B_s} \left| Q(f; B_s) - \int_{B_s} f(x)\, dx \right| \leq \sum_{B_s} E(f; B_s)
$$

$$
\leq \sum_{B_s} \frac{\mathrm{vol}(B_s)}{\mathrm{vol}(B)} \varepsilon_{\mathrm{abs}} = \varepsilon_{\mathrm{abs}}.
$$

In single-processor computers, active subregions are often processed in a fixed
order, which simplifies the management of the respective data belonging to each
subregion.

In the case of a one-dimensional integral, the usual choice of an a priori order
for processing active subintervals is from left to right. However, other orders of
processing active subintervals are conceivable. In Rice [325] it is demonstrated
that the order in which active subregions are processed in locally adaptive algorithms is not critical for the efficiency of the algorithm.

It has to be noted that splitting the tolerance by means of (12.102) is only
possible for an *absolute* error requirement $\varepsilon_{\mathrm{abs}}$. It is *not* possible to split up
a *relative* error requirement $\varepsilon_{\mathrm{rel}}$ in the same way, since cancellation may occur
when the results of different subregions are added. As a result, the overall integral
estimate may fail to satisfy the overall relative error requirement $\varepsilon_{\mathrm{rel}}$ even if the
individual relative error requirements of all subregions are satisfied.

Software, (Locally Adaptive Algorithms) The subroutine NAG/d01ahf uses a locally
adaptive subdivision strategy for the computation of univariate integrals over bounded intervals. The basic module implements a non-adaptive Patterson sequence of 1-, 3-, 7-, 15-, 31-,
63-, 127-, and 255-point formulas. The convergence of these formulas is accelerated by using
the ε-algorithm (see Section 12.3.5).

Global Subdivision Strategy

With a global subdivision strategy there are *no inactive* regions, i. e., *all* regions
are eligible for further subdivision at any time during the algorithmic procedure
(Malcolm, Simpson [281]).

The processing order of the subregions is determined dynamically, i. e., it is
not fixed in advance. The subregion B_s which possesses the largest (absolute)
error estimate at the moment of decision is selected for further subdivision:

$$e_s := \max\{e_1, e_2, \ldots, e_k\}.$$

The integration programs based on global subdivision strategies are the most efficient and reliable programs available for conventional (von Neumann) computers.

All adaptive programs in QUADPACK [22], for instance, are based on global
subdivision strategies. Many programs contained in libraries such as IMSL, NAG
(and others) use QUADPACK subroutines. So practically all adaptive integration
programs are based on this kind of subdivision. The following two sections will
describe some of these programs.

12.5.4 Software for Univariate Problems: Globally Adaptive Integration Programs

CMLIB/qag $\approx$ IMSL/MATH-LIBRARY/qdag $\approx$ NAG/d01auf $\approx$ QUADPACK/qag $\approx$
SLATEC/qag are subroutines which use a globally adaptive subdivision strategy
for the evaluation of univariate integrals over finite intervals. The subroutines
CMLIB/qage $\approx$ QUADPACK/qage $\approx$ SLATEC/qage return additional information
about the integration process (e.g., a list of the subintervals chosen by the algorithm). The user can choose between a number of basic modules which implement
different pairs of Gauss-Kronrod formulas. The subroutine NAG/d01akf (which
is based on QUADPACK/qag) provides only one pair of Gauss-Kronrod formulas.

Integrands with Unknown Singularities

The subroutines CMLIB/qags $\approx$ IMSL/MATH-LIBRARY/qdags $\approx$ NAG/d01ajf $\approx$
QUADPACK/qags $\approx$ SLATEC/qags use a subdivision strategy based on a pair of
10/21-point Gauss-Kronrod formulas for the evaluation of intervals over bounded
intervals. In addition, the ε-algorithm is used to accelerate the convergence of
integral approximations (see Section 12.3.5).

The subroutines CMLIB/qagse $\approx$ QUADPACK/qagse $\approx$ SLATEC/qagse return
additional a posteriori information about the integration process (e.g., a list
containing the subintervals actually used).

The subroutine NAG/d01atf only differs from NAG/d01ajf in the way the
integrand f is specified. The function is provided by the user in the form of a
subroutine which evaluates the integrand for several abscissa values at the same
time. If the evaluation of the integrand can be vectorized, the computation time
on vector computers can be reduced significantly in this way.

The programs TOMS/691/qxg and TOMS/691/qxgs are derived from the subroutines QUADPACK/qag and QUADPACK/qags respectively. However, instead of
using a pair of Gauss-Kronrod formulas, the basic module of those routines uses
an embedded sequence of four integration formulas Q_1, Q_2, Q_3, Q_4 with an increasing degree of accuracy. Firstly, the integral approximation and the corresponding
error estimate are determined for each subinterval using Q_1 and Q_2. If the error
estimate is large compared to the machine accuracy, then Q_3 is also evaluated
and the resulting error is estimated. If the accuracy is improved significantly by
changing from Q_2 to Q_3, then Q_4 is also determined. This strategy is based on
the following empirical result: With adaptive algorithms, formulas with large ab-

scissa numbers prove to be optimal for smooth functions, whereas formulas with smaller abscissa numbers are more efficient for functions with peaks, jump discontinuities, etc. Changing from Q_2 to Q_3 can be expected to yield a significant increase in accuracy only if the integrand is smooth. Hence Q_4, i.e., the formula with the largest number of abscissas, is calculated for these integrands only. For functions with peaks, jump discontinuities etc. only the formula Q_3—which comprises many fewer abscissas than Q_4—is evaluated.

Singularities with Known Location

The five subroutines CMLIB/qagp $\approx$ IMSL/MATH-LIBRARY/qdagp $\approx$ NAG/d01alf $\approx$ QUADPACK/qagp $\approx$ SLATEC/qagp use the same integration algorithm as CMLIB/qags; but in addition, they enable the user to specify the location of known singularities in the interior of the integration region. The programs subdivide the integration interval at those points, improving the convergence using the ε-algorithm. The subroutines CMLIB/qagpe $\approx$ QUADPACK/qagpe $\approx$ SLATEC/qagpe return additional a posteriori information about the integration process.

Algebraic-Logarithmic End-Point Singularities

The subroutines CMLIB/qaws $\approx$ IMSL/MATH-LIBRARY/qdaws $\approx$ NAG/d01apf $\approx$ QUADPACK/qaws $\approx$ SLATEC/qaws are tailor-made for the efficient evaluation of univariate integrals over bounded intervals with *algebraic* or *algebraic-logarithmic end-point singularities*. The integration module uses a pair of 7/15-point Gauss-Kronrod formulas for subintervals containing none of the endpoints of the integration interval, and for the boundary intervals themselves two modified Clenshaw-Curtis formulas, one with 13 and the other with 25 abscissas, are employed. The subroutines CMLIB/qawse $\approx$ IMSL/MATH-LIBRARY/qdawse $\approx$ QUADPACK/qawse $\approx$ SLATEC/qawse return additional a posteriori information about the integration process.

Trigonometric Weight Functions

The subroutines CMLIB/qawo $\approx$ IMSL/MATH-LIBRARY/qdawo $\approx$ NAG/d01anf $\approx$ QUADPACK/qawo $\approx$ SLATEC/qawo are used for the evaluation of integrals with the *trigonometric weight functions* $\cos(\omega x)$ or $\sin(\omega x)$ over bounded intervals. The integration module uses two modified Clenshaw-Curtis formulas, one with 13 and the other with 25 abscissas, for subintervals whose lengths exceed a certain ω-dependent value, whereas otherwise a pair of 7/15-point Gauss-Kronrod formulas is applied. In addition, the ε-algorithm is used to accelerate the convergence of the integral approximation. The equivalent subroutines CMLIB/qawoe $\approx$ IMSL/MATH-LIBRARY/qdawoe $\approx$ QUADPACK/qawoe $\approx$ SLATEC/qawoe return additional a posteriori information about the integration process.

The quadrature subroutines CMLIB/qawf $\approx$ IMSL/MATH-LIBRARY/qdawf $\approx$ NAG/d01asf $\approx$ QUADPACK/qawf $\approx$ SLATEC/qawf evaluate integrals over *semi-infinite* intervals $[a, \infty)$ with the *trigonometric weight functions* $\cos(\omega x)$ or $\sin(\omega x)$.

The globally adaptive subroutine QUADPACK/qawo (or an equivalent program) is applied to finite subintervals $[a_k, b_k] := [a + (k-1)c, a + kc]$, $k = 1, 2, \ldots$, where c is chosen so that the integrals

$$I_k f := \int_{a_k}^{b_k} w(x) f(x)\, dx$$

of a monotone function f have alternating signs. The ε-algorithm is used to accelerate the convergence of the resulting alternating series

$$I f = \sum_{k=1}^{\infty} I_k f.$$

The subroutines CMLIB/qawfe $\approx$ QUADPACK/qawfe $\approx$ SLATEC/qawfe return additional information about the integration process.

Unbounded Integration Regions

The subroutines CMLIB/qagi $\approx$ IMSL/MATH-LIBRARY/qdagi $\approx$ NAG/d01amf $\approx$ QUADPACK/qagi $\approx$ SLATEC/qagi use a globally adaptive subdivision strategy to evaluate integrals over *unbounded* intervals. The unbounded integration interval B is first transformed into $[0,1]$: for $B = [a, \infty)$ the transformation $x = \psi(\overline{x}) = (\overline{x} - a)/(\overline{x} + a)$ is used; for $B = (-\infty, \infty)$ the integration interval is divided at the origin and the aforementioned transformation is applied to both subintervals. The transformed integration problem is then solved with the slightly modified program QUADPACK/qags: the pair of 10/21-point Gauss-Kronrod formulas is replaced by a pair of 7/15-point Gauss-Kronrod formulas. The reduced number of abscissas in this pair of Gauss-Kronrod formulas proves to be more efficient for integrands with singularities, arising, for instance, from the aforementioned transformation. The subroutines CMLIB/qagie $\approx$ IMSL/MATH-LIBRARY/qdagie $\approx$ QUADPACK/qagie $\approx$ SLATEC/qagie return additional information about the integration process.

Cauchy Principal Value Integrals

CMLIB/qawc $\approx$ IMSL/MATH-LIBRARY/qdawc $\approx$ NAG/d01aqf $\approx$ QUADPACK/qawc $\approx$ SLATEC/qawc are programs especially suitable for the calculation of *Cauchy principal value integrals* with the weight function $1/(x - c)$. In each subdivision step the subinterval $[a_s, b_s]$ with the largest error estimate is divided into two subintervals at a point $c_s \in (a_s, b_s)$. For $c \notin [a_s, b_s]$ the division is made at $c_s := (a_s + b_s)/2$. Otherwise, c_s is the center of the longer of the two intervals $[a_s, c]$ and $[c, b_s]$. The division of $[a_s, b_s]$ at the critical point c is thus avoided. The basic module uses two modified Clenshaw-Curtis formulas with 13 and 25 abscissas respectively if $c \in (a_s - d, b_s + d)$, $d = (b_s - a_s)/20$, holds; otherwise, a pair of 7/15-point Gauss-Kronrod formulas is used. The subroutines CMLIB/qawce $\approx$ IMSL/MATH-LIBRARY/qdawce $\approx$ QUADPACK/qawce $\approx$ SLATEC/qawce return additional information about the integration process.

12.5.5 Software for Multivariate Problems: Globally Adaptive Integration Programs

Cubic Regions

The programs CMLIB/adapt $\approx$ NAG/d01eaf $\approx$ NAG/d01fcf $\approx$ TOMS/698/cuhre use a globally adaptive subdivision strategy for evaluating n-dimensional integrals over *cubic regions*. CMLIB/adapt works for dimensions from $n = 2$ to $n = 20$, whereas NAG/d01fcf and TOMS/698/cuhre only allow for $n = 2$ to $n = 15$. NAG/d01fcf and TOMS/698/cuhre divide the respective sub-cube with the largest error estimate along a plane $x_k = $ const into two sub-cubes in each subdivision step. The component k is determined from the fact that the fourth divided difference of the integrand in the direction of e_k is maximal. NAG/d01fcf uses a pair of formulas with degrees of accuracy $D = 5$ and $D = 7$ respectively for all dimensions. With TOMS/698/cuhre the user can choose between integration formulas with degrees of accuracy $D = 7$ and $D = 9$. For $n = 2$ and $n = 3$ additional integration formulas with degrees of accuracy $D = 13$ and $D = 11$ are provided.

Triangular Regions

CMLIB/twodq is a program for evaluating two-dimensional integrals over a finite *set of triangles*. In each subdivision step the triangle with the largest error estimate is divided into two sub-triangles along the line between the center of the longest side and the opposite vertex. The basic module comprises two pairs of formulas with the respective degrees of accuracy $D = 6/D = 8$ and $D = 9/D = 11$, one of which can be chosen by the user.

The subroutines TOMS/584/cubtri, TOMS/triex, and TOMS/706/cutri use a globally adaptive subdivision strategy for the evaluation of two-dimensional integrals over a *triangular region*, where the sub-triangle with the largest error estimate is divided into four congruent sub-triangles in each subdivision step. In TOMS/584/cubtri the basic module is based on a pair of formulas with degrees of accuracy $D = 5$ and $D = 8$ respectively. In TOMS/612/triex a pair of formulas with degrees of accuracy $D = 9$ and $D = 11$ is used. In addition the ε-algorithm is used in order to accelerate the convergence of integral approximations. In TOMS/706/cutri the basic module is derived from a formula with degree of accuracy $D = 13$. Moreover, this program can be used for an integration region consisting of a *finite number of triangles*.

Tetrahedra

The subroutine TOMS/720/cutet uses a globally adaptive subdivision strategy in order to evaluate three-dimensional integrals over a finite set of tetrahedra, where the sub-tetrahedra with the largest error estimate is divided into eight congruent sub-tetrahedra in each subdivision step. The basic module is based on a 43-point formula with a degree of accuracy $D = 8$.

12.5.6 Reliability Enhancement

Noise Detection

For the *mathematical* problem of integration, the integrand f is not assumed to be *disturbed* in any way. However, such an ideal integrand cannot be realized on a digital computer for the following reasons:

1. Since the set of machine numbers $\mathbb{F}$ is a *finite* set, the integrand can only be represented by points on a grid, whose fineness is determined by the actual floating-point number system. The deviation of the implemented function $\Box f$ from the "exact" function f is a (pseudo-)stochastic disturbance.

2. If the integrand is defined by complicated expressions, the deviation $\Box f - f$ can be significantly larger than elementary rounding errors.

   An important example of this situation is the recurrent calculation of multi-dimensional integrals using one-dimensional integration programs (e. g., for rectangular or cubic integration regions). If a quadrature program is not used at the lowest level of the recurrence, but uses the results of preceding numerical calculations as the values of the integrand, then these "function values" are most likely disturbed by significant (pseudo-)stochastic noise, which is of the same order as the quadrature error. It is a difficult task to balance the tolerance parameters ε_{abs} and ε_{rel} of these recursive quadrature program calls so as to achieve the best efficiency and reliability (Fritsch, Kahaner, Lyness [204]).

3. If the values of the integrand are the results of measurements (obtained, e. g., by using the analog-digital-converter of a real-time system), "noise", i.e., stochastic disturbances are unavoidable.

These (pseudo-)stochastic disturbances set a limit to the maximum achievable accuracy of the numerical result. If the user of an integration program specifies an accuracy requirement which cannot be achieved due to stochastic disturbances, then the program should be able to recognize this situation without unnecessary computational overhead and should terminate the procedure with an appropriate message. A *noise detector* is used for this purpose.

In general, globally adaptive programs use the monotonic behavior of error estimates to detect stochastic disturbances. A program may count, for instance, how many times a region subdivision does *not* improve the error estimate:

$$\textbf{if} \quad \sum_{l=1}^{L} e_l > e_s \quad \textbf{then} \quad \text{moveup} := \text{moveup} + 1.$$

When this number reaches a fixed upper bound, the algorithm terminates with the message *noisy integrand*. The error estimate returned in the corresponding output parameter informs the user of the *noise level* and of the accuracy which can possibly be achieved in the current situation.

Detection of Non-Integrable Functions

It cannot be neglected that the user of an integration program might try to integrate a *non-integrable* function (like $f(x) = x^{-1}$ on the interval $[0, 1]$). In principle it is not possible to decide whether or not f is an integrable function on the basis of *finite* information

$$S(f) = (f(x_1), f(x_2), \ldots, f(x_N)).$$

Example (Undetectable Pole) The (integrable) function

$$f(x) = 3x^2 + \frac{1}{\pi^4} \log[(\pi - x)^2] + 1$$

has a pole at $x = \pi$,

$$f(x) \to -\infty \quad \text{as} \quad x \to \pi.$$

Even by evaluating f for *all* machine numbers $x \in \mathbb{F}(2, 53, -1021, 1024, \textit{true})$, this pole *cannot* be detected. Figure 12.15 shows the numerical values of f in the neighborhood of $x = \pi$. It can be seen that the implementation of f is strictly positive for *all* $x \in \mathbb{F}$; even the two zeros of this function remain undetected.

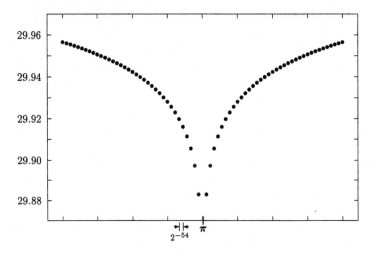

Figure 12.15: Numerical values of the function $f(x) = 3x^2 + \pi^{-4} \log[(\pi - x)^2] + 1, x \in \mathbb{F}(2, 53, -1021, 1024, \textit{true})$.

Precautionary algorithmic measures would impose

1. an *upper limit* on the *computational effort* in the form of a maximal number of function evaluations, specified by the user or by the program;

2. *lower limits* on the sizes of subregions: if the algorithmic refinement of the subregions reduces their size to the distance between machine numbers then no useful numerical integration is possible. Thus, integration programs often have a built-in termination criterion that tests the size of the smallest subregion.

Termination on account of one of the above criteria can possibly result in a premature termination of the algorithm for (improperly) integrable integrands (like $f(x) := x^{-0.99}$ on the interval $[0, 1]$), i.e., it may happen that the diagnosis *non-integrable function* is incorrect.

12.5.7 Multiple Integrands

Some applications require the computation of a large number m of integrals with the same weight function w and the same integration region B, but with different integrands $f_1, f_2, \ldots, f_m \colon B \to \mathbb{R}$:

$$\mathrm{I}f_1 = \int_B w(x) f_1(x)\, dx,$$

$$\vdots$$

$$\mathrm{I}f_m = \int_B w(x) f_m(x)\, dx.$$

Clearly, this problem can be solved by computing the m integrals one after the other. However, a substantial amount of work (required, for example, to manage data structures) can be saved by calculating the m integrals *simultaneously*.

If m integrands with sufficiently similar behavior are integrated simultaneously using one common adaptive algorithm, the norm

$$\|(\mathrm{E}(f_1; B_s), \mathrm{E}(f_2; B_s), \ldots, \mathrm{E}(f_m; B_s))^\top\|$$

of the integrand's error estimates over the subregions B_s is used to make the appropriate subdivision decisions.

Software (Multiple Integrands) The globally adaptive integration program dcuhre (Berntsen et al. [111]), which integrates several integrands simultaneously, uses the maximum norm:

$$\|(\mathrm{E}(f_1; B_s), \mathrm{E}(f_2; B_s), \ldots, \mathrm{E}(f_m; B_s))\|_\infty = \max\{\mathrm{E}(f_1; B_s), \mathrm{E}(f_2; B_s), \ldots, \mathrm{E}(f_m; B_s)\}.$$

The simultaneous computation of m integrals using adaptive algorithms can sometimes lead to serious difficulties. Adaptive algorithms choose their sequence of abscissas according to the integrand behavior observed. If m integrands are processed at once, the sequence of abscissas must (locally) comply with the requirements of the most difficult integrand. To avoid the unnecessary evaluation of (locally) simple integrands, only functions $f_1, \ldots, f_m$ with similar behavior should be integrated simultaneously when using adaptive integration programs.

Chapter 13

Systems of Linear Equations

Am Einfachen, Durchgreifenden halte ich mich and gehe ihm nach,
ohne mich durch einzelne Abweichungen irre leiten zu lassen.[1]

JOHANN WOLFGANG VON GOETHE

Although nearly all actual dependencies are *nonlinear*, linearity assumptions are widely spread throughout engineering and the natural sciences. Often they lead to the simplest models, which are preferred to equivalent, but more complex models.

Many methods (both exact as well as approximate) for analyzing and solving mathematical problems are more appropriate for linear models. This circumstance often leads to the application of linear models even when the real dependency is very different from a linear relation (e. g., in many applications in operations research). The assumption behind this decision is that not taking into account the nonlinearity of the analyzed phenomenon does not have great impact on the results, that the effects of model errors can be compensated for by choosing appropriate coefficients for the linear model, or that it is possible to improve the computed solution later (taking into account nonlinear phenomena). Thus many investigations in engineering and science—even the most complex ones—start with linear models.

The numerical solutions of difficult nonlinear problems are nearly always traced back to the solution of systems of linear equations. For instance, Newton's method is applied to systems of nonlinear algebraic equations, or finite differences methods and finite element methods are used for the numerical solution of partial differential equations. This explains why the solution of linear systems and linear least squares problems has a central position in numerics.

Example (Bearing Structure)   The basic mechanical structures of bridges, buildings, vehicles (frames) etc. are *bearing structures*, systems of rigid bodies (trusses, rods etc.) connected together at nodal points. The simplest mathematical models of such bearing structures are the perfect lattice works (cf. Fig. 13.1). The basic elements of such lattice works are straight trusses, the ends of which are connected together by perfect (frictionless) hinges. The external forces act only on hinges (nodal points) so that the trusses are exposed only to tensile or compressive stress. The results derived from such models are used in practice although some assumptions are never supported by real bearing structures. Instead of using perfect hinges, the parts of actual bearing structures are connected together rigidly.

In determining the forces in trusses (due to their dead weight and external forces), one assumes that the lattice work is in a state of equilibrium (otherwise it would move). The static

---

[1] I adhere to simple, far-reaching things, and follow them without being discouraged by a few deviations.

equilibrium condition for the nodal point i read as

$$F_i + \sum_j s_{ij} = 0, \tag{13.1}$$

where F_i is the external force acting on nodal point i and s_{ij} is the force in the member connecting the vertices i and j. In addition, the forces in the bearings must be considered; they can be interpreted as unknown forces. All the equations (13.1) can be combined into a *system of linear equations*, containing the bar forces s_{ij} and the forces in the bearings as unknowns. The known forces (external forces, weights, bar forces) can be put on the right side of the linear system. An inhomogeneous linear system of the form

$$
\begin{array}{ccccccccc}
a_{11}x_1 & + & a_{12}x_2 & + & \cdots & + & a_{1n}x_n & = & b_1 \\
a_{21}x_1 & + & a_{22}x_2 & + & \cdots & + & a_{2n}x_n & = & b_2 \\
\vdots & & \vdots & & & & \vdots & & \vdots \\
a_{n1}x_1 & + & a_{n2}x_2 & + & \cdots & + & a_{nn}x_n & = & b_n
\end{array} \tag{13.2}
$$

can thus be obtained. In the example depicted in Fig. 13.1, the unknowns x_1 and x_2 are the forces in the bearing (x and y direction). The forces in the members are described by the scalar unknowns x_3 to x_{17}, where the respective index corresponds lexicographically to the indices s_{ij} (e.g., $s_{12} \mapsto x_3$, $s_{13} \mapsto x_4$, $s_{15} \mapsto x_5$ etc.). To keep things simple, the bearing B is designed as a movable hinged bearing, and the force in y direction is the unknown x_{18}.

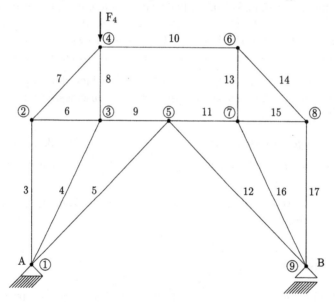

Figure 13.1: Plane lattice work.

The geometry of the lattice work is determined by the angles α, β, and γ, where α denotes the angle between the members 3 and 4 as well as 16 and 17, β denotes the angle between the members 3 and 5 as well as 12 and 17, and γ denotes the angle between the members 6 and 7 as well as 14 and 15. With this notation and the abbreviations

$$
\begin{array}{llllll}
s_\alpha & := & \sin\alpha, & s_\beta & := & \sin\beta, & s_\gamma & := & \sin\gamma, \\
c_\alpha & := & \cos\alpha, & c_\beta & := & \cos\beta, & c_\gamma & := & \cos\gamma,
\end{array}
$$

the forces $x = (x_1, \ldots, x_{18})^\top$ can be obtained from the linear system $Ax = b$, whose right-hand side is given by the vector $b = (0\ 0\ 0\ 0\ 0\ 0\ 0\ F_4\ 0\ \cdots\ 0)^\top$. The coefficient matrix is the following 18×18 matrix:

$$
\begin{pmatrix}
1 & & & s_\alpha & s_\beta & & & & & & & & & & \\
1 & 1 & & c_\alpha & c_\beta & & & & & & & & & & \\
 & & 1 & c_\gamma & & & & & & & & & \mathbf{0} & & \\
 & -1 & & s_\gamma & & & & & & & & & & & \\
 & -s_\alpha & & -1 & & 1 & & & & & & & & & \\
 & -c_\alpha & & & 1 & & & & & & & & & & \\
 & & & c_\gamma & & & -1 & & & & & & & & \\
 & & & s_\gamma & 1 & & & & & & & & & & \\
 & & -s_\beta & & & -1 & & 1 & s_\beta & & & & & & \\
 & & -c_\beta & & & & & & c_\beta & & & & & & \\
 & & & & & -1 & & & & c_\gamma & & & & & \\
 & & & & & & & & 1 & s_\gamma & & & & & \\
 & & & & & -1 & & & & & 1 & s_\alpha & & & \\
 & & & & & & & & -1 & & & c_\alpha & & & \\
 & & & & & & & & & -c_\gamma & -1 & & & & \\
 & \mathbf{0} & & & & & & & & -s_\gamma & & 1 & & & \\
 & & & & & & & s_\beta & & & & s_\alpha & & & \\
 & & & & & & & -c_\beta & & & & -c_\alpha & -1 & 1 &
\end{pmatrix}
$$

There are plenty of high-quality software products for the numerical solution of linear systems, more than for any other field of numerical data processing: the packages LAPACK, TEMPLATES, ITPACK, NSPCG, SLAP, or UMFPACK and the linear algebra parts of the IMSL and NAG libraries. Developing software for the solution of linear systems on one's own would thus be rather uneconomical.

The practical solution of linear systems is preferably achieved by applying existing software. In doing so the user has to pass through (consciously or unconsciously) three stages:

1. A *design stage*, in which a precise problem specification takes place.

2. A *realization stage* for the numerical solution of the problem using computer systems.

3. A *verification stage*, in which the solution obtained from the previous stage is critically reviewed.

13.1 Design Stage

From the beginning, a series of questions which aim at the precise problem specification have to be addressed: *What* is to be computed? *What* are the characteristics of the problem under investigation? etc. Preparing and acquiring basic facts which make it possible to choose the algorithm and/or adequate software products are also part of this stage.

13.1.1 Type of Problem

Whether or not the problem is linear or nonlinear must be considered (cf. Sections 8.5.2 and 8.5.3), as sometimes mistakes are made during the classification of problems.

Example (Polynomial Data Fitting) When fitting polynomials to prescribed data, one task might be, for instance, to adapt an optimal univariate polynomial with given degree d,

$$P_d(x; c_o, c_1, \ldots, c_d) = c_0 + c_1 x + c_2 x^2 + \cdots + c_d x^d,$$

to $k > d+1$ given data points

$$(x_1, y_1), (x_2, y_2), \ldots, (x_k, y_k)$$

with respect to the Euclidean metric. Thus the coefficients of the optimal polynomial $P_d^* \in \mathbb{P}_d$ which has a minimal distance to the data points

$$D(\Delta_k P_d^*, y) = \min \left\{ \sum_{i=1}^{k} (P_d(x_i) - y_i)^2 : P_d \in \mathbb{P}_d \right\}, \tag{13.3}$$

have to be determined.

The determination of P_d^* is a *linear* problem:

1. Each polynomial depends linearly on its parameters $c_0, c_1, \ldots, c_d$:

$$P_d(x; \lambda b_0 + \mu c_0, \lambda b_1 + \mu c_1, \ldots, \lambda b_d + \mu c_d) = \lambda P_d(x; b_0, b_1, \ldots, b_d) + \mu P_d(x; c_0, c_1, \ldots, c_d).$$

2. The search for the minimum in (13.3) is a linear problem, since the minimization of the l_2-distance (by differentiation with respect to the independent variables $c_0, c_1, \ldots, c_d$) can be recast as a linear least squares problem.

Standard Form of Systems of Linear Equations

In order to compute solutions (or to perform different types of analytical investigations) of systems of linear equations using existing software products, the system is almost always transformed into the standard form $Ax = b$.

Definition 13.1.1 (System of Linear Equations, Standard Form)

A set of m equations in n unknowns $x_1, \ldots, x_n$ of the form

$$F(x) = \begin{pmatrix} f_1(x_1, \ldots, x_n) \\ f_2(x_1, \ldots, x_n) \\ \vdots \\ f_m(x_1, \ldots, x_n) \end{pmatrix} = \begin{pmatrix} a_{11}x_1 + a_{12}x_2 + \cdots + a_{1n}x_n \\ a_{21}x_1 + a_{22}x_2 + \cdots + a_{2n}x_n \\ \vdots \quad \vdots \quad\quad \vdots \\ a_{m1}x_1 + a_{m2}x_2 + \cdots + a_{mn}x_n \end{pmatrix} = \begin{pmatrix} b_1 \\ b_2 \\ \vdots \\ b_m \end{pmatrix}$$

(or for short: $Ax = b$) in which the quantities $a_{11}, \ldots, a_{mn}$ and $b_1, \ldots, b_m$ are given, is called a system of linear equations. $A \in \mathbb{R}^{m \times n}$ is the coefficient matrix (system matrix), the vector $b \in \mathbb{R}^m$ is the right-hand side, and $x^* \in \mathbb{R}^n$ is an unknown solution vector, which must simultaneously satisfy all m equations.

Unless stated otherwise, it is assumed throughout the following sections that the elements $a_{11}, \ldots, a_{mn}$ of the matrix A and the components $b_1, \ldots, b_m$ of the vectors b are *real* numbers. Most results and methods remain valid or applicable to cases where the a_{ij} or the b_i are complex.

Quadratic and Rectangular System Matrices

According to the existence and the structure of the solution set and to the choice of suitable solution methods, three categories of matrices are distinguished:

$m = n$

This is a square coefficient matrix in $Ax = b$.

$$\boxed{m = n} \cdot \left| \begin{array}{c} \\ \\ \end{array} \right| = \left| \begin{array}{c} \\ \\ \end{array} \right|$$

A unique solution of such a system exists provided the matrix $A \in \mathbb{R}^{n \times n}$ is regular. Otherwise, depending on the specific constellation of the right hand side b, either there is no solution at all or there is a whole space of solution vectors.

$m < n$

In the case of a rectangular matrix with $m < n$, the number of equations is smaller than the number of unknowns; such systems $Ax = b$ are said to be *under-determined* systems of linear equations.

$$\boxed{m < n} \cdot \left| \begin{array}{c} \\ \\ \\ \end{array} \right| = \left| \begin{array}{c} \\ \\ \end{array} \right|$$

The solution of such a system is always a whole subspace $X \subseteq \mathbb{R}^n$ with a dimension $\dim(X) \geq n - m$.

$m > n$

With $m > n$, the number of unknowns is less than the number of equations; the system $Ax = b$ is said to be a *over-determined* system of linear equations, which in almost all cases has *no* solution.

$$\boxed{m > n} \cdot \left| \begin{array}{c} \\ \end{array} \right| = \left| \begin{array}{c} \\ \\ \end{array} \right|$$

In case of an over-determined system, one often switches over to an approximation problem, in which one seeks the minimum x^* under an l_p-norm of the residual vector $r := Ax - b$:

$$\|Ax^* - b\|_p = \min \{\|Ax - b\|_p : x \in \mathbb{R}^n\}.$$

In most cases the l_2-norm is used to determine the minimum of the residual

$$\min \{\|Ax - b\|_2^2 = \sum_{i=1}^{m} (a_{i1}x_1 + \cdots + a_{in}x_n - b_i)^2 : x \in \mathbb{R}^n\} \qquad (13.4)$$

and hence a *solution* x^* of $Ax = b$ can be found using the *least squares method*. x^* is, in this case, the solution of the linear system $A^\top Ax = A^\top b$, the so-called *normal equations* (cf. Section 10.3.1).

If other l_p-norms are used for solving the linear approximation problem, then the model remains linear, but the determination of the parameters $x_1, \ldots, x_n$ can no longer be achieved by solving a system of linear equations. In this case an iterative minimization method must be used (cf. Chapter 14).

Type and Number of Right-Hand Sides

Homogeneous systems, characterized by $b = 0$, require solution methods (cf. Section 13.7.5) other than inhomogeneous systems $(b \neq 0)$, which occur more often in practice.

If more than one linear problem

$$Ax_1 = b_1, \ Ax_2 = b_2, \ \ldots, \ Ax_k = b_k \tag{13.5}$$

with different right-hand sides, but with the same coefficient matrix A is given, then the common solution of the k equations (13.5) corresponds to the solution of *one matrix equation*

$$AX = B, \qquad A \in \mathbb{R}^{m \times n}, \quad B := [b_1, b_2, \ldots, b_k] \in \mathbb{R}^{m \times k}.$$

Special programs for the efficient solution of matrix equations are available. The above procedure requires that all data in (13.5) are present at the same time; otherwise a multi-stage approach is advisable: First, the factorization of the matrix A is computed; this part of the solution process requires $O(n^3)$ floating-point operations. Next, using this factorization, the systems are solved one by one. This stage requires $O(n^2)$ floating-point operations for each system.

Data Type

The *data types* of $a_{11}, a_{12}, \ldots, a_{mn}, \ b_1, \ldots, b_m$, and $x_1, \ldots, x_n$ are important attributes for characterizing the problem. Above all, the floating-point data types REAL, COMPLEX etc. are of importance to the choice of suitable software. If, on the other hand, an integer solution (with integer coefficients) is sought, an entirely different problem type arises.

Data Errors

An important classification of the problems under consideration can be done according to the nature and magnitude of *data errors*. In systems of linear equations, and in particular in over-determined systems, the vector b (in most cases) and sometimes the coefficients $a_{11}, a_{12}, \ldots, a_{mn}$ as well, are contaminated with inaccuracies (e. g., as a result of errors in measurements etc.). In such cases the condition of the linear system should be analyzed in the design stage to obtain information about the best possible accuracy attainable for x^*.

The solution of over-determined systems affected with errors using the *least squares method* (13.4) is useful only in cases in which the vector b is affected with significant data errors, but any perturbations of the coefficients $a_{11}, \ldots, a_{mn}$ are no greater than the order of elementary rounding errors. In such cases, the minimization of the Euclidean distance only yields an optimal solution (*maximum likelihood estimation*) if the errors in b are independently distributed random variables originating from a *single* Gaussian distribution.

If there are *outliers* among the data errors, then the distribution underlying the stochastic data perturbations is a compound. It is then more suitable to determine a solution x^* by minimizing the l_p-norm of the residual vector

$$\min\{\|Ax - b\|_p^p = \sum_{i=1}^{m}(a_{i1}x_1 + \cdots + a_{in}x_n - b_i)^p : x \in \mathbb{R}^n\} \qquad \text{with} \quad p \in (1,2)$$

(cf. Section 8.6.6). *Weighting factors*

$$\min\{\|Ax - b\|_{2,w}^2 = \sum_{i=1}^{m} w_i\,(a_{i1}x_1 + \cdots + a_{in}x_n - b_i)^2 : x \in \mathbb{R}^n\}$$

also allow the algorithm to handle data with non-uniform error distributions (accuracy) better than (13.4).

In many problems occurring in practical applications, not only the vector b but also the elements $a_{11}, a_{12}, \ldots, a_{mn}$ of the coefficient matrix are affected with errors. In such cases the determination of a solution using *orthogonal approximation* (cf. Section 8.6.7) is preferable to the least squares method.

13.1.2 Structural Properties of System Matrices

Specific structural properties of the matrices of linear systems are exploited in the development of efficient algorithms and software. Problem-oriented software requires the knowledge of specific structural characteristics of the matrix A *before* the linear system is solved.

Symmetry and Definiteness

Given a square matrix $A \in \mathbb{R}^{n \times n}$, the property of *symmetry*

$$a_{ij} = a_{ji} \qquad \text{for all} \quad i, j \in \{1, 2, \ldots, n\},$$

can easily be investigated. By choosing appropriate storage techniques, the storage requirements of such matrices can nearly be halved. If specific software for the solution of symmetric systems is used, then the computational effort (runtime) can be halved as well.

Even more advantages can be obtained if the matrix is additionally (*positive*) *definite*,

$$\langle Ax, x \rangle = \sum_{i=1}^{n}\sum_{j=1}^{n} a_{ij}x_i x_j \begin{cases} > 0 & \text{for} \quad x \neq 0 \\ = 0 & \text{for} \quad x = 0. \end{cases}$$

Unlike symmetry, positive definiteness cannot be detected directly by a simple examination of the matrix (cf. Section 13.5.3).

Percentage of Nonzero Elements and Sparsity Structures

When working with really large systems consisting of thousands of equations
and more, the number of nonzero matrix elements and their positions within the
matrix are of significance. If a great number of matrix elements are zero, the
matrix is said to be *sparse*. A linear system $Ax = b$ with a sparse coefficient
matrix $A \in \mathbb{R}^{n \times n}$ of order $n = 100\,000$ can be solved using appropriate software
(cf. Chapter 16). Systems with dense matrices (matrices in which almost all
elements $a_{ij} \neq 0$) of this order cannot be practically solved; even a supercomputer
would fail.

In the case of sparse matrices, the *sparsity structure* of the matrix is also of
great importance to the choice of algorithms and software. *Band matrices* are of
special interest, since very efficient programs are available for them.

13.1.3 Types of Solutions

Systems of linear equations are numerical problems provided with certain *desir-able* features:

(i) The problem specification is established with algebraic data (cf. Chapter 4)—
matrices and vectors.

(ii) Finitization (truncation) of the solution method is not necessary. For example, Gaussian elimination computes the solution vector of a system of linear
equations in a finite number of steps (arithmetic operations).

There is an extremely large number of arithmetic operations needed in the numerical solution of large systems of linear equations. A dense system with 1200
equations already requires a computational effort of more than 10^9 floating-point
operations (1 Gflop). As a result, rounding errors, which may affect the solution
vector considerably, occur. Thus the question arises, which numerical solutions
or solution accuracies are acceptable to the user. Therefore, even in the case of
linear systems it is necessary to distinguish between the mathematical problem
$Ax = b$ and the numerical problems arising due to the imposition of accuracy
requirements.

Linear systems with a system matrix which is not regular create an unusual
difficulty. From the mathematical-analytical point of view, the distinction between the cases

$$A \text{ is regular} \qquad or \qquad A \text{ is singular}$$

is absolutely clear. But because of data and rounding errors, poorly conditioned
problems lead to a fuzzy decision between *numerical regularity* and *numerical
singularity*. Actually, this decision can be made only if additional information
(e.g., about the type and size of data errors) is available. In the case of a numerically singular coefficient matrix A, the special position of vector b in the space
$\mathbb{R}^m$ must be taken into consideration in order to come to a useful solution of

$Ax = b$. A possible definition of a solution in the case of a numerically singular matrix is the *pseudo-normal solution* (cf. Section 13.7.4)

$$x_0 = A^+ b,$$

which is defined by means of the *generalized inverse* A^+ (cf. Section 13.7.3). In practice, the *singular value decomposition* (cf. Section 13.7) is an important aid for determining A^+ and subsequently a solution of the linear system.

If the computed solution is required to be within a small neighborhood of the true solution, then special algorithms and interval arithmetic must be used.

13.1.4 Algorithms and Software Requirements

When software for the solution of systems of linear equations is selected, the following general (see Chapter 6 and Chapter 7) as well as specific quality criteria apply.

Robustness

Programs used for solving linear systems, the characteristics of which are not completely and exactly known in advance, must be highly robust. For instance, they must be able to recognize ill-conditioned or numerically singular systems and must provide the user with adequate information (condition and/or accuracy estimates etc.).

Efficiency

Efficiency is of such a great importance when solving very large linear systems that often the practical solvability of a problem depends on the degree of exploitation of the available computer resources. For very large, sparse matrices, not only the computational efficiency but also the *storage efficiency* is of critical importance (cf. Chapter 16).

Usability

Software for solving linear systems (e.g., LAPACK programs) is generally quite user friendly. Only some kinds of problems and classes of algorithms can cause difficulties. For example, programs designed for the iterative solution of large linear systems with sparse coefficient matrices (e.g., the TEMPLATES programs from Section 16.16.2) are not suitable for black box applications, as the user must have preliminary knowledge of starting iterates, parameters of the algorithm, or preconditioning.

Portability

The highest possible degree of portability together with good efficiency distinguish the programs of LAPACK, whose excellent qualities make it the best software product for small to medium sized systems with dense coefficient matrices.

For large linear systems with sparse coefficient matrices, portability is prevented primarily by the different storage schemes. For such cases, the TEMPLATES package shows how to make generally available algorithms and programs in a portable manner (cf. Section 16.16.2).

13.2 Realization Stage

The *choice* of suitable software products and their *application* to the problem given are a part of the second stage of the numerical solution of linear systems. Before this can occur, one has to choose one of the two large classes of algorithms:

Direct methods are based on the factorization of the coefficient matrix. In the (hypothetical) absence of rounding errors, these methods produce the exact solution of a linear system after a finite number of arithmetic operations.

Iterative methods are based on the iterative determination of a fixed-point of a linear system or the minimization of quadratic functions. In most cases (even in the absence of rounding errors), they require an infinite number of operations to determine an (exact) solution.

As a guiding rule for the choice between these two alternatives, the following advice can be useful:

- Small systems are best solved using LAPACK programs (or using the corresponding descendants in the IMSL and NAG libraries), which are all based on direct methods. For large systems with band matrices, special LAPACK programs are available.

- Large, sparse systems whose coefficient matrices have a general sparsity structure are best solved using iterative methods (e.g., using programs from ITPACK, SLAP or TEMPLATES; cf. Chapter 16).

13.3 Verification Stage

In the third stage of the numerical solution of linear systems, as to whether a solution with the desired accuracy has actually been achieved has to be investigated.

In practice, for most linear systems, there is no a priori information available about the condition of the system. It is tempting to be satisfied too quickly with the computed solution if it is compared only to some vague idea of the expected solution. An uncritical acceptance of a numerical solution can lead to extremely inaccurate results. On the other hand, an objective assessment of the results obtained can be reached with reasonable additional effort. To this end, special programs have been developed which compute a condition estimate during the solution process.

13.4 Mathematical Foundations

The following section gives a brief overview of the methods used in the analysis
and solution of systems of linear equations. Detailed descriptions of the mathematical foundations can be found in Stewart [360], Strang [76], or Horn and
Johnson [59], [60].

13.4.1 Linear Spaces

In Section 8.6.1 the basic notions of *linear spaces* (*vector spaces*) are introduced,
and the linear combination of vectors $v_1, \ldots, v_k$ of the n-dimensional vector space
V_n is defined. Each vector $u \in V_n$ of the form

$$u = \alpha_1 v_1 + \cdots + \alpha_k v_k$$

with arbitrary scalars $\alpha_1, \ldots, \alpha_n \in \mathbb{R}$ is is said to be a *linear combination* of the
vectors $v_1, \ldots, v_k \in V_n$.

Definition 13.4.1 (Linear Span)   *The set*

$$\text{span}\{v_1, \ldots, v_k\} := \left\{ \sum_{j=1}^{k} \alpha_j v_j : \alpha_j \in \mathbb{R}, \ v_j \in V_n \right\}$$

*of all linear combinations of the vectors $v_1, \ldots, v_k \in V_n$ is a subspace of the vector
space V_n. This subspace is said to be spanned by $v_1, \ldots, v_k$; it is called the linear
closure of $v_1, \ldots, v_k$.*

Definition 13.4.2 (Linear Independence of Vectors)   *The k vectors
$v_1, \ldots, v_k \in V_n$ are said to be linearly independent if*

$$\sum_{j=1}^{k} \alpha_j v_j = 0 \quad \Longleftrightarrow \quad \alpha_1 = \alpha_2 = \cdots = \alpha_k = 0;$$

otherwise they are linearly dependent.

The set $M = \{v_{i_1}, \ldots, v_{i_m}\}$ is a *maximal linearly independent subset* of the set[2]
$\{v_1, \ldots, v_k\} \subset V_n$ if M consists of linearly independent vectors and if M is not
a proper subset of any other linearly independent subset of $\{v_1, \ldots, v_k\}$. If M is
maximal, then

$$\text{span}\{v_{i_1}, \ldots, v_{i_m}\} = \text{span}\{v_1, \ldots, v_k\},$$

and M is a *basis* for the linear space $\text{span}\{v_1, \ldots, v_k\}$.

If $S \subseteq V_n$ is a subspace of V_n, then there must exist linearly independent vectors
$u_1, \ldots, u_k \in S$ which span S:

$$S = \text{span}\{u_1, \ldots, u_k\}.$$

---

[2]It should be noted that no assumptions about k have been made; hence, $k > n$ can also
hold.

Definition 13.4.3 (Dimension, Coordinates) *Any basis for the subspace $S \subseteq V_n$ has the same unique number of vectors. The number k of vectors in a basis is called the dimension of the subspace and is denoted $\dim(S)$. Each vector $v \in S$ may be written uniquely as a linear combination*

$$v = \alpha_1 u_1 + \cdots + \alpha_k u_k$$

with respect to the basis $M = \{u_1, \ldots, u_k\}$. The coefficients $\alpha_1, \ldots, \alpha_k$ are called the coordinates of the vector v with respect to the basis M.

The system of the n unit vectors

$$
e_1 := \begin{pmatrix} 1 \\ 0 \\ 0 \\ \vdots \\ 0 \end{pmatrix}, \quad
e_2 := \begin{pmatrix} 0 \\ 1 \\ 0 \\ \vdots \\ 0 \end{pmatrix}, \quad \ldots, \quad
e_n := \begin{pmatrix} 0 \\ 0 \\ 0 \\ \vdots \\ 1 \end{pmatrix} \tag{13.6}
$$

is obviously a basis of V_n: each vector $x \in V_n$ with the components $\xi_1, \xi_2, \ldots, \xi_n$ can be written as

$$x = \xi_1 e_1 + \xi_2 e_2 + \cdots + \xi_n e_n.$$

In this special case, the components of a vector are equal to its coordinates.

13.4.2 Vector Norms

In Section 8.6.2 the l_p-norms (*Hölder norms*)

$$
\|u - v\|_p := \begin{cases} \left(\sum_{i=1}^{n} |u_i - v_i|^p \right)^{1/p}, & p \in [1, \infty) \\ \max\{|u_1 - v_1|, \ldots, |u_n - v_n|\}, & p = \infty, \end{cases}
$$

were introduced to define the distance between two vectors $u, v \in \mathbb{R}^n$ or $u, v \in \mathbb{C}^n$. The following discussion concerns *only* $\mathbb{R}^n$, but similar results can be derived for the vector space $\mathbb{C}^n$.

Let $\langle , \rangle$ be an inner product (cf. Section 10.2.1) in $\mathbb{R}^n$. Then

$$\|x\| := \sqrt{\langle x, x \rangle} \quad \text{or} \quad \|x\|_B := \sqrt{\langle x, x \rangle_B} = \sqrt{\langle x, Bx \rangle}$$

defines a vector norm on $\mathbb{R}^n$, provided that the matrix B is symmetric and positive definite. In case $B = I$, the resulting norm is precisely the l_2-norm (the *Euclidean norm*). But

$$\|x\|_T := \|Tx\|$$

also defines a vector norm on $\mathbb{R}^n$ if $T \in \mathbb{R}^{n \times n}$ is a regular matrix.

Equivalence of Norms

Theorem 13.4.1 (Equivalence of Norms)   *Two norms* $\| \ \|$ *and* $\| \ \|'$ *in* $\mathbb{R}^n$ *are equivalent, if real numbers* $c_2 \geq c_1 > 0$ *exist, such that*

$$c_1\|x\| \leq \|x\|' \leq c_2\|x\| \qquad \text{for all} \quad x \in \mathbb{R}^n .$$

In $\mathbb{R}^n$ *all norms are equivalent.*

Proof: Ortega, Rheinboldt [67].

For example, the vector norms $\| \ \|_1, \| \ \|_2$, and $\| \ \|_\infty$ satisfy

$$\|x\|_2 \ \leq \ \|x\|_1 \ \leq \ \sqrt{n}\|x\|_2,$$
$$\|x\|_\infty \ \leq \ \|x\|_2 \ \leq \ \sqrt{n}\|x\|_\infty,$$
$$\|x\|_\infty \ \leq \ \|x\|_1 \ \leq \ n\|x\|_\infty.$$

for all $x \in \mathbb{R}^n$. As it can be seen from the above inequalities, the constants c_1 and c_2 may depend on the dimension n.

Convergence With Respect to a Norm

As a result of the equivalence Theorem 13.4.1, the convergence of a sequence $\{x^{(k)}\}$ of vectors can be defined using an arbitrary norm:

Definition 13.4.4 (Convergence of a Sequence of Vectors)   *A sequence* $\{x^{(k)}\}$ *of vectors of the* $\mathbb{R}^n$ *converges to* $x \in \mathbb{R}^n$ *if and only if*

$$\|x^{(k)} - x\| \to 0 \qquad as \quad k \to \infty \tag{13.7}$$

holds for some norm $\| \ \|$ *on* $\mathbb{R}^n$.

As each norm is equivalent to the maximum norm $\| \ \|_\infty$, (13.7) holds if and only if the sequence $\{x^{(k)}\}$ converges component-wise:

$$\lim_{k \to \infty} x_i^{(k)} = x_i \qquad \text{for each component} \quad i = 1, 2, \ldots, n.$$

Hence the convergence of all vector components (with respect to any basis for $\mathbb{R}^n$) is equivalent to the convergence of the vector sequence with respect to any norm.

13.4.3 Orthogonality

An essential notion in linear algebra is that of the *inner product*

$$\langle u, v \rangle := u^\top v = u_1 v_1 + \cdots + u_n v_n$$

(*scalar product*) of two vectors $u, v \in \mathbb{R}^n$. If $u \neq 0$ and $v \neq 0$, then the *angle* between u and v is defined by

$$\cos \varphi := \frac{\langle u, v \rangle}{\sqrt{\langle u, u \rangle \langle v, v \rangle}}.$$

If the vectors u and v are orthogonal, i.e., if the angle φ between u and v is 90°, then it follows that $\langle u, v \rangle = 0$.

Definition 13.4.5 (Orthogonality, Orthogonal Set of Vectors)   *Two vectors $u, v \in \mathbb{R}^n$ are said to be orthogonal if $\langle u, v \rangle = 0$. The vectors $u_1, \ldots, u_k \in \mathbb{R}^n$ are said to be orthogonal if they are pairwise orthogonal. The set $\{u_1, \ldots, u_k\}$ is then said to be a set of orthogonal vectors.*

Definition 13.4.6 (Set of Orthonormal Vectors)   *An orthogonal set of vectors $u_1, \ldots, u_k$ is said to be orthonormal, if*

$$\|u_i\|_2 = \sqrt{\langle u_i, u_i \rangle} = 1, \qquad i = 1, 2, \ldots, k,$$

i. e., each vector u_i is normalized such that it has Euclidean length 1.

Orthogonal and orthonormal sets of vectors are two important cases of linearly independent sets of vectors. For example, the unit vectors (13.6) form an orthonormal basis for $\mathbb{R}^n$.

Definition 13.4.7 (Orthogonal Complement)   *The orthogonal complement $S^\perp$ of a subspace S of the $\mathbb{R}^n$ is defined as*

$$S^\perp := \{v \in \mathbb{R}^n : \langle u, v \rangle = 0 \quad \text{for all } u \in S\}.$$

$S^\perp$ itself is also a linear subspace of $\mathbb{R}^n$.

Schmidt Orthonormalization Process

Each set $\{a_1, \ldots, a_k\}$ of linearly independent vectors can be orthonormalized using the (Gram) Schmidt process. The process starts with the vector a_1, which is normalized:

$$u_1 := \frac{1}{\|a_1\|} a_1.$$

Subtracting the component in direction u_1 from a_2 yields

$$v_2 := a_2 - \langle a_2, u_1 \rangle u_1,$$

which is orthogonal to u_1 (cf. Fig. 13.2).

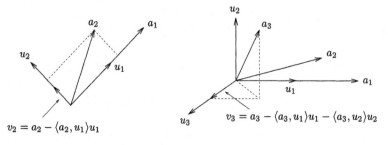

Figure 13.2: Schmidt orthonormalization process for two and three vectors.

Through normalization one obtains

$$u_2 := \frac{1}{\|v_2\|} v_2.$$

The third vector can be obtained in a similar way by

$$v_3 := a_3 - \langle a_3, u_1 \rangle u_1 - \langle a_3, u_2 \rangle u_2.$$

In general, the Schmidt process has the following algorithmic structure:

$u_1 := a_1 / \|a_1\|$
do $i = 2, 3, \ldots, k$
$\qquad v_i := a_i - \sum_{j=1}^{i-1} \langle a_i, u_j \rangle u_j$
$\qquad u_i := v_i / \|v_i\|$
end do

13.4.4 Linear Functions

Let F be a mapping from X_n to Y_m where X_n and Y_m are n- and m-dimensional vector spaces. If

1. $F(x_1 + x_2) = F(x_1) + F(x_2)$ for all $x_1, x_2 \in X_n$ (*additivity*)   and
2. $F(\alpha x) = \alpha F(x)$ for all $x \in X_n$, $\alpha \in \mathbb{R}$ (*homogeneity*),

then F is said to be a *linear transformation* (*linear function*). A *regular* linear transformation preserves the linear (in)dependence of a system of vectors. Such transformations can occur only if $n \leq m$.

Definition 13.4.8 (Null Space, Nullity)   *The set*

$$\mathcal{N}(F) := \{ x : F(x) = 0 \} \subseteq X_n$$

of all vectors which are mapped by F to the zero vector $0 \in Y_m$ is a subspace of X_n and is known as the kernel or the null space of the transformation F. The dimension of the kernel is called the nullity of F.

Definition 13.4.9 (Range Space, Rank)   *The set of all image vectors $F(x)$*

$$\mathcal{R}(F) := \{ y : \exists x \in X_n : F(x) = y \} \subseteq Y_m$$

is a subspace of Y_m and is known as the range (space) or the image (space) of the transformation F. The dimension of the range is called the rank of F.

13.4.5 Matrices

Each vector $x \in X_n$ can be expressed as a unique linear combination of basis vectors $u_1, \ldots, u_n$ of the space X_n:

$$x = \xi_1 u_1 + \xi_2 u_2 + \cdots + \xi_n u_n.$$

Each image vector $F(x) \in Y_m$ is a linear combination of basis vectors $v_1, \ldots, v_m$ of the image space Y_m. In particular, for the vectors $F(u_1), \ldots, F(u_n)$:

$$
\begin{aligned}
F(u_1) &=: a_{11}v_1 + a_{21}v_2 + \cdots + a_{m1}v_m \\
F(u_2) &=: a_{12}v_1 + a_{22}v_2 + \cdots + a_{m2}v_m \\
&\;\;\vdots \qquad\qquad \vdots \qquad\qquad\quad \vdots \\
F(u_n) &=: a_{1n}v_1 + a_{2n}v_2 + \cdots + a_{mn}v_m.
\end{aligned}
\tag{13.8}
$$

Hence it follows that

$$
\begin{aligned}
F(x) &= F\left(\sum_{j=1}^{n} \xi_j u_j\right) = \sum_{j=1}^{n} \xi_j F(u_j) = \sum_{j=1}^{n} \xi_j \sum_{i=1}^{m} a_{ij} v_i \\
&= \sum_{j=1}^{n}\sum_{i=1}^{m} a_{ij}\xi_j v_i = \sum_{i=1}^{m}\sum_{j=1}^{n} a_{ij}\xi_j v_i = \sum_{i=1}^{m} \eta_i v_i,
\end{aligned}
$$

i.e., the coordinates $\eta_1, \ldots, \eta_m$ form a representation of the image vector $F(x)$ with respect to the basis $v_1, \ldots, v_m$ of Y_m. The coordinates $\eta_1, \ldots, \eta_m$ and $\xi_1, \ldots, \xi_n$ are related in the following way:

$$
\begin{aligned}
\eta_1 &= a_{11}\xi_1 + a_{12}\xi_2 + \cdots + a_{1n}\xi_n \\
\eta_2 &= a_{21}\xi_1 + a_{22}\xi_2 + \cdots + a_{2n}\xi_n \\
&\;\;\vdots \qquad\quad \vdots \qquad\quad \vdots \qquad\qquad \vdots \\
\eta_m &= a_{m1}\xi_1 + a_{m2}\xi_2 + \cdots + a_{mn}\xi_n.
\end{aligned}
\tag{13.9}
$$

Definition 13.4.10 (Matrix of a Linear Transformation) *Corresponding to each linear transformation F there exists a unique set of coordinates of the image vectors $F(u_1), \ldots, F(u_n)$ of the basis vectors $u_1, \ldots, u_n$. They can be arranged in a matrix*

$$
A := (a_{ij}) := \begin{pmatrix}
a_{11} & a_{12} & \cdots & a_{1n} \\
a_{21} & a_{22} & \cdots & a_{2n} \\
\vdots & \vdots & & \vdots \\
a_{m1} & a_{m2} & \cdots & a_{mn}
\end{pmatrix}.
$$

It is said that A is the matrix assigned to the linear transformation F relative to the bases $\{u_1, \ldots, u_n\}$ and $\{v_1, \ldots, v_m\}$. Each linear transformation can be represented uniquely by such an $m \times n$ matrix.

It should be noted that the coefficient scheme (13.9) can be easily obtained from the scheme (13.8) by swapping rows and columns, i.e., in (13.8) the *transposed matrix $A^\top$* occurs.

The uniqueness of the relationship between linear transformations and matrices is maintained if arithmetic operations are applied to linear transformations. The matrix of $F_1 + F_2$ is $A_1 + A_2$; the matrix of the composed transformation $F_2 F_1$ is the matrix product $A_2 A_1$ (provided $F_1 : X_n \to Y_m$ and $F_2 : Y_m \to Z_k$).

Provided that bases $\{u_1, \ldots, u_n\}, \{v_1, \ldots, v_m\}$ are given, it is advisable to identify a linear function with its corresponding matrix. In particular, for the *canonical* bases $\{e_1^{(n)}, \ldots, e_n^{(n)}\} \subset \mathbb{R}^n$,

$$
e_1^{(n)} = \begin{pmatrix} 1 \\ 0 \\ \vdots \\ 0 \end{pmatrix}, \quad
e_2^{(n)} = \begin{pmatrix} 0 \\ 1 \\ \vdots \\ 0 \end{pmatrix}, \quad \ldots, \quad
e_n^{(n)} = \begin{pmatrix} 0 \\ 0 \\ \vdots \\ 1 \end{pmatrix} \in \mathbb{R}^n,
$$

and $\{e_1^{(m)}, \ldots, e_m^{(m)}\} \subset \mathbb{R}^m$ the following holds:

$$
x = \begin{pmatrix} \xi_1 \\ \xi_2 \\ \vdots \\ \xi_n \end{pmatrix}, \quad
y = F(x) = \begin{pmatrix} \eta_1 \\ \eta_2 \\ \vdots \\ \eta_m \end{pmatrix}, \quad Ax = y.
$$

Notation (Canonical Bases)   For the sake of convenience, the dimension of vectors of canonical bases will no longer be explicitly stated. Hence, e. g., instead of $e_1^{(1)}, \ldots, e_n^{(n)}$ only $e_1, \ldots, e_n$ will be used (as in (13.6)).

The set $\mathbb{R}^{m \times n}$ of real $m \times n$ matrices can be used to represent all linear transformations from the $\mathbb{R}^n$ into the $\mathbb{R}^m$ and so one can speak of a *linear transformation* $A \in \mathbb{R}^{m \times n}$.

The solution of the *linear system*

$$
Ax = b \quad \text{with} \quad A \in \mathbb{R}^{m \times n}, \; x \in \mathbb{R}^n, \; b \in \mathbb{R}^m
$$

can be interpreted as the inverse problem to the transformation A: determine all vectors $x \in \mathbb{R}^n$ which are mapped by A to the vector $b \in \mathbb{R}^m$.

This interpretation also makes it possible to apply geometrical concepts, such as those used for describing and analyzing linear transformations, to systems of linear equations.

Rank of a Matrix

Adapting the notion of the image space of a linear transformation from Definition 13.4.9 to $m \times n$ matrices $A \in \mathbb{R}^{m \times n}$ one obtains:

$$
\mathcal{R}(A) = \{y : Ax = y, \; x \in \mathbb{R}^n\} = \operatorname{span}\{a_1, a_2, \ldots, a_n\} \subseteq \mathbb{R}^m,
$$

where $a_1, \ldots, a_n$ denote the column vectors of A. This gives rise to the notion *column space* of the matrix A.

Definition 13.4.11 (Rank of a Matrix)   *The rank of a matrix is the maximum number of linearly independent column vectors; it is denoted* rank(A).

The solution of a linear system $Ax = b$ is a vector x^*, the coefficients of which represent b as a linear combination of the column vectors of A. The linear system $Ax = b$ has either zero, one, or an infinite number of solutions. If only one solution exists, the system is said to be *consistent*. The system is consistent if and only if rank($[A, b]$) = rank(A) holds. In this case the vector b is a unique linear combination of the columns of A.

Similarly, the notion of the null space from Definition 13.4.8 can be adapted to $m \times n$ matrices.

Definition 13.4.12 (Rank Deficiency of a Matrix)   *The dimension of the null space*

$$\mathcal{N}(A) = \{x : Ax = 0, \, x \in \mathbb{R}^n\} \subseteq \mathbb{R}^n$$

is called the nullity or the rank deficiency of the matrix A.

For each $m \times n$ matrix A, the following propositions are equivalent:
1. rank(A) = k.
2. There are exactly k linearly independent row vectors of A.
3. There are exactly k linearly independent column vectors of A.
4. There exists a $k \times k$ sub-matrix of A, whose determinant is non-zero, while the determinants of all $(k+1) \times (k+1)$ sub-matrices of A are zero.
5. The dimension of the range space $\mathcal{R}(A)$ is k.
6. The dimension of the null space $\mathcal{N}(A)$ is $n - k$.

Regularity of a Matrix

For each square $n \times n$ matrix A, the following propositions are equivalent:
1. A is regular. (The corresponding linear transformation is regular.)
2. A^{-1} exists.
3. rank(A) = n.
4. The row vectors of A are linearly independent.
5. The column vectors of A are linearly independent.
6. det(A) $\neq 0$.
7. The dimension of the range space $\mathcal{R}(A)$ is n.
8. The dimension of the null space $\mathcal{N}(A)$ is 0.
9. $Ax = b$ is consistent for all $b \in \mathbb{R}^n$.
10. $Ax = b$ has a unique solution for each $b \in \mathbb{R}^n$.
11. The only solution of the homogeneous system $Ax = 0$ is $x = 0$.
12. 0 is *not* a eigenvalue of A: $0 \notin \lambda(A)$ (cf. Section 13.4.7).

13.4.6 Inverse of a Matrix

If each of the linear systems

$$Ax_1 = e_1, \ Ax_2 = e_2, \ \dots, \ Ax_n = e_n$$

with $A \in \mathbb{R}^{n \times n}$ and with unit vectors (13.6) as the right hand side has unique solutions $x_1^*, \dots, x_n^*$, then these solution vectors can be interpreted as the column vectors of an $n \times n$ matrix

$$X := [x_1^*, x_2^*, \dots, x_n^*].$$

This matrix, which satisfies the equation $AX = I$, is called the *inverse* of A and is denoted A^{-1}.

Sometimes, formulas of the type

$$y = B^{-1}(I + 3A)b, \qquad z = A^{-1}(2B + I)(C^{-1} + B)b \qquad (13.10)$$

occur in mathematical and scientific publications. Computing the *inverse* matrices A^{-1}, B^{-1}, C^{-1} first, and then evaluating (13.10) seems to be the logical way to solve this problem. However, this method incurs unnecessarily high computational costs, and the accuracy of the results can be low. A more preferable approach is the stepwise solution of related linear systems, e. g.,

$$
\begin{array}{ll}
w := (I + 3A)b & \textbf{solve} \quad Cu = b \\
\textbf{solve} \quad By = w & v := (2B + I)(u^* + Bb) \\
& \textbf{solve} \quad Az = v.
\end{array}
$$

The inverse matrix, as such, is *very seldom* required in practice.

13.4.7 Eigenvalues of a Matrix

Definition 13.4.13 (Eigenvalues of a Matrix) *The eigenvalues of a matrix $A \in \mathbb{R}^{n \times n}$ are the n zeros of its characteristic polynomial*

$$
P_n(z; A) := \det(A - zI) =
\begin{vmatrix}
(a_{11} - z) & a_{12} & \cdots & a_{1n} \\
a_{21} & (a_{22} - z) & \cdots & a_{2n} \\
\vdots & \vdots & \ddots & \vdots \\
a_{n1} & a_{n2} & \cdots & (a_{nn} - z)
\end{vmatrix}.
\qquad (13.11)
$$

Definition 13.4.14 (Spectrum, Spectral Radius of a Matrix) *The set of zeros $\lambda_1, \dots, \lambda_n$ of the characteristic polynomial (13.11)*

$$\lambda(A) := \{\lambda_1, \dots, \lambda_n\}, \quad \lambda_i \in \mathbb{C},$$

is called the spectrum of the matrix A. The maximum of the absolute values of the eigenvalues of A is called the spectral radius of A,

$$\varrho(A) := \max\{|\lambda_1|, \dots, |\lambda_n|\}.$$

A more comprehensive discussion of eigenvalues and eigenvectors of a matrix can be found in Chapter 15. Algorithms and software products for their practical calculation are also presented there.

13.4.8 Matrix Norms

Mathematical analysis requires a precise formulation of the notion of neighborhood (e. g., the closeness of a matrix to a singular matrix). Therefore, a *distance function in the space of matrices* is needed. Matrix norms can be used to define such distance functions. They are valuable tools for quantifying notions such as the degree of perturbation to a matrix.

Definition 13.4.15 (Matrix Norms)   *A function* $\| \ \| \ : \ \mathbb{R}^{m \times n} \to \mathbb{R}$ *which satisfies the conditions*

$$1. \ \|A\| \geq 0,$$
$$2. \ \|A\| = 0 \quad \Longleftrightarrow \quad A = 0 \quad (definiteness),$$
$$3. \ \|\alpha A\| = |\alpha| \, \|A\|, \quad \alpha \in \mathbb{R} \quad (homogeneity) \quad and$$
$$4. \ \|A + B\| \leq \|A\| + \|B\| \quad (triangle \ inequality)$$

for all $A, B \in \mathbb{R}^{m \times n}$ *is called a matrix norm.*

The matrix norms most commonly used in numerics are the *p-norms*

$$\|A\|_p := \max_{x \neq 0} \frac{\|Ax\|_p}{\|x\|_p}, \tag{13.12}$$

which are defined using vector l_p-norms (cf. Section 13.4.2) and the *Frobenius norm (Schur norm)*

$$\|A\|_F := \sqrt{\sum_{i=1}^{m} \sum_{j=1}^{n} |a_{ij}|^2}.$$

According to

$$\|A\|_p = \max_{x \neq 0} \left\| A \left(\frac{x}{\|x\|_p} \right) \right\|_p = \max\{\|Ax\|_p : \|x\|_p = 1\} = \|Ax_{\max}\|_p$$

the *p*-norm $\|A\|_p$ of the matrix A can be interpreted as the maximum elongation possible when A is applied to a normalized vector, i.e., a point on the surface of the *n*-dimensional unit sphere $\{x \ : \ \|x\|_p = 1\}$. A vector with this maximum elongation is denoted $x_{\max}$.

Example (Norms of a Matrix)   The 2×2 matrix

$$A = \begin{pmatrix} 0.5 & 2 \\ 1.5 & 1 \end{pmatrix}$$

is applied to the vectors on the two-dimensional unit sphere

$$\{x : \|x\|_p = 1\}, \quad p = 1, 2, \infty.$$

The sets of vectors obtained

$$\{Ax : \|x\|_p = 1\}, \quad p = 1, 2, \infty,$$

are depicted in Fig. 13.3. The dotted lines in Fig. 13.3 symbolize the sphere whose radius is given by the maximum elongation.

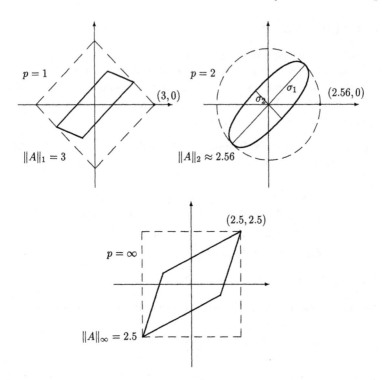

Figure 13.3: Images $\{Ax : \|x\|_p = 1\}$, $p = 1, 2, \infty$, of the unit spheres $\{x : \|x\|_p = 1\}$.

The *consistency* property

$$\|Ax\|_p \leq \|A\|_p \|x\|_p \qquad \text{for all} \quad A \in \mathbb{R}^{m \times n}, x \in \mathbb{R}^n$$

and the *sub-multiplicativity* property

$$\|AB\|_p \leq \|A\|_p \|B\|_p \qquad \text{for all} \quad A \in \mathbb{R}^{m \times n}, B \in \mathbb{R}^{n \times k}$$

of the matrix norm can be derived from (13.12).

Theorem 13.4.2 *For each matrix $A \in \mathbb{R}^{m \times n}$, the following hold:*

$$\|A\|_1 = \max\left\{ \sum_{i=1}^{m} |a_{ij}| : j = 1\ldots, n \right\}$$

$$\|A\|_\infty = \max\left\{ \sum_{j=1}^{n} |a_{ij}| : i = 1,\ldots, m \right\}$$

$$\|A\|_2 = \sqrt{\varrho(A^\top A)} = \sqrt{\varrho(AA^\top)} = \|A^\top\|_2$$

Proof: Hämmerlin, Hoffmann [52].

$\| \ \|_1$ is called the *column sum norm*, $\| \ \|_\infty$ the *row sum norm*, and $\| \ \|_2$ the *Euclidean* or *spectral norm*.

If $A \in \mathbb{R}^{n \times n}$ is symmetric, then the spectral radius of A coincides with the spectral norm of A:

$$\|A\|_2 = \sqrt{\varrho(A^T A)} = \sqrt{\varrho(A^2)} = \sqrt{\left(\varrho(A)\right)^2} = \varrho(A).$$

The calculation of $\|A\|_1$ or $\|A\|_\infty$ is algorithmically straightforward and requires only n^2 additions. The computation of the spectral norm $\|A\|_2$, on the other hand, requires an iterative process and a significantly higher computational effort. *Estimates* of $\|A\|_2$, however, can be easily derived using $\|A\|_1$ and $\|A\|_\infty$:

$$\frac{1}{\sqrt{m}} \|A\|_1 \leq \|A\|_2 \leq \sqrt{n} \|A\|_1$$

$$\frac{1}{\sqrt{n}} \|A\|_\infty \leq \|A\|_2 \leq \sqrt{m} \|A\|_\infty$$

$$\|A\|_2 \leq \sqrt{\|A\|_1 \|A\|_\infty}.$$

In some cases the *Frobenius norm*

$$\|A\|_F := \sqrt{\sum_{i=1}^{n} \sum_{j=1}^{n} |a_{ij}|^2}$$

is also used for norm estimations; it is related to the spectral norm by

$$\|A\|_2 \leq \|A\|_F \leq \sqrt{n} \|A\|_2.$$

The spectral norm and the Frobenius norm are invariant with respect to orthogonal transformations. For orthogonal matrices $Q, Z \in \mathbb{R}^{n \times n}$,

$$\|QAZ\|_2 = \|A\|_2 \quad \text{and} \quad \|QAZ\|_F = \|A\|_F.$$

13.4.9 Determinant of a Matrix

In linear algebra the determinant is an important characteristic of a matrix, used, for instance, to define the regularity of a matrix.

Definition 13.4.16 (Determinant of a Matrix) *The determinant of a square matrix $A \in \mathbb{R}^{n \times n}$ is a function*

$$\det : \mathbb{R}^{n \times n} \to \mathbb{R},$$

defined as follows:

$$\det(A) := \sum_{\text{perm}} \text{sign}(\nu_1, \ldots, \nu_n) a_{1\nu_1} a_{2\nu_2} \cdots a_{n\nu_n}. \tag{13.13}$$

The summation is over all $n!$ permutations $(\nu_1, \ldots, \nu_n)$ of the numbers $1, \ldots, n$. $\text{sign}(\nu_1, \ldots, \nu_n) = +1$ or -1, if $(\nu_1, \ldots, \nu_n)$ is an even or odd permutation respectively. An even/odd permutation is generated by an even/odd number of exchanges of two adjacent numbers.

The determinant can be obtained algorithmically as a *by-product* of solving linear systems with the use of direct methods (LU factorization with costs of $O(n^3)$), if the determinant is actually required. Definition (13.13) is not suitable for the practical determination of $\det(A)$ as the computational effort of $O(n!)$ would be prohibitive.

In practice, the value of $\det(A)$ has absolutely no impact on the solvability of the linear system or the quality of the numerical solution. The *classical* criterion $\det(A) = 0$ for the singularity of A cannot be verified due to unavoidable rounding errors, and the magnitude of $\det(A)$ does *not* give any information about possible numerical problems or numerical singularity.

Whilst with $\det(A) = 1.5 \cdot 10^{-38}$, the numerical solution can be satisfactorily accurate, in some cases the harmless looking value of $\det(A) = 1.5$ can be found in conjunction with quite useless results. These facts are based on the following determinant property:

$$\det(cA) = c^n \det(A), \qquad c \in \mathbb{R}, \quad A \in \mathbb{R}^{n \times n}.$$

Scaling a 100×100 matrix using $c = 2^{-3} = 0.125$ decreases the value of the determinant by a factor of about $2^{-300} \approx 4.9 \cdot 10^{-91}$, but this has absolutely no impact on the numerical regularity of A or on the difficulties which might arise in the course of the numerical solution of $Ax = b$.

13.5 Special Properties of Matrices

If the specific properties of the coefficient matrix can be exploited by the method chosen for solving the linear system, then both the reliability and the efficiency of a method can be improved. Therefore, the user of numerical software must be aware of any special properties of the linear problem at the design stage in order to select the appropriate software.

13.5.1 Symmetric and Hermitian Matrices

The *transposed matrix* $A^\top$ is obtained by reflecting the elements of A at the main diagonal:

$$A^\top := C \quad \text{where} \quad c_{ij} := a_{ji} \text{ for all } i, j \in \{1, 2, \ldots, n\}.$$

The linear operator corresponding to $A^\top$ is called the *adjoint* linear operator. For the inner product,

$$\langle Ax, y \rangle = \langle x, A^\top y \rangle \qquad \text{for all} \quad x, y \in \mathbb{R}^n.$$

A *symmetric matrix* coincides with its transpose, i. e.,

$$A^\top = A,$$

implying that

$$\langle Ax, y \rangle = \langle x, Ay \rangle.$$

This property is the formal condition that the linear operator corresponding to A is *self-adjoint*. Self-adjoint operators (and the symmetric matrices corresponding to them) have special features. For instance, the eigenvalues of a self-adjoint operator (or a symmetric matrix) are always real, and eigenvectors corresponding to different eigenvalues are orthogonal. Due to these properties the self-adjoint operators occupy a special position both in theory and in practice.

In the case of complex matrices $A \in \mathbb{C}^{n \times n}$, the transposition $A^\top$ or the symmetry $A^\top = A$ is less important than for real matrices. Much more important is the *conjugate transposition A^H*:

$$A^H := C \qquad \text{where} \qquad c_{ij} := \bar{a}_{ji} \quad \text{for all} \quad i, j \in \{1, 2, \ldots, n\}.$$

Matrices with the property

$$A^H = A$$

are called *Hermitian matrices*.

13.5.2 Orthogonal and Unitary Matrices

Definition 13.5.1 (Orthogonal Matrix) *An orthogonal matrix is a square matrix $Q \in \mathbb{R}^{n \times n}$ with orthonormal column vectors $q_1, \ldots, q_n$. It thus satisfies the matrix equation*

$$Q^\top Q = Q Q^\top = I. \tag{13.14}$$

Example (Orthogonal Matrix) The matrix

$$A = \begin{pmatrix} 1 & 1 \\ -1 & 1 \end{pmatrix}$$

has orthogonal column vectors. But this matrix is *not* orthogonal because the column vectors are not normalized. By normalizing A, the orthogonal matrix

$$Q = \frac{1}{\sqrt{2}} A$$

can be obtained.

From the matrix equation (13.14) it follows that

$$Q^\top = Q^{-1}.$$

In addition to this, each orthogonal matrix Q represents a length- and angle-preserving mapping from $\mathbb{R}^n$ to $\mathbb{R}^n$, as for arbitrary $x, y \in \mathbb{R}^n$

$$\langle Qx, Qy \rangle = (Qx)^\top (Qy) = x^\top Q^\top Q y = x^\top y = \langle x, y \rangle \tag{13.15}$$

holds. Each of these mappings specifies a reflection and/or a rotation of $\mathbb{R}^n$ with the origin as the center of rotation. From (13.15) it follows that

$$\|Qx\|_2^2 = \langle Qx, Qx \rangle = \langle x, x \rangle = \|x\|_2^2.$$

Multiplying a vector by an orthogonal matrix leaves its Euclidean length invariant. As a result orthogonal transformations do not increase data or rounding errors.

Definition 13.5.2 (Unitary Matrix)   *A complex matrix $Q \in \mathbb{C}^{n \times n}$ with the property*

$$Q^H Q = Q Q^H = I$$

is said to be unitary.

13.5.3 Positive Definite Matrices

Real symmetric or complex Hermitian matrices with a special property called positivity arise in many applications. Real symmetric matrices with this property are, in a certain sense, a generalization of positive real numbers.

Properties and Definition

A quadratic form $q : \mathbb{R}^n \to \mathbb{R}$

$$q(x) := \langle Ax, x \rangle = \langle x, Ax \rangle = \sum_{i=1}^{n} \sum_{j=1}^{n} a_{ij} x_i x_j \, ,$$

always implies a symmetric matrix A. The quadratic form q is said to be *positive definite* if

$$q(x) = \langle x, Ax \rangle > 0 \qquad \text{for all} \quad x \in \mathbb{R}^n \setminus \{0\}.$$

The corresponding symmetric matrix A is then also said to be *positive definite*.

Definition 13.5.3 (Positive Definite Matrix)   *A Hermitian matrix A is said to be positive definite, if*

$$x^H A x > 0 \quad \text{for all} \quad x \in \mathbb{C}^n \setminus \{0\}. \tag{13.16}$$

If only $x^H A x \geq 0$ can be proved, then A is said to be positive semidefinite.

Theorem 13.5.1 (Regularity)   *Positive definite matrices are regular.*

Proof: Supposing a singular $n \times n$ matrix A has a null space with $\dim(\mathcal{N}(A)) \geq 1$, then there exists a vector $x \neq 0$ with $Ax = 0$, which satisfies $x^H A x = 0$ in contradiction to (13.16). $\qquad\qquad\qquad\qquad\qquad\qquad\qquad\qquad\qquad\qquad\qquad\qquad$ □

Moreover, *negative definite* (and *negative semidefinite*) matrices can be defined by turning the inequalities in the definitions of positive definite (or positive semidefinite matrices) around or equivalently by making sure that $-A$ is positive definite (or semidefinite).

Definition 13.5.4 (Indefinite Matrix)   *If a Hermitian matrix is not covered by any of the above classes—i. e., $x^H A x$ can have negative and positive values depending on x—then A is said to be indefinite.*

Example (Minimization of Convex Functions)   Let $f : D \subset \mathbb{R}^n \to \mathbb{R}$ be a continuously differentiable function, and let y be an interior point of D. Then the *Taylor expansion* for a point $x \in D$ in the neighborhood of y is given by

$$f(x) = f(y) + \sum_{i=1}^{n}(x_i - y_i)\frac{\partial f}{\partial x_i}(y) + \frac{1}{2!}\sum_{i=1}^{n}\sum_{j=1}^{n}(x_i - y_i)(x_j - y_j)\frac{\partial^2 f}{\partial x_i \partial x_j}(y) + \cdots .$$

x^* is said to be a *stationary point* or a *critical point* of f if the first partial derivatives vanish at x^*. In a neighborhood of a critical point x^* it follows that

$$
\begin{aligned}
f(x) - f(x^*) &= \sum_{i=1}^{n}\sum_{j=1}^{n}(x_i - x_i^*)(x_j - x_j^*)\frac{\partial^2 f}{\partial x_i \partial x_j}(x^*) + \cdots = \\
&= (x - x^*)^\top H(f;x^*)(x - x^*) + \cdots
\end{aligned}
$$

where the $n \times n$ matrix

$$H(f;x^*) := \left(\frac{\partial^2 f}{\partial x_i \partial x_j}(x^*) \right)$$

is called the *Hesse matrix* (Hessian matrix) of f at the point x^*. The Hesse matrix is symmetric since $\partial^2 f/\partial x_i \partial x_j = \partial^2 f/\partial x_j \partial x_i$ holds. Provided that

$$z^\top H(f;x^*)z > 0 \qquad \text{for all} \quad z \in \mathbb{R}^n \setminus \{0\} \tag{13.17}$$

can be proved, then f has a *relative minimum* at x^* . If on the other hand the quadratic form in (13.17) is negative for all $z \in \mathbb{R}^n \setminus \{0\}$, then f has a *relative maximum* at x^*. If the quadratic form is indefinite, then x^* is a *saddle point*.

For $n = 1$ these criteria are equivalent to those found in the test for relative extreme values based on the sign of the second derivative. If x^* is a critical point and $f''(x^*) > 0$, then x^* is a relative minimum. If $f''(x^*)$ is negative, then x^* is a relative maximum. x^* may be a saddle point if the second derivative vanishes there.

If the quadratic form (13.17) is positive for all points in D (not only for the critical point), then f is said to be a *convex function* on D. For two distinct points $x, y \in D$ and for all $\lambda \in (0, 1)$ the following relation then holds :

$$f((1 - \lambda)x + \lambda y) \leq (1 - \lambda)f(x) + \lambda f(y).$$

Once more this is a generalization of scalar notions. When $n = 1$, f is convex if and only if $f''(x) \geq 0$ for all $x \in \mathbb{R}$.

Example (Covariance Matrices)   The *covariance matrix* of n-dimensional random vectors $X = (X_1, \ldots, X_n)^\top$ is the matrix $A = (a_{ij})$, whose elements are the expectations μ:

$$a_{ij} := \mu\left[(X_i - \mu(X_i))(X_j - \mu(X_j))\right], \qquad i = 1, 2, \ldots, n, \quad j = 1, 2, \ldots, n.$$

When $i \neq j$ these expectations are said to be covariances $\text{cov}(X_i, X_j)$, and when $i = j$ they are said to be variances $\text{var}(X_i) := \mu(X_i - \mu(X_i))^2$.

Obviously, A is symmetric, and for $z = (z_1, \ldots, z_n)^\top \in \mathbb{R}^n$ one can prove that

$$z^\top A z = \mu\left(\sum_{i=1}^{n}\sum_{j=1}^{n} z_i(X_i - \mu(X_i))z_j(X_j - \mu(X_j)) \right) = \mu\left(\left| \sum_{i=1}^{n} z_i(X_i - \mu_i) \right|^2 \right) \geq 0.$$

Example (Discretization of a Differential Equation)   A boundary value problem of the form

$$-y''(x) + \sigma(x)y(x) = f(x), \tag{13.18}$$
$$y(0) = \alpha,$$
$$y(1) = \beta$$

with $x \in [0,1]$ and $\alpha, \beta \in \mathbb{R}$, whereby α and β are constants and $f, \sigma : [0,1] \to \mathbb{R}$ are prescribed functions, can be solved as follows: A discretization of (13.18) is made by considering only the values of $y(kh)$, $k = 0, 1, \ldots, n+1$ at grid-points with the uniform spacing $h = 1/(n+1)$ for $n \in \mathbb{N}$, and the second derivative y'' is approximated by the second central difference

$$y''(x) \approx \frac{y((k+1)h) - 2y(kh) + y((k-1)h)}{h^2} = \frac{y_{k+1} - 2y_k + y_{k-1}}{h^2}.$$

As a result one obtains the linear system

$$(-y_{k+1} + 2y_k - y_{k-1})/h^2 + \sigma_k y_k = f_k, \qquad k = 1, \ldots, n, \tag{13.19}$$
$$y_0 = \alpha,$$
$$y_{n+1} = \beta$$

with $y_k := y(kh)$, $\sigma_k := \sigma(kh)$, and $f_k := f(kh)$. Boundary conditions can be imposed on the first and the last equation of (13.19) yielding the linear system

$$(2 + h^2\sigma_1)y_1 - y_2 = h^2 f_1 + \alpha$$
$$-y_{k-1} + (2 + h^2\sigma_k)y_k - y_{k+1} = h^2 f_k, \qquad k = 2, 3, \ldots, n-1$$
$$-y_{n-1} + (2 + h^2\sigma_n)y_n = h^2 f_n + \beta.$$

These equations can be transformed into the form $Ay = w$ with

$$y := (y_1, \ldots, y_n)^\top \in \mathbb{R}^n, \qquad w := (h^2 f_1 + \alpha, h^2 f_2, \ldots, h^2 f_{n-1}, h^2 f_n + \beta)^\top \in \mathbb{R}^n.$$

Thus A is an $n \times n$ matrix of the special form

$$A = \begin{pmatrix} 2 + h^2\sigma_1 & -1 & & & 0 \\ -1 & 2 + h^2\sigma_2 & -1 & & \\ & \ddots & \ddots & \ddots & \\ & & -1 & 2 + h^2\sigma_{n-1} & -1 \\ 0 & & & -1 & 2 + h^2\sigma_n \end{pmatrix}.$$

A is a real, symmetric, tridiagonal matrix; this property is independent of the function σ, although choosing an arbitrarily right-hand side for $Ay = w$ requires that $\sigma_1, \ldots, \sigma_n$ satisfy certain conditions ensuring the non-singularity of A. The real quadratic form assigned to A can easily be computed:

$$x^\top A x = \left(x_1^2 + \sum_{i=1}^{n-1} (x_i - x_{i+1})^2 + x_n^2 \right) + h^2 \sum_{i=1}^{n} \sigma_i x_i^2.$$

The expression in brackets is non-negative. It vanishes only if the components of x are all identically zero. If $\sigma(x) \geq 0$, then the last sum is non-negative and therefore

$$x^\top A x \geq \left(x_1^2 + \sum_{i=1}^{n-1} (x_i - x_{i+1})^2 + x_n^2 \right) \geq 0.$$

If A is singular, there exists a vector $\hat{x} \in \mathbb{R}^n$, $\hat{x} \neq 0$ such that $A\hat{x} = 0$ and consequently $\hat{x}^\top A \hat{x} = 0$. In this case, the expression in brackets must vanish implying that $\hat{x} = 0$. Hence, for $\sigma(x) \geq 0$ the matrix A is regular and the discretized boundary value problem can be solved for arbitrary boundary values α and β.

Criteria for the Definiteness of a Matrix

The symmetry of a matrix can be seen and verified easily. In contrast, definiteness is not such an obvious property. In some cases the positive definiteness of A is guaranteed by the user's awareness of distinctive physical interpretations of the quadratic form $x^\top Ax$ (e. g., as the kinetic energy of a system of mass points). In other situations, one of the theorems stated below can be useful in rejecting or ascertaining conjectures on the definiteness of the matrix.

For each of the following theorems on the positive definiteness of real matrices, there is a similar theorem for positive semidefinite matrices.

Theorem 13.5.2 *The diagonal elements of a positive definite matrix are positive real numbers.*

Theorem 13.5.3 *A principle sub-matrix of a positive definite matrix is positive definite. The trace (the sum of diagonal elements), the determinant, and the minors (the determinants of the principle sub-matrices) of a positive definite matrix are positive.*

Theorem 13.5.4 *The sum $A+B$ of two positive definite matrices $A, B \in \mathbb{R}^{n \times n}$ is positive definite.*

More generally, a non-negative linear combination

$$\alpha A + \beta B \qquad \text{with} \quad \alpha, \beta \geq 0$$

of positive semidefinite matrices $A, B \in \mathbb{R}^{n \times n}$ is positive semidefinite.

Theorem 13.5.5 *Let $A \in \mathbb{R}^{n \times n}$ be a positive definite matrix and $C \in \mathbb{R}^{n \times m}$. Then $C^\top AC$ is positive definite, and moreover*

$$\operatorname{rank}(C^\top AC) = \operatorname{rank}(C)$$

holds so that $C^\top AC$ is positive definite, if and only if $\operatorname{rank}(C) = m$.

Theorem 13.5.6 *A symmetric matrix $A \in \mathbb{R}^{n \times n}$ is positive semidefinite if and only if its eigenvalues are non-negative. Positive definiteness holds if and only if the eigenvalues of A are positive.*

If $A \in \mathbb{R}^{n \times n}$ is positive semidefinite, then the powers $A^2, A^3, A^4, \ldots$ are also positive semidefinite.

Theorem 13.5.7 *If $A = (a_{ij}) \in \mathbb{R}^{n \times n}$ is symmetric and strictly diagonally dominant, that is*

$$|a_{ii}| > \sum_{j \neq i} |a_{ij}|, \qquad i = 1, 2, \ldots, n,$$

and the diagonal elements a_{ii} are also positive, then A is positive definite.

For each positive real number a kth root of unity ($k = 1, 2, \ldots$) can be defined. A similar quantity can be defined for positive definite matrices.

Theorem 13.5.8   *Given a positive semidefinite matrix $A \in \mathbb{R}^{n \times n}$ and a natural number $k \geq 1$, there exists a unique positive, semidefinite, symmetric matrix $B \in \mathbb{R}^{n \times n}$ such that*

1. $B^k = A$;

2. $BA = AB$ *and there exists a polynomial $P(t)$, such that $B = P(A)$;*

3. $\mathrm{rank}(B) = \mathrm{rank}(A)$; *hence B is positive definite if and only if A is positive definite.*

The most important special case of Theorem 13.5.8 is $k = 2$: The positive (semi) definite root of the positive (semi) definite matrix A is denoted by $A^{1/2}$. Similarly, for $k = 3, 4, 5, \ldots$, $A^{1/k}$ denotes the kth positive (semi) definite root of A.

Definition 13.5.5 (Congruence of Matrices)   *Two matrices $A, B \in \mathbb{R}^{n \times n}$ are said to be congruent if there exists a regular matrix $C \in \mathbb{R}^{n \times n}$ such that*

$$A = C^\top B C.$$

Theorem 13.5.9   *A matrix A is positive definite if and only if it is congruent to the identity, or more precisely, there exists a regular matrix $C \in \mathbb{R}^{n \times n}$ with $A = C^\top C$.*

Sometimes it may be useful to further specialize the decomposition $A = C^\top C$ of a positive semidefinite matrix utilizing the principle of QR decomposition: All square matrices C have a QR decomposition $C = QR$, where Q is orthogonal (hence $Q^\top Q = I$ holds) and R is an upper triangular matrix of the same rank as C. Accordingly,

$$A = C^\top C = (QR)^\top QR = R^\top Q^\top QR = R^\top R.$$

If C is regular then R can be chosen so that its diagonal elements are positive. The corresponding decomposition of A is known as the *Cholesky factorization*.

Theorem 13.5.10 (Cholesky Factorization)   *A matrix A is positive definite if and only if there exists a regular, lower triangular matrix $L \in \mathbb{R}^{n \times n}$ with positive diagonal elements such that $A = LL^\top$.*

Let $\{v_1, \ldots, v_k\}$ be a set of vectors from vector space V, and let $\langle \, , \, \rangle$ be an inner product on V. Then the matrix

$$G = (g_{ij}) \in \mathbb{R}^{n \times n} \qquad \text{with} \quad g_{ij} := \langle v_i, v_j \rangle$$

is called the *Gramian* of the vectors $v_1, \ldots, v_k$ with respect to an inner product $\langle \, , \, \rangle$. It is a characteristic of positive definite matrices that they are always Gramians.

Theorem 13.5.11 *Let $G \in \mathbb{R}^{k \times k}$ be the Gramian of the vectors $\{w_1, \ldots, w_k\} \subset \mathbb{R}^n$ with respect to an inner product $\langle , \rangle$, and $W = [w_1 w_2 \cdots w_k] \in \mathbb{R}^{n \times k}$. Then*

1. *G is positive semidefinite;*

2. *G is regular if and only if the vectors $w_1, \ldots, w_k$ are linearly independent;*

3. *There exists a positive definite matrix $A \in \mathbb{R}^{n \times n}$ such that $G = W^\top A W$;*

4. *$\operatorname{rank}(G) = \operatorname{rank}(W)$ is the maximum number of linearly independent vectors from the set $\{w_1, \ldots, w_k\}$.*

A matrix $A \in \mathbb{R}^{n \times n}$ with $\operatorname{rank}(A) = r \leq n$ is positive semidefinite if and only if there exists a set of vectors $S = \{w_1, \ldots, w_n\} \subset \mathbb{R}^n$ containing exactly r linearly independent vectors such that A coincides with the Gramian of S with respect to the Euclidean inner product.

13.6 Special Types of Matrices

Matrices with zero elements arranged in a special structure are important both to analytical investigations and to the algorithmic solution of numerical problems in linear algebra.

13.6.1 Diagonal Matrices

An $n \times n$ matrix

$$
D := \begin{pmatrix} d_{11} & & & \mathbf{0} \\ & d_{22} & & \\ & & \ddots & \\ \mathbf{0} & & & d_{nn} \end{pmatrix}
$$

with $d_{ij} = 0$ for $i \neq j$ is said to be *diagonal*, denoted by

$$
D = \operatorname{diag}(d_{11}, \ldots, d_{nn}) \qquad \text{or} \qquad D = \operatorname{diag}(d),
$$

where d represents a vector containing the elements from the diagonal of D.

If all diagonal elements are real and positive then the diagonal matrix is said to be positive. For instance, the *identity matrix*

$$
I := \begin{pmatrix} 1 & & & \mathbf{0} \\ & 1 & & \\ & & \ddots & \\ \mathbf{0} & & & 1 \end{pmatrix}
$$

is a positive diagonal matrix. A diagonal matrix is said to be *scalar* if all diagonal elements have the same value: $D = \alpha I$ for $\alpha \in \mathbb{R}$ or $\alpha \in \mathbb{C}$. Multiplying a matrix by a scalar matrix αI is equivalent to multiplying each element of a matrix by α.

The determinant of a diagonal matrix is given by the product of its diagonal elements. A diagonal matrix is thus regular if and only if $d_{ii} \neq 0$ for all i.

There are also *rectangular*, diagonal matrices $D \in \mathbb{R}^{m \times n}$ for $m > n$ resp. $m < n$:

$$
\begin{pmatrix}
d_{11} & & & \mathbf{0} \\
& d_{22} & & \\
& & \ddots & \\
& & & d_{nn} \\
\mathbf{0} & & &
\end{pmatrix}
\quad \text{or} \quad
\begin{pmatrix}
d_{11} & & & & & \mathbf{0} \\
& d_{22} & & & & \\
& & \ddots & & & \\
\mathbf{0} & & & d_{mm} & &
\end{pmatrix},
$$

which occur, for example, in singular value decompositions (cf. Section 13.7).

13.6.2 Triangular Matrices

An $n \times n$ matrix $T = (t_{ij})$ is called an *upper triangular matrix* if $t_{ij} = 0$ for all $j < i$:

$$
T =
\begin{pmatrix}
t_{11} & t_{12} & \cdots & t_{1n} \\
& t_{22} & \cdots & t_{2n} \\
& & \ddots & \vdots \\
\mathbf{0} & & & t_{nn}
\end{pmatrix}.
$$

If $t_{ij} = 0$ for $j \leq i$, then T is called a *strict upper triangular matrix*. Lower and strict lower triangular matrices are defined in a similar manner.

As with diagonal matrices, the determinant of a triangular matrix is equal to the product of its diagonal elements. The rank is at least as great as the number of nonzero elements t_{ii} in the main diagonal.

13.6.3 Block Matrices

Partitioning a matrix into disjoint sub-matrices is referred to as *blocking*. Each element of the original matrix is contained in one and only one block (sub-matrix).

Submatrices

Let A be a $n \times n$ matrix and $\alpha, \beta \subseteq \{1, \ldots, n\}$. Then $A(\alpha, \beta)$ is defined to be a *sub-matrix* of the matrix A whereby the set α specifies the rows and β specifies the columns of the original matrix that are present in the sub-matrix.

If $\alpha = \beta$, the sub-matrix $A(\alpha, \alpha) =: A(\alpha)$ is called a *principle minor matrix*. Sometimes sub-matrices are given by specifying those rows and columns which are not contained in the sub-matrix.

The determinant of a square sub-matrix of A is called a *minor* of A.

Multiplication of Block Matrices

Provided that $\alpha_1, \ldots, \alpha_z$ is a partition (decomposition) of the set of row indices $\{1, \ldots, m\}$, and that $\beta_1, \ldots, \beta_s$ is a partition of the set of column indices $\{1, \ldots, n\}$, then the matrices $A(\alpha_i, \beta_j)$ form a partition of the given $m \times n$ matrix A. The partitioned matrix

$$A = \begin{pmatrix} A_{11} & A_{12} & \cdots & A_{1q} \\ A_{21} & A_{22} & \cdots & A_{2q} \\ \vdots & \vdots & \vdots & \vdots \\ A_{p1} & A_{p2} & \cdots & A_{pq} \end{pmatrix},$$

which consists of the sub-matrices A_{ij}, is called a *block matrix*. In particular cases, the sub-matrices A_{ij} can be 1×1 matrices, row vectors, or column vectors.

Example (Block Matrix) By partitioning the 4×5 matrix

$$A := \left(\begin{array}{cc|c|cc} 11 & 12 & 13 & 14 & 15 \\ 21 & 22 & 23 & 24 & 25 \\ 31 & 32 & 33 & 34 & 35 \\ \hline 41 & 42 & 43 & 44 & 45 \end{array} \right) = \begin{pmatrix} A_{11} & A_{12} & A_{13} \\ A_{21} & A_{22} & A_{23} \\ A_{31} & A_{32} & A_{33} \end{pmatrix}$$

a 3×3 block matrix

$$\begin{array}{ll} \alpha_1 = \{1\} & \beta_1 = \{1, 2\} \\ \alpha_2 = \{2, 3\} & \beta_2 = \{3\} \\ \alpha_3 = \{4\} & \beta_3 = \{4, 5\} \end{array}$$

can be obtained.

If the $m \times n$ matrix A and the $n \times p$ matrix B are partitioned in such a way that both partitions of $\{1, \ldots, n\}$ coincide, then it is said that the matrix partitions *conform*. In such cases

$$[AB](\alpha_i, \gamma_j) = \sum_{k=1}^{s} A(\alpha_i, \beta_k) B(\beta_k, \gamma_j),$$

holds, where $A(\alpha_i, \beta_k)$ and $B(\beta_k, \gamma_j)$ are conforming partitions of A and B.

The Inverse of a Block Matrix

Sometimes it is useful to have certain information concerning the blocks found in the inverse of a regular block matrix in order to express the inverse in partitioned form. Provided that all sub-matrices of A and A^{-1} are also regular, the inverse can be represented in different ways. Given

$$A = \begin{pmatrix} A_{11} & A_{12} \\ A_{21} & A_{22} \end{pmatrix},$$

one partitioned representation of A^{-1} might be

$$\begin{pmatrix} (A_{11} - A_{12} A_{22}^{-1} A_{21})^{-1} & A_{11}^{-1} A_{12} (A_{21} A_{11}^{-1} A_{12} - A_{22})^{-1} \\ (A_{21} A_{11}^{-1} A_{12} - A_{22})^{-1} A_{21} A_{11}^{-1} & (A_{22} - A_{21} A_{11}^{-1} A_{12})^{-1} \end{pmatrix},$$

provided that all occurring inverses exist.

Block Diagonal Matrices

Block diagonal matrices are of the form

$$
A = \begin{pmatrix} A_{11} & & & \mathbf{0} \\ & A_{22} & & \\ & & \ddots & \\ \mathbf{0} & & & A_{kk} \end{pmatrix} =: \mathrm{diag}(A_{11}, A_{22}, \ldots, A_{kk}).
$$

All sub-matrices $A_{11}, \ldots, A_{kk}$ are square shaped, but they may be of different order. Many properties of block diagonal matrices are generalizations of the respective properties of diagonal matrices, as, for example,

$$
\det(\mathrm{diag}(A_{11}, A_{22}, \ldots, A_{kk})) = \prod_{i=1}^{k} \det A_{ii}.
$$

Accordingly, A is regular if and only if all of the A_{ii} are regular, i. e.,

$$
\mathrm{rank}(\mathrm{diag}(A_{11}, A_{22}, \ldots, A_{kk})) = \sum_{i=1}^{k} \mathrm{rank}(A_{ii}).
$$

Block Triangular Matrices

Matrices of the form

$$
A = \begin{pmatrix} A_{11} & A_{12} & \cdots & A_{1k} \\ & A_{22} & \cdots & A_{2k} \\ & & \ddots & \vdots \\ \mathbf{0} & & & A_{kk} \end{pmatrix} \quad \text{or} \quad A = \begin{pmatrix} A_{11} & & & \mathbf{0} \\ A_{21} & A_{22} & & \\ \vdots & \vdots & \ddots & \\ A_{k1} & A_{k2} & \cdots & A_{kk} \end{pmatrix}
$$

are called *upper* or *lower block triangular matrices*. Again the determinant of such a matrix is given by the product

$$
\det(A) = \det(A_{11}) \det(A_{22}) \cdots \det(A_{kk}).
$$

The rank of A is at least as great as the sum of the ranks of the diagonal blocks A_{ii}.

13.6.4 Hessenberg Matrices

A matrix $A \in \mathbb{R}^{n \times n}$ is said to be an *upper Hessenberg matrix* if $a_{ij} = 0$ for $j+1 < i$:

$$
A = \begin{pmatrix} a_{11} & a_{12} & a_{13} & \cdots & a_{1,n-1} & a_{1n} \\ a_{21} & a_{22} & a_{23} & \cdots & a_{2,n-1} & a_{2n} \\ & a_{32} & a_{33} & \cdots & a_{3,n-1} & a_{3n} \\ & & a_{43} & \cdots & a_{4,n-1} & a_{4n} \\ & & & \ddots & \vdots & \vdots \\ \mathbf{0} & & & & a_{n,n-1} & a_{nn} \end{pmatrix}.
$$

A is said to be a *lower* Hessenberg matrix, if $A^\top$ is an upper Hessenberg matrix.

13.6.5 Tridiagonal Matrices

A matrix $A \in \mathbb{R}^{n \times n}$, which is simultaneously both a lower and an upper Hessenberg matrix, is called a *tridiagonal matrix*. A is tridiagonal if and only if $a_{ij} = 0$ for $|i - j| > 1$:

$$A = \begin{pmatrix} a_{11} & a_{12} & & & & \mathbf{0} \\ a_{21} & a_{22} & a_{23} & & & \\ & a_{32} & \ddots & & \ddots & \\ & & & \ddots & a_{n-1,n-1} & a_{n-1,n} \\ \mathbf{0} & & & & a_{n,n-1} & a_{nn} \end{pmatrix}$$

Symmetric Hessenberg matrices are always symmetric tridiagonal matrices.

13.6.6 Band Matrices

The nonzero elements of a *band matrix* $A \in \mathbb{R}^{n \times n}$ can be found in the main diagonal $a_{11}, \ldots, a_{nn}$ and in some adjacent side-diagonals (super- or sub-diagonals).

Example (Band Matrix) A 10×10 matrix with two upper and four lower non-vanishing side-diagonals has the following sparsity structure:

$$\begin{pmatrix} * & * & * & & & & & & & \mathbf{0} \\ * & * & * & * & & & & & & \\ * & * & * & * & * & & & & & \\ * & * & * & * & * & * & & & & \\ * & * & * & * & * & * & * & & & \\ & * & * & * & * & * & * & * & & \\ & & * & * & * & * & * & * & * & \\ & & & * & * & * & * & * & * & * \\ & & & & * & * & * & * & * & * \\ \mathbf{0} & & & & & * & * & * & * & * \end{pmatrix}$$

The symbol $*$ denotes a value which may be non-zero.

Tridiagonal matrices are special band matrices with *one* lower and *one* upper side-diagonal.

13.6.7 Permutation Matrices

A matrix $P \in \mathbb{R}^{n \times n}$ is called a *permutation matrix* if exactly one element in each row and in each column has the value 1 while all other elements are zero. The multiplication PA permutes the rows of A, while AP permutes the columns. The determinant $\det(P)$ of permutation matrices is always ± 1; thus they are always regular. They are not commutative with respect to multiplication, but the product of two permutation matrices is also a permutation matrix.

The matrix $P^{\mathsf{T}} = P^{-1}$ permutes the columns in the same way that P permutes the rows. Hence, the transformation $A \to PAP^{\mathsf{T}}$ permutes the rows *and* the columns in the same way. In the context of a system of linear equations with coefficient matrix A, this transformation is equivalent to a reordering of the variables.

13.7 Singular Value Decomposition

In Section 13.4.5 the connection was established between linear transformations (functions) $F : X_n \to Y_m$ and matrices representing the coordinates of the images of basis vectors $u_1, \ldots, u_n$ of X_n. Thus different matrices may correspond to the same linear transformation, depending on the bases

$$\{u_1, u_2, \ldots, u_n\} \subset X_n \qquad \text{and} \qquad \{v_1, v_2, \ldots, v_m\} \subset Y_m$$

with respect to which F is represented. By applying a regular $n \times n$ matrix R, the basis $\{u_1, \ldots, u_n\}$ can be transformed into the basis

$$\{Ru_1, Ru_2, \ldots, Ru_n\} \subset X_n.$$

Similarly, each regular $m \times m$ matrix T enables a basis transformation from $\{v_1, \ldots, v_m\}$ to

$$\{Tv_1, Tv_2, \ldots, Tv_m\} \subset Y_m.$$

The set of all matrices which represent a given mapping (relative to some bases) form an equivalence class.

Definition 13.7.1 (Equivalence of Matrices) *Two $m \times n$ matrices A and B are said to be equivalent if there exist regular matrices R and T with $B = RAT$. If there exist two orthogonal matrices U and V such that $B = U^\mathsf{T}AV$, then A and B are said to be orthogonally equivalent.*

The matrix representation of a linear transformation depends on what bases are chosen for the vector spaces on which it acts. The initial choice of these bases may be unsuitable to the problem at hand, and it is quite natural to change the bases to obtain a *better* representation of the linear transformation. The following theorem reveals a particularly simple representation which makes it possible to gain many insights into linear transformations.

Theorem 13.7.1 (Singular Value Decomposition) *For each matrix $A \in \mathbb{R}^{m \times n}$ there exists a pair of orthogonal matrices*

$$U = [u_1, \ldots, u_m] \in \mathbb{R}^{m \times m} \quad \text{and} \quad V = [v_1, \ldots, v_n] \in \mathbb{R}^{n \times n},$$

which enable an equivalence transformation of A to diagonal form

$$U^\mathsf{T}AV = S := \operatorname{diag}(\sigma_1, \sigma_2, \ldots, \sigma_k) \in \mathbb{R}^{m \times n}, \qquad k = \min\{m, n\} \qquad (13.20)$$

such that the diagonal values of S satisfy

$$\sigma_1 \geq \sigma_2 \geq \cdots \geq \sigma_r > \sigma_{r+1} = \cdots = \sigma_k = 0 \qquad \text{with} \quad r = \operatorname{rank}(A).$$

If $A \in \mathbb{R}^{n \times n}$ is a regular matrix, then $\sigma_1 \geq \sigma_2 \geq \cdots \geq \sigma_n > 0$.

Proof: Golub, van Loan [50].

Definition 13.7.2 (Singular Vectors)   *The matrix U consists of the orthonormal eigenvectors $u_1, \ldots, u_m$ of $AA^\top$, which are called the left singular vectors. The matrix V consists of the orthonormal eigenvectors $v_1, \ldots, v_n$ of $A^\top A$, which are called the right singular vectors.*

Definition 13.7.3 (Singular Values)   *The diagonal elements $\sigma_1, \ldots, \sigma_k$ of the diagonal matrix S, which is orthogonally equivalent to A, are called the singular values of the matrix A.*

From

$$S^\top S = (U^\top AV)^\top (U^\top AV) = V^\top A^\top U U^\top AV = V^{-1} A^\top AV$$

the orthogonal equivalence of $A^\top A$ and

$$S^\top S = \mathrm{diag}(\sigma_1^2, \sigma_2^2, \ldots, \sigma_r^2, 0, \ldots, 0)$$

can be proved. Since similar matrices have the same spectrum (cf. Section 15.1.2), it follows that

$$\sigma_i = \sqrt{\lambda_i}, \quad \lambda_i \in \lambda(A^\top A), \qquad i = 1, 2, \ldots, n.$$

The singular values of a matrix $A \in \mathbb{R}^{m \times n}$ are unique. The singular vectors on the other hand are unique only if σ_i^2 is a simple eigenvalue of $A^\top A$. For degenerate singular values, the corresponding singular vectors can be chosen as an arbitrary orthonormal basis of the associated eigenspace.

The spectral norm and the Frobenius norm of a matrix $A \in \mathbb{R}^{n \times n}$ are strongly related to the singular values of A:

$$
\begin{aligned}
\|A\|_2 &= \sigma_1, \\
\|A\|_F &= \sqrt{\sigma_1^2 + \cdots + \sigma_k^2}.
\end{aligned}
$$

The determinant $\det(A)$ of a square matrix $A \in \mathbb{R}^{n \times n}$ is also related to its singular values:

$$|\det(A)| = |\det(U)\,\det(S)\,\det(V^\top)| = \prod_{i=1}^{k} \sigma_i$$

since the determinant of an orthogonal matrix is either $+1$ or -1.

13.7.1 Geometry of Linear Transformations

If, in the linear space X defined by $x = Vx'$ and in the space Y defined by $y = Uy'$, the system of coordinates is transformed, then the linear mapping originally represented by A is represented by S:

$$y' = U^\top y = U^\top Ax = U^\top A(Vx') = (U^\top AV)x' = Sx'.$$

In the new system of coordinates, the linear mapping, originally underlying matrix A, has a very simple representation. In the case of an $n \times n$ matrix, $y' = (\eta'_1, \ldots, \eta'_n)^\top$ is connected to $x' = (\xi'_1, \ldots, \xi'_n)^\top$ by

$$\eta'_1 = \sigma_1 \xi'_1, \ldots, \eta'_r = \sigma_r \xi'_r, \quad \eta'_{r+1} = 0, \ldots, \eta'_n = 0. \tag{13.21}$$

The first coordinate-axis of the space X is mapped onto the first coordinate-axis of the space Y with the scaling factor $\sigma_1 > 0$, in the same way the second, $\ldots$, rth coordinate-axes are. The remaining coordinate-axes are mapped to the zero vector $0 \in Y$.

From (13.21) it can be seen that under $y' = Sx'$, the surface of the unit sphere

$$K_n = \{x' : \|x'\|_2 = 1\}$$

is mapped onto the r-dimensional hyper-ellipsoid

$$E(\sigma_1, \ldots, \sigma_n) = \{y' = (\eta'_1, \ldots, \eta'_n)^\top : \left(\frac{\eta'_1}{\sigma_1}\right)^2 + \cdots + \left(\frac{\eta'_r}{\sigma_r}\right)^2 = 1,$$
$$\eta'_{r+1} = \cdots = \eta'_n = 0\}.$$

In this hyper-ellipsoid $(\sigma_1, 0, \ldots, 0)^\top$ is (one of) the vector(s) with a maximum distance from the zero vector $0 \in \mathbb{R}^n$. If $r < n$, the hyper-ellipsoid $E(\sigma_1, \ldots, \sigma_n)$ contains the zero vector. If A is regular, $E(\sigma_1, \ldots, \sigma_n)$ does not contain the zero vector, and $(0, \ldots, 0, \sigma_n)^\top$ is (one of) the vector(s) with minimal distance to the zero vector. In this case

$$S^{-1} = \operatorname{diag}(1/\sigma_1, 1/\sigma_2, \ldots, 1/\sigma_n),$$

from which one can derive that $1/\sigma_1, \ldots, 1/\sigma_n$ are the singular values of A^{-1}.

In the case of a 2×2 matrix, the singular value decomposition can be illustrated geometrically. The orthogonal matrices of the decomposition $A = USV^\top$,

$$U = \begin{pmatrix} \cos\alpha & -\sin\alpha \\ \sin\alpha & \cos\alpha \end{pmatrix} \quad \text{and} \quad V^\top \overset{!}{=} \begin{pmatrix} \cos\beta & \sin\beta \\ -\sin\beta & \cos\beta \end{pmatrix},$$

specify rotations with angles α and $-\beta$ centered at the origin. The composite mapping $USV^\top$ is shown in Fig. 13.4.

13.7.2 Structure of Linear Transformations

From the singular value decomposition $A = USV^\top$, a great deal of information about the structure of the mapping represented by a matrix $A \in \mathbb{R}^{n \times n}$ can be obtained. For instance, orthogonal bases of the range space $\mathcal{R}(A)$ and of the null space $\mathcal{N}(A)$ can be gleaned directly:

$$\mathcal{R}(A) = \operatorname{span}\{u_1, \ldots, u_r\} \quad \text{and} \quad \mathcal{N}(A) = \operatorname{span}\{v_{r+1}, \ldots, v_n\}.$$

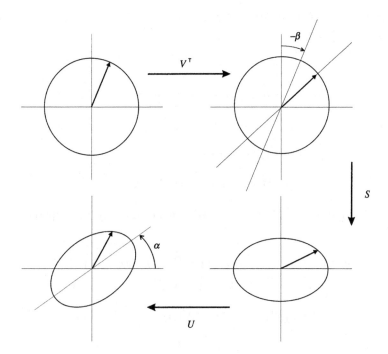

Figure 13.4: Composite mapping $USV^\top = A \in \mathbb{R}^{2\times2}$. A vector x is first mapped to $V^\top x$ then $SV^\top x$ and finally to $USV^\top x = Ax$.

Using the row vectors $z_1, \ldots, z_n$ of A, one can also express $Ax = 0$ in the form

$$z_i^\top x = \langle z_i, x \rangle = 0, \qquad i = 1, 2, \ldots, n.$$

Each vector $x \in \mathcal{N}(A)$ is thus orthogonal to each column vector of $A^\top$ and also therefore, to each vector in

$$\mathcal{R}(A^\top) = \operatorname{span}\{z_1, z_2, \ldots, z_n\}.$$

$\mathcal{R}(A^\top)$ is the orthogonal complement of the null space:

$$\mathcal{R}(A^\top) = \mathcal{N}(A)^\perp.$$

Definition 13.7.4 (Sum of two Subspaces, Direct Sum)   *If S_1 and S_2 are subspaces of $\mathbb{R}^n$, then their sum, defined by*

$$S_1 + S_2 := \{v = v_1 + v_2 : v_1 \in S_1, v_2 \in S_2\}$$

is also a subspace of $\mathbb{R}^n$; if in addition $S_1 \cap S_2 = \{0\}$, then

$$S_1 \oplus S_2 := \{v = v_1 + v_2 : v_1 \in S_1, v_2 \in S_2\}$$

is called the direct sum of S_1 and S_2.

Definition 13.7.5 (Orthogonal Projection)   *Each vector $v \in S_1 \oplus S_2$ can be uniquely written as*

$$v = v_1 + v_2, \qquad v_1 \in S_1, \quad v_2 \in S_2.$$

The vector v_i is called the orthogonal projection of v onto the subspace $S_i, i = 1, 2$.

Each linear transformation $A : U \to W$ satisfies:

$$U = \mathcal{N}(A) \oplus \mathcal{N}(A)^{\perp} = \mathcal{N}(A) \oplus \mathcal{R}(A^{\mathsf{T}})$$
$$W = \mathcal{R}(A) \oplus \mathcal{R}(A)^{\perp} = \mathcal{R}(A) \oplus \mathcal{N}(A^{\mathsf{T}}).$$

Hence, the vectors $u \in U$ and $w \in W$ can be uniquely decomposed as (cf. Fig. 13.5)

$$u = u_A + u_{\mathcal{N}} \qquad \text{with} \quad u_A \in \mathcal{R}(A^{\mathsf{T}}), \quad u_{\mathcal{N}} \in \mathcal{N}(A)$$
$$w = w_A + w_{\mathcal{N}} \qquad \text{with} \quad w_A \in \mathcal{R}(A), \quad w_{\mathcal{N}} \in \mathcal{N}(A^{\mathsf{T}}).$$

u_A is the orthogonal projection of u onto $\mathcal{N}(A)^{\perp} = \mathcal{R}(A^{\mathsf{T}})$ and

$$P_{\mathcal{R}(A^{\mathsf{T}})} : U \to \mathcal{R}(A^{\mathsf{T}})$$

denotes the corresponding projection mapping. Similarly,

$$P_{\mathcal{R}(A)} : W \to \mathcal{R}(A)$$

denotes the projection onto the range space of A.

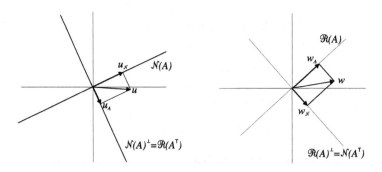

Figure 13.5: Decomposition of $u \in U$ and $w \in W$ into orthogonal components.

13.7.3 Generalized Inverse Mappings

Each mapping A can be expressed as a composite mapping

$$A = \hat{A} P_{\mathcal{R}(A^{\mathsf{T}})},$$

where $\hat{A} : \mathcal{N}(A)^{\perp} \to \mathcal{R}(A)$ is the restriction of A to $\mathcal{N}(A)^{\perp}$:

$$\hat{A}x := Ax \qquad \text{for all} \quad x \in \mathcal{N}(A)^{\perp}.$$

The mapping $\hat{A}$ is injective and has therefore an inverse

$$\hat{A}^{-1} : \mathcal{R}(A) \to \mathcal{N}(A)^{\perp}.$$

Based on $\hat{A}^{-1}$, a mapping can be defined which leads to a generalization of the inverse of a matrix.

Definition 13.7.6 (Generalized Inverse, Pseudo-Inverse)   *By defining*

$$A^{+}w := \begin{cases} \hat{A}^{-1}w & \text{for} \quad w \in \mathcal{R}(A) \\ 0 & \text{for} \quad w \in \mathcal{R}(A)^{\perp} \end{cases}$$

a composite mapping

$$A^{+} = \hat{A}^{-1} P_{\mathcal{R}(A)},$$

called the generalized inverse or pseudo-inverse A^{+} of A, is established.

The pseudo-inverse defined in this way is *uniquely* determined and satisfies the *Moore-Penrose conditions*:

$$\begin{aligned} (AA^{+})A &= A & (13.22) \\ (A^{+}A)A^{+} &= A^{+} & (13.23) \\ (AA^{+})^{\top} &= AA^{+} & (13.24) \\ (A^{+}A)^{\top} &= A^{+}A. & (13.25) \end{aligned}$$

From (13.22) and (13.24) it follows that

$$P_{\mathcal{R}(A)} = AA^{+},$$

and from (13.23) and (13.25) it follows that

$$P_{\mathcal{N}(A)^{\perp}} = A^{+}A.$$

If $A \in \mathbb{R}^{n \times n}$ is regular, then the projections $P_{\mathcal{R}(A)}$ and $P_{\mathcal{N}(A)^{\perp}}$ are identical mappings and it can be established that

$$A^{+} = \hat{A}^{-1} P_{\mathcal{R}(A)} = \hat{A}^{-1} I = A^{-1}.$$

Example (Generalized Inverse Mapping)   In the case of a 2×2 matrix A with $\text{rank}(A) = 1$, an intuitive geometric interpretation of the generalized inverse mapping can be given.

Fig. 13.6 shows a general vector u and its unique orthogonal decomposition into $u_{\mathcal{N}}$ (the projection of u onto the null space $\mathcal{N}(A)$) and u_A (the projection of u onto the orthogonal complement of $\mathcal{N}(A)$). None of the components vanish. The vector u is mapped to Au in the range space $\mathcal{R}(A)$.

Also, the vector w shown in Fig. 13.6 has a non-vanishing component w_A along $\mathcal{R}(A)$. The generalized inverse mapping A^{+} assigns w to u_A in the subspace $\mathcal{N}(A)^{\perp}$.

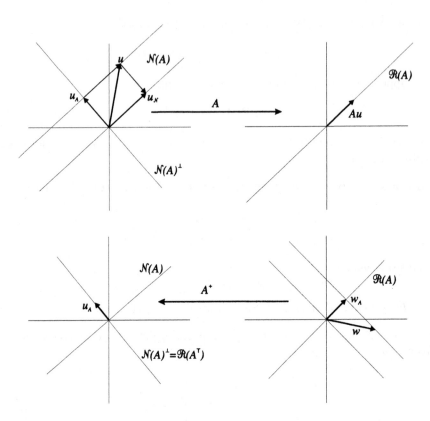

Figure 13.6: The mapping of $u \in \mathbb{R}^2$ to Au by $A \in \mathbb{R}^{2\times2}$ with rank$(A) = 1$ (top figure) and the generalized inverse mapping of $w \in \mathbb{R}^2$ by $A^+ \in \mathbb{R}^{2\times2}$ (bottom figure).

Example (Diagonal Matrices)   The pseudo-inverse of a diagonal matrix

$$A = \mathrm{diag}(\alpha_1,\ldots,\alpha_k,0,\ldots,0) \in \mathbb{R}^{m\times n}, \quad k \le \min\{m,n\}$$

with $\alpha_i \ne 0, i = 1,\ldots,k$ is given by

$$A^+ = \mathrm{diag}(1/\alpha_1,\ldots,1/\alpha_k,0,\ldots,0) \in \mathbb{R}^{n\times m}, \tag{13.26}$$

since $AA^+ = \mathrm{diag}(1,\ldots,1,0,\ldots,0) \in \mathbb{R}^{m\times m}$ and $A^+A = \mathrm{diag}(1,\ldots,1,0,\ldots,0) \in \mathbb{R}^{n\times n}$ obviously satisfy the Moore-Penrose conditions.

Example (Block Matrices)   The pseudo-inverse of a block matrix

$$A = \begin{pmatrix} B & 0 \\ 0 & 0 \end{pmatrix} \in \mathbb{R}^{m\times n}, \quad B \in \mathbb{R}^{k\times k}, \quad k \le \min\{m,n\},$$

where B is a regular $k \times k$ matrix, is given by

$$A^+ = \begin{pmatrix} B^{-1} & 0 \\ 0 & 0 \end{pmatrix} \in \mathbb{R}^{n\times m}.$$

Again a proof can be obtained by verifying the Moore-Penrose conditions.

If two matrices $A, B \in \mathbb{R}^{m \times n}$ are orthogonally equivalent, such that a relation of the form

$$A = UBV^\top$$

with orthogonal matrices $U \in \mathbb{R}^{m \times m}, V \in \mathbb{R}^{n \times n}$ holds, then the corresponding relation between the pseudo-inverses A^+ and B^+ is given by

$$A^+ = VB^+U^\top. \tag{13.27}$$

From (13.26) and (13.27) it follows that the pseudo-inverse of a general matrix $A \in \mathbb{R}^{m \times n}$ can be derived using its singular value decomposition

$$A = USV^\top \quad \text{with} \quad S = \mathrm{diag}(\sigma_1, \ldots, \sigma_k, 0, \ldots, 0)$$

because

$$A^+ = VS^+U^\top \quad \text{with} \quad S^+ = \mathrm{diag}(1/\sigma_1, \ldots, 1/\sigma_k, 0, \ldots, 0).$$

In general, A^+ does *not* continuously depend on the elements of A. The nature of the discontinuity is discussed in more detail later in the book.

Example (Discontinuity of A^+) For

$$A = \begin{pmatrix} 1 & 1 \\ 1 & 1+\varepsilon \end{pmatrix}$$

and $\varepsilon > 0$, A is regular and

$$A^+ = A^{-1} = \frac{1}{\varepsilon} \begin{pmatrix} 1+\varepsilon & -1 \\ -1 & 1 \end{pmatrix}. \tag{13.28}$$

If, on the other hand, $\varepsilon = 0$, then A is singular with the pseudo-inverse

$$A^+ = \begin{pmatrix} 1 & 0 \\ 0 & 0 \end{pmatrix},$$

which cannot be derived from (13.28) by the limiting process $\varepsilon \to 0$.

13.7.4 General Solution of Linear Systems

Using the generalized inverse, all cases occurring in the solution of inhomogeneous linear systems $Ax = b$ with $b \neq 0$ and $A \in \mathbb{R}^{n \times n}$ can be described in a concise manner.

Regular Matrix

If $A \in \mathbb{R}^{n \times n}$ is a regular matrix (and, hence, $\mathrm{rank}(A) = n$), then from $A^+ = A^{-1}$ it follows that

$$x_0 = A^{-1}b$$

solves the linear system $Ax = b$. In practice, A^{-1} will *not* be used for the determination of the solution due to inefficiency. Instead, almost all efficient methods for determining the solution are based on suitable factorizations of A, which are discussed later. Ill-conditioned regular matrices are also discussed later.

Singular Matrix

If $A \in \mathbb{R}^{n \times n}$ is a singular matrix with $\mathrm{rank}(A) = k < n$, there are two cases to be distinguished:

Case 1: $b \in \mathcal{R}(A)$

In this case, the vector $x_0 = A^+b$ is a solution of the linear system, but the same is true for each vector $v = x_0 + z$ with $z \in \mathcal{N}(A)$:

$$Av = A(x_0 + z) = Ax_0 + Az = b + 0 = b.$$

In such cases, the solution of the linear system $Ax = b$ is the *manifold*

$$X := \left\{ v : \; v = A^+b + Hy; \; y \in \mathbb{R}^{n-k} \right\},$$

where the column vectors of $H \in \mathbb{R}^{n \times (n-k)}$ form a basis of the null space $\mathcal{N}(A)$ of the matrix A. However there exists one *special* vector $x_0 \in X$ with minimal Euclidean length (cf. Fig. 13.7).

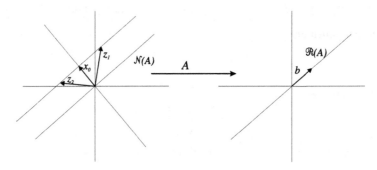

Figure 13.7: The mapping of several vectors by a singular matrix onto the vector $b = Ax_0 = Az_i \in \mathcal{R}(A)$.

Case 2: $b \notin \mathcal{R}(A)$

If the vector b *cannot* be expressed as a linear combination of the column vectors of A, then $Ax = b$ is *inconsistent*. In such cases, no vector x_0 satisfying $Ax_0 = b$ exists: the system is *not solvable*.

Depending on the actual problem, it may be useful to determine the vector x_0 (or manifold X) for which the residual $r = Ax - b$ has minimal length (norm).

Definition 13.7.7 (Normal Solution)   *A vector* $v \in \mathbb{R}^n$ *is called a normal solution of* $Ax = b$ *if* v *has minimal residual:*

$$\|Av - b\| = \min \left\{ \|Ay - b\| : \; y \in \mathbb{R}^n \right\}.$$

Hence, a normal solution of $Ax = b$ is the vector in the image space $\mathcal{R}(A)$ which has minimal distance to b. It can be determined from the orthogonal projection of b onto $\mathcal{R}(A)$

$$Ax_0 = b_A = AA^+b.$$

That is,

$$x_0 = A^+ b$$

is proved to be a normal solution. Minimizing the residual, does not uniquely determine a single solution but rather it determines a *manifold of solutions*. In addition to x_0, each vector $v = x_0 + z$ with $z \in \mathcal{N}(A)$ yields a minimal value for the residual $\|b_A - b\|_2$; the manifold of solutions is again given as

$$X := \left\{ v : v = A^+ b + H y; \ y \in \mathbb{R}^{n-k} \right\}.$$

Again, there exists a unique vector in X with minimal length.

Definition 13.7.8 (Pseudo-Normal Solution)   *A vector $x_0 \in \mathbb{R}^n$ is said to be the pseudo-normal solution of $Ax = b$, if x_0 is the unique normal solution with minimal length:*

$$\|x_0\| = \min \left\{ \|v\| : v \text{ is a normal solution} \right\}.$$

Unique Definition of Solution

The results of this section can be summarized by the following theorem:

Theorem 13.7.2   *The one and only pseudo-normal solution of $Ax = b$ is $x_0 = A^+ b$.*

13.7.5 The Solution of Homogeneous Systems

If the right-hand side of a linear system is the zero vector, then the linear system

$$Ax = 0$$

is said to be *homogeneous*. If the matrix $A \in \mathbb{R}^{n \times n}$ has full rank, then only the trivial solution $x = 0$ exists. On the other hand, if the matrix A is singular, with $\operatorname{rank}(A) = k < n$, then the manifold of solutions is given by the $(n - k)$ dimensional subspace $\mathcal{N}(A)$ of $\mathbb{R}^n$—the null space of A.

An orthogonal basis for the $(n - k)$ dimensional solution space $\mathcal{N}(A)$ can be obtained from the singular value decomposition $A = U S V^\top$. With the notation $V = [v_1, \ldots, v_n]$ and $U^\top = [u_1, \ldots, u_n]$ it can be proved that

$$A v_i = \sigma_i u_i, \quad i = 1, 2, \ldots, n,$$

i.e., $A v_i = 0$ for $i = (k + 1), \ldots, n$; the column vectors $\{v_{k+1}, \ldots, v_n\}$ of V corresponding to the singular values $\sigma_{k+1} = \cdots = \sigma_n = 0$ form an orthogonal basis of the null space $\mathcal{N}(A)$.

13.7.6 Linear Data Fitting

The design of mathematical models is often based on approximation algorithms, particularly if the available data is affected by stochastic perturbations.

For this purpose, a large amount of data is usually collected, more than is necessary for the determination of model parameters. It follows that the resulting system is over-determined (and therefore, for the most part, unsolvable).

It is therefore only possible to find the vector x_0 whose image Ax_0 has minimum distance to b:

$$\text{dist}(Ax_0, b) = \min\left\{\text{dist}(Ax, b) : x \in \mathbb{R}^n\right\}.$$

This distance can be quantified by using an l_p-norm of the residual $Ax - b$:

$$\text{dist}(Ax, b) := \|Ax - b\|_p.$$

As to which l_p-norm is used for this purpose basically depends on the problem itself. Depending on the value chosen for the parameter p, the deviations are given different weight of in proportion to the distance. For instance, the larger the value of p, the smaller the effect of small deviations (< 1). On the other hand, large errors get heavier weight. Subsequently, single *outliers* (caused by unsystematic perturbations of a measuring instrument etc.) can greatly affect the result.

The Euclidean norm is the most logical choice of a generalization of the way distance is measured in (three dimensional) space. It also requires the least computational effort for the solution of the fitting problem. Because of these facts, the l_2-norm is used for defining the distance in most practical applications. The linear data fitting problem

$$\min\left\{\|Ax - b\|_2 : x \in \mathbb{R}^n\right\}, \qquad A \in \mathbb{R}^{m \times n}, \quad b \in \mathbb{R}^m, \quad m > n,$$

can be viewed as a minimization problem. The solution of this problem can be interpreted as follows: The solution vectors are those $x \in \mathbb{R}^n$ which are mapped by A onto the orthogonal projection of b onto the image space $\mathcal{R}(A)$. If the mapping A is injective, i.e., if $\text{rank}(A) = n$, then the solution is unique. Otherwise a *zero-manifold* given by a residual class of the null space $\mathcal{N}(A)$ is obtained.

In the following sections the l_2-norm, i.e., the Euclidean norm, is always used to measure the distance, unless otherwise indicated.

Solution of Linear Least Squares Problems

Similarly to systems of linear equations the solution of linear least squares problems can be treated with the help of the pseudo-inverse.

For $A \in \mathbb{R}^{m \times n}$, the existence of the singular value decomposition $A = USV^\top$ is guaranteed by Theorem 13.7.1. Defining

$$p := V^\top x = \begin{pmatrix} p_1 \\ p_2 \end{pmatrix} \in \mathbb{R}^n \qquad \text{and} \qquad q := U^\top b = \begin{pmatrix} q_1 \\ q_2 \end{pmatrix} \in \mathbb{R}^m$$

with $p_1, q_1 \in \mathbb{R}^k$, one obtains—the Euclidean norm is preserved by orthogonal transformations—the minimization problem

$$\|b - Ax\|_2 = \|U^\top(b - USV^\top x)\|_2 = \|q - Sp\|_2 = \left\|\begin{pmatrix} q_1 - S_k p_1 \\ q_2 \end{pmatrix}\right\|_2 \longrightarrow \min$$

with $S_k := \operatorname{diag}(\sigma_1, \ldots, \sigma_k) \in \mathbb{R}^{k \times k}$.

The norm of the residual $\|r\|_2 = \|b - Ax\|_2$ is minimal for arbitrary $p_2 \in \mathbb{R}^{n-k}$ and

$$p_1 = S_k^{-1} q_1 \in \mathbb{R}^k .$$

Hence, a normal solution has been found. Obviously, for $p_2 = 0$ the Euclidean norm $\|p\|_2$ and thus $\|x\|_2 = \|Vp\|_2$ are minimal. This implies that

$$x_0 = V \begin{pmatrix} S_k^{-1} & 0 \\ 0 & 0 \end{pmatrix} U^\top b = A^+ b$$

is the unique normal solution for which the Euclidean norm $\|x_0\|_2$ is also minimal, i.e., x_0 is the pseudo-normal solution.

The results of this section can be summarized in an extension of Theorem 13.7.2 to rectangular matrices:

Theorem 13.7.3   *The unique pseudo-normal solution of a least squares problem $Ax = b$ with $A \in \mathbb{R}^{m \times n}, x \in \mathbb{R}^n, b \in \mathbb{R}^m$ and $m \geq n$ is given by*

$$x_0 = A^+ b$$

with $A^+ \in \mathbb{R}^{n \times m}$.

If A has full rank, i.e., $\operatorname{rank}(A) = n$, then

$$A^+ = (A^\top A)^{-1} A^\top.$$

In this case, the solution of the linear least squares problem coincides, as expected, with the solution of the normal equations

$$A^\top A x = A^\top b.$$

13.8 The Condition of Linear Systems

The condition of a linear system can easily be illustrated in two dimensions. The solution of

$$
\begin{array}{cccccll}
a_{11} x & + & a_{12} y & = & b_1 & \quad (\text{straight line } g_1) \\
a_{21} x & + & a_{22} y & = & b_2 & \quad (\text{straight line } g_2)
\end{array}
$$

is the intersection point S of the two lines g_1 and g_2 in the x-y plane (cf. Fig. 13.8).

In the case of a well conditioned 2×2 system, the intersection point S varies only a little if the original lines are replaced by the *perturbed* lines $\tilde{g}_1$ and $\tilde{g}_2$. Given,

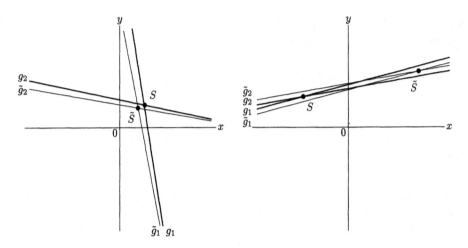

Figure 13.8: Well conditioned, and ill-conditioned 2×2 systems.

on the other hand, an ill conditioned 2×2 system, the the distance between S and $\tilde{S}$ may be very large, even if the perturbations are small; this is due to the fact that the angle between the two lines is very small (a glancing intersection).

The smaller the angle between the lines (the closer they are to being parallel) the worse the condition of the system. If the two lines are perpendicular to one another, then the system is optimally conditioned.

These graphically oriented considerations concerning the condition of a linear system can be generalized to n-dimensional spaces (Golub, Ortega [48]) though such an approach has some disadvantages, e.g., there is no way of proving statements regarding the favorable/unfavorable constellations of the right-hand side b. For this reason another approach, based on the first order condition (cf. Section 2.6), is generally accepted for the quantitative analysis of the condition of linear systems.

13.8.1 Condition of Regular Systems

The linear system $Ax = b$ may or may not be sensitive to perturbations; a more precise quantification may be gained by analyzing the system (parameterized by $t \in \mathbb{R}$)

$$(A + t \, \Delta A)x(t) = b + t \, \Delta b, \qquad x(0) = x^*$$

where $\Delta A \in \mathbb{R}^{n \times n}$ and $\Delta b \in \mathbb{R}^n$ (Golub, Van Loan [50]). Given that A is regular, the function $x(t)$ is differentiable at $t = 0$:

$$\dot{x}(0) = A^{-1}(\Delta b - \Delta A \, x^*).$$

The Taylor expansion of $x(t)$ is given by

$$x(t) = x^* + t\dot{x}(0) + O(t^2).$$

It thus follows that the absolute error can be estimated using

$$\|\Delta x(t)\| \quad = \quad \|x(t) - x^*\| \le t\|\dot{x}(0)\| + O(t^2) \le$$

$$\le \quad t\|A^{-1}\| \left(\|\Delta b\| + \|\Delta A\|\|x^*\|\right) + O(t^2)$$

and (due to $\|b\| \le \|A\|\|x^*\|$) the relative error can be estimated using

$$\frac{\|\Delta x(t)\|}{\|x^*\|} \quad \le \quad t\|A^{-1}\| \left(\frac{\|\Delta b\|}{\|x^*\|} + \|\Delta A\|\right) + O(t^2) \le$$

$$\le \quad \|A\|\|A^{-1}\| \left(t\frac{\|\Delta b\|}{\|b\|} + t\frac{\|\Delta A\|}{\|A\|}\right) + O(t^2). \qquad (13.29)$$

The error estimation (13.29) gives rise to the following definition:

Definition 13.8.1 (Condition Number of a Regular Matrix)   *The condition number of a regular matrix A, denoted $\mathrm{cond}(A)$ or $\kappa(A)$ is defined to be*

$$\mathrm{cond}(A) = \kappa(A) := \|A\|\|A^{-1}\|. \qquad (13.30)$$

By introducing the notations

$$\rho_A(t) := t\frac{\|\Delta A\|}{\|A\|} \qquad \text{and} \qquad \rho_b(t) := t\frac{\|\Delta b\|}{\|b\|}$$

for the relative errors of A and b, the error estimate (13.29) can be written as

$$\frac{\|\Delta x(t)\|}{\|x^*\|} \le \mathrm{cond}(A)\,(\rho_A + \rho_b) + O(t^2).$$

Generally, the relative data errors ρ_A of A and ρ_b of b affect the result of the perturbed system to an extent that is proportional to $\mathrm{cond}(A)$. If the matrix A has a large condition number $\mathrm{cond}(A)$, then the linear system $Ax = b$ is said to be *ill-conditioned*.

The value of the condition number $\mathrm{cond}(A)$ depends on the norm being used. However, due to norm equivalence in $\mathbb{R}^n$ it can be proved, for instance, that

$$\frac{1}{n}\kappa_\infty(A) \le \kappa_2(A) \le n\kappa_\infty(A)$$

holds for the condition numbers

$$\kappa_\infty(A) := \|A\|_\infty \|A^{-1}\|_\infty \qquad \text{and} \qquad \kappa_2 := \|A\|_2 \|A^{-1}\|_2 = \frac{\sigma_1(A)}{\sigma_n(A)}$$

where $\sigma_1(A) \ge \cdots \ge \sigma_n(A) > 0$ are the singular values of A (cf. Golub and Van Loan [50]).

It can be shown that $\kappa_p(A)$ for each p-norm. The closer $\kappa_p(A)$ comes to the minimal value 1, the better the condition of the linear system $Ax = b$. Orthogonal matrices Q have an optimal condition number, namely $\kappa_2(Q) = 1$.

Example (Effects of Ill-Conditioning)   The linear system

$$
\begin{array}{rcrcrcl}
3y_1 &+& 1.5y_2 &+& y_3 &=& 0.2 \\
1.5y_1 &+& y_2 &+& 0.75y_3 &=& 1 \\
y_1 &+& 0.75y_2 &+& 0.6y_3 &=& 1
\end{array}
\qquad (13.31)
$$

has the coefficient matrix

$$
A = \begin{pmatrix} 3.00 & 1.50 & 1.00 \\ 1.50 & 1.00 & 0.75 \\ 1.00 & 0.75 & 0.60 \end{pmatrix},
$$

whose inverse A^{-1} can easily be calculated:

$$
A^{-1} = \begin{pmatrix} 3 & -12 & 10 \\ -12 & 64 & -60 \\ 10 & -60 & 60 \end{pmatrix}.
$$

Hence $\|A\|_\infty = 5.5$, $\|A^{-1}\|_\infty = 136$, and so $\kappa_\infty(A) = 748$.

This relatively large value (large for a system consisting of only three equations and three unknowns) shows that the linear system (13.31) may be very sensitive to data perturbations. In fact, if the right-hand side $b = (0.2, 1, 1)^\top$ is perturbed by $\Delta b = (0.01, -0.01, 0.01)^\top$, then the solution vector $x = (-1.4, 1.6, 2)^\top$ changes to

$$
x + \Delta x \qquad \text{with} \qquad \Delta x = (0.25, -1.36, 1.30)^\top.
$$

Hence, it follows that

$$
\|\Delta x\|_\infty / \|x\|_\infty = 1.36/2 = 0.68 \qquad \text{and} \qquad \|\Delta b\|_\infty / \|b\|_\infty = 0.01/1 = 0.01
$$

and so the relative error is increased by a factor of 68. The value of $\kappa_\infty(A) = 748$ indicates that even more unfavorable cases can occur for this coefficient matrix.

If a stable algorithm is used for the solution of a linear system, one must be aware of the fact that at least some rounding errors resulting from the execution of the algorithm within a floating-point arithmetic will be propagated in the same way as data errors, i.e., they can possibly affect the results to an extent that is proportional to $\kappa(A)$. In the case of an ill-conditioned system, the result will be greatly affected by rounding errors.

In a given floating-point arithmetic, the single rounding errors are bounded by the relative rounding error bound *eps*. The relative effect of a single rounding error may be, in the worst case, $\kappa(A)\,eps$. A relative perturbation of order 1 means that the effect of the perturbation is of the same order as the original quantity. Hence, it can be expected that if

$$
\kappa(A)\,eps \geq 1
$$

then the numerical solution of the linear system, with the coefficient matrix A, on a computer with rounding error bound *eps* may not even be correct in the first digit. Such a linear system or the corresponding matrix is said to be *numerically singular* with respect to the floating-point arithmetic under consideration.

Generally, the solution of such a linear system is meaningless for one reason alone: that at least some of the coefficients or the components of the right-hand sides will be affected with a relative error of order *eps* due to rounding. The effect of these data inaccuracies yields a fundamental change in the solution, as indicated by the error estimate given above as well as by errors which are discussed in the following sections.

The sensitivity of linear systems, whose matrix is *not* necessarily square or regular, to perturbations is discussed in the following sections.

13.8.2 Effects of a Perturbed Right-Hand Side

If only the right-hand side of a linear system is perturbed, then the following situation occurs:

linear system		solution
original system	$Ax = b$	$x_0 = A^+ b$
perturbed system	$Ax = b + \Delta b$	$\tilde{x}_0 = A^+(b + \Delta b)$

From the definition of the perturbation of the solution it follows that

$$\Delta x_0 \ = \ \tilde{x}_0 - x_0 \ = \ A^+(b + \Delta b) - A^+ b = A^+ \Delta b$$

and further that

$$\|\Delta x_0\| \ \leq \ \|A^+\| \, \|\Delta b\|. \tag{13.32}$$

This estimate of the absolute error shows that even a slight perturbation of the right-hand side b can produce a large change in the solution x_0, if $\|A^+\|$ or, in the case of a regular matrix, $\|A^{-1}\|$ is large. But being *large* cannot be determined by an absolute criterion; a *relative* error estimate is often more easily interpreted.

From $Ax_0 = b_A$ it follows that

$$\|b_A\| \ \leq \ \|A\| \, \|x_0\|,$$

and provided that $\|A\| > 0$,

$$\|x_0\| \geq \frac{\|b_A\|}{\|A\|}. \tag{13.33}$$

Combining (13.32) and (13.33) yields

$$\frac{\|\Delta x_0\|}{\|x_0\|} \ \leq \ \frac{\|A^+\| \, \|\Delta b\|}{\|x_0\|} \ \leq \ \frac{\|A^+\| \, \|\Delta b\|}{\|b_A\|/\|A\|}.$$

By reducing the compound fraction and expanding it with $\|b\|$, the following estimate is obtained for the relative error of x_0:

$$\frac{\|\Delta x_0\|}{\|x_0\|} \ \leq \ \|A\| \, \|A^+\| \frac{\|\Delta b\|}{\|b\|} \cdot \frac{\|b\|}{\|b_A\|}$$

If

$$\varrho_b := \frac{\|\Delta b\|}{\|b\|}$$

denotes the relative perturbation of the right-hand side, the following estimate can be given for the relative error ϱ_x of x_0:

$$\varrho_x \leq \|A\| \, \|A^+\| \frac{\|b\|}{\|b_A\|} \varrho_b. \tag{13.34}$$

In (13.34) there are two factors amplifying the effects of the relative error ρ_b of the right-hand side: There is the factor $\|A\| \|A^+\|$, which depends only on the matrix A, and the factor $\|b\|/\|b_A\|$, which depends on both b and A.

Definition 13.8.1 can be generalized as follows:

Definition 13.8.2 (Condition Number of a General Matrix)   *The condition number of a rectangular matrix A, denoted by* cond(A) *or* $\kappa(A)$, *is defined to be*

$$\text{cond}(A) = \kappa(A) := \|A\| \|A^+\|.$$

In the case of regular matrices $b_A = b$ and so $\|b\|/\|b_A\| = 1$. For singular matrices $(\text{rank}(A) < n)$, the factor $\|b\|/\|b_A\|$ is also of importance. Its geometrical interpretation can be seen from the relation

$$\frac{\|b\|}{\|b_A\|} = \left| \frac{1}{\cos \varphi} \right|,$$

which can be derived using the singular value decomposition of A (cf. Fig. 13.9).

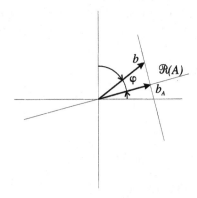

Figure 13.9: Geometric interpretation of the condition number cond$(b; A) = \|b\|/\|b_A\|$.

Definition 13.8.3 (Condition Number of a Right-Hand Side)   *The condition number of the right-hand side $b \in \mathbb{R}^m$ of a linear system with coefficient matrix $A \in \mathbb{R}^{m \times n}$ is defined to be*

$$\text{cond}(b; A) = \kappa(b; A) := \frac{\|b\|}{\|b_A\|}.$$

At best $\text{cond}(b; A) = 1$, which corresponds to $b \in \mathcal{R}(A)$. Now, the estimate (13.34) can be given as

$$\varrho_x \leq \text{cond}(A) \, \varrho_b.$$

In the case of a regular square matrix A, $A^+ = A^{-1}$ and $b_A = b$ yield the particular estimate

$$\varrho_x \leq \|A\| \, \|A^{-1}\| \, \varrho_b. \tag{13.35}$$

Therefore (consistent with Definition 13.8.1) the condition number is given by

$$\text{cond}(A) = \|A\| \|A^{-1}\|.$$

13.8.3 Effects of a Perturbed Matrix

In the case of a perturbed matrix and a non-perturbed right-hand side

linear system		solution
original system	$Ax = b$	$x_0 = A^+b$
perturbed system	$(A + \Delta A)x = b$	$\tilde{x}_0 = (A + \Delta A)^+b$

the absolute error of the solution is given by

$$\Delta x_0 = \tilde{x}_0 - x_0 = (A + \Delta A)^+b - A^+b.$$

At this point, three situations have to be considered: (i) the rank remains unchanged, (ii) the rank increases, or (iii) the rank decreases. Depending on the actual category, different effects can be observed.

Case 1: Perturbation Preserves the Rank

If $\text{rank}(A + \Delta A) = \text{rank}(A)$ or, equivalently, if the conditions

$$AA^+\Delta A \;=\; \Delta A \tag{13.36}$$
$$A^+A(\Delta A)^\mathsf{T} \;=\; (\Delta A)^\mathsf{T} \tag{13.37}$$
$$\|A^+\| \, \|\Delta A\| \;<\; 1 \tag{13.38}$$

are satisfied, then the following estimate (Wedin [377]) for the relative error

$$\varrho_x \leq \frac{\|A\| \, \|A^+\| \, \varrho_A}{1 - \|A\| \, \|A^+\| \, \varrho_A} = \frac{\kappa(A)}{1 - \kappa(A)\rho_A} \rho_A,$$

is obtained, where

$$\varrho_A := \frac{\|\Delta A\|}{\|A\|}$$

characterizes the relative perturbation of A.

Case 2: Perturbation Increases the Rank

If $\text{rank}(A + \Delta A) > \text{rank}(A)$, then only conditions (13.36) and (13.37) but not (13.38) are satisfied, i. e., a vector $u \neq 0$ exists with

$$Au = 0, \quad \text{but} \quad (A + \Delta A)u \neq 0,$$

or, in terms of vector space theory,

$$u \in \mathcal{N}(A) \quad \text{and} \quad u \notin \mathcal{N}(A + \Delta A).$$

One thence obtains

$$u = (A + \Delta A)^+ (A + \Delta A)u = (A + \Delta A)^+ Au + (A + \Delta A)^+ \Delta Au,$$

and since $Au = 0$, it follows that

$$u = (A + \Delta A)^+ \Delta Au.$$

Passing over to norms yields the estimate

$$\|u\| \leq \|(A + \Delta A)^+\| \, \|\Delta A\| \, \|u\|$$

and finally

$$\|(A + \Delta A)^+\| \geq \frac{1}{\|\Delta A\|},$$

which results in the discontinuous dependency between A and A^+. Situations with $\text{rank}(A + \Delta A) > \text{rank}(A)$ should thus always be avoided; one way of doing so is to modify the perturbed matrix $\tilde{A} = A + \Delta A$ which originates from the application problem in such a way that its rank, denoted $\text{rank}(\tilde{A})$, is reduced to a *pseudo-rank* denoted $\text{rank}_\varepsilon(\tilde{A})$. The parameter ε represents information about the size of the absolute error (the perturbation ΔA) of $\tilde{A}$.

Definition 13.8.4 (Pseudo-Rank of a Matrix)   *The pseudo-rank* $\text{rank}_\varepsilon(A)$ *of a matrix* $A \in \mathbb{R}^{m \times n}$ *with respect to a distance* $\varepsilon > 0$ *is the minimal rank which occurs among all matrices from the corresponding ε-neighborhood of A:*

$$\text{rank}_\varepsilon(A) := \min \left\{ \text{rank}(B) : B \in \mathbb{R}^{m \times n}, \|A - B\| < \varepsilon \right\}.$$

If it is known, for example, that the elements of $\tilde{A}$ originate from a measurement that only permits an absolute accuracy of ± 0.01, then it may be advisable to use $\text{rank}_{0.01}(\tilde{A})$ instead of $\text{rank}(\tilde{A})$.

Even if the coefficients of A are only affected by elementary rounding errors (corresponding to the considered floating-point number system), A must be treated as *numerically singular* as long as

$$\text{rank}_\varepsilon(A) < \min\{m, n\} \quad \text{with} \quad \varepsilon := eps\|A\|_2.$$

The pseudo-rank $\text{rank}_\varepsilon(A)$ can be determined using the singular value decomposition

$$A = USV^\top \quad \text{with} \quad S = \text{diag}(\sigma_1, \sigma_2, \ldots, \sigma_k, 0, \ldots, 0), \quad k \leq \min\{m, n\}.$$

Theorem 13.8.1 *Provided that $A \in \mathbb{R}^{m \times n}$ has the singular value decomposition $A = USV^\top$ and $j < k = \text{rank}(A)$, then, among all matrices B with rank j, the matrix*

$$A_j = \sum_{i=1}^{j} \sigma_i u_i v_i^\top$$

has minimal distance to A:

$$\min\{\|A - B\|_2 : B \in \mathbb{R}^{m \times n}, \text{rank}(B) = j\} = \|A - A_j\|_2 = \sigma_{j+1}.$$

Proof: Golub, Van Loan [50].

With

$$\bar{S} := \text{diag}(\bar\sigma_1, \bar\sigma_2, \ldots, \bar\sigma_k, 0, \ldots, 0) \qquad \bar\sigma_i := \begin{cases} \sigma_i, & \text{if } \sigma_i \geq \varepsilon \\ 0 & \text{otherwise}, \end{cases}$$

a modified matrix $\bar{A} := U\bar{S}V^\top \in \mathbb{R}^{m \times n}$ is defined, for which

$$
\begin{aligned}
\|A - \bar{A}\|_2 &= \|U(S - \bar{S})V^\top\|_2 \\
&= \|S - \bar{S}\|_2 \\
&= \|\text{diag}(0, \ldots, 0, \sigma_{j+1}, \ldots, \sigma_n)\|_2 \\
&= \sigma_{j+1} \\
&< \varepsilon.
\end{aligned}
$$

Thus $\bar{A}$ is the matrix closest to A with pseudo-rank

$$\text{rank}_\varepsilon(\bar{A}) = \begin{cases} j & \text{for } \bar{A} \neq 0 \\ 0 & \text{otherwise} \end{cases}$$

in the ε-neighborhood of A, i.e.,

$$\{B \in \mathbb{R}^{m \times n} : B = A + \Delta A, \|\Delta A\| < \varepsilon\}.$$

For the modified matrix $\bar{A}$, the estimate

$$\varrho_x = \frac{\|\Delta x_0\|}{\|x_0\|} \leq \|A^+\|\,\|A\|\varrho_A \left(1 + \frac{\|b_A - b_{\bar{A}}\|}{\|b_{\bar{A}}\|}\right) + \|A^+ A - \bar{A}^+ \bar{A}\|$$

is obtained for the relative error in x.

Case 3: Perturbation Decreases the Rank

In cases where $\text{rank}(A + \Delta A) \leq \text{rank}(A)$ and $c := \|A^+\|_2\|\Delta A\|_2 < 1$, it can be proved that (Wedin [377])

$$\|(A + \Delta A)^+ - A^+\|_2 \leq \|A^+\|_2\, c \left[1 + \frac{1}{1 - c} + \frac{1}{(1 - c)^2}\right],$$

i.e., in this case A^+ depends continuously on A. A decrease in the rank is therefore not as critical as an increased rank.

13.9 Condition of Least Squares Problems

In the linear least squares problem

$$\min\{\|Ax - b\|_2 : x \in \mathbb{R}^n\}$$

the effects of data perturbations in A and b, on the solution x and on the residual $r = Ax - b$, are analyzed in the following.

Theorem 13.9.1 *From*

$$\text{rank}(A) = n \qquad and \qquad \kappa(A)\varrho_A = \|A^+\|\|\Delta A\| < 1,$$

it follows that

$$\text{rank}(A + \Delta A) = n,$$

and the following estimates of the residual $r = Ax - b$ can be obtained:

$$\|\Delta x\| \leq \frac{\kappa(A)}{1 - \kappa(A)\varrho_A}\left(\varrho_A\|x\| + \varrho_b\frac{\|b\|}{\|A\|} + \varrho_A\kappa(A)\frac{\|r\|}{\|A\|}\right) \qquad (13.39)$$

and

$$\|\Delta r\| \leq \varrho_A\|x\|\|A\| + \varrho_b\|b\| + \varrho_A\kappa(A)\|r\|.$$

Proof: Wedin [377].

The relation $b_A = Ax$ implies that

$$\|x\| \geq \frac{\|b_A\|}{\|A\|}.$$

Combining this inequality with (13.39), assertions for the relative error in x can be obtained:

$$\frac{\|\Delta x\|}{\|x\|} \leq \frac{\kappa(A)}{1 - \kappa(A)\rho_A}(\rho_A + \kappa(b; A)\rho_b) + \frac{(\kappa(A))^2\rho_A}{1 - \kappa(A)\rho_A}\frac{\|r\|}{\|b_A\|}.$$

This error estimate shows that the sensitivity of a linear least squares problem for which $r \neq 0$ is also determined by the *second power* of the condition number of A. Even solutions of linear least squares problems which have been determined *without* using normal equations are thus influenced in their sensitivity by $(\kappa(A))^2$. Using normal equations to derive the modified problem

$$Cx = c \qquad \text{with} \quad C = A^\top A \quad \text{and} \quad c = A^\top b,$$

the condition number is always given by

$$\kappa(C) = \kappa(A^\top A) = (\kappa(A))^2.$$

If the residual vanishes, i.e., if b is an element of the image space $\mathcal{R}(A)$, the following bound for the absolute error:

$$\|\Delta x\| \leq \frac{\kappa(A)}{1 - \kappa(A)\varrho_A}\left(\varrho_A\|x\| + \varrho_b\frac{\|b\|}{\|A\|}\right)$$

is obtained. For the relative error $\|\Delta x\|/\|x\|$, the above mentioned inequality

$$\varrho_x \leq \frac{\kappa(A)}{1 - \kappa(A)\rho_A}\left(\varrho_A + \varrho_b\right)$$

is obtained from the error estimate in Section 13.8.

13.10 Condition Analysis Using the SVD

Let $Ax = b$ be a linear system. $USV^\mathsf{T}x = b$ is then an equivalent one because of the singular value decomposition $A = USV^\mathsf{T}$. Moreover,

$$SV^\mathsf{T}x = U^\mathsf{T}b.$$

With $p := V^\mathsf{T}x$ and $q := U^\mathsf{T}b$, the *decoupled* system of linear equations

$$Sp = q \tag{13.40}$$

can be obtained. Since the Euclidean norm is preserved by orthogonal transformations, it follows that

$$\|S\|_2 = \|A\|_2, \quad \|p\|_2 = \|x\|_2, \quad \|q\|_2 = \|b\|_2.$$

For the decoupled system (13.40), the condition number of the matrix is given by

$$\kappa_2(S) = \|S\|_2\,\|S^+\|_2 = \|A\|_2\,\|A^+\|_2.$$

Because $\|S\|_2 = \sigma_1$ and $\|S^+\|_2 = 1/\sigma_k$, it follows that

$$\kappa_2(A) = \kappa_2(S) = \frac{\sigma_1}{\sigma_k}.$$

The condition number of the right-hand side can be obtained as

$$\kappa_2(b; A) = \frac{\|b\|_2}{\|b_A\|_2} = \frac{\|q\|_2}{\|q_k\|_2} = \sqrt{\frac{\sum_{i=1}^n q_i^2}{\sum_{i=1}^k q_i^2}}.$$

Thus, for the estimate (13.34) of the relative error ρ_x in the solution x_0, due to the perturbation in the right-hand side, one obtains

$$\rho_x \leq \kappa(A)\,\kappa(b; A)\,\rho_b = \frac{\sigma_1}{\sigma_k}\frac{\|q\|}{\|q_k\|}\rho_b.$$

This error estimate is tight since given a matrix A, vectors b and Δb can be found to validate the inequality. The worst case—the strongest magnification of the relative error—occurs if b (or b_A) is extremely *contracted* (shrunk) by A^+, and if Δb is extremely *stretched* (elongated). This happens if b is in the direction of the eigenvector corresponding to the eigenvalue of AA^T with maximum absolute value and if Δb is in the direction of the eigenvector of AA^T which corresponds to the eigenvalue with minimal absolute value (cf. Fig. 13.10). The other extreme occurs if b and Δb have the reverse positions in the eigenspace of AA^T (cf. Fig. 13.11).

These considerations show that a large condition number $\kappa(A)$ of a matrix A does not necessarily imply an ill-conditioned linear system $Ax = b$; the direction of b relative to the eigenvectors of AA^T is also of great importance.

Since U consists of the orthonormal eigenvectors of AA^T, b is an eigenvector of AA^T if and only if the transformed right-hand side q has only *one* component not equal to zero:

$q = (q_1, 0, \ldots, 0)^T$ means that b is an eigenvector corresponding to the eigenvalue with maximal absolute value—this is the worst case (cf. Fig. 13.10).

$q = (0, \ldots, 0, q_k, 0, \ldots, 0)^T$ means that b is an eigenvector corresponding to the eigenvalue with minimal absolute value (not equal to zero)—this is the most favorable case (cf. Fig. 13.11).

13.10.1 Case Study: Condition Analysis

Given the matrix

$$A = \left(\begin{array}{cc} 780 & 563 \\ 913 & 659 \end{array} \right),$$

the singular value decomposition $A = USV^T$ with

$$U = \left(\begin{array}{cc} -0.64956 & -0.76031 \\ -0.76031 & 0.64956 \end{array} \right), \qquad V^T = \left(\begin{array}{cc} -0.81084 & -0.58526 \\ -0.58526 & 0.81084 \end{array} \right)$$

and $S = \mathrm{diag}(1481.0, 6.9286 \cdot 10^{-4})$ can be derived.

Unfavorable Right-Hand Side

The right-hand sides

$$b = \left(\begin{array}{c} -6.4956 \\ -7.6031 \end{array} \right) \qquad \text{and} \qquad \tilde{b} = b + \Delta b = \left(\begin{array}{c} -6.4994 \\ -7.5999 \end{array} \right)$$

differ by

$$\|\Delta b\| = 0.005.$$

Applying U^T to b yields

$$q = U^T b = \left(\begin{array}{c} 10.0 \\ 0 \end{array} \right),$$

the most unfavorable case for the right-hand side, and

$$\tilde{q} = U^T \tilde{b} = \begin{pmatrix} 10.0 \\ 0.005 \end{pmatrix},$$

yields the most unfavorable case for the perturbation. The solutions of the decoupled systems are

$$p = \begin{pmatrix} 6.7524 \cdot 10^{-3} \\ 0.0 \end{pmatrix} \quad \text{and} \quad \tilde{p} = \begin{pmatrix} 6.7524 \cdot 10^{-3} \\ 7.2165 \end{pmatrix},$$

and thus, by using V, both solutions

$$x = \begin{pmatrix} -5.4752 \cdot 10^{-3} \\ -3.9519 \cdot 10^{-3} \end{pmatrix} \quad \text{and} \quad \tilde{x} = \begin{pmatrix} -4.2290 \\ 5.8475 \end{pmatrix}$$

can be calculated. The relative error is given by

$$\varrho_x = \frac{\|\Delta x\|}{\|x\|} = \frac{\|\tilde{x} - x\|}{\|x\|} = 1068.7,$$

which is a value very close to the upper bound $\kappa_A \varrho_b$:

$$\varrho_x \leq \kappa_A \varrho_b = \frac{1481.0}{6.9286 \cdot 10^{-4}} \cdot \frac{0.005}{10} = 1105.2.$$

Favorable Right-Hand Side

The right-hand sides

$$b = \begin{pmatrix} -7.6031 \\ 6.4956 \end{pmatrix} \quad \text{and} \quad \tilde{b} = \begin{pmatrix} -7.6064 \\ 6.4918 \end{pmatrix}$$

differ once more by

$$\|\Delta b\| = 0.005.$$

Applying U^T to b and $\tilde{b}$ yields

$$q = \begin{pmatrix} 0.0 \\ 10.0 \end{pmatrix} \quad \text{and} \quad \tilde{q} = \begin{pmatrix} 0.005 \\ 10.0 \end{pmatrix}.$$

Solving the decoupled systems and applying V leads to

$$x = \begin{pmatrix} -8447.1 \\ 11703. \end{pmatrix} \quad \text{and} \quad \tilde{x} = \begin{pmatrix} -8447.1 \\ 11703. \end{pmatrix}.$$

In this case the solution $\tilde{x}$ of the perturbed system coincides with the solution x of the original system—the error ϱ_x vanishes.

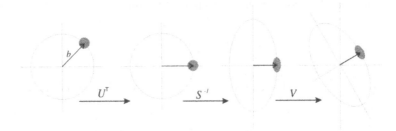

Figure 13.10: *The worst case:* the right-hand side b is the most contracted by the mapping A^{-1}. Hence, the *relative* perturbation of $x = A^{-1}b$ is greatest in this case.

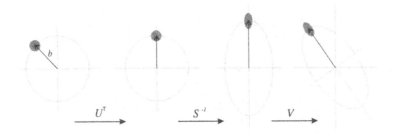

Figure 13.11: *The most favorable case:* the right-hand side b is most elongated by the mapping A^{-1}. Hence, in this case the *relative* perturbation in $x = A^{-1}b$ is smallest.

13.11 Direct Methods

In this and the following sections, the algorithmic principles behind direct methods[3] for solving linear systems with dense coefficient matrices are briefly discussed. The sections are relatively short due to the supply of outstanding software (cf. Section 13.15) which makes it possible to solve nearly all problems reliably and efficiently using existing products. If it happens that a problem cannot be solved optimally by the available software, it is best to start with LAPACK programs for which the source code is freely available and fit them to individual needs. Developing software for oneself is not advised.

13.11.1 The Elimination Principle

The *classical* method of solving linear systems is based on the procedure already described in Section 5.7.1: Since a linear combination of equations preserves the solution of the linear system, appropriate linear combinations are used to systematically eliminate the unknowns.

---

[3] *Iterative* methods, which are mostly used for large, sparse systems, can be found in Chapter 16.

By multiplying the first equation of

$$
\begin{array}{ccccccc}
a_{11}x_1 & + & a_{12}x_2 & + \cdots + & a_{1n}x_n & = & b_1 \\
a_{21}x_1 & + & a_{22}x_2 & + \cdots + & a_{2n}x_n & = & b_2 \\
\vdots & & \vdots & & \vdots & & \vdots \\
a_{n1}x_1 & + & a_{n2}x_2 & + \cdots + & a_{nn}x_n & = & b_n
\end{array}
\tag{13.41}
$$

by a_{i1}/a_{11} $(a_{11} \neq 0)$ and then subtracting this equation from the ith equation, the coefficient of x_1 in the resulting equation vanishes; the variable x_1 is eliminated. For $i = 2, 3, \ldots, n$ one thence obtains a system of $n-1$ linear equations

$$
\begin{array}{ccccc}
a_{22}^{(1)}x_2 & + \cdots + & a_{2n}^{(1)}x_n & = & b_2^{(1)} \\
\vdots & & \vdots & & \vdots \\
a_{n2}^{(1)}x_2 & + \cdots + & a_{nn}^{(1)}x_n & = & b_n^{(1)}
\end{array}
\tag{13.42}
$$

with $n-1$ unknowns $x_2, x_3, \ldots, x_n$. The coefficients of the reduced linear system are given as

$$
a_{ij}^{(1)} := a_{ij} - a_{1j}\frac{a_{i1}}{a_{11}}, \quad b_i^{(1)} := b_i - b_1\frac{a_{i1}}{a_{11}}, \quad i, j = 2, 3, \ldots, n,
$$

and together with the first, unchanged equation

$$
a_{11}x_1 + a_{12}x_2 + \cdots + a_{1n}x_n = b_1
$$

these equations are equivalent to the original system (13.41).

Provided that $a_{ii}^{(k)} \neq 0$ (cf. Section 13.11.3), this process can be continued recursively. After $n-1$ steps, the remaining system is finally reduced to a *single* equation in *one* unknown x_n, and the equivalent system has the form:

$$
\begin{array}{ccccc}
a_{11}x_1 + a_{12}x_2 + \cdots & \cdots + & a_{1n}x_n & = & b_1 \\
a_{22}^{(1)}x_2 + \cdots & \cdots + & a_{2n}^{(1)}x_n & = & b_2^{(1)} \\
\ddots & & \vdots & & \vdots \\
a_{n-1,n-1}^{(n-2)}x_{n-1} + & a_{n-1,n}^{(n-2)}x_n & = & b_{n-1}^{(n-2)} \\
& a_{nn}^{(n-1)}x_n & = & b_n^{(n-1)}.
\end{array}
\tag{13.43}
$$

The $a_{ij}^{(k)}$ and $b_i^{(k)}$ are the coefficients or elements of the right-hand side, generated by the elimination process at the kth step.

A linear system of the form (13.43) is called a *triangular system*. In other words, the corresponding coefficient matrix is an upper (or right) triangular matrix. Such linear systems can be solved recursively using *back substitution*, provided that none of the diagonal elements is zero. The last equation yields

$$
x_n = \frac{b_n^{(n-1)}}{a_{nn}^{(n-1)}},
$$

substituting x_n in the previous equation yields

$$x_{n-1} = \frac{b_{n-1}^{(n-2)} - a_{n-1,n}^{(n-2)} x_n}{a_{n-1,n-1}^{(n-2)}};$$

and so on. Finally, after $x_2, \ldots, x_n$ have been computed and after substituting them into the first line,

$$x_1 = \frac{b_1 - a_{12} x_2 - a_{13} x_3 - \cdots - a_{1n} x_n}{a_{11}}$$

is obtained. In the algorithmic description of these two processes—elimination and back substitution—the $a_{ij}^{(k)}$ do not require any additional storage. Elements no longer required in the elimination process can be overwritten, and the original matrix A and right-hand side b are destroyed.

Elimination:

> **do** $k = 1, 2, \ldots, n-1$
>   **do** $i = k+1, k+2, \ldots, n$
> $a_{ik} := a_{ik}/a_{kk}$ (13.44)
>   **do** $j = k+1, k+2, \ldots, n$
> $a_{ij} := a_{ij} - a_{ik} a_{kj}$ (13.45)
>   **end do**
> $b_i := b_i - a_{ik} b_k$
>   **end do**
> **end do**

Back substitution:

> **do** $k = n, n-1, \ldots, 1$
> $x_k := b_k$
>   **do** $i = k+1, k+2, \ldots, n$
> $x_k := x_k - a_{ki} x_i$
>   **end do**
> $x_k := x_k/a_{kk}$
> **end do**

It must be noted that when using elimination methods for solving linear systems, *no truncation error* can arise. The elimination algorithm involves only the four basic arithmetic operations in $\mathbb{R}$ or $\mathbb{C}$. If the operations can be done exactly (in a rational arithmetic), then the exact result is obtained.

Computational Effort

The elimination (13.45) in the innermost loop of the triangular decomposition requires the most floating-point operations. The total number of floating-point additions *and* multiplications of (13.45) is

$$\sum_{k=1}^{n-1}(n-k)^2 = \sum_{k=1}^{n-1}k^2 = \frac{2n^3 - 3n^2 + n}{6} \approx \frac{n^3}{3}.$$

Furthermore,

$$\sum_{k=1}^{n-1}(n-k) = \sum_{k=1}^{n-1}k = \frac{n^2}{4}$$

floating-point division operations are required for (13.44). In the case of large linear systems, this computational effort of order $O(n^2)$ can almost be ignored. Assuming that the execution of a floating-point addition takes the same amount of time as that of a floating-point multiplication (usually one cycle per operation), the total time spent by the elimination process can be characterized by the order $2n^3/3$.

Example (The Time Spent on Elimination) A workstation with a clock rate $f_c = 200\,\text{MHz}$ has a cycle time of

$$T_c = \frac{1}{200 \cdot 10^6} = 5 \cdot 10^{-9}\,\text{s} = 5\,\text{ns}.$$

For a 1200×1200 linear system, a lower bound for the total runtime is given by

$$T \geq \frac{2n^3}{3}T_c = 5.76\,\text{s}.$$

The actual runtime might be around five to twenty times longer than this minimal runtime due to inevitable overhead.

It is always important to realize that the amount of computational work for the solution of a general linear system rises with the *third power* of the number n of equations: doubling n means that the computational costs are increased by a factor of eight. The solution of large linear systems—using methods as described above—requires (even on very fast machines) an amount of computational work which cannot be ignored. This is of importance since the solution of a linear system often appears in *inner* loops of large scale application software.

13.11.2 LU Factorization

The elimination process which transforms the original system (13.41) into the triangular system (13.43) corresponds to a decomposition of the matrix A

$$A = LU,$$

whose factors are triangular matrices. This factorization is thus also called *triangular decomposition* of the matrix A. L is a lower triangular matrix with

$$l_{11} = l_{22} = \cdots = l_{nn} = 1,$$

and U is an upper triangular matrix. U is the coefficient matrix of the reduced system (13.43):

$$U = \begin{pmatrix} a_{11} & a_{12} & a_{13} & \cdots & a_{1n} \\ & a_{22}^{(1)} & a_{23}^{(1)} & \cdots & a_{2n}^{(1)} \\ & & a_{33}^{(2)} & \cdots & a_{3n}^{(2)} \\ & & & \ddots & \vdots \\ \mathbf{0} & & & & a_{nn}^{(n-1)} \end{pmatrix}.$$

The generation of the matrix L corresponds to the elimination process. The first step, which transforms (13.41) into (13.42), corresponds to a matrix multiplication $M_1 A$ with

$$M_1 := \begin{pmatrix} 1 & & & \mathbf{0} \\ -a_{21}/a_{11} & 1 & & \\ \vdots & & \ddots & \\ -a_{n1}/a_{11} & & & 1 \end{pmatrix}.$$

The reduced system (13.43) can be obtained according to

$$M A x = M b \quad \text{with} \quad M := M_{n-1} \cdots M_2 M_1.$$

The so-called *Gaussian transformations* M_k are given as

$$M_k := I - m_k e_k^\mathsf{T}, \qquad m_k := \left(0, \ldots, 0, -a_{k+1,k}^{(k-1)}/a_{kk}^{(k-1)}, \ldots, -a_{nk}^{(k-1)}/a_{kk}^{(k-1)}\right)^\mathsf{T}.$$

Each Gaussian transformation is described by a lower triangular matrix whose diagonal elements are all 1. Accordingly, the product matrix M and its inverse M^{-1} are also of the same type. Therefore, from

$$M A = U \quad \text{it follows that} \quad L = M^{-1} = M_1^{-1} M_2^{-1} \cdots M_{n-1}^{-1}.$$

On the basis of the LU factorization of A, the solution of the linear system $Ax = b$ proceeds in three steps:

    factorize $A = LU$
    solve $Ly = b$
    solve $Ux = y$

Splitting up the whole process into one factorization and two solution steps has the additional advantage that the first step, which requires most of the computational effort, has to be done only once if several linear systems with the same coefficient matrix but different right hand sides are to be solved. In practice this often happens, and it is possible to proceed as follows:

    factorize $A = LU$

    do $i = 1, 2, \ldots, m$
        solve $Ly = b_i$
        solve $Ux_i = y$
    end do

$O(mn^2)$ floating-point operations for back substitution must be added to the $O(n^3)$ amount of computational work involved in the LU factorization of A.

If the inverse should actually be needed—which happens very rarely—it is possible to proceed as indicated above. Using the unit vectors as right-hand sides,

$$b_i := e_i, \quad i = 1, 2, \ldots, n,$$

the column vectors $a'_1, a'_2, \ldots, a'_n$ of A^{-1} can be determined:

$$a'_i = x_i, \quad i = 1, 2, \ldots, n.$$

To avoid unnecessary computational effort, a linear system $Ax = b$ should never be solved by computing A^{-1} first, and then determining the solution by left multiplying the right-hand side by A^{-1}!

13.11.3 Pivot Strategies

It must be noted that the execution of the elimination process as described previously can run into obstacles.

After k steps of the elimination process, the remaining system

$$
\begin{array}{ccccc}
a^{(k)}_{k+1,k+1} x_{k+1} & + & \cdots & + & a^{(k)}_{k+1,n} x_n & = & b^{(k)}_{k+1} \\
\vdots & & & & \vdots & & \vdots \\
a^{(k)}_{n,k+1} x_{k+1} & + & \cdots & + & a^{(k)}_{n,n} x_n & = & b^{(k)}_{n}
\end{array}
\tag{13.46}
$$

consists of $n-k$ equations involving the unknowns $x_{k+1}, \ldots, x_n$. Using the first equation of this system for the elimination of x_{k+1} in the $(k+2)$nd, $(k+3)$rd, $\ldots$, nth equation, presupposes that the coefficient $a^{(k)}_{k+1,k+1}$ is non-zero. However, even a very small value of $a^{(k)}_{k+1,k+1}$ (relative to the other coefficients) is not harmless. Such a small value often results from the cancelation of leading digits, and accordingly its low accuracy perturbs the course of further computations.

Example (Elimination Without Pivot Strategy) Solving the 50×50 linear system with

$$
A = \begin{pmatrix}
6 & 1 & & & & 0 \\
8 & 6 & 1 & & & \\
& 8 & 6 & 1 & & \\
& & \ddots & \ddots & \ddots & \\
& & & 8 & 6 & 1 \\
0 & & & & 8 & 6
\end{pmatrix}
\quad \text{and} \quad
b = \begin{pmatrix}
7 \\
15 \\
15 \\
\vdots \\
15 \\
14
\end{pmatrix}
$$

using the method described in Section 13.11.1 (elimination and back substitution in single precision IEC/IEEE arithmetic), produces the numerically unsatisfying result

$$
\tilde{x} = \begin{pmatrix}
1.000 \cdot 10^0 \\
1.000 \cdot 10^0 \\
1.000 \cdot 10^0 \\
\vdots \\
2.512 \cdot 10^7 \\
-3.349 \cdot 10^7
\end{pmatrix}
\quad \text{instead of} \quad
x^* = \begin{pmatrix}
1 \\
1 \\
1 \\
\vdots \\
1 \\
1
\end{pmatrix}.
$$

The elimination process in its simplest form is *numerically unstable*.

These difficulties can be overcome by modifying the course of the algorithm: As the order of the equations in (13.46) is irrelevant to the solution, the $(k+1)$st row can be interchanged with a more suitable row. Selecting the row which yields the largest absolute value for x_{k+1} is advisable.

Hence, in the recursive transformation of (13.41) into the triangular form (13.43) each elimination step is preceded by so-called *pivoting*:

In the $(k+1)$st step, the coefficient in the first column of the remaining system (13.46), with the largest absolute value, is searched for. Next the selected *pivot row* is interchanged with the first row of (13.46) so that the denominator of the factors used in the elimination process becomes the matrix element which was determined by the pivot search. As a result, all elimination factors

$$l_{i,k+1} := a_{i,k+1}^{(k)} / a_{k+1,k+1}^{(k)}, \qquad i = k+2, \ k+3, \ldots, n$$

have an absolute value ≤ 1. By expressing each single row interchange as a permutation matrix P_k, it can easily be seen that the triangular decomposition

$$P_{n-1} P_{n-2} \cdots P_2 P_1 A = PA = LU$$

is actually computed for PA and not for A. The factorization of A is given by

$$A = P^{-1} LU = P^{\top} LU.$$

It can be shown that these row interchanges are essential for the *stability* of the elimination process. This algorithmic technique is known as *column pivoting* or *partial pivoting*.

In the absence of rounding errors, for a regular matrix the LU factorization with partial pivoting is always successful:

Theorem 13.11.1 (LU Factorization) *For each regular matrix A there exists a permutation matrix P, such that a triangular decomposition*

$$PA = LU$$

can be obtained. The matrix P can be chosen in such a way that $|l_{ij}| \leq 1$ for all elements of L.

Proof: Deuflhard, Hohmann [44].

If it becomes apparent that in the $(k+1)$st step *all* coefficients of x_{k+1} are small relative to the coefficients in the original system (13.41), then it makes no sense to continue the process any further. In general this breakdown occurs with numerically singular matrices. Such matrices can thus be discovered during the course of the computation. An example of a stopping criterion for the elimination process is given by

$$|a_{i,k+1}^{(k)}| \leq eps \, \|A\|, \qquad i = k+1, k+2, \ldots, n,$$

where $a_{i,k+1}^{(k)}$ are the coefficients of x_{k+1} in (13.46), and where $\|A\|$ is a norm of the *original* coefficient matrix.

Scaling

The partial pivoting discussed above can only guarantee numerical stability if the values of the coefficients in the equations of (13.41) are of the same order. Otherwise, each row can be forced to become the *pivot row* by multiplying the row by a sufficiently large factor.

It would be simple to divide each equation of $Ax = b$ by a weight factor

$$d_i := \sum_{j=1}^{n} |a_{ij}|, \qquad i = 1, 2, \ldots, n.$$

Then for each row of the scaled (equilibrated) coefficient matrix, the sum of the row elements taken absolutely is 1, and the row sum norm of the matrix is also 1. In general though, performing these divisions introduces additional rounding errors in all coefficients. In the case of an ill-conditioned system, these perturbations in the matrix coefficients can significantly affect the result. Therefore, the weight factors d_i are only used for comparison with the $|a_{i,k+1}^{(k)}|$ since scaling is only of importance to the partial pivoting. The elimination process itself is not affected by scaling.

13.12 Special Types of Linear Systems

Since computational costs resulting from the solution of large linear systems cannot be neglected, it is important to exploit any special structure in the coefficient matrix in order to reduce the complexity of the algorithm.

Also, the storage requirements of n^2 floating-point numbers can be reduced: For a linear system consisting of 400 equations with 400 unknowns, 160 000 matrix elements must be stored and so 640 000 bytes of memory are required if single precision arithmetic is used (1 matrix element can be held in 4 bytes). If, for example, the matrix is symmetric, then only half the storage locations are required.

13.12.1 Symmetric, Positive Definite Matrices

A special matrix property which can easily be verified is the *symmetry* of the coefficient matrix, i. e., whether or not

$$a_{ij} = a_{ji} \qquad \text{for all} \quad i, j \in \{1, 2, \ldots, n\}.$$

Among various symmetric matrices, the positive definite matrices are of particular interest (cf. Section 13.5.3); such matrices arise, for example, from the discretization of boundary value problems in ordinary and partial differential equations which often occur in physics and engineering.

For symmetric and positive definite matrices there is a symmetric form of the elimination algorithm for the computation of the *Cholesky factorization* (cf. Theorem 13.5.10)

$$A = LL^{\mathsf{T}}$$

where L is a lower triangular matrix or

$$A = LDL^\mathsf{T} \tag{13.47}$$

where L is a lower triangular matrix with diagonal elements $l_{ii} = 1$ and D is a positive diagonal matrix. One great advantage of this method is that no pivoting is required to stabilize the algorithm. Computational costs are halved compared to the LU factorization. In the case of a symmetric matrix, the storage requirements are also approximately halved, since the elements a_{ij} and a_{ji} are identical and thus only need to be stored once. This property is independent of the possible definiteness of the matrix.

If A is symmetric, but *indefinite*, there exists no factorization of the form (13.47). However, A can be factorized as

$$PAP^\mathsf{T} = L\overline{D}L^\mathsf{T}$$

where $\overline{D}$ is a block diagonal matrix with 1×1 and 2×2 diagonal blocks.

There is also a special algorithmic form of the triangular decomposition, namely the *Bunch-Kaufman algorithm* for indefinite, symmetric matrices. In this algorithm numerical stability is only guaranteed if pivoting is performed. To preserve symmetry during the course of the elimination process, the corresponding *rows* and *columns* must be interchanged. The pivot elements must therefore be chosen from the elements of the main diagonal.

The favorable stability properties of the elimination algorithms for symmetric, positive definite systems are completely independent of the condition of the coefficient matrix. However, positive definite matrices can be very ill-conditioned, as is shown in the following example:

Example (Positive Definite Matrix)   The Hilbert matrices $H_n \in \mathbb{R}^{n \times n}$, $n = 2, 3, 4, \ldots$

$$h_{ij} = \frac{1}{i+j-1}, \qquad i, j = 1, 2, \ldots, n$$

are symmetric and positive definite. The condition number of H_n grows rapidly with n:

n	2	3	4	$\cdots$	10	$\cdots$
$\kappa_2(H_n)$	19.3	524	$1.55 \cdot 10^4$	$\cdots$	$1.60 \cdot 10^{13}$	$\cdots$

One can therefore see that definiteness and condition are two independent properties of a matrix.

13.12.2 Band Matrices

Efficiency can be improved significantly if the special structure of band matrices can be exploited by the algorithm. In practice, large systems with a different kind of sparsity structure often occur: each equation of the system involves only a few unknowns so that most coefficients are zero. In contrast to band matrices though, these zero elements are spread over the whole matrix in an irregular pattern. Such general sparse matrices are discussed in Chapter 16.

The number non-null(A) of non-vanishing elements of a band matrix A is approximately

$$\text{non-null}(A) \approx n\left(1 + k_u + k_l\right),$$

provided that the values of k_l and k_u (the number of upper and lower side-diagonals) are small. If $k_u, k_l \ll n$, a considerable reduction of significant data can be gained. This must be exploited both by the storage scheme for the data and by the execution of the elimination algorithm. For instance, the loops in the algorithm on page 230 have only to run through the nonzero elements.

For fixed $k_u, k_l \ll n$ the number of floating-point operations necessary for the solution of the linear system is proportional only to n (as opposed to n^3); this is only asymptotically true for large n.

Tridiagonal matrices are extremely important in practice. They have only two side-diagonals in addition to the main diagonal which may contain nonzero elements:

$$k_u = k_l = 1.$$

In systems of linear equations with tridiagonal matrices, symmetry and definiteness often hold. The solution of such systems requires only about $5n$ arithmetic operations: It can thus be computed much more rapidly than the solution of a dense system of the same order.

The linear system $Ax = b$ with a band matrix A should by no means be solved by computing A^{-1} and then performing the multiplication $A^{-1}b$. The reason for this is that the inverse does *not* maintain the sparsity structure of the band matrix.

Example (Inverse of a Tridiagonal Matrix) The following 5×5 tridiagonal matrix A has a *dense* inverse A^{-1}:

$$A = \begin{pmatrix} 1 & -1 & & & 0 \\ -1 & 2 & -1 & & \\ & -1 & 2 & -1 & \\ & & -1 & 2 & -1 \\ 0 & & & -1 & 2 \end{pmatrix}, \qquad A^{-1} = \begin{pmatrix} 5 & 4 & 3 & 2 & 1 \\ 4 & 4 & 3 & 2 & 1 \\ 3 & 3 & 3 & 2 & 1 \\ 2 & 2 & 2 & 2 & 1 \\ 1 & 1 & 1 & 1 & 1 \end{pmatrix}.$$

13.13 Assessment of the Accuracy Achieved

The various versions of elimination methods lead directly to arithmetic algorithms which do not involve the use of iterations or similar limiting schemes which depend on accuracy parameters and thus no truncation error occurs. Direct methods (elimination algorithms) are afflicted only by rounding errors caused by the execution of the arithmetic operations within a floating-point arithmetic. The huge number of operations, however, does not make a direct estimate of rounding error propagation possible since such an estimate would be too pessimistic.

On the other hand it is crucial to know to what extent the result delivered, for instance, by a library subroutine is affected with errors. Given the hierarchical graduation of the error (cf. Chapter 2), the effects of the rounding error generated by executing programs for solving linear systems in an floating-point arithmetic should be less than the effects of the data errors already inherent in the linear system. Therefore, both the order of the effect of data errors and the total effect of rounding errors must be understood.

13.13.1 Condition Number Estimates

In order to determine the *effects of data errors*, at the very least the magnitude of the *condition number* of the linear system must be estimated (cf. Section 13.8). According to (13.30) $\|A^{-1}\|$ remains to be determined since $\|A\|$ can be calculated directly. As each p-norm has the property (13.12), it follows that

$$\|A^{-1}\| = \max_{b \neq 0} \frac{\|A^{-1}b\|}{\|b\|}.$$

In order to determine an estimate of $\|A^{-1}\|$, a right-hand side b with $\|b\| = 1$ has to be generated such that the solution $x = A^{-1}b$ obtained is as large as possible (Bischof, Tang [114], [115]). This can be done systematically using a special algorithm after A has been factorized. Such an algorithm is implemented, for instance, in the LAPACK/*con programs (cf. Table 13.1 and Section 13.17).

Table 13.1: LAPACK condition number estimates for Hilbert matrices H_2, H_3, ..., H_{15}. For values of $1/eps = 9.01 \cdot 10^{15}$ or higher the condition estimates deteriorate significantly.

n	exact condition $\kappa_\infty(H_n)$	condition estimate	
		LAPACK/dgecon	LAPACK/dpocon
2	$2.70 \cdot 10^1$	$2.70 \cdot 10^1$	$2.70 \cdot 10^1$
3	$7.48 \cdot 10^2$	$7.48 \cdot 10^2$	$7.48 \cdot 10^2$
4	$2.84 \cdot 10^4$	$2.84 \cdot 10^4$	$2.84 \cdot 10^4$
5	$9.44 \cdot 10^5$	$9.44 \cdot 10^5$	$9.44 \cdot 10^5$
6	$2.91 \cdot 10^7$	$2.91 \cdot 10^7$	$2.91 \cdot 10^7$
7	$9.85 \cdot 10^8$	$9.85 \cdot 10^8$	$9.85 \cdot 10^8$
8	$3.39 \cdot 10^{10}$	$3.39 \cdot 10^{10}$	$3.39 \cdot 10^{10}$
9	$1.10 \cdot 10^{12}$	$1.10 \cdot 10^{12}$	$1.10 \cdot 10^{12}$
10	$3.54 \cdot 10^{13}$	$3.54 \cdot 10^{13}$	$3.54 \cdot 10^{13}$
11	$1.23 \cdot 10^{15}$	$1.23 \cdot 10^{15}$	$1.23 \cdot 10^{15}$
12	$4.12 \cdot 10^{16}$	$3.80 \cdot 10^{16}$	$4.09 \cdot 10^{16}$
13	$1.32 \cdot 10^{18}$	$4.28 \cdot 10^{17}$	$3.36 \cdot 10^{18}$
14	$4.54 \cdot 10^{19}$	$5.96 \cdot 10^{18}$	$1.07 \cdot 10^{32}$
15	$1.54 \cdot 10^{21}$	$6.04 \cdot 10^{17}$	$2.27 \cdot 10^{32}$

13.13.2 Backward Error Analysis

The first insight into the size of the *rounding error* can be gained by computing the *residual* of the obtained approximate solution $\tilde{x}$, i.e., by applying $\tilde{x}$ to the linear system $Ax = b$:

$$r := A\tilde{x} - b \qquad (13.48)$$

or

$$r_i := \sum_{j=1}^{n} a_{ij}\tilde{x}_j - b_i, \qquad i = 1, 2, \ldots, n.$$

The computation of r using an accurate approximation $\tilde{x}$ leads to an extreme situation of cancelation, as the components of $A\tilde{x}$ and b, as a rule, cancel each other out almost completely (cf. Section 5.7.4). One thus attempts to satisfy

$$\tilde{r}_i \approx \Box r_i = \Box\left(\sum_{j=1}^{n} a_{ij}\tilde{x}_j - b_i\right) \qquad (13.49)$$

by using *higher precision* (more digits in the mantissa) or an *exact scalar product* (cf. Section 5.7.4). According to (13.48) $\tilde{x}$ is the exact solution of

$$A\tilde{x} = b + r,$$

i.e., a system with data perturbations of order of $\|r\|$. If the order of magnitude of $\tilde{r}$ computed according to (13.49) is substantially smaller than the order of the already existing data perturbations (on grounds of model errors and other types of errors), then in the context of the problem as a whole, the system has been solved with sufficient accuracy.

Of course, the difference between $\tilde{x}$ and the true solution x in

$$\|\tilde{x} - x\| \leq \kappa(A)\,\|x\|\,\frac{\|r\|}{\|b\|}$$

may be still of the order of the right-hand side, cf. (13.35). Sufficient accuracy means only that this error will most probably be concealed by errors originating from other sources.

13.13.3 Iterative Refinement

If the residual vector r is unacceptably large, then the information inherent in r can be exploited to improve $\tilde{x}$. An attempt to determine a correction Δx such that $\tilde{x} - \Delta x$ solves the original linear system:

$$\begin{array}{rcl}
A\tilde{x} &=& b + r \\
\underline{A(\tilde{x} - \Delta x)} &=& \underline{b} \\
A\Delta x &=& r \approx \tilde{r}
\end{array}$$

can be made. Obviously, such a correction can be obtained approximately by solving the linear system

$$A\Delta x = \tilde{r}. \qquad (13.50)$$

This system has the same coefficient matrix A, but a new right-hand side $\tilde{r}$. The expensive factorization of A can thus be reused and only the amount of work needed for back substitution has to be taken into account (cf. Section 13.11).

As in general Δx is substantially smaller than x, it is sufficient to compute Δx with moderate accuracy in order to gain an essential improvement of $\tilde{x}$. For the newly generated approximation

$$\tilde{\tilde{x}} := \tilde{x} - \Delta x \qquad (13.51)$$

again, according to (13.49), the residual can be computed.

However, as soon as

$$\frac{\|\tilde{r}\|}{\|A\|\|\tilde{x}\|} < eps, \qquad (13.52)$$

a procedure according to (13.50) is no longer meaningful since in this case the floating-point computation of the residual for the true solution of $Ax = b$ can result in a vector of length of $eps\,\|A\|\,\|x\|$. That means, if (13.52) holds then $\tilde{x}$ and x cannot be distinguished with respect to their computed residuals!

A procedure according to (13.50) and (13.51), which can be iterated theoretically is called an *iterative refinement*, although usually only the step (13.50), (13.51) is performed; no improvements would be gained by carrying out more steps.

In the case of a nearly or actually singular matrix, it can happen that $\|\tilde{r}\| \approx \|b\|$ and $\|\Delta x\| \approx \|\tilde{x}\|$. In this case a numerical solution within the floating-point arithmetic under consideration is generally not possible, because the system is too ill-conditioned.

13.13.4 Experimental Condition Analysis

The scalar quantities

$$\rho_b = \frac{\|\Delta b\|}{\|b\|} \qquad \text{and} \qquad \rho_A = \frac{\|\Delta A\|}{\|A\|}$$

describe the perturbations of the right-hand side and/or of the system matrix only in a highly simplified manner. If more precise information is available, then the sensitivity of the solution to perturbation can be analyzed using a Monte-Carlo method (cf. Section 2.7.3) whereby the perturbations are simulated by random number generators, which have been adapted to the known structure of the perturbations Δb and/or ΔA. In this way one can also examine how an *unusual* perturbation affects the result (as was done in the example of the linear system in Section 2.7.3).

13.14 Methods for Least Squares Problems

13.14.1 Normal Equations

One approach to the solution of linear least squares problems is to set up and solve the normal equations

$$A^{\mathsf{T}} A x = A^{\mathsf{T}} b \qquad (13.53)$$

using the Cholesky method, which is very advantageous with respect to computational costs. A serious drawback to this method is the deteriorated condition of the transformed problem compared to the original one:

$$\text{cond}_2(A^\top A) = (\text{cond}_2(A))^2.$$

13.14.2 QR Method

The QR method for the solution of least squares problems is less susceptible to difficulties (Björk [35]). Since it is based on the QR factorization of A, the method is applied to the original data.

Theorem 13.14.1 (QR Factorization) *Each matrix $A \in \mathbb{R}^{m \times n}$ with $m \geq n$ can be factorized as*

$$A = Q \begin{pmatrix} R \\ 0 \end{pmatrix}, \tag{13.54}$$

where $0 \in \mathbb{R}^{(m-n) \times n}$; R is an upper $n \times n$ triangular matrix and Q is an orthogonal $m \times m$ matrix.

Proof: Lawson, Hanson [65].

If the matrix A has full rank n, then the triangular matrix R is non-singular and the QR factorization (13.54) can be used to solve the linear least squares problem. This is because

$$\|Ax - b\|_2 = \left\| \begin{pmatrix} Rx - q_1 \\ q_2 \end{pmatrix} \right\|_2 \qquad \text{where} \qquad q := \begin{pmatrix} q_1 \\ q_2 \end{pmatrix} = Q^\top b.$$

Thus x solves the linear system $Rx = q_1$. It can be proved that

$$\|Ax - b\|_2 = \|q_2\|_2.$$

If A is not of full rank, or the rank of A is not known, then a QR factorization with column pivoting, or a singular value decomposition (cf. Section 13.7.6) can be performed. If $m \geq n$, the QR factorization with column pivoting is given by

$$A = Q \begin{pmatrix} R \\ 0 \end{pmatrix} P^\top,$$

where P is a permutation matrix chosen so that R has the form

$$R = \begin{pmatrix} R_{11} & R_{12} \\ 0 & 0 \end{pmatrix}.$$

R_{11} is a square, non-singular, triangular matrix. The basic solution of the linear least squares problem can be obtained by using QR factorization with partial pivoting. By applying orthogonal (unitary) transformations, R_{12} can be eliminated, and thus a *complete orthogonal factorization* can be gained, from which the solution with minimal norm can be obtained (Golub, Van Loan [50]).

13.15 LAPACK—The Fundamental Linear Algebra Package

LAPACK (*Linear Algebra Package*) is a freely available (*public domain*) software package consisting of Fortran 77 subprograms which make possible the numerical solution of standard problems in linear algebra. At present the complete LAPACK software product (version 2.0) comprises about 600 000 lines of Fortran code[4] in about 1000 routines and a user guide (Anderson et al. [3]).

LAPACK was developed to solve linear systems, linear least squares problems and eigenvalue problems, as well as to perform the factorizations of matrices, singular value decompositions and condition estimates. There are LAPACK programs for dense matrices and band matrices, but not for sparse matrices with general sparsity structure. Analogous programs are provided for real and complex matrices both in single and double precision.

LAPACK contains *black box programs* (*driver routines*) for the easy solution of the problems mentioned above and *computational routines* for the solution of a certain part of a problem. The computational routines called by the driver routines are outstanding because of their high reliability and efficiency on many high performance computers.

As the source code of the LAPACK programs is freely available and well documented, LAPACK serves as a toolbox for developers of algorithms as well as for software engineers.

For numerical problems in linear algebra with dense or band matrices, LAPACK represents the de-facto standard. For instance, the IMSL libraries and the NAG libraries have adopted many of the LAPACK subroutines (sometimes slightly modified). For this reason, other software products for these linear algebra problems will not be dealt with in this book. Software for problems involving sparse matrices is discussed in Chapter 16.

Availability of LAPACK

LAPACK is *public domain* and can be used free of charge. The complete package, or individual programs can be obtained through the Internet via:

E-mail: The procurement of LAPACK software via e-mail is restricted; only single routines, not whole packages, can be ordered.

> *procedure*: e-mail to: `netlib@ornl.gov`
> subject: `send` <*name of program*> `from lapack`

The body of the message can be left blank, or additional requests for LAPACK programs can be placed there. The string <*name of program*> has to be replaced by the name of the respective Fortran program.

---

[4]The 600 000 lines of code comprise not only the LAPACK programs, but also test routines, timing routines, BLAS programs etc. Strictly speaking, LAPACK programs currently consist of 133 000 lines of Fortran code (excluding comments).

FTP: Using FTP, not only single LAPACK programs, but also the whole LAPACK package as well as precompiled versions for a large number of popular hardware platforms (e. g., HP-PA RISC, DEC 3000/5000, SUN 4, IBM RS-6000) can be obtained.

procedure: FTP to	host:	`netlib.att.com`
	user:	`ftp`
	password:	*your e-mail address*
	directory:	`/netlib/lapack`

WWW: There is also a *World-Wide Web* (WWW) site for NETLIB which provides an easy interface for the procurement of LAPACK software.

procedure: Use MOSAIC (or any other WWW *client*) to connect to

URL: `http://netlib.att.com/netlib/lapack/index.html`

Suggestions, comments, questions, and the like can be sent to the LAPACK team either via conventional mail

LAPACK Project c/o J. J. Dongarra
Computer Science Department, University of Tennessee
Knoxville, Tennessee 37996-1301, USA

or by e-mail to `lapack@cs.utk.edu`.

13.15.1 The History of LAPACK

During the sixties, some 20 different authors published algorithms (written in Algol 60) for solving algebraic eigenvalue problems in the journal *Numerische Mathematik*.

During the early seventies, intensive research was launched in the US to develop high quality software under the name NATS (cf. Section 7.2.4). One of the first things the NATS project achieved was the development of a Fortran 66 program package based on the algorithms published in *Numerische Mathematik* to solve algebraic eigenvalue problems. In 1976 the accomplishment of this effort was released—the EISPACK software package (Smith et al. [354]). Later an extension was published (Garbow et al. [16]).

In 1975, the development of an efficient and portable software package for solving systems of linear equations was commenced at the *Argonne National Laboratory* (Argonne, Illinois, USA). In 1977, the programs developed so far (mainly by J. J. Dongarra, J. R. Bunch, C. B. Moler, and G. W. Stewart), including test programs, were sent to 26 different institutions for testing and assessment. After corrections, improvements, and renewed tests, the software product LINPACK was finished and released early in 1979 (Dongarra et al. [11]).

LINPACK and EISPACK have provided high-quality portable software for the solution of problems in dense linear algebra for a long time, but on modern high-performance computers they often achieve only a fraction of the *peak performance* of the machines.

The LAPACK project was initiated in the eighties by J.J. Dongarra and J. Demmel. The goal was to supersede the widely used software packages LINPACK and EISPACK, principally by achieving far greater efficiency—but at the same time also adding extra functionality, using some new or improved algorithms, and integrating the two original sets of algorithms into a unified package. Great emphasis was placed on the use of special design techniques which ensure the applicability of the LAPACK algorithms and programs to modern high performance computers. For instance, all performance-critical LAPACK programs were carefully structured in order to minimize data transfers within the memory hierarchy (Bischof [113]).

13.15.2 LAPACK and BLAS

An essential feature of LINPACK is the use of specific subroutines (so-called kernels), which for the first time were applied to solve elementary subtasks. The kernels, named BLAS (*Basic Linear Algebra Subroutines*), were defined concurrently with the LINPACK project and published as Fortran 66 programs in 1979 (Lawson et al. [266]).

The purpose of using BLAS programs for elementary vector operations is twofold:

1. They are designed in such a way as to achieve good floating-point performance (e. g., using loop unrolling) on many computers.

2. They can easily be replaced by individually optimized versions (e. g., written in an assembler language). This does not affect portability, because the Fortran version is still available.

Only vector operations (e. g., determining the norm of a vector) or vector-vector operations (e. g., scalar products of vectors) have been used in LINPACK; these basic operations have been standardized in the BLAS-1 package.[5] Hence, the greater part of the algorithms had to be implemented using Fortran statements. Only the BLAS-2 package for matrix-vector operations, which was published in 1988 (Dongarra et al. [163]), made it feasible to implement many algorithms in numerical linear algebra in such a way that the main part of the computation was performed using BLAS programs.

In order to exploit the potential performance of computers with large memory hierarchies, it is indispensable to reduce the number of data transfers between different levels of memory to a minimum. It is desirable to maximize the ratio of floating-point operations to memory references and to reuse data as much as possible while it is stored in the higher levels of the memory hierarchy (cf. Chapter 6). This goal cannot be accomplished by using only vector and matrix-vector operations in the implementation of the algorithms; extensive use of elementary matrix-matrix operations (matrix-matrix multiplication etc.) must also

---

[5]The BLAS package presented in 1979 was later called BLAS-1.

be made. In 1990, the BLAS-3 package (Dongarra et al. [161]), which offers the necessary software modules, was published. The machine specific implementation of these subprograms may yield an even greater performance increase on modern computers than the (exclusive) use of optimized BLAS-1 and BLAS-2 programs.

LAPACK programs are written in such a way that the major portion of the computational work is performed by calls to the BLAS. Specially tuned and highly effective implementations of the BLAS are available for many modern high-performance computers. The BLAS enable LAPACK routines to achieve high performance *and* portability.

The BLAS are not strictly speaking part of LAPACK, but the Fortran 77 code for the BLAS is distributed with LAPACK or can be obtained separately from NETLIB. The code for these model implementations is not tuned to any specific hardware platform, i.e., it is completely portable. However, in order to achieve a very high degree of efficiency, the code must be individually adapted and optimized.

13.15.3 Block Algorithms

LINPACK Block Algorithms Using BLAS-1 Programs

The *classical* elimination methods for solving linear systems are based on a sequence of floating-point operations applied to single scalar data elements. LINPACK was the first package in which algorithms were implemented in such a way that elementary operations are also applied to vectors. The principle of such a program design is illustrated by the program `LINPACK/spofa`, which factorizes a symmetric positive definite matrix into $A = U^T U$ (Cholesky factorization). From the matrix-vector form of the Cholesky factorization (a_j is a vector of length $j-1$)

$$\begin{pmatrix} A_{11} & a_j & A_{13} \\ \cdot & a_{jj} & a_j^T \\ \cdot & \cdot & A_{33} \end{pmatrix} = \begin{pmatrix} U_{11}^T & 0 & 0 \\ u_j^T & u_{jj} & 0 \\ U_{13}^T & \mu_j & U_{33}^T \end{pmatrix} \begin{pmatrix} U_{11} & u_j & U_{13} \\ 0 & u_{jj} & \mu_j^T \\ 0 & 0 & U_{33} \end{pmatrix};$$

the equations

$$a_j = U_{11}^T u_j$$
$$a_{jj} = u_j^T u_j + u_{jj}^2$$

can be obtained. Supposing U_{11} has already been computed, the vector u_j and the scalar u_{jj} can be obtained from the equations:

$$U_{11}^T u_j = a_j$$
$$u_{jj}^2 = a_{jj} - u_j^T u_j.$$

This algorithm has been implemented in the program `LINPACK/spofa` using a call to the BLAS-1 program `BLAS/sdot` (for the dot product of two vectors) as follows:

```
      DO 30 J = 1, N
         INFO = J
         S = 0.0E0
         JM1 = J - 1
         IF (JM1 .LT. 1) GO TO 20
         DO 10 K = 1, JM1
            T = A(K,J) - SDOT (K-1, A(1,K), 1, A(1,J), 1)
            T = T/A(K,K)
            A(K,J) = T
            S = S + T*T
   10    CONTINUE
   20    CONTINUE
         S = A(J,J) - S
C     ......EXIT
         IF (S .LE. 0.0E0) GO TO 40
         A(J,J) = SQRT (S)
   30 CONTINUE
```

LAPACK Block Algorithms Using BLAS-2 Programs

The following program (which can be obtained by applying simple program transformations to the code above) performs exactly the same computations but is
coded in a typical LAPACK style. It uses the BLAS-2 program BLAS/strsv which
solves a triangular system of equations. For reasons described later, this is not
the code actually used in LAPACK.

```
      DO 10 j = 1, n
         CALL strsv ('upper', 'transpose', 'non-unit', j-1, a, lda, a(1,j), 1)
         s = a(j,j) - SDOT (j-1, a(1,j), 1, a(1,j), 1)
         IF (s .LE. zero) GO TO 20
         a(j,j) = SQRT (s)
   10 CONTINUE
```

This program replaces the k-loop of the algorithm implemented in LINPACK/spofa
with a call to the BLAS-2 program BLAS/strsv. This change alone is sufficient to
significantly increase the performance of the code on many machines. For example, using a single processor on a Cray Y-MP for solving systems with 500×500
matrices, the floating-point performance increased from 72 to 251 Mflop/s. As
this is already 81 percent of the optimal performance of matrix-matrix multiplication operations, the use of BLAS-3 programs does not promise further significant
gains in performance.

On an IBM 3090E VF, the program transformation above *does not* yield a
performance increase; both programs achieve the same performance of about 23
Mflop/s. For a machine with a peak performance of 75 Mflop/s for matrix-matrix
multiplication, this is anything but satisfactory. The special architecture of the
IBM 3090 computer series only makes peak performance computation possible
through the use of matrix-matrix operations. Thus in such a case, performance
gains are possible only through the use of BLAS-3 programs.

LAPACK Block Algorithms Using BLAS-3 Programs

As already mentioned, some computers (e. g., the IBM 3090) obtain a satisfying efficiency only if BLAS-3 programs are used. In order to exploit the higher speed of the BLAS-3 programs, the algorithm must undergo a deeper level of restructuring and be recast as a block algorithm which operates on sub-matrices (blocks) of the original matrix.

The mechanism for deriving block algorithms is illustrated by the Cholesky factorization of a symmetric, positive definite matrix. The defining equation, in partitioned form, of the Cholesky factorization is as follows:

$$\begin{pmatrix} A_{11} & A_{12} & A_{13} \\ \cdot & A_{22} & A_{23} \\ \cdot & \cdot & A_{33} \end{pmatrix} = \begin{pmatrix} U_{11}^\mathsf{T} & 0 & 0 \\ U_{12}^\mathsf{T} & U_{22}^\mathsf{T} & 0 \\ U_{13}^\mathsf{T} & U_{23}^\mathsf{T} & U_{33}^\mathsf{T} \end{pmatrix} \begin{pmatrix} U_{11} & U_{12} & U_{13} \\ 0 & U_{22} & U_{23} \\ 0 & 0 & U_{33} \end{pmatrix}.$$

Hence:

$$\begin{aligned} A_{12} &= U_{11}^\mathsf{T} U_{12} \\ A_{22} &= U_{12}^\mathsf{T} U_{12} + U_{22}^\mathsf{T} U_{22}. \end{aligned}$$

Supposing that U_{11} has already been computed, U_{12} can be derived from the equation

$$U_{11}^\mathsf{T} U_{12} = A_{12}$$

using the BLAS-3 program BLAS/strsm. U_{22} can then be obtained from the equation

$$U_{22}^\mathsf{T} U_{22} = A_{22} - U_{12}^\mathsf{T} U_{12}.$$

Hence, first the symmetric sub-matrix A_{22} must be updated by a call to the BLAS-3 program BLAS/ssyrk, and then its Cholesky factorization is computed. As Fortran 77 does not support recursion, the program LAPACK/spotf2 must be used. Thus, successive blocks of columns of the matrix U are computed.

In typical LAPACK programming style, the code for the block algorithm could look like this:

```
      DO 10 j = 1, n, nb
         jb = MIN( nb, n-j+1 )
         CALL strsm( 'left', 'upper', 'transpose', 'non-unit', j-1, jb,
     $               one, a, lda, a(1,j), lda )
         CALL ssyrk( 'upper', 'transpose', jb, j-1, -one, a(1,j), lda,
     $               one, a(j,j), lda )
         CALL spotf2( jb, a(j,j), lda, info )
         IF( info .NE. 0 ) GO TO 20
   10 CONTINUE
```

This program achieves 49 Mflop/s on an IBM 3090E VF. This is more than twice the performance of the LINPACK/spofa program. However, on a single processor of a Cray Y-MP, very little performance gain can be observed. In a configuration with 8 processors, however, a significant increase is obtained (cf. Table 13.2).

There are also other ways of increasing the performance of the program given above. Hence, the actual implementation of `LAPACK/spotrf` differs from the code fragment shown above. As mentioned in Chapter 6, for many linear algebra problems, there are three vectorizable variants of the algorithms, the so-called i, j and k variants.

LINPACK and all the previous examples in this section use the j variant. This variant is based on the solution of (triangular) systems of linear equations and achieves significantly poorer performance on many computers than the i variant, which is based on matrix-matrix multiplication. Hence, the i variant was chosen for the implementation of `LAPACK/spotrf`.

Table 13.2: The floating-point performance [Mflop/s] and the efficiency [%] of several variants (using specifically optimized BLAS programs) of the Cholesky factorization of a 500×500 matrix.

Machine Number of processors	IBM 3090 VF 1		Cray Y-MP 1		Cray Y-MP 8	
j variant: LINPACK	23	(21 %)	72	(22 %)	72	(3 %)
j variant using BLAS-2	24	(22 %)	251	(75 %)	378	(14 %)
j variant using BLAS-3	49	(45 %)	287	(86 %)	1225	(46 %)
i variant using BLAS-3	50	(46 %)	290	(87 %)	1414	(53 %)
Peak performance	108	(100 %)	333	(100 %)	2644	(100 %)

Floating-Point Performance of Block Factorizations

The block algorithms for LU and Cholesky factorization are easy to derive and neither additional floating-point operations nor extra working storage is required.

Table 13.3 illustrates the floating-point performance of the two programs `LAPACK/sgetrf` (in single precision on Cray machines) and `LAPACK/dgetrf` (in double precision on IBM machines) for the LU factorization of a real matrix. This corresponds to a 64-bit floating-point arithmetic on all computers tested. A block size of 1 means that the unblocked algorithm is used since it is faster or at least as fast as the corresponding blocked algorithm. It can be seen from Table 13.3 that the advantage of a block factorization becomes more apparent as the order of the matrix is increased.

Table 13.3: Floating-point performance [Mflop/s] of `LAPACK/sgetrf` resp. `dgetrf`.

	Number of processors	Block size	Dimension n					Peak performance
			100	200	300	400	500	
IBM RS/6000-530	1	32	19	25	29	31	33	50
IBM 3090J VF	1	64	23	41	52	58	63	108
Cray 2	1	64	110	211	292	318	358	488
Cray Y-MP	1	1	132	219	254	272	283	333
Cray Y-MP	8	64	195	556	920	1188	1408	2644

LAPACK programs for Cholesky factorization have been tested under the same conditions. The results are listed in Table 13.4.

Table 13.4: Floating-point performance [Mflop/s] of LAPACK/spotrf resp. LAPACK/dpotrf with uplo = 'u'.

	Number of processors	Block size	Dimension n					Peak performance
			100	200	300	400	500	
IBM RS/6000-530	1	32	21	29	34	36	38	50
IBM 3090J VF	1	48	26	43	56	62	67	108
Cray 2	1	64	109	213	294	318	362	488
Cray Y-MP	1	1	126	219	257	275	285	333
Cray Y-MP	8	32	146	479	845	1164	1393	2644

Blocked QR Factorization

The traditional algorithm for QR factorization is based upon the use of elementary Householder matrices $H = I - \tau u u^{\mathsf{T}}$, where u is a column vector and τ is a scalar (cf. Section 15.1.4). On most machines this algorithm comes close to reaching peak performance even with the invocation of BLAS-2 routines. When transforming it to block form, the product of k elementary Householder matrices of order n must be represented as a block form of a Householder matrix (cf. Section 13.19.5):

$$H_1 H_2 \ldots H_k = I - UTU^{\mathsf{T}},$$

where U is an $n \times k$ matrix whose columns correspond to the vectors $u_1, u_2, \ldots, u_k$ and T is an upper triangular matrix of order k.

13.15.4 Structure of LAPACK

LAPACK contains three different types of programs:

Black box or **driver routines** solve standard linear algebra problems such as the solution of a system of linear equations or determining the eigenvalues of a matrix. All driver routines offer a very comfortable and simple user interface (all calls to computational and auxiliary routines are done by the driver routine itself), the user is well advised to use a driver routine whenever there is one that meets his requirements.

Computational routines perform certain computational tasks such as the computation of LU factorizations or the reduction of matrices to diagonal form.

Auxiliary routines can be classified as follows:

- programs for general low-level functions, such as scaling a matrix, computing a matrix norm, or generating an elementary Householder matrix;

- extensions to the BLAS, such as programs for matrix-vector operations with complex, symmetric matrices;
- programs which execute subtasks of block algorithms.

Data Types and Precision

LAPACK provides the same functionality for *real* and *complex data*. For most computations there are both real and complex program variants, but there are a few exceptions. For example, corresponding to the routines for real symmetric, indefinite systems of linear equations, there are routines for complex Hermitian matrices $(A = A^H)$ as well as for complex symmetric matrices $(A = A^T)$, because both types of complex systems occur in practical applications.

For all LAPACK programs there are *single* and *double precision* versions. Double precision programs for complex matrices require the data type DOUBLE COMPLEX, which is *not* defined in the Fortran 77 standard, but is nevertheless available on most modern computers.

As a result of the number of supported data types, there are *four* versions of each LAPACK program. If the programming language supports generic names (as does Fortran 90, but not Fortran 77), then in most cases all four variants can be combined into a single program.

Naming Scheme

The name of each LAPACK routine is a coded specification of its function.

All driver and computational routines have names consisting of 5 or 6 letters.

The *first letter* indicates the Fortran 77 data type as follows:

s   REAL
d   DOUBLE PRECISION
c   COMPLEX
z   DOUBLE COMPLEX   or   COMPLEX*16   (both non-standard).

When referring to an LAPACK routine generically, regardless of the data type, the first letter is replaced by a star *. Thus, for instance, *gesv refers to *any or all* of the routines sgesv, cgesv, dgesv, and zgesv.

The *second* and *third* letter indicate the type of the matrix. Most of these two-letter codes apply to both real and complex matrices; a few apply specifically to one or the other, as indicated in Table 13.5.

When referring to a class of routines that performs the same function on different types of matrices, the first *three* letters are replaced by a star *. Thus, for instance, *sv refers to *all* routines for systems of linear equations that are listed in Table 13.6 (sv is shorthand for *solve*).

The last three letters indicate the computation performed. Their definitions are explained in Sections 13.16 and 13.17. For example, sgebrd is a single precision

Table 13.5: Matrix Type—second and third letter in the LAPACK naming scheme.

ge	general matrix (in some cases rectangular)
gg	general matrix in generalized eigenvalue problem
gb	general band matrix
gt	general tridiagonal matrix
he	Hermitian matrix
hp	Hermitian matrix, *packed storage*
hs	lower Hessenberg matrix
hg	lower Hessenberg matrix in generalized eigenvalue problem
or	orthogonal matrix
op	orthogonal matrix, *packed storage*
po	symmetric (or Hermitian), positive definite matrix
pp	symmetric (or Hermitian), positive definite matrix, *packed storage*
pb	symmetric (or Hermitian), positive definite band matrix
pt	symmetric (or Hermitian), positive definite tridiagonal matrix
sy	symmetric matrix
sp	symmetric matrix, *packed storage*
sb	symmetric band matrix
st	symmetric tridiagonal matrix
tb	triangular band matrix
tg	triangular matrix in generalized eigenvalue problem
tp	triangular matrix, *packed storage*
tr	triangular matrix or block triangular matrix
bd	bidiagonal matrix
un	unitary matrix
up	unitary matrix, *packed storage*

routine that performs a bidiagonal reduction of a real general matrix (brd is shorthand for *bidiagonal reduction*).

The names of auxiliary routines follow a similar scheme except that the second and third characters are usually la (e. g., slascl or clarfg). There are two kinds of exceptions: Auxiliary routines which implement an unblocked version of a block algorithm have similar names to the routines which perform the block algorithm, with the sixth character being "2" (for example, sgebd2 is the unblocked version of sgebrd); a few routines that may be regarded as extensions of BLAS are named according to the BLAS naming scheme (for example, crot or csyr).

13.16 LAPACK Black Box Programs

This section is a survey of the LAPACK driver routines for the solution of linear systems and linear least squares problems[6]. Further details on the terminology and the numerical operations they perform are given in Section 13.17.

There are two basic types of driver routines:

---

[6]The LAPACK driver routines for the solution of eigenvalue problems can be found in Section 15.7.

Simple driver routines are black box routines delivering only the solution of the problem.

Expert driver routines deliver additional information (such as the condition number, or error bounds) or perform additional computations (e. g., improving the accuracy of a numerical solution).

LAPACK expert driver routines follow the naming scheme *x and so the last character of their name is an **x**.

13.16.1 Linear Equations

Two types of driver routines for solving systems of linear equations (in matrix terms, $AX = B$) are provided:

Simple driver routines: The matrix A is overwritten by its factors (e. g., the matrices L and U) and the right-hand side $B = (b_1, \ldots, b_k)$ by the solution vector $X = (x_1, \ldots, x_k)$. The names of simple driver routines follow the scheme *sv (shortened form of *solve*).

Expert driver routines: In addition to the factorization of A and the back substitution for determining X, the following functions are performed:

- estimating the condition number cond(A) and checking for numerical singularity;
- refining the solution and computing error bounds;
- (optionally) balancing the system if A is poorly scaled.

Expert drivers require roughly twice the storage of simple drivers in order to perform these extra functions. The names of all LAPACK expert drivers contain *svx (shortened form of *solve expert*).

Table 13.6 lists the different driver routines. Some of them take advantage of special properties of, or storage schemes for, the matrix A (cf. Section 13.19).

13.16.2 Linear Least Squares Problems

Given a matrix $A \in \mathbb{R}^{m \times n}$ and a vector $b \in \mathbb{R}^m$, the *linear least squares problem* serves to determine a vector $x^* \in \mathbb{R}^n$ such that the residual is minimal:

$$\|Ax^* - b\| = \min\{\|Ax - b\|_2 : x \in \mathbb{R}^n\}. \tag{13.55}$$

If $m > n$, the system $Ax = b$ is said to be *over-determined*, if $m < n$, it is said to be *under-determined*. In most cases, $m \geq n$ and rank(A) $= n$ and so there is a unique solution to the problem (13.55). When $m < n$ or $m \geq n$ and rank(A) $< n$, there is no longer a single solution; rather there is an infinite number of solutions. In such cases the particular (unique) solution x which minimizes $\|x^*\|_2$ is determined.

Table 13.6: LAPACK driver routines for linear equations.

Type of Matrix (Storage)	driver	REAL	COMPLEX
general matrix	simple	sgesv	cgesv
	expert	sgesvx	cgesvx
general band matrix	simple	sgbsv	cgbsv
	expert	sgbsvx	cgbsvx
general tridiagonal matrix	simple	sgtsv	cgtsv
	expert	sgtsvx	cgtsvx
symmetric/Hermitian, positive definite matrix	simple	sposv	cposv
	expert	sposvx	cposvx
symmetric/Hermitian, positive definite matrix (packed storage)	simple	sppsv	cppsv
	expert	sppsvx	cppsvx
symmetric/Hermitian, positive definite band matrix	simple	spbsv	cpbsv
	expert	spbsvx	cpbsvx
symmetric/Hermitian, positive definite tridiagonal matrix	simple	sptsv	cptsv
	expert	sptsvx	cptsvx
symmetric/Hermitian, indefinite matrix	simple	ssysv	chesv
	expert	ssysvx	chesvx
symmetric/Hermitian, indefinite matrix (packed storage)	simple	sspsv	chpsv
	expert	sspsvx	chpsvx
complex symmetric matrix	simple		csysv
	expert		csysvx
complex symmetric matrix (packed storage)	simple		cspsv
	expert		cspsvx

If the matrix A is of full rank, then the driver routine *gels can be used to solve the problem (13.55) using a QR or LQ factorization of A. The driver routines *gelsx and *gelss also solve the problem (13.55) for a rank-deficient matrix A. Table 13.7 lists the driver routines for linear least squares problems.

Table 13.7: LAPACK driver routines for linear least squares problems.

Type of Matrix	Method	driver	REAL	COMPLEX
general matrix	QR or LQ factorization	simple	sgels	cgels
	complete orthogonal factorization	simple	sgelsx	cgelsx
	singular value decomposition (SVD)	simple	sgelss	cgelss

13.17 LAPACK Computational Routines

This section gives a survey of LAPACK's computational routines used for the solution of linear systems and least squares problems. Computational routines for eigenvalue problems are covered in Section 15.7.

With direct methods for the solution of the linear system $Ax = b$ or the matrix equation $AX = B$ (where the columns of B are the individual right-hand sides, and the columns of X are the corresponding solutions), the main task is the factorization of the matrix A.

The type of factorization chosen depends on the properties of the matrix A. LAPACK provides routines for the following types of matrices:

General Matrices (LU Factorization With Partial Pivoting):

$$A = PLU;$$

Symmetric, Positive Definite Matrices (Cholesky Factorization):

$$A = U^\mathsf{T} U \qquad \text{or} \qquad A = LL^\mathsf{T};$$

Symmetric, Indefinite Matrices (Symmetric, Indefinite Factorization):

$$A = P^\mathsf{T} U^\mathsf{T} DUP \qquad \text{or} \qquad A = PLDL^\mathsf{T} P^\mathsf{T}.$$

U is an upper triangular matrix; L is a lower triangular matrix; P is a permutation matrix; and D is a block diagonal matrix with blocks of order 1 or 2.

The following list describes the tasks of the computational routines corresponding to the last three characters of their name (cf. Tables 13.8 and 13.9):

LAPACK/*trf are programs for the factorization of the matrix A (obviously not needed for triangular matrices).

LAPACK/*trs are programs for solving $AX = B$ with back substitution using an existing factorization of A (or a triangular matrix A). They require the factorization computed by LAPACK/*trf.

LAPACK/*con estimates the reciprocal κ^{-1} of the condition number $\kappa(A) = \|A\| \, \|A^{-1}\|$. A modification of the Hager method (Higham [224]) is used to estimate $\|A^{-1}\|$, except for symmetric, positive definite, tridiagonal matrices for which A^{-1} can be computed directly with reasonable computational effort (Higham [222], [223]). The argument norm makes it possible to switch between κ_1 and κ_∞. The programs require the norm of the original matrix A and the factorization returned by LAPACK/*trf.

Table 13.8: LAPACK computational routines for linear systems.

Type of Matrix (Storage)	Operation	REAL	COMPLEX
general matrix	factorize	sgetrf	cgetrf
	solve	sgetrs	cgetrs
	condition number	sgecon	cgecon
	error bounds	sgerfs	cgerfs
	invert	sgetri	cgetri
	balance	sgeequ	cgeequ
general band matrix	factorize	sgbtrf	cgbtrf
	solve	sgbtrs	cgbtrs
	condition number	sgbcon	cgbcon
	error bounds	sgbrfs	cgbrfs
	balance	sgbequ	cgbequ
general tridiagonal matrix	factorize	sgttrf	cgttrf
	solve	sgttrs	cgttrs
	condition number	sgtcon	cgtcon
	error bounds	sgtrfs	cgtrfs
symmetric/Hermitian, positive definite matrix	factorize	spotrf	cpotrf
	solve	spotrs	cpotrs
	condition number	spocon	cpocon
	error bounds	sporfs	cporfs
	invert	spotri	cpotri
	balance	spoequ	cpoequ
symmetric/Hermitian, positive definite matrix (packed storage)	factorize	spptrf	cpptrf
	solve	spptrs	cpptrs
	condition number	sppcon	cppcon
	error bounds	spprfs	cpprfs
	invert	spptri	cpptri
	balance	sppequ	cppequ
symmetric/Hermitian, positive definite band matrix	factorize	spbtrf	cpbtrf
	solve	spbtrs	cpbtrs
	condition number	spbcon	cpbcon
	error bounds	spbrfs	cpbrfs
	balance	spbequ	cpbequ
symmetric/Hermitian, positive definite tridiagonal matrix	factorize	spttrf	cpttrf
	solve	spttrs	cpttrs
	condition number	sptcon	cptcon
	error bounds	sptrfs	cptrfs

Table 13.9: LAPACK computational routines for linear systems (continued).

Type of Matrix (Storage)	Operation	REAL	COMPLEX
symmetric/Hermitian, indefinite matrix	factorize	ssytrf	chetrf
	solve	ssytrs	chetrs
	condition number	ssycon	checon
	error bounds	ssyrfs	cherfs
	invert	ssytri	chetri
symmetric/Hermitian, indefinite matrix (packed storage)	factorize	ssptrf	chptrf
	solve	ssptrs	chptrs
	condition number	sspcon	chpcon
	error bounds	ssprfs	chprfs
	invert	ssptri	chptri
complex symmetric matrix	factorize		csytrf
	solve		csytrs
	condition number		csycon
	error bounds		csyrfs
	invert		csytri
complex symmetric matrix (packed storage)	factorize		csptrf
	solve		csptrs
	condition number		cspcon
	error bounds		csprfs
	invert		csptri
triangular matrix	solve	strtrs	ctrtrs
	condition number	strcon	ctrcon
	error bounds	strrfs	ctrrfs
	invert	strtri	ctrtri
triangular matrix (packed storage)	solve	stptrs	ctptrs
	condition number	stpcon	ctpcon
	error bounds	stprfs	ctprfs
	invert	stptri	ctptri
triangular band matrix	solve	stbtrs	ctbtrs
	condition number	stbcon	ctbcon
	error bounds	stbrfs	ctbrfs

Example (Condition Estimate) For test purposes, a one parameter family of matrices $A_n(p) \in \mathbb{R}^{n \times n}$ has been defined as

$$[A_n(p)]_{ij} = \binom{p+j-1}{i-1}, \quad i,j = 1, 2, \ldots, n. \tag{13.56}$$

The inverse $A_n^{-1}(p)$ is given by

$$[A_n^{-1}(p)]_{ij} = (-1)^{i+j} \sum_{l=0}^{n-j} \binom{p+l-1}{l} \binom{l+j-1}{i-1} \quad \text{with} \quad \binom{r}{s} := 0 \quad \text{for} \quad r < s.$$

The condition number $\kappa(A_n(p))$ increases rapidly with increasing n and p (cf. Fig. 13.12). Since the inverse is known explicitly in this test case, the condition number can be easily derived from $\kappa(A_n) = \|A_n\| \|A_n^{-1}\|$. If a condition estimate is found using LAPACK/sgecon, then the following becomes apparent: As long as the condition number $\kappa(A_n)$ is smaller than $1/eps \approx 1.68x10^7$, the estimated value is in perfect agreement with the actual condition number. For matrices with a higher condition number, an *under-estimation* of the actual condition can be observed (cf. Fig. 13.13).

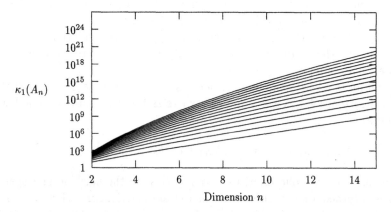

Figure 13.12: Condition numbers of the matrices (13.56) for $p = 1, 2, \ldots, 15$; the lowermost curve corresponds to $p = 1$, the uppermost curve corresponds to $p = 15$.

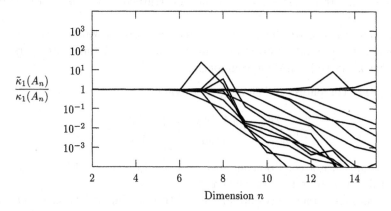

Figure 13.13: Accuracy of the condition number estimate produced by LAPACK/sgecon.

LAPACK/*rfs are used for determining error bounds for the solution delivered by LAPACK/*trs and for refining this solution (cf. also next section). The programs require the original matrices A and B, the factorization of A computed by LAPACK/*trf and the solution X from LAPACK/*trs.

LAPACK/*tri are programs for computing the inverse A^{-1} using the factorization of the matrix A. The programs require the factorization of A from LAPACK/*trf.

LAPACK/*equ are used for determining the scaling factors of the matrix A.

13.17.1 Error Bounds

LAPACK/*rfs (*refine solution*) programs compute error bounds for numerical solution using iterative refinement. This iterative refinement of the solution is
performed in the same precision as the input data. In particular, the residual is
not computed with extra precision, as was done earlier.

Let $\hat{x}$ be the numerically computed result, x the exact solution of $Ax = b$,
and $r = b - A\hat{x}$ the residual of $\hat{x}$. Let $|b|$ ($|A|$) denote the vector (the matrix) of
absolute values of b (A): $|b|_k := |b_k|$ ($|A|_{ij} := |a_{ij}|$).

Except for in a very small number of cases, $\hat{x}$ is the exact solution of a slightly
perturbed system $(A + \Delta A)\hat{x} = b + \Delta b$, where ΔA and Δb satisfy the following
inequalities:

$$|\Delta a_{ij}| \leq \omega_c |a_{ij}| \quad \text{and} \quad |\Delta b_k| \leq \omega_c |b_k| \qquad \text{for all} \quad i, j, k.$$

ω_c is the error in the solution $\hat{x}$; this error is close to the machine precision *eps*
and it is returned from a subroutine in the argument **berr**. Hence, $\hat{x}$ is the
exact solution of a system which differs from the original system only by a few
rounding errors in each entry. In case the elements of A themselves are affected
with rounding errors, the solution of the unperturbed system is as accurate as
the data allow.

The rare cases in which ω_c is not $O(eps)$ are cases in which A is so ill-conditioned that the iteration does not converge, or where A and x are sparse and
$|A| |x|$ has a (nearly) vanishing component.

LAPACK/*rfs computes the following error bound:

$$\frac{\|x - \hat{x}\|_\infty}{\|\hat{x}\|_\infty} = \frac{\|A^{-1}r\|_\infty}{\|\hat{x}\|_\infty} \leq \frac{\| |A^{-1}| |r| \|_\infty}{\|\hat{x}\|_\infty}.$$

The programs use a condition number estimate to compute $\| |A^{-1}| |r| \|_\infty$ and
return this estimate in the argument **ferr**.

ferr and **berr** themselves are affected with rounding and other errors. The
absolute error in **berr** does not exceed $(n+1)\,eps$, where n is the order of A. **ferr**
can be affected with errors due to the computation of $|r|$ and the condition number
estimate. The possible underestimation of $|r|$ is eliminated by increasing it by
the maximal rounding error which may occur in its computation. A detailed
discussion of the error estimation techniques used in LAPACK can be found in
Arioli, Demmel, Duff [90].

13.17.2 Orthogonal Factorizations

LAPACK provides a number of routines for performing the orthogonal (or unitary
in the case of complex matrices) factorization of rectangular $m \times n$ matrices (cf.
Table 13.10). In addition to the QR factorization, programs for LQ, QL, as well
as RQ factorizations are provided. These may be useful if $m < n$, or if a lower
triangular matrix L is needed instead of an upper triangular matrix R. In fact

Table 13.10: LAPACK computational routines for orthogonal factorizations.

	Operation	REAL	COMPLEX
QR factorization	factorization with pivoting	sgeqpf	cgeqpf
	factorization *without* pivoting	sgeqrf	cgeqrf
	generation of Q	sorgqr	cungqr
	multiplication of matrix by Q	sormqr	cunmqr
LQ factorization	factorization *without* pivoting	sgelqf	cgelqf
	generation of Q	sorglq	cunglq
	multiplication of matrix by Q	sormlq	cunmlq
QL factorization	factorization *without* pivoting	sgeqlf	cgeqlf
	generation of Q	sorgql	cungql
	multiplication of matrix by Q	sormql	cunmql
RQ factorization	factorization *without* pivoting	sgerqf	cgerqf
	factorization *without* pivoting	stzrqf	ctzrqf
	generation of Q	sorgrq	cungrq
	multiplication of matrix by Q	sormrq	cunmrq

all four factorization routines accept arbitrary values for m and n, so in some cases R or L are trapezoidal rather than triangular. Note that the routine for QR factorization is the only one which performs pivoting.

The orthogonal matrix Q is returned by the factorization routines as a product of Householder matrices (cf. Section 13.19.5). There are additional routines for generating Q (or part of Q) explicitly or for computing matrix products of the form QC, $Q^H C$, CQ, or CQ^H.

13.17.3 Singular Value Decomposition (SVD)

The LAPACK computation routines for the singular value decomposition of a rectangular matrix $A \in \mathbb{R}^{m \times n}$ or $A \in \mathbb{C}^{m \times n}$ are based on two stages:

1. The matrix A is transformed into a bidiagonal form:

$$A = U_1 B V_1^\top \qquad \text{resp.} \qquad A = U_1 B V_1^H.$$

Here U_1 and V_1 are orthogonal (unitary if A is complex) matrices, and B is a real bidiagonal matrix. The program LAPACK/*gebrd performs these computations; U_1 and V_1 are stored as products of Householder matrices. The matrices U_1 and V_1 may be computed explicitly using the routine LAPACK/*orgbr (LAPACK/*ungbr if A is complex).

2. The singular value decomposition of the bidiagonal matrix B is computed as

$$B = U_2 S V_2^\top.$$

Here again, U_2 and V_2 are orthogonal matrices and S is a diagonal matrix containing the singular values of A. The singular vectors of A are then given by the columns of $U := U_1 U_2$ and $V := V_1 V_2$.

Table 13.11: LAPACK computation routines for singular value decomposition.

Type of Matrix	Operation	REAL	COMPLEX
general matrix	bidiagonal reduction	sgebrd	cgebrd
orthogonal/ unitary matrix	generate factor matrices after bidiagonal reduction	sorgbr	cungbr
bidiagonal matrix	singular values/singular vectors	sbdsqr	cbdsqr

13.18 LAPACK Documentation

Each LAPACK program is fully documented in the LAPACK User's Guide [3]. The documentation of each program comprises the corresponding SUBROUTINE or FUNCTION statement, followed by the type declaration and dimension of the arguments, the *purpose* of the routine, and a description of its *parameters*.

13.18.1 Parameters

The parameters of all LAPACK programs appear in the following order: the specification of options, the *dimension* of the problem, arrays or scalars storing *input data* (some of which may be overwritten by results), arrays or scalars storing *output data*, working storage arrays (and their array dimensions), and finally the parameter `info` for diagnostics.

Parameters specifying options are of type CHARACTER*1. The possible values of these parameters are expressed as one-letter abbreviations for each option (e. g., 't' for transposition). In order to improve readability, a longer string may be passed as an actual parameter, but only the *first* character is *significant*:

```
CALL  sgetrs ('transpose', . . . )
```

Dimension of the Problem

It is permissible to specify a *zero* problem dimension. In such cases the according computation (or part of it) will be skipped. However, negative dimensions are regarded as incorrect input.

Array Parameters

Vectors and matrices are stored in one- or two-dimensional arrays in the conventional manner. That is, if a one-dimensional array x of order (length) n stores the vector $x = (x_1, \ldots, x_n)^\top \in \mathbb{R}^n$, then the array element x(i) stores the vector component x_i for $i = 1, 2, \ldots, n$. If a two-dimensional array a of order (lda, n) stores the $m \times n$ matrix A, then the array element a(i,j) stores the matrix element a_{ij} for $i = 1, 2, \ldots, m$ and $j = 1, 2, \ldots, n$.

Each two-dimensional array parameter is immediately followed in the parameter list by a parameter specifying the *leading dimension* of the array; the name of this parameter comes in the form:

$$ld < array\text{-}name > .$$

LAPACK is written in Fortran 77. There are, therefore, no assumed-shape arrays, no automatic arrays and no dynamic arrays available. In LAPACK programs, formal array parameters are usually declared as assumed-size arrays. In the case of one-dimensional arrays (vectors), the range of the array is determined by the actual parameter. In the case of two-dimensional arrays (matrices), the declaration

```
REAL  a(lda, *)
```

determines the range of the first dimension explicitly using the index lda; only the range of the second dimension (denoted by the star) remains variable. Nevertheless, LAPACK documentation always gives the dimension as (lda, n) because this form also specifies the *minimal* value of the last dimension.

Example (Array Parameters) The matrix $A \in \mathbb{R}^{10 \times 10}$ and the right-hand side $b \in \mathbb{R}^{10}$ are passed to the program LAPACK/sposv as actual parameters:

```
...
INTEGER      lda
PARAMETER    (lda = 100)
REAL         a (lda, lda),  b(lda)
...
n = 10
DO i = 1, n
   DO j = 1, n
      a(i,j) = 1E0/REAL(i + j - 1)
   END DO
   b(i) = 1E0
END DO
...
CALL  sposv (uplo, n, 1, a, lda, b, lda, info)
...
```

13.18.2 Error Handling

All LAPACK programs have a diagnostic parameter info which indicates the success or failure of the computation:

info = 0: The algorithm has *successfully* (without detecting any errors) terminated;

info < 0: The value of one or more of the input parameters is *illegal*.

info > 0: An *error* has been detected during the course of computation.

All LAPACK programs check that the actual parameters have permitted values. If an illegal value of the ith argument is detected, the routine sets info = -i, and then the error-handling routine xerbla is called. In such cases no assignments to output parameters (except info) are performed.

The standard version of the program `xerbla` issues an error message and *halts* execution, so that no LAPACK program returns to the calling program with `info` < 0. This can occur only if a non-standard (user modified) version of `xerbla` is used.

13.19 LAPACK Storage Schemes

LAPACK supports the following four storage schemes for matrices:

- conventional storage in a two-dimensional array;
- packed storage for symmetric, Hermitian or triangular matrices;
- band storage for band matrices;
- 2 or 3 one-dimensional arrays (vectors) for bidiagonal and tridiagonal matrices respectively.

The examples below illustrate only the relevant part of the array; array arguments may of course have additional (unused) rows or columns, according to the usual rules for passing array arguments in Fortran 77.

13.19.1 Conventional Storage

A matrix $A \in \mathbb{R}^{n \times n}$ is usually stored in a two-dimensional array `a`, with the matrix element a_{ij} stored in array element $a(i,j)$. The Fortran storage concept is used: the matrix elements are stored *column-wise* (*column major order*).

If a matrix is triangular, only the elements of the relevant triangular part are accessed. The remaining elements of the array need not be set. Such elements are indicated by "." in the following example of a 4×4 matrix. *Upper* and *lower* triangular matrices are distinguished by the argument `uplo`.

uplo	Triangular matrix A				Storage in array a			
'u'	a_{11}	a_{12}	a_{13}	a_{14}	a_{11}	a_{12}	a_{13}	a_{14}
		a_{22}	a_{23}	a_{24}	·	a_{22}	a_{23}	a_{24}
			a_{33}	a_{34}	·	·	a_{33}	a_{34}
	0			a_{44}	·	·	·	a_{44}
'l'	a_{11}			**0**	a_{11}	·	·	·
	a_{21}	a_{22}			a_{21}	a_{22}	·	·
	a_{31}	a_{32}	a_{33}		a_{31}	a_{32}	a_{33}	·
	a_{41}	a_{42}	a_{43}	a_{44}	a_{41}	a_{42}	a_{43}	a_{44}

Also, if the matrix is symmetric or Hermitian, then either the upper or lower triangular part of the matrix must be stored; the remaining elements need not be set.

uplo	Hermitian matrix A	Storage in array a
'u'	$\begin{pmatrix} a_{11} & a_{12} & a_{13} & a_{14} \\ \bar{a}_{12} & a_{22} & a_{23} & a_{24} \\ \bar{a}_{13} & \bar{a}_{23} & a_{33} & a_{34} \\ \bar{a}_{14} & \bar{a}_{24} & \bar{a}_{34} & a_{44} \end{pmatrix}$	$\begin{matrix} a_{11} & a_{12} & a_{13} & a_{14} \\ \cdot & a_{22} & a_{23} & a_{24} \\ \cdot & \cdot & a_{33} & a_{34} \\ \cdot & \cdot & \cdot & a_{44} \end{matrix}$
'l'	$\begin{pmatrix} a_{11} & \bar{a}_{21} & \bar{a}_{31} & \bar{a}_{41} \\ a_{21} & a_{22} & \bar{a}_{32} & \bar{a}_{42} \\ a_{31} & a_{32} & a_{33} & \bar{a}_{43} \\ a_{41} & a_{42} & a_{43} & a_{44} \end{pmatrix}$	$\begin{matrix} a_{11} & \cdot & \cdot & \cdot \\ a_{21} & a_{22} & \cdot & \cdot \\ a_{31} & a_{32} & a_{33} & \cdot \\ a_{41} & a_{42} & a_{43} & a_{44} \end{matrix}$

13.19.2 Packed Storage

Symmetric, Hermitian, and triangular matrices can be stored more compactly than they can be in conventional storage schemes, so that only half the storage is required. The relevant part of the matrix (specified by uplo) is packed by columns in a one-dimensional array (vector).

By convention, the names of LAPACK arrays containing matrices in packed storage end with the letter 'p':

- uplo = 'u': In this case the part of the matrix A *above* the main diagonal holds all relevant information. The matrix element a_{ij} is stored (column-wise) for $i \leq j$ in the one-dimensional array ap at position $i + j(j-1)/2$.

- uplo = 'l': The part *below* the main diagonal of A holds all relevant information. The matrix element a_{ij} is stored (column-wise) for $j \leq i$ in the one-dimensional array ap at position $i + (2n - j)(j - 1)/2$.

uplo	Triangular matrix A	Packed storage in array ap
'u'	$\begin{pmatrix} a_{11} & a_{12} & a_{13} & a_{14} \\ & a_{22} & a_{23} & a_{24} \\ & & a_{33} & a_{34} \\ 0 & & & a_{44} \end{pmatrix}$	$a_{11}\ \underbrace{a_{12}\ a_{22}}\ \underbrace{a_{13}\ a_{23}\ a_{33}}\ \underbrace{a_{14}\ a_{24}\ a_{34}\ a_{44}}$
'l'	$\begin{pmatrix} a_{11} & & & 0 \\ a_{21} & a_{22} & & \\ a_{31} & a_{32} & a_{33} & \\ a_{41} & a_{42} & a_{43} & a_{44} \end{pmatrix}$	$\underbrace{a_{11}\ a_{21}\ a_{31}\ a_{41}}\ \underbrace{a_{22}\ a_{32}\ a_{42}}\ \underbrace{a_{33}\ a_{43}}\ a_{44}$

13.19.3 Storage of Band Matrices

An $n \times n$ band matrix A with k_l sub-diagonals and k_u super-diagonals can be stored more compactly in a two-dimensional array ab of $k_l + k_u + 1$ rows and n columns. The columns of the matrix are stored in the according array columns and the diagonals in the array rows. The matrix element a_{ij} is stored in

$$\mathrm{ab}(k_u + 1 + i - j, j), \qquad \max(1, j - k_u) \leq i \leq \min(n, j + k_l).$$

For reasons of efficiency this storage scheme should be used in practice only for band matrices with $k_l, k_u \ll n$, even though LAPACK programs work properly for *all* values k_l and k_u.

By convention, the names of LAPACK arrays containing matrices in band storage form end with the letter 'b'.

For example, when $n = 5$, $k_l = 2$ and $k_u = 1$.

Band matrix A	Band storage in array ab
$\begin{pmatrix} a_{11} & a_{12} & & & \mathbf{0} \\ a_{21} & a_{22} & a_{23} & & \\ a_{31} & a_{32} & a_{33} & a_{34} & \\ & a_{42} & a_{43} & a_{44} & a_{45} \\ \mathbf{0} & & a_{53} & a_{54} & a_{55} \end{pmatrix}$	$\begin{array}{ccccc} \cdot & a_{12} & a_{23} & a_{34} & a_{45} \\ a_{11} & a_{22} & a_{33} & a_{44} & a_{55} \\ a_{21} & a_{32} & a_{43} & a_{54} & \cdot \\ a_{31} & a_{42} & a_{53} & \cdot & \cdot \end{array}$

For symmetric or Hermitian band matrices with k_d sub-diagonals and super-diagonals, only the upper or lower triangular part of the matrix (according to the value of uplo) has to be stored:

uplo = 'u': a_{ij} is stored in $ab(k_d + 1 + i - j, j)$ for $\max(1, j - k_d) \le i \le j$;

uplo = 'l': a_{ij} is stored in $ab(1 + i - j, j)$ for $j \le i \le \min(n, j + k_d)$.

The following example illustrates the scheme for a Hermitian 5×5 matrix with $k_d = 2$.

uplo	Hermitian band matrix A	Band storage in array ab
'u'	$\begin{pmatrix} a_{11} & a_{12} & a_{13} & & \mathbf{0} \\ \bar{a}_{12} & a_{22} & a_{23} & a_{24} & \\ \bar{a}_{13} & \bar{a}_{23} & a_{33} & a_{34} & a_{35} \\ & \bar{a}_{24} & \bar{a}_{34} & a_{44} & a_{45} \\ \mathbf{0} & & \bar{a}_{35} & \bar{a}_{45} & a_{55} \end{pmatrix}$	$\begin{array}{ccccc} \cdot & \cdot & a_{13} & a_{24} & a_{35} \\ \cdot & a_{12} & a_{23} & a_{34} & a_{45} \\ a_{11} & a_{22} & a_{33} & a_{44} & a_{55} \end{array}$
'l'	$\begin{pmatrix} a_{11} & \bar{a}_{21} & \bar{a}_{31} & & \mathbf{0} \\ a_{21} & a_{22} & \bar{a}_{32} & \bar{a}_{42} & \\ a_{31} & a_{32} & a_{33} & \bar{a}_{43} & \bar{a}_{53} \\ & a_{42} & a_{43} & a_{44} & \bar{a}_{54} \\ \mathbf{0} & & a_{53} & a_{54} & a_{55} \end{pmatrix}$	$\begin{array}{ccccc} a_{11} & a_{22} & a_{33} & a_{44} & a_{55} \\ a_{21} & a_{32} & a_{43} & a_{54} & \cdot \\ a_{31} & a_{42} & a_{53} & \cdot & \cdot \end{array}$

13.19.4 Tridiagonal and Bidiagonal Matrices

A non-symmetric, $n \times n$, tridiagonal matrix is stored in three separate one-dimensional arrays (vectors), where one array of length n contains the diagonal elements and two arrays of length $n - 1$ contain the sub- and super-diagonals.

Bidiagonal matrices or symmetric, tridiagonal matrices are analogously stored in *two* one-dimensional arrays (vectors).

13.19.5 Orthogonal or Unitary Matrices

In LAPACK, real orthogonal or complex unitary matrices are often represented as the product of *Householder matrices* (*elementary reflectors*):

$$Q = H_1 H_2 \cdots H_k.$$

An elementary reflection specified by an $n \times n$-Householder matrix H corresponds to an orthogonal (unitary) matrix of the form

$$H = I - \tau v v^H, \tag{13.57}$$

where τ is a scalar and v is a vector with $|\tau|^2 \, \|v\|_2^2 = 2 \operatorname{Re}(\tau)$.

The representation (13.57) is redundant. As a consequence, the representation actually used in LAPACK (which differs from the one used in LINPACK or EISPACK) sets the first component of the vector to unity, i.e., $v_1 = 1$: v_1 thus does *not* have to be stored, and in the case of a real matrix H, it follows that $1 \leq \tau \leq 2$ or $\tau = 0$ (if $H = I$).

13.20 Block Size for Block Algorithms

LAPACK routines implementing block algorithms must have a mechanism for determining a suitable block size. It was a design decision of the LAPACK group, on the one hand, to hide this process from the user whenever possible, but, on the other hand, not to impede machine specific tuning of LAPACK routines.

LAPACK programs use the auxiliary program `ilaenv` which returns the optimal block size. The standard version included in the LAPACK package returns reasonable *default values* which have proved to be good on many test machines. However, in order to achieve optimal floating-point performance, the program `ilaenv` should be adapted individually to the machine being used (cf. LAPACK installation handbook for details).

Moreover, the optimal block size may depend on the particular routine and the combination of option arguments as well as the order of the problem.

`LAPACK/ilaenv` returns a reasonable value for the block size according to the particular routine, the order of the problem and optional parameters. If `LAPACK/ilaenv` returns a block size 1, then the respective LAPACK program will execute an *unblocked* algorithm.

13.21 LAPACK Variants and Extensions

The problems covered by LAPACK (e. g., the solution of linear systems and least squares problems, or the computation of eigenvalues and eigenvectors) are of exceptional importance in practical numerical data processing. It is, however, unpleasant to be confronted with various restrictions and weaknesses of LAPACK programs when using them in practice. For this reason efforts have been made to develop LAPACK variants and extensions without those shortcomings.

The main restrictions and weaknesses of LAPACK are described below. The most important current LAPACK update projects are subsequently discussed.

Restrictions and Weaknesses of LAPACK

Three basically different levels of reflection need to be considered when assessing the quality of LAPACK:

- At the *problem level* the question is as to whether LAPACK can indeed treat all essential problems concerning linear algebra models in practice.

- At the *algorithm level* the question is as to whether LAPACK algorithms comply with all the essential quality criteria for numerical software (accuracy, efficiency etc.; cf. Chapter 5).

- Similarly, at the *implementation level*, the quality of the actual implementation is to be assessed according to standard criteria (cf. Chapter 6).

The worst shortcoming of LAPACK at the *problem level* is the missing support for general sparse matrices. At present sparse matrices can be efficiently handled by LAPACK programs only when they have a band structure. The lack of support is primarily due to the inadaptability of all currently available methods for handling sparse matrices, the quality of which (numerical stability, efficiency etc.) greatly depends on the respective matrix (cf. Chapter 16). For instance, an iterative method for the solution of linear systems, when applied to a certain matrix, may converge very rapidly. When applied to another matrix however, it may converge very slowly or not at all. Consequently, an extension of LAPACK which would allow for the processing of general sparse matrices has not yet been proposed.

Three problems with the *implementation* of LAPACK appear. The first one arises from the fact that LAPACK is implemented in Fortran 77. Although Fortran 77 is still the most widely used programming language in the field of scientific (engineering) computation, there is a growing interest in using other programming languages, especially C and C++. However, for many computer systems, the integration of Fortran routines into C programs still proves to be rather difficult.

Another disadvantage is the complexity of many interfaces to LAPACK subroutines due to the restrictions of Fortran 77. This circumstance affects user friendliness. For instance, each work array used in LAPACK programs must be declared in the user program and passed to specific LAPACK subroutines because dynamic storage allocation is not a feature of Fortran 77. Furthermore, it is irritating to have a confusing variety of non-mnemonic subprogram names, because subprograms executing the *same* function on *different* data types must have different names since *generic* subprograms are not supported by Fortran 77 (in contrast to Fortran 90).

The third implementation problem emerges as a result of the design of LAPACK, primarily the extensive use of BLAS. By appropriately implementing BLAS, satisfactory performance can be achieved on workstations, vector and multiprocessor computers with shared memory, but not on computers with distributed memory. The potential for applying LAPACK to very large linear systems, which require substantial computational resources, is therefore reduced considerably.

CLAPACK

The software package CLAPACK, a version of LAPACK implemented in the programming language C, is available from NETLIB (cf. Section 7.3.7). CLAPACK can be found in NETLIB/clapack/clapack.tar. In contrast to the original Fortran 77 version, it is not possible to obtain single subprograms of CLAPACK; the complete package must always be transferred.

The use of CLAPACK does not require a Fortran 77 compiler, as it is a *complete* C-implementation of LAPACK. Consequently, purchase costs can be reduced if a system does not yet have a Fortran 77 compiler.

CLAPACK was generated automatically (for the most part) from LAPACK programs using the converting tool f2c (Feldman et al. [188]). It is not, therefore, a *genuinely* new-implementation. Specifically, no use is made of the potentially favorable features of C, such as dynamic storage allocation. Furthermore, it may well be that the code, generated by the converting tool, does not perform with optimal efficiency.

LAPACK Programs in The NAG Fortran 90 Library

The Fortran 90 software library *fl90* developed by NAG Ltd. is based on the same algorithms as the original Fortran library (written in Fortran 77). However, the use of Fortran 90 as implementation language allows a far reaching simplification of the user interface.

For instance, the number of the different subprogram names can be reduced by the use of generic subprograms. Furthermore, dynamic storage allocation makes passing of work arrays unnecessary. In addition the use of assumed-shape arrays made the passing of array dimensions obsolete, and the use of *dummy* parameters can be replaced by using optional parameters. The current release 1 of the *fl90* library does not yet contain all the algorithms found in the NAG Fortran library. Moreover, there are still several programs missing (e. g., programs for band matrices) from the LAPACK section.

LAPACK++

LAPACK++ (Dongarra, Pozo, Walker [169]) is an object-oriented, C++ extension to the Fortran LAPACK package.[7] The design goals for LAPACK++ include:

- maintaining speed and efficiency of native Fortran 77 codes while allowing the programmers to capitalize on the software engineering benefits of object-oriented programming,

---

[7] CLAPACK can be used as a basis for LAPACK++ instead of LAPACK.

- providing LAPACK with a simple interface that hides the implementation details of various matrix storage schemes and their corresponding factorization structures,

- providing a universal interface and open system design for integration into user-defined data structures and third-party matrix packages,

- replacing the static work array limitations of Fortran 77 with more flexible dynamic memory allocation schemes,

- utilizing function and operator *overloading* in C++ to simplify and reduce the number of entry points to LAPACK, and

- providing more meaningful naming conventions for variables and functions (e. g., names are no longer limited to six alpha-numerical characters, etc.).

The current version of LAPACK++ (Dongarra, Pozo, Walker [168]) offers

- C++ driver routines for the solution of sets of linear systems, linear least squares problems and eigenvalue problems,

- an object-oriented interface to the BLAS, and

- suitable classes for the different matrix types supported by LAPACK (general rectangular matrices, band matrices, etc.).

Example (Solving a Linear System Using LAPACK++)   Using LAPACK++, a given system of linear equations $Ax = b$ with a general $n \times n$ matrix A can be solved numerically as indicated in the following code sequence:

```
#include <lapack++.h>
...
LaGenMatFloat A(n,n);       /* declaration of the matrix A           */
LaVectorFloat x(n), b(n);   /* declaration of the vectors x and b    */
...
LaLinSolve(A,x,b);          /* solution of the linear system Ax = b  */
```

Firstly, A is declared as an $n \times n$ matrix using LaGenMatFloat, and x and b are declared as vectors of length n using LaVectorFloat. The linear system $Ax = b$ is then solved using LaLinSolve.

If a linear system with a tridiagonal matrix A is to be solved, the following code sequence can be used:

```
#include <lapack++.h>
...
LaTridiagMatFloat A(n,n);   /* declaration of the tridiagonal matrix A */
LaVectorFloat x(n), b(n);   /* declaration of the vectors x and b      */
...
LaLinSolve(A,x,b);          /* solution of the linear system Ax = b    */
```

The only modification is the declaration of the tridiagonal matrix A using LaTridiagMatFloat. In particular, the linear system has been solved in both cases by the *same* generic function LaLinSolve; the function has been overloaded to perform different tasks depending on the type of the input matrix A. The parameter list of LaLinSolve is restricted to the immediate problem data A, x, and b. By using LAPACK, two different routines, sgesv and sgtsv, would have been required. The names of these routines do not reveal very much information as to the function of the routine, and the parameter list contains a large number of variables not connected to the problem itself but rather to the internal storage management. LAPACK++ is obviously much more user-friendly than the original LAPACK.

Extensive tests (Dongarra, Pozo, Walker [168]) have shown that the overhead caused by the C++ interface programs is negligible.

SCALAPACK

Recently, more and more parallel and distributed multiprocessor computer systems have been used for solving linear systems, least squares problems, and eigenvalue problems with very large coefficient matrices (of order $n \geq 5000$).

While on parallel computer systems with shared memory a satisfying performance can be achieved by using a proper parallel implementation of BLAS, this is not the case with *distributed memory systems*. If an LAPACK routine is executed sequentially on a certain processor and a parallel BLAS routine is called, then the respective matrix operands must be transferred to the available processors and the computed results must be transferred back to the calling processor. These data transfers cause a communication overhead which is generally proportional to the number of processors. Hence with a growing number of processors the time an algorithm spends on communication increases dramatically—the algorithm is not *scalable*.

In order to efficiently exploit the multiprocessor hardware of distributed memory systems SCALAPACK (*Scalable* LAPACK) has been developed (Choi et al. [137]). A C++ version, namely SCALAPACK++ is also available.

The current version of SCALAPACK (Choi et al. [7], [137]) contains subroutines for LU factorization, QR factorization, and the Cholesky decomposition of dense matrices and for the solution of the corresponding systems of linear equations; in the case of LU factorization and Cholesky decomposition, parallel routines for condition number estimates, error bounds, and iterative refinement are also offered. Subroutines for the parallel reduction of matrices to Hessenberg form, tridiagonal form, and bidiagonal form (all of which are relevant for solving eigenvalue problems) are available (cf. Chapter 15).

The parallelization strategy used in SCALAPACK is based on a *static block cyclic* distribution of the given $m \times n$ matrix A onto P processors. Firstly, the number P is decomposed into two factors Q and R, and the different processors are represented by ordered pairs

$$(q, r), \qquad q = 1, 2, \ldots, Q, \quad r = 1, 2, \ldots, R.$$

Next, the matrix A is decomposed into $m_b \times n_b$ sub-matrices:

$$A = (A_{ij}), \qquad i = 1, 2, \ldots, m/m_b, \quad j = 1, 2, \ldots, n/n_b.$$

The assignment of the elements of A to the different processors is done by a block cyclic distribution determined by the parameters m_b, n_b, Q and R; the sub-matrix A_{ij} is assigned to processor (q, r) with

$$q = 1 + i \bmod Q \qquad \text{and} \qquad r = 1 + j \bmod R. \tag{13.58}$$

Example (Block Cyclic Distribution)   Fig. 13.14 graphically depicts a block cyclic distribution of a 16×16 matrix onto 4 processors with the parameters $m_b = 2$, $n_b = 4$ and $Q = R = 2$. Single matrix elements are indicated by a dotted box, sub-matrices are bounded by full lines. The assignment of the single sub-matrices to the different processors is symbolized by the shading of the boxes.

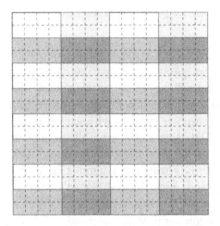

Figure 13.14: Block cyclic distribution of a 16×16 matrix onto 4 processors.

For a given data distribution, the computational work of LAPACK routines is distributed to different processors according to the *owner computes rule*. This implies that the operations on the right-hand side of the assignment have to be done by the processor to which the variable on the left hand side has been assigned. If one or more operands on the right-hand side are related to different processors, then these operands must be transferred to the processor before executing the assignment.

Example (Owner Computes Rule)   The addition operation in

    z = x + y

must be done by the processor p_z to which the variable z has been assigned. If, for example, the processor p_x is not the same as p_z, then the processor p_x must send the actual value of x to p_z in order to make the addition operation possible.

The implementation of SCALAPACK is based on the modular concept outlined in Fig. 13.15. In addition to the sequential BLAS and LAPACK programs, SCALA-PACK also uses the following modules:

BLACS (Basic Linear Algebra *Communication* Subprograms) is a message passing library designed for linear algebra. The computational model consists of a one or two dimensional grid of processes, where each process stores matrices and vectors. The BLACS include routines for sending a matrix or a submatrix from one process to another, for broadcasting sub-matrices to many

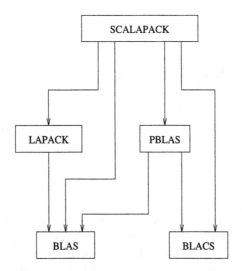

Figure 13.15: Modular structure of SCALAPACK.

processes, or for computing global reductions (sums, maxima, minima) (Dongarra et al. [12], Whaley [380]). The BLACS execute machine dependent communication routines while performing the operations, but the interface and the functionality of the BLACS itself are completely machine independent. Implementations of the BLACS are available on all relevant multiprocessor computers.

PBLAS (*Parallel* BLAS) is a package of multiprocessor implementations of BLAS routines. As the distribution of the matrix operands is somehow restricted, storage access and communication operations are optimized to an extent not possible for generally distributed matrices (Choi et al. [7], [136], [137]).

In order to simplify the design of SCALAPACK, the interfaces of the PBLAS routines have been built as similarly to the BLAS as possible. The SCALAPACK code is therefore quite similar and sometimes almost identical to the corresponding LAPACK code.

For a given problem on a certain computer system, the attainable performance of the SCALAPACK routines depends critically on the parameters m_b, n_b, Q and R of the block cyclic data distribution.

In principle, the communication overhead decreases with increasing block size $m_b n_b$. At the same time, however, the load on the different processors becomes more unbalanced, resulting in a longer overall runtime. In general there is an optimal value for $m_b = n_b$ that itself depends on the values of Q and R.

Chapter 14

Nonlinear Equations

Je weiter sich das Wissen ausbreitet,
desto mehr Probleme kommen zum Vorschein.[1]

JOHANN WOLFGANG VON GOETHE

Nearly all functional relations occurring in practice are *nonlinear*. However, many nonlinear relationships can be described *locally* (in a more or less restricted domain) by linear models, with sufficient accuracy. Still there are important phenomena such as saturation, solution branching, chaos etc. which can *only* be described by nonlinear models.

Example (Deformations of Solid Materials) Every solid object changes its shape under the influence of external forces. The relationship between the external tension (pressure), the cause, and the change of shape, the effect, can be described to a certain extent by a linear model. This model is adequate up to a certain level of tension. In solid body mechanics this linear relation is called *Hooke's law* (see, e.g., Ziegler [386]). Hooke's law is the fundamental basis for mathematically demonstrating the strength of technical constructions in mechanical engineering, civil engineering, etc.

In order to test the characteristics of materials, rod-shaped samples are exposed to an experiment. Tractive forces are exerted on the material longitudinally and are increased until the sample breaks. A plot of the measured tractive force versus the change in the length of the sample lead to the *tension dilation diagram*. The elastic and plastic features of the material as well as other features can be deduced from such a diagram.

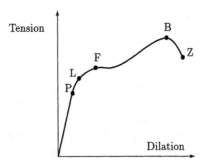

Figure 14.1: Prototype of a tension dilation diagram (for a metallic material with distinct cold-flowing properties).

In Fig. 14.1 the validity of Hooke's law is evident from 0 to the *limit of proportionality* P. At L the *limit of elasticity* (irreversible dilation 0.01 %) and at F the *limit of flowability* or *dilation* (irreversible dilation 0.2 %) is attained. For materials with distinct flowability the internal

---

[1]The more you know the more problems come to the surface.

structure changes beyond the point F. The sample then behaves like a viscous liquid. At B
the *ultimate breaking strength* is attained. Beyond this point the sample starts to constrict and
finally, at Z, the breaking point is reached.

The *theory of plasticity* develops and investigates nonlinear models of the behavior of the
material beyond the limit of elasticity (where the linear model is no longer adequate). The
theory also allows a better exploitation of the weight-bearing capacity of building-elements, for
example.

In the *theory of viscous elasticity* the temporal behavior of the dilation is investigated
and modeled. At higher temperatures metals, and polymers of a high molecular weight, tend
to increase their dilation when exposed to constant tension for a long period of time. This
process is called *creeping*. An adequate model of this phenomenon requires nonlinear functions
(materials with *nonlinear memory*).

Example (The Gas Equation) The thermodynamic state of a gas is determined by the
volume V, the pressure p, and the absolute temperature T. For an ideal gas

$$\frac{pV}{T} = R = \text{const} \tag{14.1}$$

(*ideal gas equation*). For a very high pressure the model (14.1) becomes inaccurate for real
gases. As a result corrections to (14.1) have been developed. For high pressures the equation
of Beattie and Bridgeman

$$\frac{pV}{T} = R\left[1 + \beta/V + \gamma/V^2 + \delta/V^3\right] \tag{14.2}$$

provides a more accurate description of the state of gas. The parameters

$$\begin{aligned}
\beta &= B_0 - A_0/RT - c/T^3, \\
\gamma &= -B_0 b + A_0 a/RT - cB_0/T^3, \text{ and} \\
\delta &= cB_0 b/T^3
\end{aligned}$$

of this equation depend on the temperature. The constants A_0, B_0, a, b, c occurring in these
parameter specifications depend on the gas. The *ratio of compressibility*

$$\kappa := \frac{pV}{RT} = 1 + \beta/V + \gamma/V^2 + \delta/V^3$$

signifies the deviation of the actual state of a gas from the ideal gas model ($\kappa = 1$ corresponds
to an ideal gas). In investigating the validity of equation (14.2) the following procedure is possible: Experimentally determined values of the compressibility are compared with numerically
calculated values of κ. To enable such a comparison the volume V is determined using (14.2)
for given values of p and T. Then $\kappa = pV/RT$ is determined. To do so nonlinear equations of
the form

$$V = \frac{RT}{p}\left[1 + \beta/V + \gamma/V^2 + \delta/V^3\right]$$

must be solved numerically (cf. Fig. 14.2).

Nonlinear models generally require the solution of nonlinear equations. If the
mathematical model comprises differential equations it has to be discretized first.
Besides particular simple cases[2], solutions of nonlinear equations can neither be
found in closed form nor determined after a limited number of algorithmic steps.
To sum up, it can be said that there are no methods for solving nonlinear problems
which correspond to the direct methods for solving systems of linear equations.
The solution of nonlinear equations can only be calculated *numerically*, using
iterative methods.

---

[2]Nonlinear *algebraic equations* $a_0 + a_1 x + \cdots + a_d x^d = 0$ with degree $d \leq 4$ can be solved
by formulas (see Section 14.2.10).

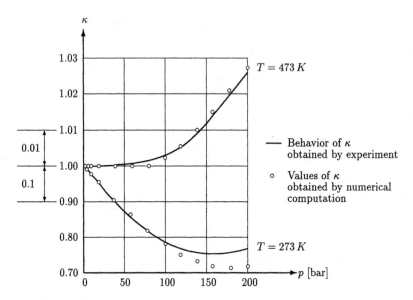

Figure 14.2: Ratio of compressibility $\kappa = pV/RT$ of methane.

Terminology (Algebraic Equations)   To distinguish them from integral equations, differential equations, and other types of equations, systems of transcendental equations discussed in this chapter are often called *algebraic* equations. This designation is *not* correct in the strict mathematical sense, since the term algebraic equations is only true in connection with polynomials and the root-finding of polynomials (see Section 14.2.10).

If nonlinear models deviate from linear models (e. g., from a regular system of linear equations) only by a small margin, then many important characteristics (e. g., the unique solvability) of the linear model apply to the nonlinear model as well. Even the iterative methods for solving systems of linear equations can be applied to systems of *weakly* nonlinear equations (see Section 14.3.1) after they have been suitable generalized. *Strongly* nonlinear equations have properties which are not preserved in the course of linearization. Nonlinear equations for instance, may not have any real solution at all (e. g., $e^{-x} - \sin x + 1 = 0$), or may have infinitely many isolated solutions (e. g., $e^{-x} - \sin x = 0$).

Definition 14.0.1 (Isolated Solution) x^* *is called an isolated solution of a nonlinear equation, if* x^* *is a solution of the equation and there are no other solutions in a neighborhood* $U(x^*)$ *of* x^*.

The occurrence of several isolated solutions often corresponds to various states of the investigated process. In such cases it must be checked as to whether or not a numerically determined solution actually is an approximation of the relevant solution of the problem under consideration.

In general it is not possible to have an overall view of the global behavior of a system of strongly nonlinear equations. Thus, to determine a relevant approximate solution, it is necessary to begin from an initial point which is located sufficiently close to the desired solution. Determining where to begin may require more effort and consideration (depending on the actual problem) than the iterative improvement of the initial approximation itself. This can often be carried out using standard software (e. g., subroutines from the IMSL or the NAG library).

The exact analysis of the particular situation is generally inevitable in order to select an effective method for solving nonlinear problems. The careless use of *black box* software rarely leads to the desired solution, particularly for systems of strongly nonlinear equations.

Example (Magnetic Disk Storage)   The read/write head of a disk drive does not touch the magnetic surface of the disk but is separated from it by a gap of approximately $1\mu m$ (see Fig. 14.3).

Since the disk rotates at high speed, air is dragged into the gap, which leads to an increase of the air pressure and, hence, guarantees the separation of the read/write head from the disk. In this way, direct contact and damage to the magnetic film can be avoided.

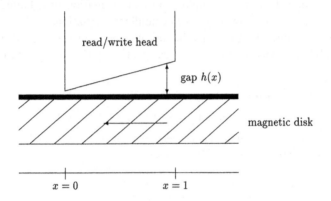

Figure 14.3: Schematic representation of the read/write head of a magnetic disk drive.

The air pressure $p(x)$ in the gap can be modeled using a differential equation:

$$p'' = \left[\frac{(p')^2}{p} + \frac{3h'h^2p'}{p} + \frac{k(p'h + ph')}{ph^3}\right] =: g(x, p, p'), \tag{14.3}$$

where the pressure p_0 of the ambient air is prevalent at the two edges $x = 0$ and $x = 1$:

$$p(0) = p(1) = p_0. \tag{14.4}$$

The constant k depends primarily on the speed of the disk and on characteristics of the ambient air (e.g. the air temperature).

For initial value problems of the form

$$y' = f(x, y), \quad y(0) = y_0 \quad \text{with} \quad y : \mathbb{R} \to \mathbb{R}^n, \quad f : \mathbb{R}^{n+1} \to \mathbb{R}^n$$

there is more standard software available (e. g., in ODEPACK) than for second order boundary value problems such as (14.3), (14.4). However, the second order differential equation (14.3) can be transformed into a *system* of *two* first order differential equations

$$\begin{pmatrix} y_1' \\ y_2' \end{pmatrix} = \begin{pmatrix} y_2 \\ g(x, y_1, y_2) \end{pmatrix}$$

which, given the initial values

$$y(0) = \begin{pmatrix} y_1(0) \\ y_2(0) \end{pmatrix} = \begin{pmatrix} y_{10} \\ y_{20} \end{pmatrix} = y_0,$$

can be solved numerically using computer programs for the solution of first order initial value problems. Due to the condition $p(0) = p_0$, the only choice of y_{10} is $y_{10} = p_0$. Suppose ODESOLVER(x, y_{20}) denotes the *numerically* obtained approximate value $y_1(x) = p(x)$ with respect to the initial value $y_2(0) = y_{20} = p'(0)$. The problem is to find the solution of the nonlinear equation

$$\text{ODESOLVER}\,(1, y_{20}) = p_0, \tag{14.5}$$

and—more or less as a by-product—to determine a numerical approximation of the function p. Equation (14.5) is a nonlinear equation which is defined in terms of an ordinary differential equation initial value problem. In this case then, the nonlinear equation *cannot* be described by a formula; it can only be evaluated numerically for a given value of y_{20}. This is sufficient for the use of methods and software designed to solve nonlinear equations.

The method applied in this example—the transformation of a boundary value problem into the solution of a system of nonlinear equations which in turn defines the initial values of an initial value problem—is called the *shooting method*. The applicability of this method depends greatly on the underlying differential equation.

Example (Eigenvalue Problems)   The algebraic eigenvalue problem

$$Ax = \lambda x \qquad \text{with} \qquad A \in \mathbb{C}^{n \times n},\ x \in \mathbb{C}^n,\ \lambda \in \mathbb{C}$$

can be interpreted—using a suitable normalizing condition for x—as a *nonlinear* problem for determining eigenvalues λ and eigenvectors x:

$$F(x) := \begin{pmatrix} x^\mathsf{T} x - 1 \\ (A - \lambda I)x \end{pmatrix} = 0.$$

Nevertheless it would be extremely inefficient to solve eigenvalue problems numerically using software designed for solving systems of nonlinear equations. There are many special algorithms and efficient software packages for solving eigenvalue problems, which are discussed in Chapter 15. The most comprehensive collection of software for algebraic eigenvalue problems can be found in LAPACK [3].

Software (Interactive Program Systems)   The interactive program system TK SOLVER was developed especially for the solution of both linear and nonlinear systems of algebraic equations (and also of differential equations). It allows the problem to be specified in terms of a combination of declarative and procedural language elements.

The Standard Form F(x) = 0

For theoretical purposes (e. g., for analyzing the convergence of solution algorithms), and as a preparatory step preceding the use of numerical software, systems of nonlinear equations or single nonlinear equations are often transformed into the *canonical form* (*zero form*)

$$\begin{pmatrix} f_1(x_1, x_2 \ldots, x_n) \\ f_2(x_1, x_2 \ldots, x_n) \\ \vdots \\ f_m(x_1, x_2 \ldots, x_n) \end{pmatrix} = \begin{pmatrix} 0 \\ 0 \\ \vdots \\ 0 \end{pmatrix} \qquad \text{or} \qquad F(x) = 0. \tag{14.6}$$

$x = (x_1, \ldots, x_n)^\top \in \mathbb{R}^n$ denotes the vector of unknowns and

$$F : B \subseteq \mathbb{R}^n \to \mathbb{R}^m$$

the function defining the system of nonlinear equations, which is composed of the *component functions*

$$f_i : B \subseteq \mathbb{R}^n \to \mathbb{R}, \quad i = 1, 2, \ldots, m.$$

The mathematical problem can now be specified as follows: A vector $x^* \in \mathbb{R}^n$ is to be determined such that *all* component equations

$$f_1(x^*) = 0, \; f_2(x^*) = 0, \ldots, f_m(x^*) = 0$$

are solved simultaneously. Hence an n-tuple $(x_1^*, \ldots, x_n^*)$ is sought that is a zero of *all* component functions $f_1, \ldots, f_m$.

It is conventional to consider this particular problem whenever a system of nonlinear equations is given. Other problem types may be useful as well, such as the determination of n-tuples that satisfy at least *one* of the component equations.

If the n-tuples are viewed as points in $\mathbb{R}^n$, then the set of points (given that it is non-empty) which satisfy *one* of the equations represents a multi-dimensional surface. In this geometrical interpretation, the determination of a solution of a system of m equations is equivalent to determining the intersection of m such surfaces (whereas the other—equally reasonable—type of problem refers to the determination of their union).

The relationship between the number m of equations and the number n of unknowns leads to the distinction of the following cases:

m = n

If $m = n$ then, in most practical situations, the system of nonlinear equations (14.6) has one or more *isolated* solutions x^*. These solutions are also known as the *zeros* of the function F. A specific case is $m = n = 1$, i.e., a *scalar* equation with one *scalar* unknown. Section 14.2 is devoted entirely to this particular case.

m < n

If there are fewer equations than unknowns, then (14.6) is said to be *underdetermined*. Such systems generally have an $(n-m)$-dimensional family of solutions. Underdetermined systems arise when, for example, an additional parameter $\lambda \in \mathbb{R}$ is introduced into a system with $m = n$, i.e., for a *parameter-dependent system*

$$F(x, \lambda) = 0 \tag{14.7}$$

with $F : \mathbb{R}^{n+1} \to \mathbb{R}^n$. As a rule, the *set* of solutions of the system of equations (14.7) is a one-parameter family of solutions in $\mathbb{R}^{n+1}$.

m > n

If there are more equations than unknowns, then (14.6) is said to be *overdetermined* and is generally inconsistent, i.e., there are no solutions. In this case the transition to a nonlinear data fitting (approximation) problem is appropriate:

$$\min \{\|F(x)\| \ : \ x \in B \subseteq \mathbb{R}^n\}.$$

For this purpose the most commonly used method is the minimization of the Euclidean vector norm or its squared form:

$$\min \left\{ \|F(x)\|_2^2 = F^{\mathsf{T}}(x)F(x) = \sum_{i=1}^{m}\big(f_i(x_1,\ldots,x_n)\big)^2 \ : \ x \in B \subseteq \mathbb{R}^n \right\}. \quad (14.8)$$

Accordingly, in this case the "solution" of (14.6) is determined using the *least squares method* (see Section 14.4).

Moreover, the minimization of (14.8) may also be appropriate, for methodical reasons, for systems (14.6) with $m = n$ (see Section 14.4.1).

The Numerical Problem F(x) = 0

As already mentioned, the mathematical problem $F(x) = 0$ *cannot* be solved exactly using a finite number of algorithmic steps. However, every practical method should lead to (at least) one approximate value of the solution(s) within a limited number of operations. In order to achieve this goal, a *numerical problem* (mathematical problem + tolerance) must first be defined. This can be done using two different criteria:

The Error Criterion: The absolute or relative deviation of the approximate solution $\tilde{x}$ from the exact solution x^* must be smaller than a given tolerance:

$$\begin{aligned} \|\tilde{x} - x^*\| &\leq \ \tau_{\mathrm{abs}}, \\ \|\tilde{x} - x^*\| &\leq \ \tau_{\mathrm{rel}}\|x^*\|. \end{aligned} \quad (14.9)$$

The Residual Criterion: The residual $F(\tilde{x})$ of the approximate solution $\tilde{x}$ must stay below a given tolerance:

$$\|F(\tilde{x})\| \leq \tau_f.$$

These two criteria are not independent of one another; but for each individual application, it must be decided which kind of tolerance should be chosen, which computationally determined quantity should be substituted in (14.9) for the unknown solution x^*, and how the actual values τ_{abs}, τ_{rel}, or τ_f should be specified.

Inequality (14.9) guarantees the correctness of $\lfloor -\log_{10}(\tau_{\mathrm{rel}})\rfloor$ decimal places of those components of the vector $\tilde{x}$ which are of the same magnitude as $\|x^*\|$. Criterion (14.9) does not, however, provide information about the accuracy of those components of $\tilde{x}$ whose absolute values are much smaller than $\|x^*\|$. If guaranteed accuracy is required for *all* components of the solution, then a *componentwise* criterion of accuracy has to be specified or the system (14.6) has to be scaled properly prior to its numerical solution.

The Influence of Floating-Point Arithmetic

In order to solve a system of nonlinear equations, the function $F : \mathbb{R}^n \to \mathbb{R}^n$ must be implemented in the form of a computer (sub) program. It follows that zeros of the implementation $\tilde{F} : \mathbb{F}^n \to \mathbb{F}^n$ are determined, rather than those of the mathematically defined function F. The term "zero" loses its common meaning even in the scalar case ($n = 1$), because the equation

$$\tilde{f}(x) = 0 \quad \text{with} \quad \tilde{f} : \mathbb{F} \to \mathbb{F}$$

may either have *several* solutions (zeros) or *no* solution at all in the neighborhood of an isolated zero of the original function f.

Example (No Zeros) The function

$$f(x) = 3x^2 + \frac{1}{\pi^4} \ln(\pi - x)^2 + 1 \tag{14.10}$$

has a pole at $x = \pi$; $f(x) \to -\infty$ as $x \to \pi$. From $f(3.14) \approx 30.44$ and $f(3.15) \approx 30.67$ it follows that f has two zeros in the interval $[3.14, 3.15]$. But these zeros *cannot* be determined numerically, since

$$f(x) > 0$$

holds for all $x \in \mathbb{F}(2, 53, -1021, 1024, true)$ for which (14.10) can be evaluated. The two zeros of f,

$$x_1^* \approx \pi - 10^{-647} \quad \text{and} \quad x_2^* \approx \pi + 10^{-647},$$

are located too close to π, and $f(x) \leq 0$ is valid only in an interval located between two machine numbers (see Fig. 14.4). The implementation $\tilde{f} : \mathbb{F} \to \mathbb{F}$ thus has *no* zeros at all.

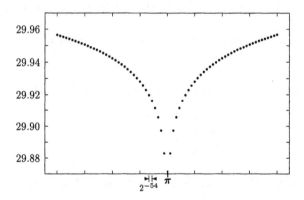

Figure 14.4: The function $f(x) = 3x^2 + \pi^{-4} \ln(\pi - x)^2 + 1$ has *no* zeros when numerically calculated in $\mathbb{F}(2, 53, -1021, 1024, true)$ (analytically, however, there are *two*).

Example (Large Number of Zeros) The polynomial

$$\begin{aligned}
P_7(x) &= x^7 - 7x^6 + 21x^5 - 35x^4 + 35x^3 - 21x^2 + 7x - 1 = \\
&= ((((((x - 7) + 21)x - 35)x + 35)x - 21)x + 7)x - 1 = (x - 1)^7
\end{aligned}$$

has its only zero at $x^* = 1$. The implementation $\tilde{P}_7 : \mathbb{F} \to \mathbb{F}$ based on the Horner scheme has thousands of zeros in the neighborhood of $x = 1$ (evaluated in $\mathbb{F}(2, 24, -125, 128, true)$), i.e., in a single precision IEC/IEEE-arithmetic); for $x > 1$ there are even thousands of points with a *negative* function value due to cancellation (see Fig. 4.10), although $P_7(x) > 0$ in this region.

14.1 Iterative Methods

In order to solve a system of nonlinear equations with an iterative algorithm, the transformation of $F(x) = 0$ into an equivalent *fixed-point form*

$$\begin{pmatrix} x_1 \\ \vdots \\ x_n \end{pmatrix} = \begin{pmatrix} t_1(x_1,\ldots,x_n) \\ \vdots \\ t_n(x_1,\ldots,x_n) \end{pmatrix}, \tag{14.11}$$

i.e., $x = T(x)$ with a vector function $T(x) = \left(t_1(x),\ldots,t_n(x)\right)^\top \in \mathbb{R}^n$, is often useful.

Definition 14.1.1 (Fixed-Point) $x^* = (x_1^*,\ldots,x_n^*)^\top$ *is called a fixed-point of equation (14.11) if $x^* = T(x^*)$.*

Definition 14.1.2 (Equivalence of Systems of Equations)   *The two systems (14.6) and (14.11) are said to be equivalent if and only if every zero x^* of (14.6) is also a fixed-point of (14.11) and vice versa.*

Occasionally systems of nonlinear equations are already given in fixed-point form. In most cases, however, the equations are given in the form $F(x) = 0$ and an appropriate fixed-point form (14.11) has to be found (see Section 14.3.1). As the following examples demonstrate, this fixed-point form is far from unique.

Example (Quadratic Equation)   The nonlinear algebraic equation

$$P_2(x) = 1.25 - x - 0.25x^2 = 0$$

can be transformed into an equivalent fixed-point form $x = t(x)$ in the following ways:

$$\begin{aligned}
x &= 1.25 - 0.25x^2 \quad \text{or} & (14.12)\\
x &= \sqrt{|5 - 4x|} \quad \text{or} &\\
x &= -4 + 5/x. & (14.13)
\end{aligned}$$

Example (Logarithmic Equation)   The nonlinear transcendental equation

$$f(x) = x + \ln x = 0$$

can be transformed into an equivalent fixed-point problem in the following ways:

$$\begin{aligned}
x &= -\ln x \quad \text{or} & (14.14)\\
x &= e^{-x} \quad \text{or} &\\
x &= \frac{\alpha x + e^{-x}}{\alpha + 1}, \quad \alpha \in \mathbb{R}\setminus\{-1\}. & (14.15)
\end{aligned}$$

Note   Most examples in this section and in other sections as well, deal with *one* nonlinear function in *one* variable $f : \mathbb{R} \to \mathbb{R}$. This way the examples are clearer and the reader is able to reproduce the results easily (even on pocket calculators).

14.1.1 Fixed-Point Iteration

The simplest iteration for determining a fixed-point is obtained by inserting the previous approximation in $T(\cdot)$:

$$x^{(k+1)} := T(x^{(k)}), \quad k = 0, 1, 2, \ldots . \tag{14.16}$$

Whether or not this method leads to the determination of a fixed-point depends not only on the function T, but also on the selection of the initial value $x^{(0)} \in \mathbb{R}^n$ (see also Section 14.1.4).

Definition 14.1.3 (Iterative Process)   *The fixed-point iteration (14.16) defines an iterative process if there is a nonempty set $D^* \subset \mathbb{R}^n$ such that for all $x^{(0)} \in D^*$ the infinite sequence $\{x^{(k)}\}$ can be determined, i. e., if all points $x^{(1)}$, $x^{(2)}$, $x^{(3)}, \ldots$ are located in the domain of T.*

Example (Quadratic Equation)   For $x^{(0)} = 1.25$ the iteration

$$x^{(k+1)} := -4 + 5/x^{(k)},$$

which corresponds to the fixed-point problem (14.13), terminates (provided the calculation is carried out without rounding error) at $x^{(1)} = 0$. $x^{(2)}$ cannot be determined and, as a result $x^{(3)}, x^{(4)}, \ldots$ cannot be determined either.

Example (Logarithmic Equation)   The iteration

$$x^{(k+1)} := -\ln x^{(k)},$$

which corresponds to (14.14), can actually only be started for $x^{(0)} > 0$. The calculation of $x^{(2)}$ is only possible for $x^{(0)} \in (0,1)$; $x^{(3)}$ can only be obtained for $x^{(0)} \in (1/e, 1) \approx (0.3679, 1)$; $x^{(4)}$ only for $x^{(0)} \in (1/e, e^{-1/e}) \approx (0.3679, 0.6922)$ and so on. Since $D^* = \{x^*\}$, the desired infinite sequence $\{x^{(k)}\}$ is actually only defined if the initial point $x^{(0)}$ is equal to the fixed-point $x^* \approx 0.567$.

Definition 14.1.4 (The Convergence of an Iterative Method)   *In the context of an iterative process (14.16) with $x^{(0)} \in D^*$, convergence (of the sequence $x^{(0)}, x^{(1)}, x^{(2)}, \ldots$) to a solution x^* means that*

$$\lim_{k \to \infty} x^{(k)} = x^*.$$

The convergence of vectors $x^{(k)} \in \mathbb{R}^n$ is defined in terms of the norm convergence

$$\lim_{k \to \infty} \|x^{(k)} - x^*\| = 0$$

or the equivalent component-wise convergence:

$$\lim_{k \to \infty} x_i^{(k)} = x_i^*, \quad i = 1, 2, \ldots, n.$$

Each iterative process yields (according to Definition 14.1.3) an *infinite* sequence $\{x^{(k)}\}$ for any initial value $x^{(0)} \in D^*$. But the iteration (14.16) is only useful for solving the system of nonlinear equations $F(x) = 0$ if a sufficiently large set $E \subset D^* \subset \mathbb{R}^n$ exists such that the sequence $x^{(0)}, x^{(1)}, x^{(2)}, \ldots$ *converges* to a solution x^* for all initial values $x^{(0)} \in E$.

Definition 14.1.5 (Region of Attraction, Local and Global Convergence)
*The set $E \subseteq D \subseteq \mathbb{R}^n$ is known to be the region of attraction of the solution x^*
if*

$$\lim_{k \to \infty} x^{(k)} = x^* \qquad \text{for all} \quad x^{(0)} \in E.$$

*Moreover, if $E = D$, i. e., if the entire domain of T is the region of attraction
of an iteration, then the convergence is referred to as global convergence. If, on
the other hand, the region of attraction only contains a neighborhood $U(x^*)$ of the
solution, then the convergence is referred to as local convergence.*

For a nonlinear equation with several isolated solutions, it may of course happen
that not all solutions have a (non-trivial) region of attraction, i. e., only certain
solutions can be approximately determined by fixed-point iteration.

Example (Quadratic Equation)   The polynomial $P_2(x) = 1.25 - x - 0.25x^2$ has zeros at
$x_1^* = 1$ and $x_2^* = -5$. With fixed-point iteration based on (14.12),

$$x^{(k+1)} := T(x^{(k)}) = 1.25 - 0.25(x^{(k)})^2, \tag{14.17}$$

the sequence $\{x^{(k)}\}$ is convergent for all $x^{(0)} \in [-5, 5]$. The region of attraction of x_1^* is the
entire open interval $(-5, 5)$, whereas x_2^* can only be reached from the trivial initial values
$x^{(0)} = 5$ and $x^{(0)} = -5$, (which require the knowledge of x_2^*). For starting points with $|x^{(0)}| > 5$
the sequence defined by (14.17) *diverges.*

No general statement can be made about the size of the region of attraction. For
many iterative methods, faster convergence in a neighborhood of the solution can
be obtained at the expense of a smaller region of attraction.

Definition 14.1.6 (Stationary Iterative Method)   *An iterative process*

$$x^{(k+1)} = T(x^{(k)})$$

*is called stationary if the iteration function T is independent of the actual iteration
step. With non-stationary methods the iteration rule can change from step to
step (depending on the course of the iteration):*

$$x^{(k+1)} = T_k(x^{(k)}).$$

The polyalgorithms in Section 14.2.9 and the *damped Newton method* (14.61) are
examples of non-stationary iterative methods.

Definition 14.1.7 (One-Step Method, Multi-Step Method)   *In*

$$x^{(k+1)} = T(x^{(k)})$$

the next computed value depends only on one *previous value. This method is
therefore referred to as a one-step method. If, on the other hand,*

$$x^{(k+1)} = T(x^{(k)}, x^{(k-1)}, \ldots, x^{(k-s)})$$

then such an iterative scheme is said to be an $(s+1)$-step method.

The secant method (cf. Section 14.2.5) is an example of a two-step method.
Müller's method (cf. Section 14.2.6) is a three-step method.

14.1.2 Convergence of Iterative Methods

The main question concerning iterative methods is that of the conditions under which they converge. Convergence criteria can be derived from the contraction theorem, which ensures the existence of a fixed-point if T is a *contraction mapping*.

Definition 14.1.8 (Contraction Mapping)   *A function $T : D \subset \mathbb{R}^n \to \mathbb{R}^n$ is said to a contraction on a set $D_0 \subset D$ if it satisfies a Lipschitz condition*

$$\|T(x) - T(y)\| \leq L\|x - y\| \qquad \text{for all} \quad x, y \in D_0 \tag{14.18}$$

with a Lipschitz constant (contraction constant) $L < 1$.

When inserting $x = x^{(k)}$ and $y = x^*$ in (14.18), the inequality

$$\|x^{(k+1)} - x^*\| \leq L\|x^{(k)} - x^*\| < \|x^{(k)} - x^*\|$$

is obtained because of $T(x^{(k)}) = x^{(k+1)}$ and $T(x^*) = x^*$. Hence, the iteration $x^{(k+1)} = T(x^{(k)})$ reduces the distance of the iterates $x^{(k)}, x^{(k+1)}, \ldots$ from x^* at each step, provided T is a contraction.

Note   A *linear* mapping $A : \mathbb{R}^n \to \mathbb{R}^n$ is a contraction on $D_0 = \mathbb{R}^n$ if and only if $\|A\| < 1$. The norm dependency of contractivity is apparent.

Theorem 14.1.1 (The Contraction Theorem)   *If $T : D \subset \mathbb{R}^n \to \mathbb{R}^n$ is a contraction on the closed set $D_0 \subset D$ and if*

$$TD_0 \subset D_0 \tag{14.19}$$

then it follows that:

1. *T has exactly one fixed-point x^* in D_0;*

2. *$x^{(k+1)} = T(x^{(k)})$ converges to x^* for every initial value $x^{(0)} \in D_0$.*

Proof: Due to (14.19), the sequence

$$x^{(0)}, x^{(1)}, x^{(2)}, \ldots \qquad \text{with} \qquad x^{(k+1)} = T(x^{(k)})$$

is defined for any initial value $x^{(0)} \in D_0$, and all points $x^{(k)}$ are located in D_0. Since T is assumed to be a contraction mapping, the following estimate holds:

$$\begin{aligned} \|x^{(k+1)} - x^{(k)}\| &= \|T(x^{(k)}) - T(x^{(k-1)})\| \\ &\leq L\|x^{(k)} - x^{(k-1)}\| \leq \cdots \\ &\leq L^k\|x^{(1)} - x^{(0)}\|. \end{aligned}$$

It follows that

$$
\begin{aligned}
\|x^{(k+m)} - x^{(k)}\| &\leq \|x^{(k+m)} - x^{(k+m-1)}\| + \|x^{(k+m-1)} - x^{(k+m-2)}\| + \cdots \\
&\quad \cdots + \|x^{(k+1)} - x^{(k)}\| \\
&\leq (L^{m-1} + L^{m-2} + \cdots + 1)\|x^{(k+1)} - x^{(k)}\| \\
&\leq \frac{1 - L^m}{1 - L} L^k \|x^{(1)} - x^{(0)}\| \\
&\leq \frac{L^k}{1 - L} \|x^{(1)} - x^{(0)}\|.
\end{aligned}
\tag{14.20}
$$

So for each $\varepsilon > 0$ there exists a $K(\varepsilon) \in \mathbb{N}$ such that

$$
\|x^{(k+m)} - x^{(k)}\| < \varepsilon \qquad \text{for all} \quad k > K(\varepsilon), \quad m \in \mathbb{N}.
$$

The sequence $\{x^{(k)}\}$ is therefore a Cauchy sequence with a limit $\bar{x} \in D_0$ (the set D_0 is closed, and $\mathbb{R}^n$ is complete). For this limit

$$
\|\bar{x} - T(\bar{x})\| \leq \|\bar{x} - x^{(k)}\| + \|x^{(k)} - T(\bar{x})\| \leq \|\bar{x} - x^{(k)}\| + L\|x^{(k-1)} - \bar{x}\|,
$$

and because $\|x^{(k)} - \bar{x}\| \to 0$ it follows that

$$
\|\bar{x} - T(\bar{x})\| = 0,
$$

i. e., $\bar{x}$ is a fixed-point. Moreover, $\bar{x}$ is the *only* fixed-point, since the assumption of the existence of a second fixed-point $\bar{\bar{x}}$ leads to

$$
\|\bar{x} - \bar{\bar{x}}\| = \|T(\bar{x}) - T(\bar{\bar{x}})\| \leq L\|\bar{x} - \bar{\bar{x}}\|,
$$

which can only hold for $\bar{x} = \bar{\bar{x}}$.   □

If the assumptions of this theorem hold with $D_0 = D = \mathbb{R}^n$, then T is a contraction on the whole of $\mathbb{R}^n$. In this case the contraction theorem is a *global* statement which guarantees the convergence of the sequence $\{x^{(k)}\}$ to the unique fixed-point x^* of T for any initial value $x^{(0)} \in \mathbb{R}^n$. If the mapping T is a contraction only on a proper subset $D_0 \neq \mathbb{R}^n$, then the contraction theorem guarantees only *local* convergence.

Note (The Importance of the Contraction Theorem) The contraction theorem is one of the most important theorems in applied mathematics. It remains valid for any complete, normed linear space. For nonlinear inverse problems (e. g., partial differential equations) it is one of the most important tools for proving the existence and uniqueness of solutions as well as for the derivation of iterative methods for the approximate determination of solutions.

Convergence and Floating-Point Operations

On a computer, the values occurring in an iterative method can only be (vectors of) floating-point numbers $\mathbb{F}$ (see Section 4.4.3). Since these numbers—unlike the real numbers $\mathbb{R}$—do not constitute a continuum, convergence in the sense of classical analysis is *not* possible. Instead, the following alternatives are encountered: The sequence $\{\tilde{x}^{(k)}\}$ of numerical approximations becomes either

stationary, i.e., for a certain $K \in \mathbb{N}$

$$\tilde{x}^{(K)} = \tilde{x}^{(K+1)} = \tilde{x}^{(K+2)} = \cdots, \qquad \text{or} \qquad (14.21)$$

periodic with period I

$$\tilde{x}^{(K+i)} = \tilde{x}^{(K+i+jI)}, \qquad i = 0, 1, \ldots, I-1, \quad j = 1, 2, 3, \ldots \qquad (14.22)$$

or, in particular, with period 2 (*alternating*)

$$\tilde{x}^{(K)} = \tilde{x}^{(K+2)} = \cdots \quad \text{and} \quad \tilde{x}^{(K+1)} = \tilde{x}^{(K+3)} = \cdots.$$

This behavior can be explained as follows: For every iteration that is convergent in the domain of real numbers, the *contraction effect*

$$\|x^{(k+1)} - x^*\| < \|x^{(k)} - x^*\| \qquad (14.23)$$

occurs close to the limit x^*. If T is evaluated in a floating-point arithmetic, rounding errors are inevitable, and only perturbed values $\tilde{x}^{(k+1)}$ are obtained instead of $x^{(k+1)} = T(x^{(k)})$. The magnitude of the rounding error is bounded by an $\varepsilon \in \mathbb{R}$:

$$\|\tilde{x}^{(k+1)} - x^{(k+1)}\| = \|\tilde{x}^{(k+1)} - T(x^{(k)})\| \leq \varepsilon.$$

In the worst case the (vector) rounding error may even divert an iterate from its *goal* x^*. However, as long as the contraction effect (14.23) is stronger than that of rounding errors, the sequence $\{\tilde{x}^{(k)}\} \subset \mathbb{F}^n$ continues to approach x^*. If, on the other hand, $\tilde{x}^{(k)}$ is so close to x^* that the contraction effect is countered by the scatter effect of rounding errors—then the sequence of iterates no longer converges to x^*. With a Lipschitz constant L and $\delta = \varepsilon/(1 - L)$ the following cases can be distinguished:

1. As long as $\|\tilde{x}^{(k)} - x^*\| > \delta$, the contraction effect dominates the effect of rounding errors and

$$\|\tilde{x}^{(k+1)} - x^*\| < \|\tilde{x}^{(k)} - x^*\|, \qquad (14.24)$$

   i.e., the sequence of floating-point iterates approaches the solution x^* of the nonlinear equations (14.11) monotonically (with respect to the norm $\| \ \|$).

2. If, on the other hand, $\|\tilde{x}^{(k)} - x^*\| \leq \delta$, then it can only be concluded that

$$\|\tilde{x}^{(k+1)} - x^*\| \leq \delta,$$

   i.e., the sequence of floating-point iterates remains within a δ-neighborhood of the solution x^*.

As a result of (14.24) and the finite distance between the floating-point numbers, the relationship $\|\tilde{x}^{(k)} - x^*\| \leq \delta$ must eventually hold for sufficiently large $k \in \mathbb{N}$. Beyond this index k the sequence of iterates stays within the neighborhood $\{x : \|x - x^*\| \leq \delta\}$ of x^*, which contains only a *finite* number of floating-point numbers (vectors). After a finite number of iterations a particular iterate $\tilde{x}^{(i)}$ occurs a second time. But from this moment on every stationary iteration becomes *periodic*, since the process is deterministic.

These considerations show that convergent iterative methods can be executed in a floating-point arithmetic without suffering great damage. The stationarity of the sequence $\{\tilde{x}^{(k)}\}$ (in the sense of (14.21) or (14.22)) is the discrete counterpart to convergence in analysis.

For the sake of efficiency, in most practical situations the iteration is not continued until it becomes stationary. Consequently, rounding errors are not dealt with in this chapter.

Error Estimates

As $m \to \infty$, the estimate (14.20) leads to the *a priori error estimate*

$$\|x^* - x^{(k)}\| \leq \frac{L^k}{1-L}\|x^{(1)} - x^{(0)}\|. \tag{14.25}$$

If a Lipschitz constant L of T is known, then it is possible to determine a bound on the maximum number of steps $k_{\max}$ necessary to meet the accuracy requirement

$$\|x^* - x^{(k)}\| \leq \varepsilon_{\mathrm{abs}}$$

after the first iteration:

$$k_{\max} = \left\lceil \frac{\log\left(\varepsilon_{\mathrm{abs}}(1-L)\,/\,\|x^{(1)} - x^{(0)}\|\right)}{\log L} \right\rceil. \tag{14.26}$$

Interpreting $x^{(k-1)}$ as a new starting point $x^{(0)}$, inequality (14.25) enables the *a posteriori error estimate*

$$\|x^* - x^{(k)}\| \leq \frac{L}{1-L}\|x^{(k)} - x^{(k-1)}\|, \tag{14.27}$$

which makes it possible to estimate the absolute error after k iterations.

Lipschitz Constants

An essential prerequisite for the applicability of the contraction theorem and the error estimates (14.25) and (14.27) is the knowledge of a good (as small as possible) Lipschitz constant L for the function T. Sometimes optimal Lipschitz constants can be obtained using computer algebra systems.

If T is continuously differentiable on a bounded and convex set D_0, then T satisfies a Lipschitz condition with Lipschitz constant $L = \|T'\|_p$ on that set. T' denotes the *functional matrix (Jacobian matrix)* of T:

$$
T'(x) := \begin{pmatrix}
\dfrac{\partial t_1(x)}{\partial x_1} & \dfrac{\partial t_1(x)}{\partial x_2} & \cdots & \dfrac{\partial t_1(x)}{\partial x_n} \\[2ex]
\dfrac{\partial t_2(x)}{\partial x_1} & \dfrac{\partial t_2(x)}{\partial x_2} & \cdots & \dfrac{\partial t_2(x)}{\partial x_n} \\[1ex]
\vdots & \vdots & & \vdots \\[1ex]
\dfrac{\partial t_n(x)}{\partial x_1} & \dfrac{\partial t_n(x)}{\partial x_2} & \cdots & \dfrac{\partial t_n(x)}{\partial x_n}
\end{pmatrix}
$$

and

$$
\|T'\|_p = \max\{\|T'(x)\|_p : x \in D_0\}.
$$

Example (Logarithmic Equation) There are various ways to transform the logarithmic equation $x + \ln x = 0$ into an equivalent fixed-point problem $x = t(x)$. The resulting fixed-point problems can be better or worse suited for the iteration method.

A neighborhood D_0 of the fixed-point x^* on which t is a contraction exists only if $|t'(x^*)| < 1$.

Variant 1: For $x^{(k+1)} = -\ln x^{(k)}$ the following estimate holds:

$$
|t'(x^*)| = \frac{1}{x^*} \approx 2.
$$

It follows that the contraction theorem cannot be applied to show that the iteration converges to x^* in this case.

Variant 2: For $x^{(k+1)} = e^{-x^{(k)}}$ the following estimate holds:

$$
|t'(x^*)| = e^{-x^*} \approx 0.6,
$$

i.e., for an appropriately chosen $x^{(0)}$ the fixed-point iteration converges to x^*. In the interval $D_0 = [0.4, 0.6]$,

$$
L = \max\{|t'(x)| : x \in D_0\} = \max\{e^{-x} : x \in [0.4, 0.6]\} = e^{-0.4} \approx 0.67.
$$

Using (14.26),

$$
k_{\max} = \left\lceil \frac{\log(\varepsilon_{\text{abs}}\, 0.33/0.2)}{\log 0.67} \right\rceil = \left\lceil \frac{\log(1.65\, \varepsilon_{\text{abs}})}{\log 0.67} \right\rceil,
$$

and for, $\varepsilon_{\text{abs}} = 10^{-6}$ the maximum number of iterations is $k_{\max} = 34$. Consequently, for every initial point $x^{(0)} \in [0.4, 0.6]$, the fixed-point iteration achieves an absolute accuracy of 10^{-6} after 34 steps at most.

In this case the a posteriori error estimate (14.27) is of the form

$$
\|x^* - x^{(k)}\| \le 2.033\|x^{(k)} - x^{(k-1)}\|.
$$

Hence, if the fixed-point iteration (starting with $x^{(0)} \in D_0$) is terminated as soon as

$$
\|x^{(k)} - x^{(k-1)}\| \le \varepsilon_{\text{abs}}/2.033,
$$

then it is guaranteed that the accuracy requirement $\|x^* - x^{(k)}\| \le \varepsilon_{\text{abs}}$ is met.

Example (Quadratic Equation) For the iteration rule (14.17) it holds that $t'(x) = -0.5x$, i.e., $|t'(x)| < 1$ is only valid for $|x| < 2$. The contraction theorem only ensures the convergence of the iteration (14.17) to the fixed-point $x^* = 1$ (which is unique in $(-2, 2)$) for initial values $x^{(0)} \in (-2, 2)$. Thus, the convergence conditions of the contraction theorem are *sufficient* but not *necessary*, as a comparison with the actual region of attraction $(-5, 5)$ shows (see page 282).

14.1.3 Rate of Convergence

The efficiency of an iterative method is closely related to the number of iterations which are needed to achieve a given accuracy requirement. A universal characterization of the convergence rate of an iterative method is *not* possible, not even with rigorous restrictions as to the class of functions to which T may belong. The convergence behavior of the sequence $\{x^{(k)}\}$ can only be described in a certain neighborhood of x^* (which is often very small) by characterizing the *manner* in which the absolute error $e_k = x^{(k)} - x^*$ decreases as $k \to \infty$.

Definition 14.1.9 (Order of Convergence, Convergence Factor) *A sequence $\{x^{(k)}\} \subset \mathbb{R}^n$ that converges to x^* has convergence order p and convergence factor a if*

$$\lim_{k \to \infty} \frac{\|x^{(k+1)} - x^*\|}{\|x^{(k)} - x^*\|^p} = \lim_{k \to \infty} \frac{\|e_{k+1}\|}{\|e_k\|^p} = a > 0.$$

Note Definitions of a convergence order (Ortega, Rheinboldt [67]), much more general than the one above are possible. Such definitions are in any case *asymptotic* statements which characterize the error behavior as $k \to \infty$. For practical values of k they may actually be wrong or may just hold approximately.

In general, a sequence with a higher convergence *order* converges faster than a sequence of lower order. Two particular cases are exceptionally important:

$$\boldsymbol{p = 1} \quad \textit{linear convergence,}$$
$$\boldsymbol{p = 2} \quad \textit{quadratic convergence.}$$

The convergence order p is not necessarily an integer. For example, the secant method (see Section 14.2.5) has the convergence order

$$p = (1 + \sqrt{5})/2 \approx 1.618\ldots .$$

The convergence *factor* is mainly used for comparing different methods with the same convergence order.

Example (Comparison of Methods) Two iterative methods are to be compared. Method $\mathcal{M}_1$ is linearly convergent with convergence factor

$$a = \lim_{k \to \infty} \frac{\|e_{k+1}\|}{\|e_k\|} = 0.6.$$

Method $\mathcal{M}_2$ is quadratically convergent with the same convergence factor

$$\bar{a} = \lim_{k \to \infty} \frac{\|\bar{e}_{k+1}\|}{\|\bar{e}_k\|^2} = 0.6.$$

For the sake of simplicity, suppose that for practically relevant values of k

$$\frac{\|e_{k+1}\|}{\|e_k\|} \approx 0.6 \quad \text{and} \quad \frac{\|\bar{e}_{k+1}\|}{\|\bar{e}_k\|^2} \approx 0.6.$$

For method $\mathcal{M}_1$ this means that

$$\|e_k\| \approx 0.6\|e_{k-1}\| \approx 0.6^2\|e_{k-2}\| \approx \cdots \approx 0.6^k\|e_0\|,$$

and for method $\mathcal{M}_2$

$$\|\bar{e}_k\| \approx 0.6\|\bar{e}_{k-1}\|^2 \approx 0.6\Big[0.6\|\bar{e}_{k-2}\|^2\Big]^2 = 0.6^3\|\bar{e}_{k-1}\|^4 \approx \cdots \approx (0.6)^{2^k-1}\|\bar{e}_0\|^{2^k}.$$

Assume that both methods have the same initial error, i.e., $x^{(0)} = \bar{x}^{(0)}$, with an initial error $\|e_0\| = \|\bar{e}_0\| = 0.1$. To achieve the absolute accuracy 10^{-8}, i.e.,

$$\|e_k\| = 0.6^k\|e_0\| = 0.6^k \cdot 0.1 \leq 10^{-8} \quad \text{and}$$
$$\|\bar{e}_k\| = (0.6)^{2^k-1}\|\bar{e}_0\|^{2^k} = 0.6^{2^k-1} \cdot 0.1^{2^k} \leq 10^{-8},$$

the iteration has to be continued until

$$k \geq \left\lceil \frac{-\log_{10} 0.1 - 8}{\log_{10} 0.6} \right\rceil = 32$$

for method $\mathcal{M}_1$, and using method $\mathcal{M}_2$ it suffices that

$$2^k \geq \left\lceil \frac{-\log_{10} 0.6 - 8}{\log_{10} 0.6 + \log_{10} 0.1} \right\rceil = 7,$$

i.e., only three iterations are needed. The computational effort (in terms of the number of iterations needed) of the quadratically convergent method is lower than that of the linearly convergent method by an order of magnitude .

In the case of one scalar equation the order p can be interpreted as the factor by which the number of correct decimal places of $x^{(k)}$ is increased at each iteration.

Suppose the error of the kth iterate $x^{(k)}$ is $\|e_k\| = 10^{-d}$, then in a neighborhood of x^* the relation

$$\|e_{k+1}\| \approx a\|e_k\|^p = a(10^{-d})^p = a10^{-pd} \tag{14.28}$$

holds. For quadratic convergence the number of correct decimal places is approximately *doubled* at every iteration (in a small neighborhood of the solution). For a linearly convergent sequence, the error reduction is characterized only by the convergence factor a. An increase of the exponent in (14.28), and consequently an increase in the number of correct decimal places, does not take place since $p = 1$.

Theorem 14.1.2 (Rate of Convergence, Fixed-Point Iteration) *If the function T of the fixed-point iteration (14.16) has continuous second partial derivatives, then the rate of convergence can be estimated by*

$$\|x^{(k+1)} - x^*\| \leq \|T'(x^*)\|\|x^{(k)} - x^*\| + \frac{L_2}{2}\|x^{(k)} - x^*\|^2,$$

where $L_2 = \|T''\|$ is a Lipschitz constant for T'.

1. *If $T'(x^*) \neq 0$ then the error decreases (asymptotically) by the factor*

$$a = \|T'(x^*)\| < 1$$

at each iteration step, i.e., the sequence $\{x^{(k)}\}$ is linearly convergent with convergence factor a.

2. If $\|T'(x^*)\| = 0$ then the convergence is quadratic with convergence factor $a = L_2/2$.

Example (Logarithmic Equation)   The iteration

$$x^{(k+1)} := \left(\alpha x^{(k)} + e^{-x^{(k)}}\right)/(\alpha+1),$$

which corresponds to (14.15), has an iteration function with the following derivatives:

$$t'(x) = \left(\alpha - e^{-x}\right)/(\alpha+1) \qquad \text{and} \qquad t'(x^*) = \left(\alpha - e^{-x^*}\right)/(\alpha+1).$$

The iteration converges most rapidly for $\alpha \approx 1/\exp(x^*)$, since, in this case, the convergence factor $|t'(x^*)| \approx 0$ is nearly optimal.

14.1.4 Determination of the Initial Values

The results of the previous section show that iterative methods with $T'(x^*) = 0$ (or at least $T'(x^*) \approx 0$) have a particularly favorable convergence rate. For many systems of nonlinear equations, however, rapid convergence is only achieved if the initial iterate is chosen from a small subset of the region of attraction. This subset is generally located close to the desired solution x^*. Thus, in connection with the practical solution of nonlinear equations, it is most crucial to find an appropriate initial value $x^{(0)}$. Difficulty in determining initial values grows in proportion to the dimension of the problem (*"curse of dimensionality"*).

Random Search

In general, rapidly convergent iterative methods only have small regions of attraction. A *uniform random search* can be made prior to using a fixed-point iteration in order to determine an initial value $x^{(0)}$ that is located—with high probability— in the region of attraction of the desired solution x^*. In a random search it is usually assumed that the desired point x^* is located in an n-dimensional hypercube

$$D = [a_1, b_1] \times [a_2, b_2] \times \cdots \times [a_n, b_n] \subset \mathbb{R}^n.$$

Generating n independent random numbers $z_1, \ldots, z_n$, located in the intervals $[a_i, b_i]$, according to a continuous uniform distribution[3] leads to a *random point*

$$z = (z_1, \ldots, z_n)^\top \in \prod_{i=1}^n [a_i, b_i] = D \subset \mathbb{R}^n.$$

A sequence of many n-dimensional random points is generated in this way and used to evaluate F. Finally, the point $\bar{z}$ with the smallest residual $\|F(\bar{z})\|$ of all points of the sequence is chosen as the desired initial iterate.

---

[3] Generators of random numbers with continuous uniform distribution can be found, for example, in the IMSL and the NAG program libraries (see Chapter 17).

Example (Curse of Dimensionality) The effect of the dimensionality can be demonstrated in a simple model. If, for example, the region of attraction is the spherical neighborhood

$$K_n = \{x : \ (x_1 - x_1^*)^2 + \cdots + (x_n - x_n^*)^2 \le 1\}$$

of the solution x^*, and if a hypercube

$$Q_n = \{x : \ |x_i - x^*| \le 5, \ i = 1, 2, \ldots, n\}$$

is chosen as the domain of a uniform random search, then the probability that a randomly chosen n-dimensional vector $x^{(0)} \in Q_n$ is an appropriate initial value (i.e., $x^{(0)}$ is inside the spherical region of attraction) is given by

$$P_n = \frac{\text{volume}\,(K_n)}{\text{volume}\,(Q_n)}.$$

For dimensions $n = 1, 50$ and 100 the following values are obtained:

n	volume (K_n)	volume (Q_n)	P_n
1	2	10^1	$2.00 \cdot 10^{-1}$
50	$1.73 \cdot 10^{-13}$	10^{50}	$1.73 \cdot 10^{-63}$
100	$2.37 \cdot 10^{-40}$	10^{100}	$2.37 \cdot 10^{-140}$

Under the assumptions made above (which are in fact characteristic of certain nonlinear problems), the probability of obtaining an appropriate initial value using a random search converges exponentially to zero as the dimension increases. It is as probable that an appropriate initial value for a seven-dimensional problem can be obtained in this way as it is for an Austrian lottery ticket to have the winning numbers:

$$\frac{1}{\dbinom{45}{6}} \approx 1.68 \cdot 10^{-7}.$$

Determining an appropriate initial iterate for problems with $n = 100$ is as complex as searching the entire universe for *one* specific molecule.

This example demonstrates the necessity of using all available information for determining the initial values for solving higher-dimensional problems. Pure random searching in practical situations is hopeless.

Example (Random Search) Points

$$x \in [a_1, b_1] \times [a_2, b_2] \times \cdots \times [a_{20}, b_{20}] \subset \mathbb{R}^{20}$$

with very small residual $\|F(x)\|_2$ were searched for with the help of uniformly distributed random numbers, which satisfy a system of 20 nonlinear equations as well as possible. In Fig. 14.5 the course of this random search is depicted in the form of steps which correspond to the currently smallest residual found in the random search. The graph conveys an idea as to how fast the computational effort—measured by the number of function values to be calculated—increases as the residual decreases.

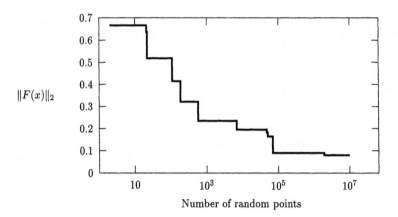

$\|F(x)\|_2$

Figure 14.5: The minimization of the residual of a function $F : \mathbb{R}^{20} \to \mathbb{R}^{20}$ using a random search. Newly generated random points whose residuals are larger than the smallest residual obtained by then are not depicted.

Homotopy

If no adequate initial approximation for starting an iterative method is available then a *homotopy* or *continuation method* often leads to a a sufficiently accurate approximate solution. For that purpose a parameter inherent to the problem or an artificially introduced parameter $\lambda \in \mathbb{R}$ (usually $\lambda \in [0, 1]$) is used to transform a nonlinear system of n equations in n unknowns into a family of problems

$$H(x, \lambda) = 0 \qquad\qquad (14.29)$$

whose solutions $x^*(\lambda)$ depend continuously on λ under certain conditions (see e. g., Allgower, Georg [85], Rheinboldt [324], Schwetlick [73]). In this case the problem family (14.29) has a one-dimensional manifold (a space curve) in $\mathbb{R}^{n+1}$ as its set of solutions.

In many practical situations the problem inherently depends on a parameter λ. If this is not the case, then

$$H(x, \lambda) := F(x) + (\lambda - 1)F(x^{(0)})$$

can be used to define a solution curve

$$H\big(x(\lambda), \lambda\big) = 0, \qquad \lambda \in [0, 1],$$

the endpoints of which are $x(0) = x^{(0)}$ and $x(1) = x^*$. In order to close the gap between $x^{(0)}$ and x^*, the interval $[0, 1]$ can be subdivided by intermediate points

$$0 = \lambda_0 < \lambda_1 < \cdots < \lambda_K = 1,$$

and the corresponding sequence of nonlinear equations

$$H(x, \lambda_k) = 0, \qquad k = 1, 2, \ldots, K,$$

can be solved one after the other, where the solution $x^{(k)}$ of the kth system of nonlinear equations can be used as the initial iterate for solving the $(k + 1)$st system. If the distance $\lambda_{k+1} - \lambda_k$ is chosen small enough, then $x^{(k)}$ is a suitable initial approximation for the iterative determination of $x^{(k+1)}$.

Example (Transportation Network) The flow through the edges of a nonlinear transportation network is zero as long as there are no transportation activities. The gradual increase in demand for transportation (possibly until saturation is reached) leads, in a natural way, to a sequence of nonlinear problems.

Software (Homotopy Method) In the software package HOMPACK (see Watson, Billups, Morgan [376] and Rheinboldt [324]) three different *globally* convergent homotopy algorithms are implemented. HOMPACK is available via NETLIB/hompack.

The software package PITCON is also suitable for parametrized systems, where the parameter λ is inherent in the problem. PITCON is available via NETLIB/contin.

Minimization

Another possible approach to getting close to the solution from an inadequate initial value is to minimize the function $\Phi := \|F\| : \mathbb{R}^n \to \mathbb{R}$. In general, the region of attraction of the slowly convergent gradient methods, used for the minimization of Φ, is larger than that of the rapidly convergent equation solvers (see Section 14.4.1).

Software (Minimization) Programs designed to perform minimization can be found in the two comprehensive mathematical libraries: IMSL and NAG. In addition, there are a number of special software products (see Moré, Wright [20]).

14.1.5 The Termination of an Iteration

In mathematical terms, an iterative process yields an *infinite* sequence $\{x^{(k)}\}$ for each initial value $x^{(0)} \in D^*$. For the practical solution of systems of nonlinear equations though, only algorithms which lead to a result for all initial values, after a *finite* number of steps, are useful.

For both the developer and the user of software for solving nonlinear equations, the selection of appropriate termination criteria and values for the respective program parameters (τ_{abs}, τ_{rel}, τ_f etc.) is important and difficult. A trade-off between accuracy and efficiency has to be made:

- The iteration must *not* be terminated *too early* (i. e., at a larger distance from the desired solution than required).

- The iteration should *not* be terminated *too late* (i. e., after an unnecessarily high accuracy has been achieved with an unnecessarily high computational effort).

For a convergent sequence $\{x^{(k)}\}$, a reliable program should be able to recognize whether or not the latest approximate value $x^{(k)}$ is of the required accuracy. This decision is based on a *convergence test*. For inadequate initial values $x^{(0)} \notin D^*$ or for problems which actually have no solution, mechanisms which recognize non-convergence and terminate the iteration with a corresponding message to the user must be provided. Tests for *non*-convergence are as important as tests for convergence.

Termination Criteria

There are two termination criteria for an iterative equation solver:

1. The numerical problem is solved by determining an approximate value $\tilde{x}$ for x^* that complies with a given tolerance, specified, for example, by one of the following inequalities:

$$\|\tilde{x} - x^*\| \leq \tau_{\text{abs}}, \tag{14.30}$$
$$\|\tilde{x} - x^*\| \leq \tau_{\text{rel}} \|x^*\| \tag{14.31}$$

or

$$\|F(\tilde{x})\| \leq \tau_f.$$

2. The numerical problem cannot be solved at all, because, for example, there is no zero of the nonlinear function F.

Note In strict mathematical terms, *neither* of the previous decisions can be made on a computer, since only a *limited* number of discrete function values $F(x^{(1)}), \ldots, F(x^{(k)})$, and possibly several values of the derivatives, are available as information about F. Moreover, the solution x^* itself is not available (except for artificial examples) in order to check the validity of (14.30) and (14.31).

The Error Criterion

In most software products estimates of the error $\|e\| = \|\tilde{x} - x^*\|$ are used as a termination criterion. The iterative process is terminated as soon as, for example, one of the inequalities

$$\|x^{(k+1)} - x^{(k)}\| \leq \tau_{\text{abs}},$$
$$\|x^{(k+1)} - x^{(k)}\| \leq \tau_{\text{rel}} \|x^{(k+1)}\|$$

is satisfied. Often the combination,

$$\|x^{(k+1)} - x^{(k)}\| \leq \tau_{\text{abs}} + \tau_{\text{rel}} \|x^{(k+1)}\|, \tag{14.32}$$

is used to avoid difficulties when $x^{(k+1)} \approx 0$. If the sequence $\{x^{(k)}\}$ converges *rapidly* to x^*, i.e., if

$$\|x^{(k+1)} - x^*\| \ll \|x^{(k)} - x^*\|,$$

then

$$\|x^{(k+1)} - x^{(k)}\| \approx \|x^{(k)} - x^*\|,$$

and $\|x^{(k+1)} - x^{(k)}\| \leq \tau_{\text{abs}}$ implies $\|x^{(k+1)} - x^*\| \ll \tau_{\text{abs}}$. If, on the other hand though, the sequence $\{x^{(k)}\}$ converges very *slowly* to x^*, i. e., if

$$\|x^{(k+1)} - x^*\| \approx \|x^{(k)} - x^*\|,$$

then $\|x^{(k+1)} - x^{(k)}\|$ can be very small even if the required accuracy has not yet been achieved (see Fig. 14.6). In this case the above termination criteria may cause a premature termination.

The Residual Criterion

The *residual* $F(\tilde{x})$ yields a less critical termination criterion: $F(\tilde{x})$ is possibly perturbed by cancellation effects, as $F(\tilde{x}) \approx 0$ for $\tilde{x}$ close to the solution x^*. For functions with a large derivative $\|F'(x^*)\| \gg 1$, the residual may also be very large, $\|F(\tilde{x})\| \gg \tau_f$, even if the error $\|\tilde{x} - x^*\|$ is very small (see Fig. 14.7). Therefore, it is advisable not to use the residual condition

$$\|F(x^{(k)})\| \leq \tau_f \tag{14.33}$$

as the only termination criterion. Most library programs use convergence tests which are a combination of both criteria, (14.32) *and* (14.33).

Criteria for Non-Convergence

The test for *divergence* of a sequence $\{x^{(k)}\}$ is commonly made by checking

$$\|x^{(k)}\| \geq x_{\max} \quad \text{or} \quad \|F(x^{(k)})\| \geq f_{\max}, \tag{14.34}$$

or after repeated increase of the residuals, e. g.,

$$\|F(x^{(k)})\| > \|F(x^{(k-1)})\| > \|F(x^{(k-2)})\|. \tag{14.35}$$

Provided $F(x) = 0$ actually has a solution with a non-trivial (non-degenerate) region of attraction, (14.35) indicates that the initial value $x^{(0)}$ was inadequate. In such a case it is advisable to restart the iterative method from a new initial value, determined, for example, by a random search.

Apart from (14.34) and (14.35), *non-convergence* can possibly become evident in the form of systematic or unsystematic *oscillations* of the sequence $\{x^{(k)}\}$. Recognizing this behavior would be far too costly for a library program, for which efficiency is an important quality criterion. Thus, iterations $\{x^{(k)}\}$ with unspecific non-convergent behavior or unacceptably slow convergence are terminated by introducing a user-chosen limit on the maximum number of iterations

$$k \leq k_{\max}. \tag{14.36}$$

Criterion (14.36) may also stop the iteration if the tolerance parameters τ_{abs}, τ_{rel}, τ_f are chosen *too small*. In some programs a separate test checks as to whether

$$\|x^{(k+1)} - x^{(k)}\| \approx eps\|x^{(k)}\|$$

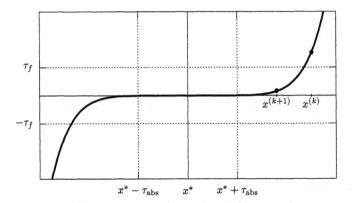

Figure 14.6: The iterate $x^{(k+1)}$ meets the residual criterion, but *not* the error criterion.

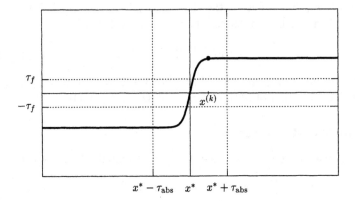

Figure 14.7: The iterate $x^{(k)}$ meets the error criterion, but *not* the residual criterion.

and, if necessary, terminates the iteration with a corresponding message.

However, if the values of F are afflicted with inaccuracies above the level of rounding errors (*noisy functions*) or if the tolerance parameters demand an accuracy that cannot be achieved, then the limit (14.36) terminates the iteration.

Another reason for terminating an iteration may be that certain subalgorithms cannot be applied. For example, if a system of linear equations, which is to be solved in Newton's method, contains a (numerically) singular equation, then the entire solution process may be terminated.

14.2 Nonlinear Scalar Equations

A scalar, nonlinear equation of the form

$$f(x) = 0 \qquad \text{where} \quad f : \mathbb{R} \to \mathbb{R} \tag{14.37}$$

is a special case ($m=n=1$) of the higher-dimensional system (14.6).

There are several reasons for separately discussing the case of *one* equation with *one* scalar unknown:

1. Some techniques (e. g., bisection) can only be used for solving scalar, nonlinear equations (14.37).

2. The determination of the roots of a univariate polynomial is an important special case of (14.37), for which many software products are available.

3. Methods for solving scalar *and* higher dimensional equations (e. g., Newton's method), can be geometrically depicted much more easily in the one-dimensional case. Moreover, analyses and theoretical considerations are often much simpler for one-dimensional problems.

4. One-dimensional iterative methods are used as auxiliary algorithms in many methods for *systems* of nonlinear equations (see e. g., Section 14.3.1).

Notation (Scalar Nonlinear Equations) In the study of scalar equations, functions are denoted by small letters (mostly with f). The only exceptions to this convention are polynomials, for which capital letters are used in this book. In most cases polynomials are denoted P.

14.2.1 The Multiplicity of a Zero

A zero x^* of a function $f : \mathbb{R} \to \mathbb{R}$ can be characterized more precisely if its multiplicity is known.

Definition 14.2.1 (Multiplicity of a Zero) *If f can be factorized in a neighborhood of x^* in the form*

$$f(x) = (x - x^*)^m \varphi(x), \tag{14.38}$$

where φ is continuous in a neighborhood of x^* and $\varphi(x^*) \neq 0$, then x^* is a zero of multiplicity m. If $m = 1$, x^* is called a simple zero.

Example (Multiplicity of a Zero) Definition (14.38) allows for zeros of *non*-integer multiplicities. Thus, $x^* = 0$ is a simple zero ($m = 1$) of the function

$$f(x) = x\sqrt{1 - x},$$

and $x^* = 1$ is a zero of multiplicity $m = 1/2$.

Theorem 14.2.1 (Integer Multiplicity of a Zero) *If f is m times continuously differentiable in a neighborhood $U(\{x^*\})$ of a zero x^*, then it follows from $f \in C^m\big(U(\{x^*\})\big)$,*

$$f'(x^*) = f''(x^*) = \cdots = f^{(m-1)}(x^*) = 0, \quad \text{and} \quad f^{(m)}(x^*) \neq 0, \tag{14.39}$$

that the zero x^* is of multiplicity $m \in \mathbb{N}$.

Proof: Expanding f in a Taylor series around x^* leads to

$$f(x) = f(x^*) + (x - x^*)f'(x^*) + \frac{(x - x^*)^2}{2}f''(x^*) + \cdots + \frac{(x - x^*)^m}{m!}f^{(m)}(\xi_x).$$

From (14.39) it follows that

$$f(x) = \frac{(x - x^*)^m}{m!}f^{(m)}(\xi_x).$$

The auxiliary function

$$\varphi(x) = \frac{f^{(m)}(\xi_x)}{m!}$$

is continuous in a neighborhood of x^* and, according to assumption (14.39), it must be that $\varphi(x^*) \neq 0$. It thus follows from (14.38) that x^* is a zero of multiplicity m. □

In particular, $x^* \in (a, b)$ is a simple zero (regular zero or zero of first order) of $f \in C^1[a, b]$ if

$$f(x^*) = 0 \quad and \quad f'(x^*) \neq 0.$$

In this case therefore, the curve $y = f(x)$ intersects the x-axis at x^* at an angle different from zero.

Nonlinear equations with simple zeros are well-conditioned: If f is perturbed (e. g., by data errors or rounding errors), then the perturbed function $\tilde{f}$ has a zero which is near x^*. This is not the case for zeros of *even* multiplicity m.

Example (Double Zero) Let $f(x) = x^2 - 2x + 1$. Then $x^* = 1$ is a zero of f of multiplicity two, i.e., x^* is a double zero. The perturbed function $\tilde{f}(x) = f(x) + \varepsilon$ with $\varepsilon > 0$ has *no* real zero at all.

Suppose that x^* is a zero of f with integral multiplicity $m \geq 2$. Then the x-axis is a tangent to the graph of f at x^* because $f'(x^*) = 0$. For *odd* multiplicities m, the function f changes its sign at x^*; for *even* multiplicities m, the graph of f is included completely in $U \times [0, \infty)$ or in $U \times (-\infty, 0]$, where U is a neighborhood of x^*, i.e., f does *not* change its sign on U.

The numerical determination of multiple zeros is more complicated than that of simple zeros:

1. The attainable accuracy $|\tilde{x}^* - x^*|$ is reduced significantly due to the ill-posed-ness or extreme ill-conditioning of nonlinear problems with multiple zeros (see Section 14.2.2).

2. Most iterative methods used for determining multiple zeros have a very low efficiency (convergence rate) (as will be shown in the example on page 305), and sometimes they fail completely.

Note (Systems of Equations) *Systems* of nonlinear equations may also have multiple zeros. However, such cases are not often encountered in practice.

Modification of the Nonlinear Equation

If f' is available as well as f, the modified problem

$$u(x) = 0 \quad \text{with} \quad u(x) := \frac{f(x)}{f'(x)} \tag{14.40}$$

can be solved instead of the original equation $f(x) = 0$. If x^* is of multiplicity m, then it follows from (14.38) that

$$\frac{f(x)}{f'(x)} = (x - x^*)\frac{\varphi(x)}{m\varphi(x) + (x - x^*)\varphi'(x)} = (x - x^*)\psi(x).$$

From $\psi(x^*) = 1/m \neq 0$ it follows that x^* is a *simple* zero of $u = f/f'$. Thus, due to the difficulties mentioned above, it seems reasonable to determine multiple zeros of f by solving the modified problem (14.40). However, practical difficulties arise at points x_p where $f'(x_p) = 0$ but $f(x_p) \neq 0$, i.e., at the poles of the function u.

14.2.2 The Condition of a Nonlinear Equation

When solving a nonlinear equation numerically, all available data is provided by the function values of f. It is a reasonable assumption that the subprogram used to calculate $f(x)$ returns a perturbed value $\tilde{f}(x)$ instead of the exact function value due to data errors, rounding errors and the like. If an upper bound on the perturbation

$$|\Delta f(x)| = |\tilde{f}(x) - f(x)| \leq \varepsilon \tag{14.41}$$

is known, then it is possible to estimate the error effect $|x^* - \tilde{x}^*|$. If x^* is a zero of multiplicity m, then it follows that

$$
\begin{aligned}
f(\tilde{x}^*) &= f(x^*) + (\tilde{x}^* - x^*)f'(x^*) + \frac{(\tilde{x}^* - x^*)^2}{2}f''(x^*) + \cdots \\
&\quad \cdots + \frac{(\tilde{x}^* - x^*)^{m-1}}{(m-1)!}f^{(m-1)}(x^*) + \frac{(\tilde{x}^* - x^*)^m}{m!}f^{(m)}(\xi_{\tilde{x}^*}) \\
&= \frac{(\tilde{x}^* - x^*)^m}{m!}f^{(m)}(\xi_{\tilde{x}^*}) \\
&\approx \frac{(\tilde{x}^* - x^*)^m}{m!}f^{(m)}(x^*).
\end{aligned}
$$

From (14.41) it also follows that $|f(\tilde{x}^*)| \leq \varepsilon$, i.e., the maximum value of $f(\tilde{x}^*)$ is $\pm\varepsilon$, where

$$\pm\varepsilon \approx \frac{(\tilde{x}^* - x^*)^m}{m!}f^{(m)}(x^*),$$

whence

$$|\tilde{x}^* - x^*| \approx \varepsilon^{1/m}\left|\frac{m!}{f^{(m)}(x^*)}\right|^{1/m}. \tag{14.42}$$

For a multiplicity $m \gg 1$ and a small error bound ε, the factor $\varepsilon^{1/m}$ is much larger than ε. A dramatic deterioration of the accuracy must therefore be expected, even though the other factor possibly has a compensating effect. In general, problems with zeros of high multiplicity are ill-conditioned with respect to the inaccuracies in f.

Example (Multiple Root of a Polynomial)   At $x^* = 1$ the polynomial $P_7(x) := (x-1)^7$ has a zero of multiplicity seven (see Fig. 14.8).

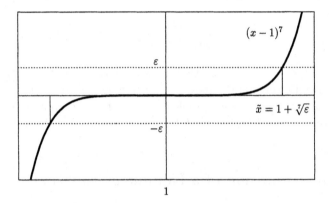

Figure 14.8: The polynomial $P_7(x) = (x-1)^7$ is represented in a neighborhood of $x = 1$.

The zero of the perturbed function

$$\tilde{P}_7(x) = (x-1)^7 - \varepsilon$$

is $\tilde{x}^* = 1 + \varepsilon^{1/7}$; thus, in accordance with (14.42),

$$|\tilde{x}^* - x^*| = \sqrt[7]{\varepsilon}.$$

The (absolute) condition number with respect to the data perturbation $\Delta \mathcal{D} = \tilde{\mathcal{D}} - \mathcal{D}$,

$$\kappa_{\mathrm{abs}} = \frac{\|\tilde{x}^* - x^*\|}{\|\tilde{\mathcal{D}} - \mathcal{D}\|} = \frac{\varepsilon^{1/7}}{\varepsilon} = \varepsilon^{-6/7},$$

depends on the size of the perturbation. This kind of dependence is typical for nonlinear problems.

ε	10^{-1}	10^{-7}	10^{-14}
κ_{abs}	7.2	10^6	10^{12}

Thus, due to the large condition number, a result with only *one* correct decimal place can be expected when using a single-precision IEC/IEEE-arithmetic. A double-precision arithmetic yields a result with *three* correct decimal places. Practical calculations (cf. the examples on pages 303, 304, and 306) demonstrate the correctness of this reasoning.

Even the determination of *simple* zeros may be ill-conditioned: several simple zeros located very close to each other lead to difficulties similar to those encountered when dealing with a multiple zero.

Example (Ill-Conditioned Simple Roots of a Polynomial)   The polynomial

$$P_2(x) = (x - 2)^2 - 10^{-6}$$

has the two simple zeros 2 ± 10^{-3}. A perturbation $\varepsilon = 10^{-6}$ changes the values of the zeros by $\pm 10^{-3}$. Hence, the condition number is 10^3.

However, large condition numbers are not only confined to zeros of higher multiplicities or many simple zeros situated close together.

Example (Wilkinson Polynomial)   A famous example was formulated by Wilkinson [381]: The polynomial

$$P_{20}(x) = (x - 1)(x - 2) \cdots (x - 19)(x - 20) = x^{20} - 210x^{19} + \cdots \qquad (14.43)$$

has the isolated simple zeros $1, 2, 3, \ldots, 20$. Calculating the (integer) coefficients of the polynomial in an INTEGER arithmetic (with sufficient precision) and rounding the obtained values to single-precision IEC/IEEE floating-point numbers leads to a perturbed polynomial which has 16 conjugate *complex* zeros, whose imaginary parts differ from zero significantly (see Fig. 14.9).

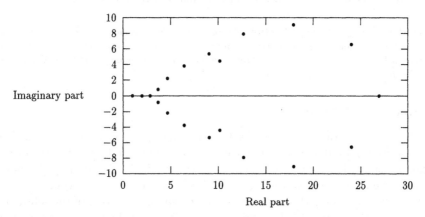

Figure 14.9: The zeros of the polynomial (14.43) are complex with significant imaginary parts when single-precision IEC/IEEE arithmetic is used to calculate the coefficients.

In this example, the single-precision IEC/IEEE arithmetic is obviously not sufficient for computing results that comply with the characteristics of the problem. A significant improvement can be achieved, however, using a multiple-precision arithmetic since the data—the coefficients of the polynomial P_{20}—are integers.

14.2.3 The Bisection Method

A special property of one-dimensional equations with continuous functions f is the following: If the signs of the function values at the end points of an interval are different, then it can be concluded that this interval contains a zero of f.

Theorem 14.2.2   *If the function* $f : [a, b] \rightarrow \mathbb{R}$ *is continuous, i.e., if* $f \in C[a, b]$, *then* f *takes each value between* $f(a)$ *and* $f(b)$ *at least once on* $[a, b]$.

If the values $f(a)$ and $f(b)$ of a function $f \in C[a, b]$ have opposite signs, then $f(x^*) = 0$ for at least one point $x^* \in (a, b)$. Thus, the existence of a zero is guaranteed.

In the *bisection algorithm* (*continuous binary search*), interval bisections are made until the interval known to contain a zero of f is as small as required:

initial interval: $[x_{\text{left}}, x_{\text{right}}]$ with $f(x_{\text{left}})f(x_{\text{right}}) < 0$

do $k = 1, 2, 3, \ldots$
     $x_{\text{center}} := (x_{\text{left}} + x_{\text{right}})/2;$
     if $f(x_{\text{left}})f(x_{\text{center}}) \leq 0$ **then** $x_{\text{right}} := x_{\text{center}}$
                        else $x_{\text{left}} := x_{\text{center}}$
   if *termination criterion is met* **then exit**
end do

The bisection algorithm uses an *adaptive* discretization to acquire information about f: At which point the next evaluation of the function f has to be made can only be decided on the basis of all information available so far.

Note (Optimality of the Bisection Algorithm) For continuous functions with sign changes, the bisection algorithm is *optimal*: in terms of a *worst-case* analysis it yields a zero of the required accuracy with the smallest number of function evaluations of all relevant algorithms (Sikorski [349]).

Convergence Behavior

The bisection algorithm is guaranteed to converge if two starting points whose function values are different with respect to their signs are found. On the other hand, no similarly useful criteria exist for the methods discussed in the next sections (Newton's method, the secant method etc.). There is no way of ensuring that for these methods $x^{(0)}$ is located in the region of attraction of the solution.

The bisection method only requires the continuity of the function f (but *not* its differentiability). As a trade-off, it converges only *linearly* with the convergence factor $1/2$. After k evaluations of f in the interior of the given interval, the actual interval length $l = x_{\text{right}} - x_{\text{left}}$ is reduced by a factor of 2^k relative to the initial interval. Thus, the following absolute accuracies can be guaranteed:

k	10	20	40	80
ε_{abs}	$10^{-3} \, l_{\text{initial}}$	$10^{-6} \, l_{\text{initial}}$	$10^{-12} \, l_{\text{initial}}$	$10^{-24} \, l_{\text{initial}}$

The number of steps in the bisection method has to be *doubled* in order to guarantee the doubling of the number of correct decimal places. On the other hand, a quadratically convergent method (like Newton's method, see Section 14.2.4) only requires a single additional step to improve the accuracy to the same extent.

$l_{\text{initial}}/2^k$ is an upper bound on the absolute error in $x^{(k)}$, which may, however, be far too pessimistic.

Example (Non-Monotonic Convergence of Bisection) For $f(x) = xe^{-x} - 0.06064$,

$$f(0) = -0.06064 < 0 \quad \text{and} \quad f(1) = 0.30723944\ldots > 0.$$

Since f is a continuous function, at least one zero can be found in the interval $[0,1]$ using the bisection algorithm. After $k = 12$ bisection steps with $x_{\text{center}} = 6.469727 \cdot 10^{-2}$, a value with the absolute error $4.63 \cdot 10^{-6}$ is achieved. The error bound would guarantee such an accuracy not before $k = 18$. It is remarkable that in the next bisection step ($k = 13$) the absolute error of $x_{\text{center}} = 6.457520 \cdot 10^{-2}$ is deteriorated to $-1.17 \cdot 10^{-4}$, a value which is closer to the respective error bound $2^{-13} = 1.22 \cdot 10^{-4}$. This example demonstrates that the error behavior of the bisection iteration is generally *not monotonic*.

Example (Effects of Rounding Errors) The polynomial

$$P_7(x) = x^7 - 7x^6 + 21x^5 - 35x^4 + 35x^3 - 21x^2 + 7x - 1 \tag{14.44}$$

has its only zero (of multiplicity 7) at $x^* = 1$. (14.44) is just another way of writing $(x - 1)^7$. The bisection algorithm is used in order to find x^* starting with the initial interval $[0,5]$ and the values $f(x_{\text{left}}) = -1$ and $f(x_{\text{right}}) = 16384$. The accuracy increases during the first n_{max} steps, while the interval length decreases to $5 \cdot 2^{-n}$ n_{max} depends on the machine arithmetic. For example, with a single-precision program on a personal computer, the best value possible is $x_{\text{center}} = 1.13372$; in a double-precision arithmetic this value is $x_{\text{center}} = 1.005094$. With the function values $f(x_{\text{center}}) = 8.34 \cdot 10^{-7}$ and $f(x_{\text{center}}) = 2.91 \cdot 10^{-16}$, these abscissas are *numerical zeros* of the function (14.44) with respect to the machine arithmetic. Nevertheless, the errors with respect to the exact zero $x^* = 1$ (0.13 and $5.1 \cdot 10^{-3}$ respectively) are very high. In Fig. 14.8 (on page 300) it can be seen that the values $\varepsilon = 8.34 \cdot 10^{-7}$ and $\varepsilon = 2.91 \cdot 10^{-16}$ correspond to the arguments $\tilde{x} = 1.13539$ and $\tilde{x} = 1.00603$ respectively.

Software (Bisection Algorithms) The bisection method, along with other methods, is often incorporated in a polyalgorithm, as in the subprograms IMSL/MATH-LIBRARY/zbren, NAG/c05adf, NAG/c05agf or NAG/c05azf.

Unlike Newton's method and the secant method, the bisection method *cannot* be transferred to *systems* of nonlinear equations.

14.2.4 Newton's Method

In Newton's method,[4] the nonlinear function $f : \mathbb{R} \to \mathbb{R}$, whose zero(s) x^* cannot be determined directly, is replaced by *linear model functions* l_k,

$$l_k(x) := a_k + b_k x \approx f(x),$$

the zeros $x_k^* = -a_k/b_k$ of which are used as approximations of the zero x^* of f. The functions l_k are obtained, for example, by truncating the Taylor expansion about $x^{(k)}$ after the linear term:

$$l_k(x) := f(x^{(k)}) + (x - x^{(k)})f'(x^{(k)}) \approx f(x),$$

---

[4]Sir Isaac Newton published his root-finding method specially for the solution of the Kepler equation $f(x) = x - a\sin x - b = 0$. J. Raphson reformulated Newton's method so that it more closely resembles its current form. This method is therefore also referred to as the *Newton-Raphson method*.

i.e., $a_k := f(x^{(k)}) - x^{(k)} f'(x^{(k)})$ and $b_k := f'(x^{(k)})$ and, hence,

$$x_k^* = x^{(k)} - \frac{f(x^{(k)})}{f'(x^{(k)})}.$$

On the basis of the assumption that x_k^* is a better approximation to x^* than $x^{(k)}$, a sequence $\{x^{(0)}, x^{(1)}, x^{(2)}, \dots \}$ is defined as follows:

$$x^{(k+1)} = x^{(k)} - \frac{f(x^{(k)})}{f'(x^{(k)})}, \qquad k = 0, 1, 2, \dots, \tag{14.45}$$

which converges to x^* under certain conditions, e.g., if the initial value $x^{(0)}$ is sufficiently close to x^*.

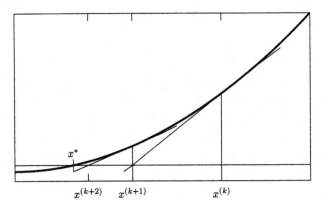

Figure 14.10: The tangent of f in the point $(x^{(k)}, f(x^{(k)}))$ intersects $y = 0$. This intersection is used as the new approximate value $x^{(k+1)}$.

The geometrical interpretation of Newton's method is shown in Fig. 14.10.

Example (Newton's Method) For $f(x) = xe^{-x} - 0.6064$, the first derivative is $f'(x) = (1 - x)e^{-x}$; thus, the corresponding Newton sequence is given by

$$x^{(k+1)} := x^{(k)} - \frac{x^{(k)} e^{-x^{(k)}} - 0.6064}{(1 - x^{(k)})e^{-x^{(k)}}}.$$

Starting at $x^{(0)} = 0$ the very rapidly convergent sequence

k	1	2	3
$x^{(k)}$	0.0604	0.0646757	0.0649263
absolute error	$-4.05 \cdot 10^{-3}$	$-1.69 \cdot 10^{-5}$	$-4.63 \cdot 10^{-7}$

is obtained. However, for $x^{(0)} = 0.99$ ($x^{(0)} = 1$ is not possible), the accuracy $1.2 \cdot 10^{-7}$ is not achieved until $k = 91$. Until $k = 89$, *not a single decimal place* of the approximations is correct!

Example (Root of a Polynomial) The zero $x^* = 1$ of the polynomial (14.44) is determined numerically using Newton's method. With $x^{(0)} = 5$, the most exact approximate values are $x^{(23)} = 1.043442$ for a single-precision arithmetic and $x^{(84)} = 0.99902912\dots$ for a double-precision arithmetic. In both cases, the sequence $\{x^{(k)}\}$ starts to oscillate for larger k.

Software (Newton's Method) The program NETLIB/TOMS/681 implements Newton's method for univariate nonlinear equations (along with a bisection algorithm).

Convergence Behavior

Since Newton's method is based on an iterative process $x^{(k+1)} = t(x^{(k)})$, it follows that

$$t(x) = x - \frac{f(x)}{f'(x)} \quad \text{and} \quad t'(x) = \frac{f(x)f''(x)}{[f'(x)]^2}.$$

In a neighborhood U of a simple zero x^*, the inequality $|t'(x)| < 1$ holds due to $f(x^*) = 0$, and therefore the function t is a *contraction* on every closed set $U_0 \subset U$. Theorem 14.1.1 (the contraction theorem) guarantees that Newton's method converges to the simple zero x^* for every initial point $x^{(0)}$ sufficiently close to x^*.

From (14.45) it follows that

$$x^{(k+1)} - x^* = x^{(k)} - x^* - \frac{f(x^{(k)})}{f'(x^{(k)})}. \tag{14.46}$$

If x^* is a *simple* zero, i.e., if $f(x^*) = 0$ and $f'(x^*) \neq 0$, then, because of (14.46) and the expansion

$$f(x^*) = f(x^{(k)}) + (x^* - x^{(k)})f'(x^{(k)}) + \frac{(x^* - x^{(k)})^2}{2} f''\big(x^{(k)} + \vartheta(x^* - x^{(k)})\big)$$

with $\vartheta \in (0,1)$, it follows that

$$x^{(k+1)} - x^* = \frac{(x^* - x^{(k)})^2}{2} \frac{f''\big(x^{(k)} - \vartheta(x^* - x^{(k)})\big)}{f'(x^{(k)})},$$

and therefore

$$\lim_{k \to \infty} \frac{|e_{k+1}|}{|e_k|^2} = \lim_{k \to \infty} \frac{|x^{(k+1)} - x^*|}{|x^{(k)} - x^*|^2} = \frac{1}{2} \left| \frac{f''(x^*)}{f'(x^*)} \right| =: a. \tag{14.47}$$

Thus, for simple zeros and sufficiently differentiable f Newton's method is (at least) *quadratically* convergent.

Multiple Zeros

Because $t'(x^*) = 1 - 1/m < 1$, Newton's method *converges* from any starting point $x^{(0)}$ which is sufficiently close to x^*. This is true, even if x^* is a zero of multiplicity $m \in \{2, 3, 4, \ldots\}$, i.e., even if (14.39) holds. However, the the rate of convergence is only *linear* for a zero of higher multiplicity.

One way to *avoid* multiple zeros is to solve the modified equation

$$u(x) := \frac{f(x)}{f'(x)} = 0$$

instead of $f(x) = 0$ (see (14.40)). The sequence to be used for this modified problem is

$$x^{(k+1)} = x^{(k)} - \frac{u(x^{(k)})}{u'(x^{(k)})} = \frac{f(x^{(k)})f'(x^{(k)})}{[f'(x^{(k)})]^2 - f(x^{(k)})f''(x^{(k)})}. \tag{14.48}$$

Since x^* is always a *simple* zero of u, (14.48) always converges quadratically. The only disadvantage of (14.48) is the additionally required second derivative of f and the slightly higher cost for the determination of $x^{(k+1)}$. However, practical difficulties arise from the fact that the denominator of (14.48) approaches zero for $x^{(k)} \to x^*$ at double zeros of f and zeros of higher multiplicity, i.e., the size of the denominator becomes very small in a neighborhood of a multiple zero.

Example (Multiple Root of a Polynomial) Unlike the non-modified Newton method (where the most precise value is obtained after 23 steps), (14.48) converges rapidly for the polynomial (14.44). For both single- and double-precision arithmetic $x^{(2)} = 0.977087$ is the most accurate approximation for $x^* = 1$. More accurate values cannot be obtained due to the size of the condition number of this nonlinear equation.

Newton's Method for Multiple Zeros

Quadratic convergence for zeros of higher multiplicities can be achieved not only by modifying the problem, but also by modifying the method. In a neighborhood of a zero x^* with multiplicity m, the relation

$$f(x) = (x - x^*)^m \varphi(x) \approx (x - x^*)^m c \tag{14.49}$$

holds. This leads to

$$\frac{f(x)}{f'(x)} \approx \frac{x - x^*}{m} \qquad \text{or} \qquad x^* \approx x - m\frac{f(x)}{f'(x)}.$$

The accordingly modified sequence

$$x^{(k+1)} := x^{(k)} - m\frac{f(x^{(k)})}{f'(x^{(k)})}, \qquad k = 0, 1, 2, \dots, \tag{14.50}$$

converges quadratically even at multiple zeros, provided the correct value of the multiplicity m is used in (14.50) (Hämmerlin, Hoffmann [52]).

Example (Multiple Root of a Polynomial) Depending on the value m used in the modified Newton sequence (14.50), the convergence of the iteration is generally extensively accelerated compared to the unmodified Newton method.

m	6.5	6.9	6.99	7	7.01	7.1	7.5
k	2	6	1	1	1	1	2
$\|e_{abs}\|$	$2.5 \cdot 10^{-2}$	$6.9 \cdot 10^{-3}$	$5.7 \cdot 10^{-3}$	0	$5.7 \cdot 10^{-2}$	$5.7 \cdot 10^{-2}$	$2.0 \cdot 10^{-2}$

The efficiency of the Newton variant (14.50) depends critically on the correctness of the value m used. If this value cannot be determined analytically, then a good *estimate* should at least be used.

Estimation of the Multiplicity of a Zero

Provided that

$$|x^{(k)} - x^*| < |x^{(k-1)} - x^*| \quad \text{and} \quad |x^{(k)} - x^*| < |x^{(k-2)} - x^*|,$$

$x^{(k)}$ can be substituted for x^* in (14.49):

$$f(x^{(k-1)}) \approx (x^{(k-1)} - x^{(k)})^m c$$
$$f(x^{(k-2)}) \approx (x^{(k-2)} - x^{(k)})^m c.$$

Then this system is solved with respect to m:

$$m \approx \frac{\log[f(x^{(k-1)})/f(x^{(k-2)})]}{\log[(x^{(k-1)} - x^{(k)})/(x^{(k-2)} - x^{(k)})]}.$$

This estimate of the multiplicity can be used, for example, in (14.50).

14.2.5 The Secant Method

Using Newton's method to solve a nonlinear equation $f(x) = 0$ requires explicit knowledge of the first derivative f' of the function f. However, in many practical cases f' is not available as an explicit representation, or it can only be obtained at high computational cost. In these cases, $f'(x^{(k)})$ can be approximated using a differential quotient:

$$f'(x^{(k)}) \approx \frac{f(x^{(k)} + h_k) - f(x^{(k)})}{h_k} =: d_k.$$

If the approximate value d_k is substituted for $f'(x^{(k)})$ in the Newton iteration (14.45) then the *discretized Newton methods* are obtained:

$$x^{(k+1)} := x^{(k)} - \frac{f(x^{(k)})}{d_k}, \qquad k = 0, 1, 2, \ldots .$$

With the step length $h_k := x^{(k-1)} - x^{(k)}$, the important special case of the *secant method* is obtained:

$$x^{(k+1)} := x^{(k)} - f(x^{(k)}) \frac{x^{(k-1)} - x^{(k)}}{f(x^{(k-1)}) - f(x^{(k)})}, \qquad k = 0, 1, 2, \ldots .$$

In geometrical terms this means that a secant, namely the linear interpolation between the two points

$$(x^{(k-1)}, f(x^{(k-1)})) \quad \text{and} \quad (x^{(k)}, f(x^{(k)})),$$

is used as a linear model for f instead of the tangent line of Newton's method (see Fig. 14.11).

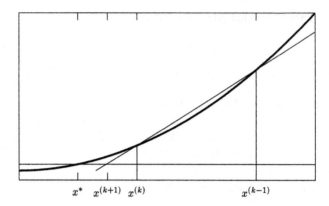

Figure 14.11: The intersection of a secant and the x-axis is used as the new approximation $x^{(k+1)}$ for x^*.

Software (Secant Method) The secant method for solving nonlinear equations is used in the programs `NAG/c05ajf` and `NAG/c05axf`.

Terminology (Regula Falsi) The iteration

$$d_k := \frac{f(\bar{x}) - f(x^{(k)})}{\bar{x} - x^{(k)}}$$

with an *unchanging* point $\bar{x}$ is called the *regula falsi method* by some authors (e. g., by Ortega and Rheinboldt [67]). Other authors, however, use the term *regula falsi* as a synonym for the secant method (see e. g., Isaacson, Keller [61]).

Convergence Behavior

Unlike the Newton method, the secant method does *not* have the simple structure $x^{(k+1)} = t(x^{(k)})$. It is not a stationary one-step method, but is rather a *two-step method* of the form $x^{(k+1)} = t(x^{(k)}, x^{(k-1)})$. Thus, special methods for the examination of its convergence properties are required (Ortega, Rheinboldt [67], Hämmerlin, Hoffmann [52]). When applying these methods, it turns out that the secant method is of the

$$\text{convergence order} \quad p = \frac{1 + \sqrt{5}}{2} = 1.618\ldots.$$

In a neighborhood of a zero x^* of f the estimate

$$e_{k+1} \approx \frac{1}{2}\left|\frac{f''(x^*)}{f'(x^*)}\right| e_k e_{k-1}$$

is valid for the secant method, whereas for Newton's method the relation

$$e_{k+1} \approx \frac{1}{2}\left|\frac{f''(x^*)}{f'(x^*)}\right| e_k^2$$

holds (cf. (14.47)).

14.2.6 Muller's Method

Müller's method (Müller [294]), in the English language literature referred to as *Muller's method*, is a generalization of the secant method: A *quadratic model* (a parabola) is used instead of a linear model of f.

With the secant method, the line interpolating the points

$$(x^{(k-1)}, f(x^{(k-1)})) \quad \text{and} \quad (x^{(k)}, f(x^{(k)}))$$

intersects the x-axis $y = 0$ at $x^{(k+1)}$, thus defining the $(k+1)$st point of the iteration. In Muller's method, $x^{(k+1)}$ is found by intersecting the parabola interpolating the three points

$$(x^{(k-2)}, f(x^{(k-2)})), \quad (x^{(k-1)}, f(x^{(k-1)})) \quad \text{and} \quad (x^{(k)}, f(x^{(k)})),$$

and the x-axis. Provided that there are two real zeros of the quadratic interpolation polynomial, the one which is closer to $x^{(k)}$ is used as $x^{(k+1)}$.

Unlike Newton's method and the secant method, Muller's method can be used to determine *complex* zeros of a function as well. This possibility has to be ruled out if only real zeros of f are desired.

However, Muller's method offers one advantage over other algorithms used to determine complex zeros: It is possible to start with real initial values and, if necessary, to continue the algorithm using complex approximations.

Software (Muller's Method) The program IMSL/MATH-LIBRARY/zreal is used to determine the real zeros of real, nonlinear functions $f : \mathbb{R} \to \mathbb{R}$, whereas the program IMSL/MATH-LIBRARY/zanly can be used to determine the zeros of a complex function $f : \mathbb{C} \to \mathbb{C}$.

Convergence Behavior

For Muller's method, which is a *three-step method* of the form

$$x^{(k+1)} = t(x^{(k)}, x^{(k-1)}, x^{(k-2)}),$$

the following relation holds:

$$e_{k+1} \approx \frac{1}{6} \left| \frac{f'''(x^*)}{f'(x^*)} \right| e_k e_{k-1} e_{k-2}.$$

The convergence order is $p = 1.839\ldots$ (Hildebrand [58]) and hence Muller's method converges superlinearly and more rapidly than the secant method.

14.2.7 Efficiency Assessment

In order to compare the efficiency of different nonlinear equation solvers, not only does the *number of iterations* have to be considered, but also the *computational cost* per each iteration, which consists of two parts:

1. the effort which is needed for the *evaluation* of the function f (calling a function procedure) and, if required, of derivatives of f, as well as

2. the *overhead* which is necessary to calculate the next iteration value $x^{(k+1)}$, to test for termination criteria etc.

In most cases the overhead does not play an important role in comparing different methods, because either it consists of a negligible number of arithmetic operations (needed to calculate $x^{(k+1)}$), or the computational cost of the overhead is approximately the same for each method (termination criteria etc.). The efficiency of a method can therefore be measured by the increase in accuracy per function evaluation.[5]

For example, two methods $\mathcal{M}_1$ and $\mathcal{M}_2$ are to be assessed: Method $\mathcal{M}_1$ has convergence order p and requires three function evaluations per iteration; method $\mathcal{M}_2$ has convergence order $\bar{p}$ and requires only one function evaluation per iteration, i.e., three iterations of $\mathcal{M}_2$ require the same computational effort as one iteration of $\mathcal{M}_1$.

What convergence order $\bar{p}$ must method $\mathcal{M}_2$ have, so that three iterations of $\mathcal{M}_2$ yield the same accuracy as one iteration of $\mathcal{M}_1$? Assuming that $a = \bar{a} = 1$ and $e_k = \bar{e}_k$, it follows that, approximately,

$$e_{k+1} = (e_k)^p, \qquad \bar{e}_{k+1} = (\bar{e}_k)^{\bar{p}},$$
$$\bar{e}_{k+2} = (\bar{e}_{k+1})^{\bar{p}} = \left((\bar{e}_k)^{\bar{p}}\right)^{\bar{p}} = (\bar{e}_k)^{\bar{p}^2},$$
$$\bar{e}_{k+3} = (\bar{e}_{k+2})^{\bar{p}} = \left(\left((\bar{e}_k)^{\bar{p}}\right)^{\bar{p}}\right)^{\bar{p}} = (\bar{e}_k)^{\bar{p}^3}.$$

For $e_{k+1} = \bar{e}_{k+3}$ then this means $p = \bar{p}^3$ or $\bar{p} = \sqrt[3]{p}$.

Thus, the number $\sqrt[w]{p}$ can be used as an *efficiency index* for an iterative method, where p denotes the convergence order of the method and w the number of function evaluations per iteration:

Method	Order of Convergence	Efficiency Index	
Bisection	1		1
Newton	2	$\sqrt{2} \approx$	1.414
Secant	1.618		1.618
Muller's	1.839		1.839

Note that a higher convergence order does not necessarily lead to a higher efficiency index. For example, a version of Newton's method with cubic convergence ($p = 3$) can be constructed using three function evaluations (f, f', and f'') per iteration. But its efficiency index $\sqrt[3]{3} \approx 1.442$ is only slightly larger (by 2 %) than that of the classical Newton method and much lower than that of the secant method or Muller's method.

---

[5] For the sake of simplicity, it is assumed that both the evaluations of f and of f' require the same computational effort.

14.2.8 The Acceleration of Convergence

A slowly converging sequence $\{x^{(k)}\}$ often requires a considerable computational effort to determine an accurate approximation to its limit. There are several ways to *accelerate the convergence* of the sequence in order to reduce this cost, i. e., to raise the efficiency of the method. One way is to use extrapolation (as discussed in Chapter 12 in the context of numerical integration); another method, which can be applied to the iterates of nonlinear equation solvers under certain conditions, is the Δ^2 *method*.

If $\{x^{(k)}\}$ is a linearly convergent sequence with limit x^* and a convergence factor $a < 1$, then it follows that

$$\lim_{k \to \infty} \frac{|x^{(k+1)} - x^*|}{|x^{(k)} - x^*|} = a < 1.$$

Assume that the signs of the errors $x^{(k)} - x^*$, $x^{(k+1)} - x^*$, and $x^{(k+2)} - x^*$ are all the same and that k is large enough for

$$\frac{x^{(k+1)} - x^*}{x^{(k)} - x^*} \approx a \approx \frac{x^{(k+2)} - x^*}{x^{(k+1)} - x^*} \tag{14.51}$$

to hold. Then,

$$(x^{(k+1)} - x^*)^2 \approx (x^{(k+2)} - x^*)(x^{(k)} - x^*),$$

$$(x^{(k+1)})^2 - 2x^{(k+1)}x^* + (x^*)^2 \approx x^{(k+2)}x^{(k)} - (x^{(k)} + x^{(k+2)})x^* + (x^*)^2,$$

$$(x^{(k+2)} - 2x^{(k+1)} + x^{(k)})x^* \approx x^{(k+2)}x^{(k)} - (x^{(k+1)})^2,$$

and

$$\begin{aligned}
x^* &\approx \frac{x^{(k+2)}x^{(k)} - (x^{(k+1)})^2}{x^{(k+2)} - 2x^{(k+1)} + x^{(k)}} \\
&= \frac{(x^{(k)})^2 + x^{(k)}x^{(k+2)} - 2x^{(k)}x^{(k+1)} + 2x^{(k)}x^{(k+1)} - (x^{(k)})^2 - (x^{(k+1)})^2}{x^{(k+2)} - 2x^{(k+1)} + x^{(k)}} \\
&= \frac{x^{(k)}(x^{(k+2)} - 2x^{(k+1)} + x^{(k)}) - ((x^{(k)})^2 - 2x^{(k)}x^{(k+1)} + (x^{(k+1)})^2)}{x^{(k+2)} - 2x^{(k+1)} + x^{(k)}} \\
&= x^{(k)} - \frac{(x^{(k+1)} - x^{(k)})^2}{x^{(k+2)} - 2x^{(k+1)} + x^{(k)}}.
\end{aligned}$$

Definition 14.2.2 (Δ^2 Method)   *Based on a sequence $\{x^{(k)}\}$, a new sequence $\{\bar{x}^{(k)}\}$ can be constructed according to*

$$\bar{x}^{(k)} := x^{(k)} - \frac{(x^{(k+1)} - x^{(k)})^2}{x^{(k+2)} - 2x^{(k+1)} + x^{(k)}}; \tag{14.52}$$

this technique is called the Δ^2 method.

This name is derived from the representations of $x^{(k+1)} - x^{(k)}$ and $x^{(k+2)} - 2x^{(k+1)} + x^{(k)}$ in terms of the difference operators

$$
\begin{aligned}
\Delta x^{(k)} &:= x^{(k+1)} - x^{(k)} \quad \text{and} \\
\Delta^2 x^{(k)} &:= \Delta\left(x^{(k+1)} - x^{(k)}\right) = \Delta x^{(k+1)} - \Delta x^{(k)} = x^{(k+2)} - 2x^{(k+1)} + x^{(k)},
\end{aligned}
$$

which makes it possible to write (14.52) as

$$
\bar{x}^{(k)} := x^{(k)} - \frac{\left(\Delta x^{(k)}\right)^2}{\Delta^2 x^{(k)}}.
$$

The Δ^2 method is based on the expectation that the sequence $\{\bar{x}^{(k)}\}$ defined by (14.52) converges to x^* *faster* than the original sequence $\{x^{(k)}\}$. The following theorem confirms this conjecture:

Theorem 14.2.3   *Let $\{x^{(k)}\}$ be a sequence which converges to the limit x^*, and for which*

$$
e_{k+1} = (a + \varepsilon_k)e_k
$$

with $e_k \neq 0$, $|a| < 1$, and $\varepsilon_k \to 0$ as $k \to \infty$. Then the sequence $\{\bar{x}^{(k)}\}$ defined by (14.52) converges to x^* faster than the original one, i. e.,

$$
\frac{\bar{x}_k - x^*}{x_k - x^*} \to 0 \quad as \quad k \to \infty.
$$

Proof: Henrici [56].

Definition 14.2.3 (Steffensen's Method)   *To apply Δ^2 acceleration not all values of the sequence $\{x^{(k)}\}$ have to be calculated in advance. The Steffensen method applies the Δ^2 acceleration whenever possible, i. e., as soon as the first three values of the original sequence are available.*

14.2.9 Polyalgorithms

When developing numerical software for solving nonlinear equations, both the reliability and the efficiency of the programs should be as high as possible. The combination of two or more algorithms is one possible way to achieve both aims. For example, the bisection algorithm yields maximal reliability as an equation solver for continuous functions with sign changes. In a small neighborhood of a zero, however, Newton's method or the secant method may be much more efficient. The combination of more than one algorithm yields what is called a *polyalgorithm*.

Software (Polyalgorithms) Examples of polyalgorithms for the solution of nonlinear (scalar) equations can be found in the publications of Bus, Dekker [130], and Brent [124]. The polyalgorithm of Brent is implemented in the program IMSL/MATH-LIBRARY/zbren.

14.2.10 Roots of Polynomials

Determining the zeros of a univariate polynomial $P_d \in \mathbb{P}_d$ of degree d whose coefficients are complex or real leads to the classical problem of finding the *roots* (solutions, zeros) of the *algebraic equation*

$$P_d(x) = a_0 + a_1 x + \cdots + a_d x^d = 0, \quad a_i \in \mathbb{C}. \tag{14.53}$$

The *fundamental theorem of algebra* is the most important discovery dealing with the roots of equation (14.53).

Theorem 14.2.4 (Fundamental Theorem of Algebra) *The algebraic equation (14.53) has exactly d solutions, provided each zero is counted as many times as its multiplicity.*

Thus, there exist unique real or complex numbers

$$x_1^*, x_2^*, \ldots, x_k^* \in \mathbb{C} \quad \text{and} \quad m_1, m_2, \ldots, m_k \in \mathbb{N}$$

(with $m_1 + m_2 + \cdots + m_k = d$), which make it possible for the polynomial P_d to be factorized as

$$P_d(x) = a_d (x - x_1^*)^{m_1} (x - x_2^*)^{m_2} \ldots (x - x_k^*)^{m_k}.$$

If the coefficients $a_0, a_1, \ldots, a_d$ are real, then for every zero x^* the complex conjugate $\bar{x}^*$ is also a zero of the same multiplicity. As a consequence, every algebraic equation (14.53) of odd degree (and with real coefficients) has at least one real solution.

For polynomials $P_d \in \mathbb{P}_d$ of degree $d \leq 4$, there exist *formulas* for the determination of the roots of (14.53). In general, equation (14.53) can be solved only approximately for degrees $d \geq 5$ (Abel's theorem). In practice, however, iterative methods are employed even for solving polynomial equations of third and fourth degree.

All iterative methods used to determine the solutions of general nonlinear equations can be applied to algebraic equations. Since P_d' can be derived easily, Newton's method is a particularly favorable choice for solving (14.53). However, that Newton's method—unlike Muller's method—leads to a *real* sequence $\{x^{(k)}\}$ for *real* initial values has to be taken into consideration. The determination of complex zeros using Newton's method requires complex initial values and complex arithmetic.

There are many special methods for the numerical determination of particular zeros or all the zeros of a polynomial, like the *Bernoulli method*, which makes use of the characteristics of the difference equation

$$a_d u_{i+d} + a_{d-1} u_{i+d-1} + \cdots + a_0 u_i = 0$$

(with coefficients $a_0, \ldots, a_d$). A variant of this method was combined with other algorithms and made into a polyalgorithm with three stages, the *Jenkins-Traub polyalgorithm* (Jenkins, Traub [242], Jenkins [241]).

Software (Jenkins-Traub Method) A version of the Jenkins-Traub polyalgorithm which finds the roots of polynomials with real coefficients is implemented in the program `IMSL/MATH-LIBRARY/zporc`, and a version dealing with polynomials with complex coefficients can be found in the program `IMSL/MATH-LIBRARY/zpocc`.

Other equation solvers which determine the roots of polynomials were developed by Laguerre, Lin, Graeffe, Bairstow, Lehmer, Rutishauser, and others (Householder [229]).

Software (Laguerre's Method) The program `IMSL/MATH-LIBRARY/zplrc` can be used to find the zeros of polynomials with real coefficients based on the Laguerre method. A version with an expanded parameter list is implemented in `NAG/c02agf` (real coefficients) and in `NAG/c02aff` (complex coefficients).

14.3 Systems of Nonlinear Equations

In general, the numerical methods to solve *systems* of nonlinear equations

$$F(x) = 0$$

with $F : \mathbb{R}^n \to \mathbb{R}^n$ cannot be derived in a straightforward manner from the one-dimensional methods for the following reasons:

1. Certain methods (e. g., the bisection method) can be applied *only* to scalar equations.

2. Some one-dimensional methods (e. g., the secant method) can be extended to n dimensions in different ways.

3. There are classes of methods which are not known in the one-dimensional case. They are useful only for $n \geq 2$ (e. g., methods based on minimization algorithms).

4. Due to the computational cost required to solve a system of equations, the generalized one-dimensional methods may have been modified significantly for the sake of efficiency (e. g., by using the Broyden method).

5. The *scaling* of the problem becomes important for $n \geq 2$: As a rule, the variables $x_1, \ldots, x_n$ should be scaled in such a way that they are approximately of the same size.

14.3.1 Generalized Linear Methods

Systems of linear equations are usually written in the standard form

$$Ax = b. \tag{14.54}$$

(14.54) has to be transformed to a fixed-point representation in order to obtain an iterative method

$$x^{(k+1)} := T(x^{(k)}).$$

For systems of *linear* equations, this transformation is straightforward (see Chapter 16). For a system of nonlinear equations

$$f_1(x_1, \ldots, x_n) = 0$$
$$\vdots \qquad \vdots$$
$$f_n(x_1, \ldots, x_n) = 0$$

analogous transformations can be executed by solving nonlinear scalar equations. To do so every coordinate function f_i is solved with respect to x_i:

$$x_1 = t_1(x_2, x_3, \ldots, x_n)$$
$$\vdots \qquad \vdots$$
$$x_i = t_i(x_1, \ldots, x_{i-1}, x_{i+1}, \ldots, x_n) \qquad (14.55)$$
$$\vdots \qquad \vdots$$
$$x_n = t_n(x_1, \ldots, x_{n-2}, x_{n-1}).$$

This is possible, for example, if

$$\frac{\partial f_i}{\partial x_i}(x_1, \ldots, x_n) \neq 0, \qquad i = 1, 2, \ldots, n.$$

The successive transformation of the coordinate functions leads to the desired fixed-point representation $x = T(x)$. There are various algorithmic approaches to obtain (14.55):

The Nonlinear Jacobi Method

do $k = 0, 1, 2, \ldots$
    do for $i \in \{1, 2, \ldots, n\}$
        solve $f_i(x_1^{(k)}, \ldots, x_{i-1}^{(k)}, u, x_{i+1}^{(k)}, \ldots, x_n^{(k)}) = 0$ (*calculate* $u \in \mathbb{R}$);
        $x_i^{(k+1)} := u$
    end do
    if *termination criterion is met* **then exit**
end do

The Nonlinear Gauss-Seidel Method

do $k = 0, 1, 2, \ldots$
    do $i = 1, 2, \ldots, n$
        solve $f_i(x_1^{(k+1)}, \ldots, x_{i-1}^{(k+1)}, u, x_{i+1}^{(k)}, \ldots, x_n^{(k)}) = 0$ (*calculate* $u \in \mathbb{R}$);
        $x_i^{(k+1)} := u$
    end do
    if *termination criterion is met* **then exit**
end do

The Nonlinear SOR Method

do $k = 0, 1, 2, \ldots$
    do $i = 1, 2, \ldots, n$
        solve $f_i(x_1^{(k+1)}, \ldots, x_{i-1}^{(k+1)}, u, x_{i+1}^{(k)}, \ldots, x_n^{(k)}) = 0$ (*calculate* $u \in \mathbb{R}$);
        $x_i^{(k+1)} := x_i^{(k)} + \omega(u - x_i^{(k)})$
    end do
    if *termination criterion is met* **then exit**
end do

Of course these methods only converge if the function $F = (f_1, \ldots, f_n)^\top$ and the initial value $x^{(0)}$ satisfy certain conditions (cf. Section 14.1.2).

Detailed investigations of the convergence of generalized linear methods can be found, for example, in Ortega, Rheinboldt [67].

Generalized linear methods are very suitable for use on parallel computers (e. g., see Bertsekas, Tsitsiklis [34]); this is indicated, for instance, by the FORALL-loop (the i-loop) of the nonlinear Jacobi iteration.

14.3.2 Newton's Method

Newton's method also works with *systems* of equations. Firstly, the function $F : \mathbb{R}^n \to \mathbb{R}^n$ is replaced by suitable linear model functions

$$L_k(x) := a_k + J_k(x - x^{(k)}) \approx F(x), \qquad a_k \in \mathbb{R}^n, \quad J_k \in \mathbb{R}^{n \times n},$$

the roots x_k^* of which are used as approximations of the zero x^* of F. The coefficients (vectors and matrices) of these affine mappings are obtained, for example, by linearizing F using the Taylor expansion about $x^{(k)}$:

$$a_k := F(x^{(k)}), \qquad J_k := F'(x^{(k)}).$$

$J_k \in \mathbb{R}^{n \times n}$ is the *Jacobian matrix (functional matrix)* of F at $x^{(k)}$:

$$F'(x^{(k)}) = \begin{pmatrix} \dfrac{\partial f_1}{\partial x_1}(x^{(k)}) & \dfrac{\partial f_1}{\partial x_2}(x^{(k)}) & \cdots & \dfrac{\partial f_1}{\partial x_n}(x^{(k)}) \\[2mm] \dfrac{\partial f_2}{\partial x_1}(x^{(k)}) & \dfrac{\partial f_2}{\partial x_2}(x^{(k)}) & \cdots & \dfrac{\partial f_2}{\partial x_n}(x^{(k)}) \\[2mm] \vdots & \vdots & & \vdots \\[2mm] \dfrac{\partial f_n}{\partial x_1}(x^{(k)}) & \dfrac{\partial f_n}{\partial x_2}(x^{(k)}) & \cdots & \dfrac{\partial f_n}{\partial x_n}(x^{(k)}) \end{pmatrix}. \tag{14.56}$$

One iteration of Newton's method for nonlinear systems consists of the following substeps:

1. $L_k(x) = 0$ is solved, i. e., the solution of the system of linear equations with the matrix $J_k = F'(x^{(k)})$ and the right-hand side $-F(x^{(k)})$,

$$F'(x^{(k)})\Delta x^{(k)} = -F(x^{(k)}), \qquad (14.57)$$

is determined in order to obtain the update vector $\Delta x^{(k)} = x - x^{(k)}$.

2. $x^{(k+1)}$ is calculated:

$$x^{(k+1)} := x^{(k)} + \Delta x^{(k)}. \qquad (14.58)$$

Newton's method for solving nonlinear systems uses a linear model of F, which can be illustrated geometrically using tangent planes. The system $F(x) = 0$ consists of n equations

$$f_j(x) = 0, \qquad \text{where} \qquad f_j : \mathbb{R}^n \to \mathbb{R}, \quad j = 1, 2, \ldots, n.$$

$y = f_j(x)$ defines a *surface* in $\mathbb{R}^{n+1}$, whose *tangent (hyper-)plane* at the position $x^{(k)} \in \mathbb{R}^n$ is given by

$$f_j(x^{(k)}) + \sum_{i=1}^{n} \frac{\partial f_j}{\partial x_i}(x^{(k)})(x_i - x_i^{(k)}) = f_j(x^{(k)}) + f_j'(x^{(k)})(x - x^{(k)}). \qquad (14.59)$$

The equation

$$f_j(x^{(k)}) + f_j'(x^{(k)})(x - x^{(k)}) = 0$$

defines the intersection of the tangent hyperplane (14.59) and the hyperplane $y = 0$. This intersection is an affine subspace of dimension $n-1$. Thus, the system of linear equations (14.57) defines the intersection of n affine subspaces.

Note (Systems of Equations and Scalar Equations) For $n = 1$ Newton's method for solving *systems* of equations is reduced to Newton's method for solving *one* scalar equation. Nevertheless, there are many other n-dimensional methods which reduce to the one-dimensional Newton method when they are applied to one scalar equation, such as, for example,

$$x^{(k+1)} = x^{(k)} - F'(x^{(k)})^{-1}F(x^{(k)}) + (n - 1)\Phi(x^{(k)}),$$

where $\Phi : \mathbb{R}^n \to \mathbb{R}^n$ can denote *any* function. Newton's method, however, has a special position among these methods due to its convergence order $p \geq 2$.

Operation Counts

If the system of linear equations (14.57) is solved by applying one of the common direct methods (e. g., LU decomposition and back substitution), then the computation of *one* Newton iteration requires

$$
\begin{array}{ll}
n^2 + n & \text{component function evaluations,} \\
2n^3/3 + O(n^2) & \text{arithmetic operations, and} \\
n^2 + O(n) & \text{store operations.}
\end{array}
$$

In this list, the evaluation of $F'(x^{(k)})$ is counted as the evaluation of n^2 component functions. Thus, the component evaluations (which may comprise a large number of operations) and, above all, solving the system of linear equations incur the most significant costs.

Economizing the Computational Effort

A straightforward approach to reducing the number of operations schedules the evaluation of the Jacobian matrix J_k and its LU decomposition only every I steps:

> **do** $k = 0, 1, 2, \ldots$
> $u^{(1)} := x^{(k)}$;
>   **decompose** $J_k = F'(x^{(k)})$ (*evaluation and LU decomposition*)
>   **do** $i = 1, 2, \ldots, I$
>   **solve** $J_k \Delta u^{(i)} = -F(u^{(i)})$ (*back substitution*)
> $u^{(i+1)} := u^{(i)} + \Delta u^{(i)}$
>   **end do**
> $x^{(k+1)} := u^{(I+1)}$
> **end do**

This variation is based on the idea that $J(x^{(k)})$ is a good approximation for $J(u^{(i)})$ if the Jacobian matrix does not change too rapidly. Therefore, the simplified method can almost be considered to be Newton's method. However, this assumption is often wrong, especially if the distance from $x^{(k)}$ to x^* is too great. The simplified algorithm with $I > 1$ thus possibly leads to a *divergent* sequence $\{x^{(k)}\}$ while the sequence given by Newton's method ($I = 1$) may converge.

Another way to reduce the computational cost of Newton's method takes into account that the *exact* solution of the linear system of equations (14.57) is not required as long as $x^{(k)}$ is not close to x^*. The package NITSOL, for example, does not determine $\Delta x^{(k)}$ by using (14.57), but ensures that $\Delta x^{(k)}$ satisfies the inequality

$$\|F(x^{(k)}) + F'(x^{(k)})\Delta x^{(k)}\|_2 \leq \tau_k \|F(x^{(k)})\|_2 \qquad \text{with} \qquad \tau_k \in (0,1). \qquad (14.60)$$

An apt $\Delta x^{(k)}$ can be found by solving (14.57) using an iterative method that terminates as soon as (14.60) is satisfied.

Software (Large Systems of Equations)   The software package NITSOL was developed by H. F. Walker (Utah State University; E-mail: walker@math.usu.edu) particularly for solving very large systems of nonlinear equations.

The efficiency of Newton's method can be improved if the convergence rate is increased by taking into account certain second derivatives $\partial^2 F/\partial x_i \partial x_j$.

Software (Second Derivative Methods)   The software product TENSOLVE, developed by R. B. Schnabel (University of Colorado; E-mail: bobby@cs.colorado.edu), implements a tensor method which exceeds Newton's method in that second derivatives of F are also used.

The User's Contribution

When using Newton's method, the user has to write a subprogram that determines the values of the Jacobian matrix. This, however, makes the differentiation step more prone to errors.

Software (Jacobian Matrix) In order to check the correctness of a subprogram which evaluates a Jacobian matrix obtained by manual differentiation, the values returned by the program can be compared to the differential quotients of the component functions of F. For example, the program IMSL/MATH-LIBRARY/chjac executes such an automatic check.

A subprogram for evaluating F' can be generated automatically using automatic differentiation with appropriate software systems (e. g., MATHEMATICA).

Convergence Behavior

Any mathematical statement about the convergence behavior of Newton's method for solving *systems* of nonlinear equations requires a lot of mathematical tools. The more practical the statements are intended to be, the more sophisticated these tools must become. A detailed discussion of the convergence of Newton's method and its variants can be found in Ortega, Rheinboldt [67] or Schwetlick [73].

One of the simplest convergence theorems deals with regular zeros.

Definition 14.3.1 (Regular Zero) *Let $x^* \in \mathbb{R}^n$ be a zero of the function $F : \mathbb{R}^n \to \mathbb{R}^n$. If the Jacobian matrix $F'(x^*)$ is regular, and if F has continuous second partial derivatives in a neighborhood*

$$S_\varepsilon(x^*) := \{x \in \mathbb{R}^n : \|x - x^*\| \le \varepsilon\}$$

of x^*, then x^* is called a regular zero.

Theorem 14.3.1 *If the Newton iteration is started close enough to a regular zero x^*, then the Jacobian matrices $F'(x^{(k)})$, $k = 0, 1, 2, \ldots$, are regular, and the sequence $\{x^{(k)}\}$ converges quadratically to x^*, i. e.,*

$$\|x^{(k+1)} - x^*\| \le c \|x^{(k)} - x^*\|^2.$$

Proof: Ortega, Rheinboldt [67].

This convergence theorem is *local*, i. e., it is valid only for suitable starting vectors $x^{(0)}$. Nevertheless, a qualitative characterization of the convergence of Newton's method can be derived from this theorem: In a (possibly very small) neighborhood of a regular zero *quadratic* convergence always occurs.

The Extension of the Region of Attraction

Experience with systems of strongly nonlinear equations shows that the regions of attraction of the zeros are often very small when using Newton's method. Hence, the *damped Newton method* is used in practice. As opposed to the usual Newton step, the iteration

$$x^{(k+1)} := x^{(k)} + \lambda_k \Delta x^{(k)}, \qquad 0 < \lambda_k < 1, \tag{14.61}$$

is executed, where the damping factors λ_k are chosen in such a way that

$$\|F(x^{(k+1)})\| < \|F(x^{(k)})\|, \qquad k = 0, 1, 2, \ldots, \tag{14.62}$$

for some norm $\| \; \|$. If $F'(x)$ is regular at $x^{(k)}$, then for a sufficiently small λ_k the inequality (14.62) must hold. The choice of the step length in (14.61) can also be carried out according to the *trust region* principle (see Section 14.4.1). According to this principle, the value $x^{(k+1)}$ is restricted to a neighborhood of $x^{(k)}$ (*trust region*) in which the linear model function L_k represents an acceptable approximation of F (Sorensen [357], Dennis, Schnabel [43]).

Another modification of Newton's method deals with the system of linear equations that must be solved at each iteration. Instead of (14.57) the equation

$$\left(F'(x^{(k)}) + \lambda_k I \right) \Delta x^{(k)} = -F(x^{(k)})$$

is solved; again, the factors λ_k are chosen in such a way that (14.62) holds and the matrix $F'(x^{(k)}) + \lambda_k I$ is regular ($I \in \mathbb{R}^{n \times n}$ denotes the identity matrix).

14.3.3 The Secant Method

With the *discrete*, scalar Newton method, i.e., the scalar secant method (see Section 14.2.5), $x^{(k+1)}$ is obtained as the solution of the linear equation

$$\bar{l}_k(x) = f(x^{(k)}) + (x - x^{(k)}) \frac{f(x^{(k)} + h_k) - f(x^{(k)})}{h_k} = 0.$$

In this equation the linear function $\bar{l}_k$ can be interpreted in two ways:

1. $\bar{l}_k$ is an approximation of the tangent equation

$$l_k(x) = f(x^{(k)}) + (x - x^{(k)}) f'(x^{(k)});$$

2. $\bar{l}_k$ is the linear interpolation of f between the points $x^{(k)}$ and $x^{(k)} + h_k$.

By extending the scalar secant method to n dimensions, *different* methods for solving systems of nonlinear equations are obtained depending on the interpretation of $\bar{l}_k$. The first interpretation leads to the discrete Newton method, the second one to interpolation methods.

Definition 14.3.2 (Discrete Newton's Method)   *If Newton's method is modified in such a way that in equation (14.57) the Jacobian matrix $F'(x)$ is replaced by a (discrete) approximation $A(x, h)$, the discrete Newton method is obtained.*

The partial derivatives in the Jacobian matrix (14.56) are replaced by (forward) differences

$$A(x, h) e_i := [F(x + h_i e_i) - F(x)]/h_i, \qquad i = 1, 2, \ldots, n, \qquad (14.63)$$

where $e_i \in \mathbb{R}^n$ is the ith unit vector and $h_i = h_i(x)$ is the step length of the discretization. A possible choice of the step length is

$$h_i := \begin{cases} \varepsilon |x_i| & \text{if } x_i \neq 0, \\ \varepsilon & \text{otherwise,} \end{cases}$$

with $\varepsilon := \sqrt{eps}$, where *eps* denotes the machine epsilon of the floating-point number system $\mathbb{F}$ (see Section 4.7.2).

Linear Interpolation

In linear interpolation, each of the tangent planes (14.59) is replaced by a (hyper-)plane which interpolates the component function f_i at $n+1$ given points $x^{k,j}$, $j = 0, 1, \ldots, n$, located in a neighborhood of $x^{(k)}$. Vectors $a^{(i)}$ and scalars α_i are thus chosen in such a way that for

$$L_i(x) := \alpha_i + a^{(i)\top} x, \qquad i = 1, 2, \ldots, n \qquad (14.64)$$

the following relations hold:

$$L_i(x^{k,j}) = f_i(x^{k,j}), \qquad i = 1, 2, \ldots, n, \quad j = 0, 1, \ldots, n.$$

The next iterate $x^{(k+1)}$ is obtained by intersecting the n hyperplanes (14.64) and the hyperplane $y = 0$ in $\mathbb{R}^{n+1}$. $x^{(k+1)}$ is the solution of the system of linear equations

$$L_i(x) = 0, \qquad i = 1, 2, \ldots, n. \qquad (14.65)$$

Depending on the selection of the interpolation points $x^{k,j}$, numerous different methods are derived.

Brown's Method and Brent's Method

In numerical programs two variants of the higher-dimensional secant methods are usually implemented: *Brown's method* and *Brent's method*. Brown's method (see, for example, the article of K. M. Brown in Byrne, Hall [133]) combines the processes of approximating F' and that of solving the system of linear equations (14.65) using Gaussian elimination. Brent's method [124] uses a QR factorization for solving the equations (14.65). Both methods are quadratically convergent (like Newton's method) but require only $(n^2 + 3n)/2$ function evaluations per iteration.

In a comparative study Moré and Cosnard [293] found that Brent's method is often better than Brown's method, and that the discrete Newton method is usually the most efficient if the F-evaluations do not require too much computational effort.

Software (Brent's Method, Brown's Method) Brent's method and Powell's hybrid algorithm (see Section 14.4.3) are implemented in MINPACK.

An inefficient implementation of Brown's method (with $O(n^4)$ arithmetic operations per step) was removed from a former IMSL release, and was replaced by two programs based on Powell's algorithm (IMSL/MATH-LIBRARY/neqnf and IMSL/MATH-LIBRARY/neqnj).

14.3.4 Modification Methods

Iterative methods of exceptional efficiency can be constructed by using an approximation A_k of $F'(x^{(k)})$ which is derived from the preceding approximation A_{k-1} by a *rank-1 modification*, i.e., by adding a matrix of rank 1:

$$A_{k+1} := A_k + u^{(k)} [v^{(k)}]^\top, \qquad u^{(k)}, v^{(k)} \in \mathbb{R}^n, \qquad k = 0, 1, 2, \ldots.$$

According to the *Sherman-Morrison formula* (Ortega, Rheinboldt [67])

$$(A + uv^\top)^{-1} = A^{-1} - \frac{1}{1 + v^\top A^{-1} u} A^{-1} uv^\top A^{-1},$$

the recursion

$$B_{k+1} := B_k - \frac{B_k u^{(k)} [v^{(k)}]^\top B_k}{1 + [v^{(k)}]^\top B_k u^{(k)}}, \qquad k = 0, 1, 2, \ldots,$$

holds for $B_{k+1} := A_{k+1}^{-1}$, provided that $1 + [v^{(k)}]^\top A_k^{-1} u^{(k)} \neq 0$. Thus, the necessity of solving a system of linear equations in every iteration is avoided; a matrix-vector multiplication operation suffices. Accordingly, a reduction of the computational effort from $O(n^3)$ to $O(n^2)$ is achieved. There is, however, a major disadvantage: The convergence is no longer quadratic (as it is in Newton's method and the methods of Brent and Brown); it is only *superlinear*:

$$\lim_{k \to \infty} \frac{\|x^{(k+1)} - x^*\|}{\|x^{(k)} - x^*\|} = 0. \tag{14.66}$$

Broyden's Method

The vectors $u^{(k)}$ and $v^{(k)}$ can be chosen using the principle of the secant approximation, as explained in the following. If f is a scalar function, then the differential quotient a_k used in the secant method as an approximation for $f'(x^{(k)})$ is uniquely defined by

$$a_{k+1}(x^{(k+1)} - x^{(k)}) = f(x^{(k+1)}) - f(x^{(k)}).$$

However, for $n > 1$ the matrix $A_{k+1} \in \mathbb{R}^{n \times n}$ is not unique if defined by

$$A_{k+1}(x^{(k+1)} - x^{(k)}) = F(x^{(k+1)}) - F(x^{(k)}). \tag{14.67}$$

This equation is called the *quasi-Newton equation* since (14.67) also holds for every other matrix of the form

$$\bar{A}_{k+1} := A_{k+1} + pq^\top$$

where $p, q \in \mathbb{R}^n$ and $q^\top(x^{(k+1)} - x^{(k)}) = 0$. On the other hand,

$$y_k := F(x^{(k)}) - F(x^{(k-1)}) \qquad \text{and} \qquad s_k := x^{(k)} - x^{(k-1)}$$

only convey information about the partial derivative of F in the direction of s_k, not about the partial derivative in directions orthogonal to s_k. A_{k+1} should thus be equivalent to A_k in that

$$A_{k+1}q = A_k q \qquad \text{for all} \quad q \in \{v : v \neq 0; \; v^\top s_k = 0\}. \tag{14.68}$$

Starting from an initial approximation $A_0 \approx F'(x^{(0)})$ (which can be obtained, for example, by using the differential quotients (14.63)), the sequence $A_1, A_2, \ldots$ is uniquely determined by (14.67) and (14.68) (Broyden [129], Dennis, Moré [156]).

For the corresponding sequence $B_0 = A_0^{-1} \approx [F'(x^{(0)})]^{-1}$, $B_1, B_2, \ldots$ the Sherman-Morrison formula can be used to obtain the recursion

$$B_{k+1} := B_k + \frac{(s_{k+1} - B_k y_{k+1})s_{k+1}^\top B_k}{s_{k+1}^\top B_k y_{k+1}}, \qquad k = 0, 1, 2, \ldots,$$

which only requires matrix-vector multiplication operations and thus only $O(n^2)$ computational work. With the matrices B_k the iteration

$$x^{(k+1)} := x^{(k)} - B_k F(x^{(k)}), \qquad k = 0, 1, 2, \ldots,$$

can be defined; it is called *Broyden's method*. This method converges superlinearly in terms of (14.66) if the update vectors s_k converge (as $k \to \infty$) to the update vectors of the Newton method. Thus, Broyden's method is a good example of the significance of local linearization for solving nonlinear equations.

Software (Broyden's Method) Broyden's method is employed in many software products, often without being mentioned explicitly. Moreover, it is part of Powell's polyalgorithm (see Section 14.4.3), which is implemented in many software products, e. g., IMSL/MATH-LIBRARY/neqnf.

14.3.5 Large Nonlinear Systems

Solving large systems of nonlinear equations, i. e., systems of the form $F(x) = 0$, where $F : \mathbb{R}^n \to \mathbb{R}^n$ and $n \gg 1$, is a very hard problem. Large nonlinear systems can be solved only with difficulty because the determination of suitable initial values (see Section 14.1.4) is very complicated and because of the very high overall computational cost.

The difficulty of finding a suitable starting value of the iteration can be avoided in those exceptional cases where good a priori estimates of x^* are available.

Example (Stiff Ordinary Differential Equations) The initial value problem for *stiff differential equations* (a special class of ordinary differential equations) can be solved numerically by using an implicit method. In each step of such an algorithm, a system of nonlinear equations has to be solved. To that end, a good initial approximation is available from the previous step. Special software packages for solving stiff differential equations are e. g., LARKIN (by Deuflhard, Bader, Nowak) or FACSIMILE (by A. Curtis).

Special (direct or iterative) methods developed for large systems of linear equations with a sparse matrix can be applied to large systems of nonlinear equations with a sparse Jacobian matrix F' (see Chapter 16). The sparsity structure of F' does not change when using the discrete Newton method (see e. g., Powell, Toint [318]).

Software (Large Nonlinear Systems) Software for solving large systems of nonlinear equations and nonlinear data fitting problems can be found in the Harwell Subroutine Library. The program HARWELL/ns02 is an implementation of the Powell method (see Section 14.4.3) for sparse Jacobian matrices. HARWELL/ns03 solves systems of nonlinear equations of the special form

$$F(x) + Ax = 0$$

where the Jacobian matrix F' and the (constant) matrix A are sparse. It also performs data fitting using the Levenberg-Marquardt method (see Section 14.4.2).

The software package NITSOL, already mentioned on page 318, was developed particularly for solving systems of nonlinear equations with large numbers of unknowns.

There is a version of the package TENSOLVE (see page 318) used for efficiently solving large systems of nonlinear equations with a sparse Jacobian matrix.

14.4 Nonlinear Data Fitting

In general, an overdetermined system of nonlinear equations—in which the number m of equations is larger than the number n of unknowns—has no solution. The only possibility is to look for vectors $x^* \in \mathbb{R}^n$ which comply with $F(x) = 0$ "as well as possible", i. e.,

$$\|F(x^*)\|_p = \min\{\|F(x)\|_p : x \in D\}. \tag{14.69}$$

Thus, instead of solving $F(x) = 0$, a *nonlinear data fitting problem* is solved. For overdetermined systems of equations, the minimization of the residual (14.69) is often equivalent to the estimation of a parameter vector in a nonlinear model.

In principle, the minimization problem (14.69) can be solved using one of the minimization methods from Section 14.4.1. However, there are many special algorithms for nonlinear data fitting with respect to the squared Euclidean norm $\| \ \|_2^2$, i. e., for the *least squares method*

$$\|F(x)\|_2^2 = F^\top F(x) = \sum_{i=1}^{m} [f_i(x)]^2.$$

As an example, the widely used Gauss-Newton method is discussed below.

Algorithms for the minimization with respect to other l_p-norms (where $p \neq 2$) can be found, for example, in Fletcher, Grant, Hebden [190].

The Gauss-Newton Method

In order to obtain an iterative method for the minimization of $\|F(x)\|_2^2$, the function F can be replaced, as in Newton's method, by linear functions

$$L_k(x) := F(x^{(k)}) + F'(x^{(k)})(x - x^{(k)}).$$

The next iterate $x^{(k+1)}$ is determined using *linear* data fitting:

$$\|L_k(x)\|_2^2 = L_k^\top L_k(x) \longrightarrow \min. \tag{14.70}$$

This is called the *Gauss-Newton method*.

Provided that the rectangular Jacobian matrix

$$J_k := F'(x^{(k)}) = \left(\frac{\partial f_i}{\partial x_j}(x^{(k)}), \ i = 1, 2, \ldots, m, \ j = 1, 2, \ldots, n\right)$$

is of full rank—i. e., rank$(J_k) = n$—then the linear data fitting problem (14.70) has a unique solution that can be determined using special methods. For example, the vector $s_k := x^{(k+1)} - x^{(k)}$ can be determined from the equations

$$(J_k^{\top} J_k)s_k = -J_k^{\top} F(x^{(k)}) \tag{14.71}$$

using Cholesky decomposition or QR decomposition.

As with the minimization methods (see Section 14.4.1), the class of Gauss-Newton methods can be extended by the *trust region* principle. To this end the step length is limited to a neighborhood of $x^{(k)}$ where the linear model L_k is a sufficiently accurate approximation for F. The relevant quality criterion is the reduction of the target function:

$$\|L_k(x^{(k)})\|_2 - \|L_k(x^{(k+1)})\|_2 \approx \|F(x^{(k)})\|_2 - \|F(x^{(k+1)})\|_2.$$

Software (The Gauss-Newton Method)   The Gauss-Newton algorithm—with extensions to enhance its reliability and efficiency—is the basis for nonlinear data fitting programs in the NAG library, the MATLAB optimization toolbox, the OPTIMA library, and the software packages TENSOLVE and DFNLP (Moré, Wright [20]).

14.4.1 Minimization Methods

Newton's method and its variants converge rapidly as soon as the iterates $x^{(k)}$ are sufficiently close to the solution x^*. At larger distances from x^* the convergence is often unsystematic and slow; moreover, the determination of a starting point $x^{(0)}$ located in the region of attraction of x^* may be extremely difficult.

Algorithms used to minimize $\|F\|$ (discussed in this section) have much larger regions of attraction but converge only *linearly*. They are therefore often used in polyalgorithms to determine an intermediate approximation, which serves as a starting point of a more rapidly converging Newton variant.

The mathematical problem discussed in this section is the *minimization* of functions $f : \mathbb{R}^n \to \mathbb{R}$, i.e., the search for points $x^* \in \mathbb{R}^n$ where f has a minimum.

Definition 14.4.1 (Minimum of a Function)   *If in an open neighborhood* $K_\varepsilon(x^*) = \{x : \|x - x^*\| < \varepsilon\}$ *of* x^*

$$f(x^*) < f(x) \qquad \text{for all} \quad x \in K_\varepsilon(x^*) \setminus \{x^*\}, \tag{14.72}$$

then $f : \mathbb{R}^n \to \mathbb{R}$ *has a strong local minimum at the point* $x^* \in \mathbb{R}^n$. *If the strict inequality in (14.72) is replaced by* $f(x^*) \le f(x)$, *then* x^* *is said to be a weak local minimum of* f. *If (14.72) holds in the entire domain of* f *(e. g.,* $D = \mathbb{R}^n$*), then* x^* *is referred to as a global minimum of* f *(see Fig. 14.12).*

All considerations remain valid for maxima in an analogous way if f is replaced by $-f$. Accordingly, attention is restricted to minimization.

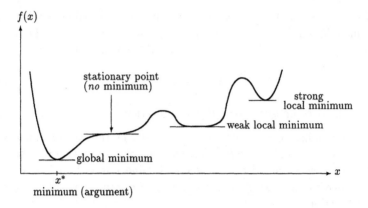

Figure 14.12: Different types of minima of a function $f : \mathbb{R} \to \mathbb{R}$.

Terminology (Minimum)   The expression *minimum* is used not only for the minimal value of a function, but also for the point $x^* \in \mathbb{R}^n$ at which the corresponding *function value* $f(x^*)$ is minimal. The more precise term *local/global minimizer* could be used as an alternative as well. Both terms are common.

The minimization of functions and the determination of zeros of nonlinear equations are linked by the fact that each solution x^* of $F(x) = 0$ is a minimum of the function

$$\|F\| : \mathbb{R}^n \to \mathbb{R}.$$

Thus, the zeros of $F = (f_1, \ldots, f_n)^\top$ correspond to the *global* minima of the function

$$\Phi(x) := \|F(x)\|_2^2 = [F(x)]^\top F(x) = \sum_{i=1}^{n} [f_i(x_1, \ldots, x_n)]^2. \qquad (14.73)$$

The opposite is not true! Not every *local* minimum of (14.73) corresponds to a zero of F. This fact has to be considered whenever a minimization method is used for determining zeros, because (except for special cases) *all* minimization methods used in practice are only capable of finding *local* minima.

   If a differentiable function $f : \mathbb{R} \to \mathbb{R}$ has a minimum at $x^* \in \mathbb{R}$, then its tangent is horizontal there, i.e., $f'(x^*) = 0$. However, this is only a *necessary* but not a *sufficient* condition for extrema (see Fig. 14.12). The criterion of a horizontal tangent can be generalized to the higher-dimensional case $f : \mathbb{R}^n \to \mathbb{R}$ using the gradient $\nabla f : \mathbb{R}^n \to \mathbb{R}^n$.

Definition 14.4.2 (The Gradient of a Function)   *If* $f : \mathbb{R}^n \to \mathbb{R}$ *is a differentiable function then*

$$\nabla f(x) := \left(\frac{\partial f}{\partial x_1}(x), \frac{\partial f}{\partial x_2}(x), \ldots, \frac{\partial f}{\partial x_n}(x) \right)^\top$$

is called the gradient of f at the point x.

Notation (Gradient) As is often found in the literature on nonlinear optimization problems, the notation $g(x)$ will also sometimes be used for the gradient $\nabla f(x)$ in this book.

A differentiable function $f : \mathbb{R}^n \to \mathbb{R}$ has a minimum at x^* only if the gradient function vanishes there, i.e.,

$$\nabla f(x^*) = 0.$$

At every local minimum $\bar{x}$ of a function $\Phi := F^{\mathsf{T}} F$ the relation $\nabla \Phi(\bar{x}) = 0$ holds:

$$\nabla \Phi(\bar{x}) = \nabla (F^{\mathsf{T}} F)(\bar{x}) = 2 J^{\mathsf{T}}(\bar{x}) F(\bar{x}) = 0. \tag{14.74}$$

As a result, at every local minimum of $F^{\mathsf{T}} F$, where $\bar{x}$ is *not* a zero $(F(\bar{x}) \neq 0)$, the Jacobian matrix $J(\bar{x})$ must be *singular*.

Classification of Minimization Methods

Minimization methods can be classified according to the kind of information they use about the function $f : \mathbb{R}^n \to \mathbb{R}$ which is to be minimized:

1. Only function values $f(x)$ are used; such minimization methods are also suitable for non-smooth functions.

2. Function values and values of the gradient function $g(x) := \nabla f(x)$ are used.

3. Function values, gradient values and values of the second derivatives, i.e., of the *Hessian matrix* $H_f := \nabla^2 f = g'(x)$,

$$H_f(x) := \begin{pmatrix} \dfrac{\partial^2 f}{\partial x_1 \partial x_1}(x) & \dfrac{\partial^2 f}{\partial x_1 \partial x_2}(x) & \cdots & \dfrac{\partial^2 f}{\partial x_1 \partial x_n}(x) \\[2ex] \dfrac{\partial^2 f}{\partial x_2 \partial x_1}(x) & \dfrac{\partial^2 f}{\partial x_2 \partial x_2}(x) & \cdots & \dfrac{\partial^2 f}{\partial x_2 \partial x_n}(x) \\[2ex] \vdots & \vdots & & \vdots \\[2ex] \dfrac{\partial^2 f}{\partial x_n \partial x_1}(x) & \dfrac{\partial^2 f}{\partial x_n \partial x_2}(x) & \cdots & \dfrac{\partial^2 f}{\partial x_n \partial x_n}(x) \end{pmatrix}$$

are used.

Direct search methods are minimization methods that only use function values; no approximations of ∇f or H_f (e.g., using difference quotients) are computed. These methods converge slowly and are inefficient. In practice they are only used for problems with small n.

Uniform and sequential random search methods as well as the nonlinear simplex method[6] of Nelder and Mead [297] belong to this category.

---

[6]The *nonlinear* simplex method is used for minimizing nonlinear functions; it has nothing in common with the simplex algorithm in linear programming.

Software (Nonlinear Simplex Method)   The nonlinear simplex method of Nelder and Mead is implemented, for example, in the program IMSL/MATH-LIBRARY/bcpol. It is particularly useful if ∇f cannot be determined without difficulty or if f is subject to random perturbations.

Gradient methods are minimization methods that utilize explicit information about the gradient ∇f (e.g., in the form of a subprogram) in addition to function values. The application of gradient methods makes sense only if $f \in C^1$. Gradient methods build a locally linear model of f (valid in a small neighborhood of the current approximation of the desired minimum), which is used to derive the direction in which the next iterate is sought.

Quasi Newton methods approximate $\nabla^2 f$ using values f and ∇f. They produce a sequence of *quadratic models* of f, the respective minima of which constitute the sequence of iterates.

Newton's method and its variants require the explicit knowledge of f, ∇f, *and* $\nabla^2 f$. Of course, the appropriate differentiability of f is a prerequisite. The minimum of the quadratic function

$$q_k(s) := f(x^{(k)}) + \nabla f(x^{(k)})^\top s + \frac{1}{2} s^\top \nabla^2 f(x^{(k)}) s,$$

modeling the function f at the point $x^{(k)}$, defines the next iterate $x^{(k+1)}$. Given that the Hessian matrix $H_f(x^{(k)}) = \nabla^2 f(x^{(k)})$ is positive definite, the function $q_k : \mathbb{R}^n \to \mathbb{R}$ has a unique global minimum that can be computed by solving the system of linear equations

$$\nabla^2 f(x^{(k)}) s_k = -\nabla f(x^{(k)}).$$

The next iterate is then given by $x^{(k+1)} := x^{(k)} + s_k$.

In the case of problems where $m = n$, i.e., systems of nonlinear equations, Newton's method can also be applied to $F(x) = 0$ directly. For data fitting problems ($m > n$) Newton's method or one of its variants is particularly attractive if the second derivative of f is easily obtained.

Gradient Methods

Since *sufficient* criteria for the existence of a local minimum (e.g., $\nabla f(x^*) = 0$ and $H_f(x^*)$ positive definite, if $f \in C^2$) are difficult to check, it is common to search only for points x^* that satisfy the *necessary* criterion of stationarity.

Definition 14.4.3 (Stationary Point) x^* *is called a critical point or stationary point of f if the gradient ∇f vanishes at x^*:*

$$\nabla f(x^*) = 0.$$

In general, a practical minimization method should yield a sequence $\{x^{(k)}\}$ whose corresponding function values gradually decrease:

$$f(x^{(0)}) > f(x^{(1)}) > f(x^{(2)}) > \cdots .$$

The question therefore arises as to in which directions from $x^{(k)}$ points are located, whose function values are smaller than $f(x^{(k)})$? The expansion

$$f(x + hd) = f(x) + h\langle g(x), d\rangle + o(h)$$

with $h \in \mathbb{R}$, $d \in \mathbb{R}^n$, $\|d\| = 1$, $g(x) := \nabla f(x)$ and the Euclidean inner product $\langle , \rangle$, shows that the *local variation* of f in a neighborhood of x is largest if the direction vector d is a multiple of g, the gradient of f. Therefore the local direction of steepest descent in the Euclidean norm is $-g(x)$, the negative of the gradient vector. This fact was first recognized by Cauchy 150 years ago. He used the gradient direction in his method of steepest descent.

Definition 14.4.4 (Method of Steepest Descent)   *The gradient method*

$$x^{(k+1)} := x^{(k)} - h_k g(x^{(k)}), \qquad k = 0, 1, 2, \ldots ,$$

for which the step length h_k is chosen such that

$$f(x^{(k)} - h_k g(x^{(k)})) \leq f(x^{(k)} - hg(x^{(k)})) \qquad \text{for all} \quad h \in (0, \bar{h}],$$

is called the method of steepest descent.

Every step length h_k is determined by a *one-dimensional* minimization in the negative gradient direction $-g(x^{(k)})$ starting from $x^{(k)}$.

The advantage of this method is that the region of attraction of the minimum x^* is often very large; whence convergence may occur even for unfavorable initial points. The disadvantage of this method is its very slow convergence and, as a consequence, its high computational effort (which is mainly required for the evaluation of the functions f and g).

Methods with a Variable Metric

The direction of steepest descent clearly depends on the inner product $\langle , \rangle$. If through a symmetric positive definite matrix B a new inner product

$$\langle u, v \rangle_B := \langle u, Bv \rangle = \sum_{i=1}^n u_i (Bv)_i$$

is introduced (and hence also a new norm and a new *metric*), then

$$
\begin{aligned}
f(x + hd) &= f(x) + h\langle g(x), d\rangle + o(h) \\
&= f(x) + h\langle B^{-1}g(x), Bd\rangle + o(h) \\
&= f(x) + h\langle B^{-1}g(x), d\rangle_B + o(h).
\end{aligned}
$$

Now the vector $-B^{-1}g$ specifies the direction of steepest descent with respect to the new inner product $\langle\,,\,\rangle_B$. As a result, every method of the form

$$x^{(k+1)} := x^{(k)} + h_k d_k,$$

whose direction vectors d_k satisfy

$$\langle d_k, g(x^{(k)}) \rangle < 0,$$

i.e., guaranteeing locally decreasing function values, can be considered as a method of steepest descent with respect to some inner product $\langle\,,\,\rangle_{B_k}$. This inner product changes from iterate to iterate and so does the symmetric positive definite matrix B_k.

Such minimization methods are thus called *methods with variable metric*. With $A_k := B_k^{-1}$, their iteration rule is

$$x^{(k+1)} := x^{(k)} - h_k A_k g(x^{(k)}), \qquad A_k \text{ symmetric positive definite.}$$

The quality of a minimization method can be assessed with model test functions. The simplest test functions $f : \mathbb{R}^n \to \mathbb{R}$ with a finite minimum x^* are the quadratic functions

$$f(x) := \langle (x - x^*), B(x - x^*) \rangle. \tag{14.75}$$

The level sets of these functions are (hyper-) ellipsoids,

$$\{x : \langle (x - x^*), B(x - x^*) \rangle = \text{ const} \}$$

and, with respect to the inner product $\langle x, y \rangle_B := \langle x, By \rangle$, the level sets

$$\{x : \langle (x - x^*), B(x - x^*) \rangle = \langle x - x^*, x - x^* \rangle_B = \text{ const} \}$$

are spheres. No matter where the initial value $x^{(0)}$ is chosen, the direction of steepest descent always points towards the center x^* of the sphere. Which method with variable metric corresponds to this particular case? With

$$A_k = B^{-1} = [g']^{-1} = H_f^{-1},$$

the sought method turns out to be *Newton's method*

$$x^{(k+1)} := x^{(k)} - h_k [g'(x^{(k)})]^{-1} g(x^{(k)})$$

for $g(x) = 0$. With step size $h_0 = 1$, the minimum x^* of (14.75) can be reached in a single step from any initial point $x^{(0)}$.

Thus, for quadratic functions, Newton's method is the most efficient variable metric method for solving $g(x) = 0$. This is due to the explicit use of $g' = H_f$.

The Quasi-Newton Method

Generally, methods with variable metric which have access only to the values of f and ∇f need more than one step to find the minimum of a quadratic function (14.75). Methods which try to approximate the inverse of the Hessian matrix $H_f(x^{(k)})$ as accurately as possible, in an attempt to reduce the number of iterates, are called *quasi-Newton methods*. Most efficient are those with

$$A_k \to H_f^{-1}(x^{(k)}) \qquad \text{as } k \to \infty.$$

Conjugate Gradient (CG) Methods

The highest efficiency possible using the values of f and ∇f is achieved by algorithms that reach the point x^* of the quadratic functions (14.75) after no more than n steps. The behavior of such algorithms is called *quadratic convergence*, which must not be mistaken for *convergence of order $p=2$*.

The basic principle of such algorithms is geometrically obvious. It is clear that for a diagonal quadratic function, i.e., a function (14.75) with

$$B := \operatorname{diag}(\lambda_1, \lambda_2, \ldots, \lambda_n), \quad \lambda_i > 0,$$

successive one-dimensional minimization along the n orthogonal coordinates will produce the minimum after n steps at most (see Fig. 14.13). Attention must be paid to the fact that an explicit knowledge of B is *not* required!

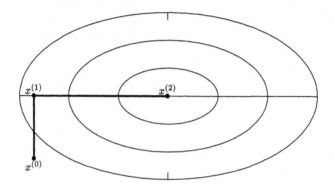

Figure 14.13: Minimization of a diagonal quadratic function along the coordinate axes.

For general quadratic functions (14.75), a coordinate transformation $x := Pu$ yields a diagonalization

$$P^\mathsf{T} B P = \operatorname{diag}(\lambda_1, \lambda_2, \ldots, \lambda_n). \tag{14.76}$$

The function

$$\begin{aligned}
\bar{f}(u) &:= f(Pu) \\
&= \langle (Pu - x^*), B(Pu - x^*) \rangle \\
&= u^\mathsf{T} P^\mathsf{T} B P u - {x^*}^\mathsf{T} B P u - u^\mathsf{T} P^\mathsf{T} B x^* + {x^*}^\mathsf{T} B x^*
\end{aligned}$$

has a diagonal quadratic part

$$u^\mathsf{T} P^\mathsf{T} B P u = u^\mathsf{T} \operatorname{diag}(\lambda_1, \lambda_2, \ldots, \lambda_n) u,$$

i.e., successive one-dimensional minimization of $\bar{f}$ along the directions of the coordinate axes will produce the minimum x^* in, at most, n steps. The search for the minimum x^* along the coordinate axes e_j for $\bar{f}$ corresponds to the search along the column vectors p_j of P for f:

$$\begin{aligned}
\bar{f}(u - he_j) &= f(P(u - he_j)) \\
&= f(Pu - hPe_j) \\
&= f(x - hp_j).
\end{aligned}$$

According to (14.76) these vectors satisfy

$$p_j^\top B p_i = 0 \qquad \text{for all} \quad i \neq j,$$

i.e., the vectors $p_1, \ldots, p_n$ are *orthogonal* with respect to the inner product $\langle\,,\,\rangle_B := \langle\,,\,B\rangle$. These vectors are therefore said to be B-orthogonal or *conjugate vectors*, and the minimization methods described above are referred to as *conjugate direction methods*.

The biggest difficulty in implementing conjugate direction methods is the procurement of conjugate vectors. They could be obtained, for example, by applying an orthogonalization procedure. However, this would require explicit knowledge of the matrix B.

With an initial point $x^{(0)}$ and a direction of steepest descent p_0, the direction p_1 can be obtained as a linear combination of p_0 and the gradient direction $g(x^{(1)})$ such that p_0 and p_1 constitute conjugate directions with respect to the matrix B:

$$p_0 = g(x^{(0)}) = Bx^{(0)} - b, \qquad x^{(1)} = x^{(0)} - \alpha_0 p_0.$$

Here the step length α_0 is the result of a one-dimensional minimization with respect to α such that

$$f(x^{(0)} - \alpha_0 p_0) \leq f(x^{(0)} - \alpha p_0).$$

The algorithm is continued with

$$p_1 = g(x^{(1)}) - \beta_0 p_0, \qquad x^{(2)} = x^{(1)} - \alpha_1 p_1, \ \ldots \,.$$

Methods of this kind, which may differ in the selection of the weights β_k, are called *conjugate gradient methods* (CG methods).

The Fletcher-Reeves Method

The following method was derived by Fletcher and Reeves [191]:

$p_0 := g(x^{(0)})$
do $k = 0, 1, 2, \ldots$
    compute α_k such that $f(x^{(k)} - \alpha_k p_k) \leq f(x^{(k)} - \alpha p_k)$;
    $x^{(k+1)} := x^{(k)} - \alpha_k p_k$;
    $\beta_k := -\|g(x^{(k+1)})\|_2^2 / \|g(x^{(k)})\|_2^2$
    $p_{k+1} := g(x^{(k+1)}) - \beta_k p_k$;

    if *termination criterion is met* **then exit**
end do

In this algorithm the matrix B is *not* explicitly referred to. The Fletcher-Reeves method can thus be applied to arbitrary (sufficiently differentiable) nonlinear functions.

In practical implementations, the method is restarted after n iterations in order to improve efficiency; for this purpose simply $\beta_n := 0$, $\beta_{2n} := 0, \ldots$ is set.

One important advantage of the Fletcher-Reeves method is its modest storage requirement, which is made possible because no approximation of H_f^{-1} has to be stored. This method is thus also appropriate for the solution of very large problems.

The Trust Region Principle

In many minimization methods, the function $f : \mathbb{R}^n \to \mathbb{R}$ which is to be minimized, is locally replaced by a *quadratic* model function $q_k : \mathbb{R}^n \to \mathbb{R}$ (valid near the current iterate $x^{(k)}$). The *trust region principle* requires that the next iterate $x^{(k+1)}$ satisfies the following two conditions:

$$
\begin{aligned}
&1. \quad x^{(k+1)} \text{ minimizes the model function } q_k, \text{ and} \\
&2. \quad \|x^{(k+1)} - x^{(k)}\| \le \delta_k.
\end{aligned}
\tag{14.77}
$$

The second condition ensures that the minimization steps do not leave the *trust region*; i.e., the domain in which q_k represents a sufficiently accurate approximation of f. The *trust radius* δ_k is kept as long as the relation

$$
f(x^{(k)}) - f(x^{(k+1)}) \approx q_k(x^{(k)}) - q_k(x^{(k+1)})
$$

ensures that q_k is a useful model function. Otherwise δ_k is made smaller and a new iterate $x^{(k+1)}$ is computed.

Software (Trust Region Principle)   All MINPACK subprograms for the minimization of nonlinear functions are based on the trust region principle.

The spheric trust region (14.77) is inappropriate for solving ill-scaled problems. Therefore, in practice, the trust region

$$
\|D_k(x^{(k+1)} - x^{(k)})\| \le \delta_k
$$

is used, where $D_k \in \mathbb{R}^{n \times n}$ is a regular *scaling matrix*. D_k is often chosen to be a diagonal matrix (Dennis, Schnabel [43]).

14.4.2 The Levenberg-Marquardt Method

As Newton's method often requires the initial application of a more slowly converging method with a larger region of attraction, the same may be necessary for the Gauss-Newton method. In such an initial phase, minimization along the gradient direction

$$
d_{k+1}^g := -J_k^\top F(x^{(k)})
$$

may be useful. As soon as the iterates come close to the solution x^*, a switch to the Gauss-Newton direction

$$d_{k+1}^{\mathrm{N}} := -(J_k^{\mathrm{T}} J_k)^{-1} J_k^{\mathrm{T}} F(x^{(k)})$$

(derived from (14.71)) takes place. The *Levenberg-Marquardt method* (see, for example, Marquardt [283]) uses the direction vector

$$d_{k+1}^{\mathrm{LM}} := -(J_k^{\mathrm{T}} J_k + \lambda I)^{-1} J_k^{\mathrm{T}} F(x^{(k)}),$$

where $d_{k+1}^{\mathrm{LM}} = d_{k+1}^{\mathrm{N}}$ for $\lambda = 0$, and $d_{k+1}^{\mathrm{LM}} \to d_{k+1}^{\mathrm{g}}$ as $\lambda \to \infty$. The method starts with a large value of λ, which gradually diminishes once a suitable neighborhood of the solution has been reached.

The step length h_k in

$$x^{(k+1)} := x^{(k)} + h_k d_{k+1}^{\mathrm{LM}}$$

is chosen so that a persistent reduction of the residuals is guaranteed:

$$\|F(x^{(k+1)})\|_2 < \|F(x^{(k)})\|_2, \qquad k = 0, 1, 2, \ldots.$$

Software (Levenberg-Marquardt Method) The ODRPACK programs (which is available via NETLIB) use the Levenberg-Marquardt method combined with the trust region principle.

14.4.3 The Powell Method

Many computer programs for the solution of systems of nonlinear equations are based on a polyalgorithm of Powell [317].

After an initial phase, in which an initial approximation $A_0 \approx F'(x^{(0)})$ and its inverse $B_0 = A_0^{-1}$ are determined using difference quotients (14.63), both the quasi Newton direction

$$d_{k+1}^{\mathrm{N}} := -B_k F(x^{(k)})$$

and the direction of the steepest descent

$$d_{k+1}^{\mathrm{g}} := -A_k^{\mathrm{T}} F(x^{(k)})$$

(cf. (14.74)) are determined at every step. The direction that is actually used is d_{k+1}, a *linear combination* of d_{k+1}^{N} and d_{k+1}^{g}, whose weights are chosen depending on the behavior of the sequence of residuals

$$\|F(x^{(0)})\|, \ldots, \|F(x^{(k-1)})\|, \|F(x^{(k)})\|.$$

In the first phase of the iteration, the direction d_{k+1} is chosen closer to the gradient direction than to the quasi-Newton direction and in close vicinity to x^*, the quasi-Newton direction is preferred.

The matrices A_{k+1} and B_{k+1} are determined by Broyden's method (see Section 14.3.4). Particular care is taken in the Powell algorithm that these matrices do *not* become rank-deficient (unless this is a property of the problem itself) to ensure that the iterates $\{x^{(i)}\}$ do not get stuck in some unsuitable subspace.

Software (Powell Method) There are many implementations of the Powell method available, e. g., IMSL/MATH-LIBRARY/neqnf, IMSL/MATH-LIBRARY/neqnj, NAG/c05nbf, NAG/c05ncf, NAG/c05ndf, NAG/c05pbf, NAG/c05pcf, NAG/c05pdf. The original program, written by Powell himself, is included in MINPACK.

14.4.4 Special Functions

If the *structure* of the function $F : \mathbb{R}^n \to \mathbb{R}^m$ is known, then special algorithms can be developed that are more efficient than universally applicable algorithms. For example, there are special methods for fitting combined linear/nonlinear approximation functions $g : \mathbb{R} \to \mathbb{R}$

$$g(t; c_0, \ldots, c_J, d_1, \ldots, d_J) := c_0 + \sum_{j=1}^{J} c_j g_j(t; d_1, \ldots, d_J) \tag{14.78}$$

to data $(t_1, y_1), \ldots, (t_m, y_m)$. Specially devised algorithms with *variable projection* techniques (Golub, Pereyra [212], Kaufmann [249]) reduce the original minimization problem

$$r(c, d; t, y) := \sum_{i=1}^{m} \left(y_i - c_0 - \sum_{j=1}^{J} c_j g_j(t_i; d_1, \ldots, d_J) \right)^2$$

to the minimization of $\bar{r}(d; t, y)$. Further information can be found in the literature.

Software (Linear/Nonlinear Approximation Functions) Programs designed to solve linear/nonlinear data fitting problems are, for example, NETLIB/MISC/varp2, varpra and NETLIB/PORT/nsf, nsg.

Important particular cases of (14.78) are the exponential sums

$$g(t; a_0, \ldots, a_J, \alpha_1, \ldots, \alpha_J) = a_0 + \sum_{j=1}^{J} a_j \exp(\alpha_j t).$$

For data fitting problems with exponential sums many special algorithms have been devised (see e. g., Ruhe [332]).

Chapter 15

Eigenvalues and Eigenvectors

Eigenheiten, die werden schon haften;
Kultiviere deine Eigenschaften![1]

JOHANN WOLFGANG VON GOETHE

The (Standard) Eigenvalue Problem

The mathematical problem dealt with in this chapter is the *linear eigenvalue problem*, where real or complex numbers $\lambda_1, \lambda_2, \ldots$ and corresponding non-zero vectors $x_1, x_2, \ldots$ which satisfy the equation

$$Ax = \lambda x \qquad \text{where} \quad A \in \mathbb{R}^{n \times n} \text{ or } A \in \mathbb{C}^{n \times n}, \qquad (15.1)$$

are to be determined. The numbers λ_i are called the *eigenvalues* of the matrix A and the corresponding vectors $x_i \neq 0$ are called the *eigenvectors* of A. Taken together they are called *eigenpairs*. To be more precise: every non-zero vector x_i is called a *right eigenvector* if it satisfies the equation $Ax_i = \lambda_i x_i$ for one of the λ_i, and *left eigenvector* if it satisfies

$$x_i^\top A = \lambda_i x_i^\top \qquad \text{or} \qquad x_i^H A = \lambda_i x_i^H.$$

In some applications *all* eigenvalues and corresponding eigenvectors are desired; these are called *complete eigenvalue problems*. Often the determination of the smallest or greatest eigenvalue or of eigenvalues in a given interval suffices. Such problems are called *partial eigenvalue problems*.

Example (General Solution of Ordinary Differential Equations) For a system of ordinary linear differential equations

$$\frac{dy}{dt} = Ay \qquad \text{with constant coefficients} \quad A \in \mathbb{R}^{n \times n} \qquad (15.2)$$

the solution is of the form

$$y(t) = e^{\lambda t} x. \qquad (15.3)$$

It contains an unknown vector $x \in \mathbb{R}^n$ and an unknown scalar λ. The formula (15.3) leads to

$$\frac{dy}{dt} = \lambda e^{\lambda t} x = A\, e^{\lambda t} x,$$

and because $e^{\lambda t} \neq 0$, to the equation

$$Ax = \lambda x.$$

---

[1] Characteristics won't disappear, cultivate rather the qualities.

(15.3) is therefore a solution of the differential equation (15.2) if λ is an eigenvalue of A and x is a corresponding eigenvector. If the matrix A possesses a complete set of eigenvectors, i. e., n linearly independent eigenvectors (see Theorem 15.1.3), then

$$y_1(t) = e^{\lambda_1 t}x_1, \; y_2(t) = e^{\lambda_2 t}x_2, \; \ldots, y_n(t) = e^{\lambda_n t}x_n$$

is a complete set of linearly independent solutions of the differential equation (15.2). Accordingly, any solution of (15.2) can be written in the form

$$y(t) = \sum_{i=1}^{n} \alpha_i y_i(t) = \sum_{i=1}^{n} \alpha_i e^{\lambda_i t}x_i,$$

where the coefficients $\alpha_1, \ldots, \alpha_n$ are determined, for example, by initial conditions. Thus, the general solution of a system of linear differential equations with constant coefficients can be obtained by solving a complete eigenvalue problem.

The term *"linear* eigenvalue problem" suggests that characteristic numbers (and vectors) of a *linear* function are desired. However, considered as a numerical problem, (15.1) is a *nonlinear* inverse problem: The eigenvalue problem (15.1) can be written as a system of $n+1$ nonlinear (real or complex) equations in $n+1$ unknown variables. If the requirement $x \neq 0$ is formulated as the normalization condition $x^{\top}x = 1$ or $x^H x = 1$ and if it is combined with the eigenvalue equation (15.1), this leads to

$$F(x, \lambda) = \begin{pmatrix} Ax - \lambda x \\ x^{\top}x - 1 \end{pmatrix} = 0, \tag{15.4}$$

i. e., a system of $n+1$ nonlinear equations for λ and the n unknown components of x. This formulation shows that the character of matrix eigenproblems is nonlinear. Nevertheless, a very inefficient and unreliable solution process results if the algorithms and programs from Chapter 14 for general systems of nonlinear equations are used to solve (15.4). For the practical solution of eigenvalue problems, it is far better to use algorithms and programs specially devised for finding eigenvalues and eigenvectors (see Section 15.7).

The Generalized Eigenvalue Problem

If not only one matrix but a pair of matrices A, B is involved:

$$Ax = \lambda Bx \quad \text{where} \quad A, B \in \mathbb{R}^{n \times n} \text{ or } A, B \in \mathbb{C}^{n \times n}, \tag{15.5}$$

then the problem is referred to as a *generalized linear eigenvalue problem*.

Terminology (Standard Eigenvalue Problems)   To distinguish between the eigenvalue problems (15.1) and (15.5), problems such as (15.1) are also called *special eigenvalue problems* or *standard eigenvalue problems*. In general the term "eigenvector" stands for *right* eigenvector. Left eigenvectors will be specially indicated.

Example (Boundary Value Problems of Ordinary Differential Equations)   For the boundary value problem

$$-y'' = \lambda y, \qquad y(0) = 0, \quad y(1) = 0, \tag{15.6}$$

the scalar λ can take on an infinite number of values, which are nontrivial solutions of (15.6). These values

$$\lambda_k := k^2\pi^2, \quad k = 1, 2, 3, \ldots,$$

are called *eigenvalues of the boundary value problem* (15.6). The corresponding functions

$$y_k(x) := \sin k\pi x, \quad k = 1, 2, 3, \ldots,$$

are referred to as *eigenfunctions* of (15.6). If (15.6) is extended by an additional positive function c,

$$-y'' = \lambda c y, \qquad y(0) = 0, \quad y(1) = 0, \tag{15.7}$$

then it is generally not possible to determine eigenvalues and eigenfunctions of (15.7) in closed form—they have to be determined numerically. With this in mind, subdividing of the interval $[0, 1]$ into grid points

$$x_0 := 0, \ x_1 := h, \ x_2 := 2h, \ \ldots, x_{n+1} := 1 \qquad \text{with} \quad h := 1/(n+1),$$

makes it possible to discretize the boundary-value problem (15.7), with $c_i := c(x_i)$, $y_0 = y_{n+1} = 0$, and $y_i \approx y(x_i)$, as follows:

$$(-y_{i+1} + 2y_i - y_{i-1})/h^2 = \lambda c_i y_i, \quad i = 1, 2, \ldots, n. \tag{15.8}$$

Thus, a generalized matrix eigenvalue problem

$$Ay = \lambda By, \tag{15.9}$$

with

$$A = \begin{pmatrix} 2 & -1 & & & \mathbf{0} \\ -1 & 2 & \ddots & & \\ & \ddots & \ddots & & \\ & & & 2 & -1 \\ \mathbf{0} & & & -1 & 2 \end{pmatrix}$$

and $B = \operatorname{diag}(h^2 c_1, h^2 c_2, \ldots, h^2 c_n)$ is obtained. As a result of the requirement $c(x) > 0$, the reformulation of (15.9) into a standard eigenvalue problem is possible:

$$B^{-1}Ay = \lambda y \qquad \text{with} \quad B^{-1} = \operatorname{diag}(h^{-2}/c_1, \ldots, h^{-2}/c_n).$$

The transformation

$$Ax = \lambda Bx \qquad \longmapsto \qquad B^{-1}Ay = \lambda y \tag{15.10}$$

is normally not recommended as a basis for the practical solution of (15.5), even in cases where the matrix B is regular. Special algorithms and software are much more useful for solving the originally specified generalized eigenvalue problem. The transformation (15.10) should be avoided, especially if B is ill-conditioned or if A and B are both symmetric, since the symmetry of $B^{-1}A$ does *not* automatically follow from the symmetry of A and B. Only in the special case of a symmetric, positive definite matrix B is it possible to obtain a symmetric standard eigenvalue problem

$$Cy = \lambda y \qquad \text{with} \quad C := L^{-1}AL^{-\top} \quad \text{and} \quad y := L^\top x$$

by utilizing the Cholesky factorization $B = LL^\top$.

15.1 Mathematical Foundations

The formulation of an eigenvalue problem as

$$(A - \lambda I)x = 0 \qquad \text{or} \qquad (A - \lambda B)x = 0 \tag{15.11}$$

makes it clear that (15.1) and (15.5) are homogeneous systems of linear equations for the n unknown components of the eigenvector x, provided that a value $\lambda \in \mathbb{R}$ or $\lambda \in \mathbb{C}$ is known. If the matrix $A - \lambda I$ or $A - \lambda B$ is regular, then $x = 0$ is the only solution of the linear system (15.11).

15.1.1 The Characteristic Polynomial

Non-trivial solutions of the special and the generalized eigenvalue problem exist if and only if, the coefficient matrix of the system of equations (15.11) is *singular*. This property can be used to characterize the eigenvalues.

Definition 15.1.1 (Characteristic Equation) *The equation*

$$P_n(z; A) := \det(A - zI) = 0$$

or

$$P_n(z; A, B) := \det(A - zB) = 0$$

is called the characteristic equation of the matrix A or the matrices A and B. The function P_n is a polynomial of degree n in z, which is called the characteristic polynomial of the matrix A or the matrices A and B.

The eigenvalues of the matrix A are the roots of the characteristic polynomial $P_n(z; A)$. According to the fundamental theorem of algebra, a polynomial of degree n possesses exactly n real or complex roots $\lambda_1, \ldots, \lambda_n$. Multiple eigenvalues are counted according to their multiplicity.

Definition 15.1.2 (Spectrum of a Matrix) *The set of $k \leq n$ distinct eigenvalues*

$$\lambda(A) := \{\lambda_1, \ldots, \lambda_k\} = \{z \in \mathbb{C} : \det(A - zI) = 0\}$$

is called the spectrum of A.

Using $\lambda(A)$ the following factorization of the characteristic polynomial $P_n(z; A)$ is possible:

$$P_n(z; A) = \prod_{i=1}^{k} (z - \lambda_i)^{n_i}.$$

Definition 15.1.3 (Algebraic Multiplicity) *The integer n_i is called the algebraic multiplicity of the eigenvalue λ_i. If $n_i = 1$, then n_i is called a simple eigenvalue.*

From a theoretical point of view, the problem of determining the eigenvalues of a matrix is solved when the characteristic polynomial is determined and its roots are calculated. However, the condition of the polynomial root finding problem, especially if $n_i > 1$, is often significantly worse than that of the original eigenvalue problem. Hence, the characteristic polynomial is *not* a useful means for determining the eigenvalues numerically, except in special cases (e. g., tridiagonal matrices).

In some cases, especially when it is easy to obtain the determinant, the eigenvalues can, of course, be determined using the roots of the characteristic polynomial. For instance, the spectrum of triangular matrices e. g.,

$$
L := \begin{pmatrix} l_{11} & & \mathbf{0} \\ \vdots & \ddots & \\ l_{n1} & \cdots & l_{nn} \end{pmatrix} \qquad \text{or} \qquad R := \begin{pmatrix} r_{11} & \cdots & r_{1n} \\ & \ddots & \vdots \\ \mathbf{0} & & r_{nn} \end{pmatrix},
$$

is obtained without any calculation:

$$
\lambda(L) = \{l_{11}, l_{22}, \ldots, l_{nn}\} \qquad \text{or} \qquad \lambda(R) = \{r_{11}, r_{22}, \ldots, r_{nn}\}.
$$

Theorem 15.1.1 *For an upper block triangular matrix*

$$
T = \begin{pmatrix} T_{11} & T_{12} \\ 0 & T_{22} \end{pmatrix}
$$

$\lambda(T) = \lambda(T_{11}) \cup \lambda(T_{22})$ *holds, where* $T_{11} \in \mathbb{C}^{p \times p}$, $T_{12} \in \mathbb{C}^{p \times q}$, $T_{22} \in \mathbb{C}^{q \times q}$.

Proof: Golub, Van Loan [50].

This theorem is a generalization of the fact that the elements $r_{11}, \ldots, r_{nn}$ in the diagonal of an upper triangular matrix R are its eigenvalues.

15.1.2 Similarity

The substitutions $y := Px$ and $c := Pb$, where P is a regular $n \times n$ matrix, lead to the transformed system of linear equations

$$
AP^{-1}y = P^{-1}c \qquad \text{or} \qquad PAP^{-1}y = c
$$

whose coefficient matrix is PAP^{-1}.

Definition 15.1.4 (Similar Matrices) *Let X be a non-singular $n \times n$ matrix. Two $n \times n$ matrices A and B, related by*

$$
B = X^{-1}AX,
$$

are said to be similar. The transformation $A \mapsto X^{-1}AX$ is called a similarity transformation.

A change from one vector space basis to another changes the matrix representation of a linear mapping. If A is the matrix of a linear mapping relative to *one* particular basis, then the set

$$\{X^{-1}AX : X \text{ is a } regular \ n \times n \text{ matrix}\} \tag{15.12}$$

consists of *all* matrices representing this mapping. It is the set of all matrices which are *similar* to the matrix A.

Similar matrices represent the same linear operator with respect to different bases. Thus, every linear mapping is equivalent to a whole class of matrices, since similarity is an equivalence relation.

Certain properties of a linear operator do not depend on the particular basis used for its matrix representation. Properties of this kind are therefore *invariant* under similarity transformations; they hold for the entire equivalence class of matrices (15.12). In the context of matrix eigenproblems, a very important invariance is that of the eigenvalues of a matrix under similarity transformations.

Theorem 15.1.2 *Let A and B be two similar matrices. Then A and B have the same characteristic polynomial:*

$$P_n(z; A) \equiv P_n(z; B).$$

Proof: Horn, Johnson [59].

The invariance of the eigenvalues of a matrix under similarity transformations is the foundation of many numerical algorithms used to solve eigenvalue problems.

Unlike eigenvalues, the *eigenvectors* of similar matrices A and $B = X^{-1}AX$ are *not* identical. There is, however, a simple connection: If y is an eigenvector of B corresponding to the eigenvalue $\lambda_i \in \lambda(B)$, then $x := Xy$ is an eigenvector of A corresponding to λ_i.

Diagonalizability

Because of their simple structure, diagonal matrices are of great importance. The question thus arises as to which type, or types of matrices are equivalent to a diagonal matrix.

Definition 15.1.5 (Diagonalizability) *If a matrix A is similar to a diagonal matrix, then it is said to be diagonalizable.*

Theorem 15.1.3 *An $n \times n$ matrix A is diagonalizable if and only if there exists a set $\{x_1, \ldots, x_n\}$ of n linearly independent vectors, each of which is an eigenvector of A.*

Proof: Horn, Johnson [59].

If a matrix A is diagonalizable by a similarity transformation, then the diagonal elements of each resulting diagonal matrix are the eigenvalues of A. The column vectors of the transformation matrix X are eigenvectors of A.

Theorem 15.1.4 *If all n eigenvalues of an $n \times n$ matrix A are distinct, then A is diagonalizable.*

Proof: Horn, Johnson [59].

Therefore, if an $n \times n$ matrix A does *not* possess n linearly independent eigenvectors, it necessarily has multiple eigenvalues. However, a matrix may have n linearly independent eigenvectors, even it has multiple eigenvalues (as, for example, in the case of the identity matrix).

15.1.3 Eigenvectors

To every eigenvalue λ_i belongs at least one eigenvector because $\det(A - \lambda_i I) = 0$ and accordingly the homogeneous system of linear equations $(A - \lambda_i I)x = 0$ has non-trivial solutions because its matrix is singular.

Every eigenvector x defines a one-dimensional subspace of $\mathbb{R}^n$ because every multiple cx is also an eigenvector of A. This subspace is invariant with respect to premultiplication by A.

Definition 15.1.6 (Invariant Subspace) *A subspace $S \subseteq \mathbb{C}^n$ with the property*

$$x \in S \quad \Rightarrow \quad Ax \in S,$$

is called invariant (under A).

If

$$AX = XB, \qquad A, B, X \in \mathbb{C}^{n \times n}, \tag{15.13}$$

then the space spanned by the column vectors of X is invariant under A, and

$$By = \lambda y \quad \Rightarrow \quad A(Xy) = \lambda(Xy).$$

If X is a regular matrix, for which (15.13) holds, then it follows that $\lambda(A) = \lambda(B)$, i.e., the spectrum of similar matrices is identical. The multiplicities of their eigenvalues also agree, as can be seen from Theorem 15.1.2.

Multiple eigenvalues can have several linearly independent, corresponding eigenvectors.

Definition 15.1.7 (Eigenspace) *Let $\lambda_i \in \lambda(A)$. The subspace*

$$\{x \in \mathbb{C}^n : Ax = \lambda_i x\}$$

of $\mathbb{C}^n$ invariant under A, is called the eigenspace of A associated with λ_i.

The maximum number m_i of linearly independent eigenvectors corresponding to an eigenvalue $\lambda_i \in \lambda(A)$ is equal to the dimension of the eigenspace associated with λ_i.

Definition 15.1.8 (Geometric Multiplicity) *The geometric multiplicity of the eigenvalue λ_i is the dimension*

$$m_i := \dim(\mathcal{N}(A - \lambda_i I)) = n - \mathrm{rank}(A - \lambda_i I),$$

i. e., the maximum number of linearly independent eigenvectors associated with λ_i.

Terminology (Multiplicity) Whenever the expression "multiplicity" is used without any further specification, *algebraic* multiplicity is meant.

For the geometric multiplicity m_i and the algebraic multiplicity n_i of every eigenvalue $\lambda_i \in \lambda(A)$ the relation

$$1 \leq m_i \leq n_i$$

holds. If the geometric and the algebraic multiplicity of all eigenvalues of a matrix $A \in \mathbb{C}^{n \times n}$ are equal, then the eigenvectors of A are a basis of $\mathbb{C}^n$. In this case such a matrix is said to have a *complete set of eigenvectors*.

Definition 15.1.9 (Defective Eigenvalue) *If the geometric multiplicity of an eigenvalue is smaller than its algebraic multiplicity, i.e.*

$$m_i < n_i \qquad \text{for some} \quad \lambda_i \in \lambda(A),$$

then λ_i is said to be a defective eigenvalue. A matrix with at least one defective eigenvalue is referred to as a defective matrix.

As a result of Theorem 15.1.3, an $n \times n$ matrix A is non-defective, i. e., a matrix has a complete set of eigenvectors, if and only if A is diagonalizable.

If λ_i is a defective eigenvalue, then the "missing" $n_i - m_i$ eigenvectors can be replaced by principal vectors.

Definition 15.1.10 (Principal Vector) *A non-zero vector x is said to be a principal vector of grade m_i associated with the eigenvalue λ_i if it is a solution of the following m_i equations:*

$$(A - \lambda_i I)^m x = 0, \qquad m = 1, 2, \ldots, m_i.$$

An eigenvector x associated with λ_i is therefore a principal vector of grade one. If $x_{m_i} := x$ is a principal vector of grade m_i associated with the eigenvalue λ_i, then the vectors in the *chain of principal vectors*

$$
\begin{aligned}
x_{m_i-1} &:= (A - \lambda_i I)x_{m_i} \\
x_{m_i-2} &:= (A - \lambda_i I)x_{m_i-1} \\
&\vdots \\
x_1 &:= (A - \lambda_i I)x_2
\end{aligned}
$$

are linearly independent. They span the *principal vector subspace* for $\lambda_i \in \lambda(A)$.

Eigenvectors or principal vectors belonging to different eigenvalues are linearly independent. The whole space $\mathbb{C}^n$ is the direct sum of the subspaces spanned by the principal vectors of a matrix $A \in \mathbb{C}^{n \times n}$. A set of n linearly independent principal vectors of an $n \times n$ matrix forms a *principal vector basis* for $\mathbb{C}^n$.

15.1.4 Unitary Similarity

In Section 15.1.2 similarity transformations

$$B = X^{-1}AX \tag{15.14}$$

with general, regular matrices X were introduced. However, for the stable numerical solution of eigenproblems, similarity transformations

$$B = Q^{-1}AQ \tag{15.15}$$

with unitary or orthogonal matrices (cf. Definition 13.5.1 and Definition 13.5.2) are of much greater importance than transformations of the form (15.14). Unitary or orthogonal matrices have the profitable property that their inverses

$$Q^{-1} = Q^H \quad \text{or} \quad Q^{-1} = Q^\mathsf{T}$$

can be obtained without further calculation.

Definition 15.1.11 (Unitary/Orthogonal Similarity of Matrices)   *Let Q be a unitary or orthogonal $n \times n$ matrix. Two $n \times n$ matrices A and B with*

$$B = Q^H AQ \quad \text{or} \quad B = Q^\mathsf{T}AQ$$

are said to be unitarily or orthogonally similar.

Premultiplication by a unitary matrix $(y = Qx)$ does not change the Euclidean norm of a vector:

$$\|y\|_2 = \sqrt{\langle y, y \rangle} = \sqrt{y^H y} = \sqrt{x^H Q^H Qx} = \sqrt{x^H I x} = \sqrt{x^H x} = \sqrt{\langle x, x \rangle} = \|x\|_2.$$

Also, the Frobenius norm of a matrix

$$\|A\|_F = \sum_{i=1}^{n}\sum_{j=1}^{n} |a_{ij}|^2$$

is invariant under unitary similarity transformations:

Theorem 15.1.5   *For two unitarily similar matrices $A, B \in \mathbb{C}^{n \times n}$*

$$\|A\|_F = \|B\|_F.$$

Proof: Horn, Johnson [59].

A unitary, similarity transformation represents, as does any other similarity transformation, a change from one basis used for a matrix representation of the underlying linear operator to another basis. In the unitary case, however, it is an exchange of orthonormal bases.

   The simplest unitary, or orthogonal, similarity transformations are rotations and reflections. Many numerically stable algorithms for solving eigenproblems are based on the successive use of such transformations.

Plane Rotations

It is a result from plane analytic geometry, that under a rotation of the coordinate axes through an angle φ (see Fig. 15.1) the coordinates of a point are transformed as follows:

$$\begin{pmatrix} x_i \\ x_j \end{pmatrix} = \begin{pmatrix} \cos\varphi & -\sin\varphi \\ \sin\varphi & \cos\varphi \end{pmatrix} \begin{pmatrix} \hat{x}_i \\ \hat{x}_j \end{pmatrix}, \qquad \begin{pmatrix} \hat{x}_i \\ \hat{x}_j \end{pmatrix} = \begin{pmatrix} \cos\varphi & \sin\varphi \\ -\sin\varphi & \cos\varphi \end{pmatrix} \begin{pmatrix} x_i \\ x_j \end{pmatrix}.$$

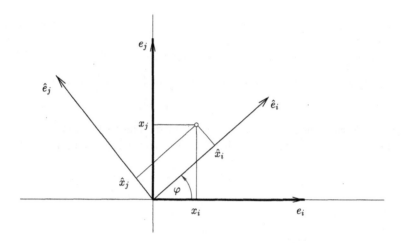

Figure 15.1: Plane rotation (in the i-j plane) through an angle φ.

In $\mathbb{R}^n$ a rotation in the i-j plane, spanned by the unit vectors $\hat{e}_i$ and $\hat{e}_j$, is obtained by a premultiplication with the transposed *rotation matrix* (*Givens matrix*, *Jacobi matrix*) $G(\varphi; i, j)$. This matrix is defined to be the identity matrix subject to the following alterations:

$$g_{ii} = g_{jj} = \cos\varphi, \quad -g_{ij} = g_{ji} = \sin\varphi \qquad \text{where} \quad |\varphi| \le \pi$$

$$G(\varphi; i, j) = \begin{pmatrix} 1 & & & & & & & & & \\ & \ddots & & & & & & & 0 & \\ & & 1 & & & & & & & \\ & & & \cos\varphi & & & -\sin\varphi & & & \\ & & & & 1 & & & & & \\ & & & & & \ddots & & & & \\ & & & & & & 1 & & & \\ & & & \sin\varphi & & & \cos\varphi & & & \\ & & & & & & & 1 & & \\ & 0 & & & & & & & \ddots & \\ & & & & & & & & & 1 \end{pmatrix}$$

A plane rotation $G^\top(\varphi; i, j)A$ affects only the ith and jth row of the matrix A. On the other hand, a postmultiplication $A\,G(\varphi; i, j)$ affects the ith and jth column of A. Accordingly, the similar matrices A and

$$B := G^{\top}(\varphi; i, j) \, A \, G(\varphi; i, j)$$

differ only in their ith and jth rows and columns.

Householder Reflections

The definition of a special unitary matrix

$$H_w := I - 2ww^H,$$

where $\|w\|_2 = 1$, $w \in \mathbb{C}^n$, gives

$$y := H_w x = (I - 2ww^H)x = x - 2\langle w, x \rangle w$$

for the product $H_w x$. If the scalar product of this formula is written as

$$\langle w, x \rangle = \|x\|_2 \cos\langle w, x \rangle,$$

then a geometrical interpretation can be gleaned: The vector y is obtained by a reflection of x on the hyperplane whose normal vector is w (see Fig. 15.2).

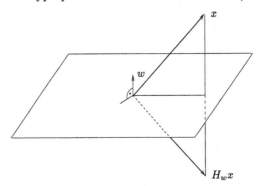

Figure 15.2: Reflection of a vector x on the plane with the normal vector w.

Unitary Similarity to Triangular Matrices

One of the most important results of matrix theory concerning eigenvalue problems is Schur's theorem, which says that every matrix $A \in \mathbb{C}^{n \times n}$ is unitarily similar to an upper triangular matrix T. The diagonal elements of such a triangular matrix are the eigenvalues of A.

Theorem 15.1.6 (Schur Decomposition)   *Given any matrix $A \in \mathbb{C}^{n \times n}$, there exists a unitary matrix $Q \in \mathbb{C}^{n \times n}$, such that*

$$Q^H A Q = T = \operatorname{diag}(\lambda_1, \lambda_2, \dots, \lambda_n) + U \tag{15.16}$$

where U is a strictly upper triangular matrix. Q can be chosen in such a way that the eigenvalues λ_i appear in any order in the diagonal matrix.

Proof: Golub, Van Loan [50].

The Schur decomposition (15.16) is *not* unique. This does not only refer to the order of the eigenvalues in the diagonal of the triangular matrix T. The elements above the diagonal grouped in the matrix U can also differ in a significant way, as the following example shows:

Example (Unitarily Similar Triangular Matrices) The two triangular matrices

$$T_1 = \begin{pmatrix} 1 & 1 & 4 \\ 0 & 2 & 2 \\ 0 & 0 & 3 \end{pmatrix} \quad \text{and} \quad T_2 = \begin{pmatrix} 2 & -1 & 3\sqrt{2} \\ 0 & 1 & \sqrt{2} \\ 0 & 0 & 3 \end{pmatrix}$$

are unitarily similar.

Theorem 15.1.6 holds analogously for unitary similarity to *lower* triangular matrices (with different transformation matrices).

The column vectors $q_1, \ldots, q_n$ of the unitary matrix Q of a Schur decomposition are called *Schur vectors*. Comparing the matrix equation $AQ = QT$ column by column, it can be seen that

$$Aq_i = \lambda_i q_i + \sum_{j=1}^{i-1} u_{ji} q_j, \qquad i = 1, 2, \ldots, n. \tag{15.17}$$

Consequently, the subspaces

$$S_k := \text{span}\{q_1, \ldots, q_k\}, \qquad k = 1, 2, \ldots, n$$

are invariant under A. Since the eigenvalues in (15.16) can be written in any order, according to Theorem 15.1.6, there is at least one k-dimensional invariant subspace associated with each set of k eigenvalues.

The relation (15.17) shows that the Schur vector q_i is an eigenvector of A if and only if all elements in the ith column of U are zero. This holds for all columns of U whenever $A^H A = AA^H$.

Definition 15.1.12 (Normal Matrices) *Complex or real quadratic matrices with*

$$A^H A = AA^H \quad \text{or} \quad A^T A = AA^T$$

are said to be normal.

Hermitian or symmetric matrices, for example, are normal.

Orthogonal Similarity to Quasi-Triangular Matrices

If a matrix $A \in \mathbb{R}^{n \times n}$ has all real eigenvalues, then the Schur decomposition can be obtained by applying an orthogonal matrix $Q \in \mathbb{R}^{n \times n}$:

$$Q^T AQ = T = \text{diag}(\lambda_1, \lambda_2, \ldots, \lambda_n) + U.$$

If A is normal (or even symmetric), then $U = 0$, i.e.,

$$Q^T AQ = \text{diag}(\lambda_1, \lambda_2, \ldots, \lambda_n).$$

If a matrix $A \in \mathbb{R}^{n \times n}$ has some conjugate complex pairs of eigenvalues, then an orthogonal similarity transformation into a real triangular matrix is impossible. Of course, A can be transformed into a *complex* triangular matrix according to Theorem 15.1.6. Any real matrix A is, however, orthogonally similar to a *real* Hessenberg matrix:

Theorem 15.1.7 (Real Schur Decomposition) *Let $A \in \mathbb{R}^{n \times n}$, then there exists an orthogonal matrix $Q \in \mathbb{R}^{n \times n}$, such that*

$$Q^{\top} A Q = \begin{pmatrix} R_{11} & R_{12} & \cdots & R_{1k} \\ & R_{22} & \cdots & R_{2k} \\ & & \ddots & \vdots \\ \mathbf{0} & & & R_{kk} \end{pmatrix}, \tag{15.18}$$

where each diagonal block R_{ii} is either a real 1×1 matrix or a real 2×2 matrix with complex conjugate eigenvalue pairs of A. Q can be chosen so that the diagonal blocks may appear in any order.

Proof: Golub, Van Loan [50].

The difference between decomposition (15.18) and a real triangular matrix is as small as possible. A Hessenberg matrix in the special form (15.18) is said to be a *quasi-triangular matrix*.

Unitary Similarity to Diagonal Matrices

Because of their simple structure, diagonal matrices, in the equivalence class of all matrices similar to A, are of special interest:

Theorem 15.1.8 *A matrix $A \in \mathbb{C}^{n \times n}$ is unitarily similar to a diagonal matrix*

$$Q^{H} A Q = \mathrm{diag}(\lambda_1, \lambda_2, \ldots, \lambda_n)$$

if and only if it is normal, i. e., if $A^{H} A = A A^{H}$.

Proof: Golub, Van Loan [50].

Thus, the matrix U of the Schur decomposition (15.16) is the zero matrix only if A is normal. The quantity $\Delta(A) := \|U\|_F$, which is independent of Q, is a measure of the deviation from normality of the matrix A:

$$\Delta^2(A) = \|U\|_F^2 = \|A\|_F^2 - \sum_{i=1}^{n} |\lambda_i|^2.$$

For real matrices it is only possible to achieve similarity to *quasi*-diagonal matrices.

Theorem 15.1.9   *A matrix $A \in \mathbb{R}^{n \times n}$ is orthogonally similar to a quasi-diagonal matrix*

$$Q^\top A Q = \text{diag}(D_{11}, D_{22}, \ldots, D_{kk})$$

if and only if it is normal, i. e., if $A^\top A = A A^\top$. Each block D_{ii} is either a real 1×1 matrix or a real 2×2 matrix of the form

$$D_{ii} := \begin{pmatrix} \alpha_i & \beta_i \\ -\beta_i & \alpha_i \end{pmatrix}.$$

Proof: Horn, Johnson [59].

If $A \in \mathbb{R}^{n \times n}$ is not only normal, but also symmetric ($A^\top = A$), then A is actually orthogonally similar to a diagonal matrix.

Theorem 15.1.10 (Symmetric Real Schur Decomposition)   *If $A \in \mathbb{R}^{n \times n}$ is symmetric, then there exists an orthogonal matrix $Q \in \mathbb{R}^{n \times n}$ such that*

$$Q^\top A Q = \text{diag}(\lambda_1, \lambda_2, \ldots, \lambda_n).$$

Proof: Golub, Van Loan [50].

Q can be chosen so that the λ_i appear in any order.

15.1.5 Similarity to (Quasi) Diagonal Matrices

If a similarity transformation of a *non-normal* matrix A is sought, such that the difference between the transformed matrix $X^{-1} A X$ and a diagonal matrix is as small as possible, then *non-unitary* transformation matrices have to be used.

Theorem 15.1.11 (Block Diagonal Form)   *For each matrix $A \in \mathbb{C}^{n \times n}$, there exists a regular matrix $X \in \mathbb{C}^{n \times n}$ such that*

$$X^{-1} A X = \text{diag}(T_{11}, T_{22}, \ldots, T_{kk}). \qquad (15.19)$$

Each block $T_{ii} \in \mathbb{C}^{n_i \times n_i}$ is an upper triangular matrix whose diagonal elements are all equal to the eigenvalue λ_i. The dimension n_i of T_{ii} is given by the algebraic multiplicity of λ_i.

Proof: Horn, Johnson [59].

Suppose X of (15.19) is partitioned as follows:

$$X = [X_1, X_2, \ldots, X_k] \qquad \text{with} \quad X_i \in \mathbb{C}^{n \times n_i}.$$

It then follows that the space $\mathbb{C}^n$ is the direct sum of invariant subspaces:

$$\mathbb{C}^n = \text{span}(X_1) \oplus \text{span}(X_2) \oplus \cdots \oplus \text{span}(X_k). \qquad (15.20)$$

Moreover, if appropriate bases for the invariant subspaces in (15.20) are chosen, then $X^{-1} A X$ has quasi-diagonal form:

Theorem 15.1.12 (Jordan Decomposition) *For each $A \in \mathbb{C}^{n \times n}$ there exists a regular matrix $X \in \mathbb{C}^{n \times n}$ such that*

$$X^{-1}AX = \mathrm{diag}(J_1, J_2, \ldots, J_l).$$

The blocks J_i are quadratic bidiagonal matrices of the form:

$$J_i := \begin{pmatrix} \lambda_i & 1 & & & \mathbf{0} \\ & \lambda_i & 1 & & \\ & & \ddots & \ddots & \\ & & & \lambda_i & 1 \\ \mathbf{0} & & & & \lambda_i \end{pmatrix}.$$

The bidiagonal matrix $\mathrm{diag}(J_1, J_2, \ldots, J_l)$ is called the Jordan canonical form of A, and is unique apart from the ordering of the blocks along the diagonal.

Proof: Horn, Johnson [59].

The matrices $J_1, \ldots, J_l$ are referred to as *Jordan blocks*. Every eigenvalue of A is associated with at least one Jordan block. If it is a simple eigenvalue, then the associated Jordan block is a 1×1 matrix

$$J_j = (\lambda_i) \in \mathbb{C}^{1 \times 1}.$$

If λ_i is a non-defective eigenvalue with the multiplicity $n_i = m_i$, then there exist n_i Jordan blocks of dimension 1 with λ_i as their element. The number of Jordan blocks containing the very same eigenvalue λ_i is equal to the number of linearly independent eigenvectors associated with λ_i.

Each column vector of the transformation matrix X in (15.19) is an eigenvector or principal vector associated with the eigenvalue in the main diagonal of the respective Jordan block.

Despite its importance in theoretical investigations, the Jordan canonical form is not used very often in numerical methods. The reason for this is the bad condition of non-unitary similarity transformations and also the difficulties caused by the numerical determination of Jordan blocks. The numerically unstable computation of ranks has a critical influence on the result.

In contrast to the Jordan decomposition, the Schur decomposition of a matrix can be obtained using numerically stable algorithms. That is why in LAPACK (see Section 15.7) and in mathematical software libraries only programs for the Schur decomposition are included.

15.1.6 Bounds for the Eigenvalues

In some cases the application of iterative methods for calculating the eigenvalues of A is not necessary. Qualitative information about the location of the eigenvalues, for example, in the form of a bounded region of the complex plane, is sufficient.

Example (Asymptotic Stability) If all solutions of a system of ordinary differential equations with constant coefficients

$$y' = Ay, \qquad A \in \mathbb{R}^{n \times n} \tag{15.21}$$

approach zero as $t \to +\infty$, then the system (15.21) is said to be *asymptotically stable*. This type of stability occurs if and only if

$$\mathrm{Re}(\lambda_i) < 0 \quad \text{for all} \quad \lambda_i \in \lambda(A),$$

i.e., if all eigenvalues of A are found in the left half of the complex plane (Coppel [142]).

Example (B-Convergence) The convergence analysis of numerical methods for solving initial value problems of stiff ordinary differential equations $y' = f(t, y)$ requires the *one-sided Lipschitz constant*

$$m := \max\{\mu(f_y(t, y)) : (t, y) \in B\}, \qquad \text{where} \quad \mu(A) := \lambda_{\max}((A + A^\top)/2),$$

to characterize the problem (Frank, Schneid, Ueberhuber [195], [196], [197]). In this case qualitative information about $\mu(A)$ is often sufficient.

Example (Convergence of Iterative Methods) An iterative method of the form

$$x^{(k+1)} := Ax^{(k)} + c, \quad k = 0, 1, 2 \ldots$$

converges to a fixed-point x^* if and only if

$$|\lambda_i| < 1 \quad \text{for all} \quad \lambda_i \in \lambda(A).$$

In order to guarantee the convergence for any initial vector $x^{(0)}$, it is sufficient to demonstrate that the *spectral radius* satisfies $\varrho(A) < 1$.

The eigenvalue definition $Ax = \lambda x$ leads directly to the following bound:

Theorem 15.1.13 *Every p-norm of a matrix A is a bound for the magnitude of its eigenvalues:*

$$|\lambda_i| \leq \|A\|_p, \qquad i = 1, 2, \ldots, n.$$

All eigenvalues of A are thus located in

$$K := \{z : |z| \leq \|A\|_p\},$$

i.e., a disk with center 0 and radius $\|A\|_p$.

Several eigenvalue algorithms use a sequence of similarity transformations:

$$X_1^{-1} A X_1, \quad X_2^{1-} A X_2, \quad X_3^{-1} A X_3, \ldots.$$

The results approximate a diagonal form to a greater and greater degree. To answer the question: How well do the diagonal elements of a matrix approximate its eigenvalues, the following procedure is useful: Let

$$r_i := |a_{i1}| + |a_{i2}| + \cdots + |a_{i,i-1}| + |a_{i,i+1}| + \cdots + |a_{in}| \qquad i = 1, 2, \ldots, n$$

be the radii of the n disks

$$K_i := \{z \in \mathbb{C} : |z - a_{ii}| \leq r_i\}, \qquad i = 1, 2, \ldots, n \tag{15.22}$$

in the complex plane. The disks K_i convey information about the eigenvalues of A as prescribed by the following theorem:

Theorem 15.1.14 (Gerschgorin Circle Theorem)   *The union of the disks*
K_i *contains all eigenvalues of A:*

$$\lambda(A) \subseteq \bigcup_{i=1}^{n} K_i. \qquad (15.23)$$

If a union of k of these disks is disjoint from the remaining $n - k$ disks, then it contains exactly k eigenvalues of A. Multiple eigenvalues are to be counted as many times as their algebraic multiplicity.

Proof: Ortega [305].

The disks (15.22) are often said to be *Gerschgorin disks*, their union (15.23) is the *Gerschgorin domain*.

If additional information about the matrix A is available, then Theorem 15.1.14 can often be stated more precisely. If A is a Hermitian matrix, for example, then all its eigenvalues are known to be real, i.e., they are in the union of the real closed intervals $\mathbb{R} \cap K_i$.

Since A and $A^\top$ have the same eigenvalues, a result like Theorem 15.1.14 can be obtained if the radii of modified Gerschgorin disks

$$K_i' := \{z \in \mathbb{C} : |z - a_{ii}| \le r_i'\}, \qquad i = 1, 2, \ldots, n$$

are defined with the *column*-wise sums:

$$r_j' := |a_{1j}| + |a_{2j}| + \cdots + |a_{j-1,j}| + |a_{j+1,j}| + \cdots + |a_{nj}| \qquad j = 1, 2, \ldots, n.$$

Often better eigenvalue estimates than (15.23) are obtained by

$$\lambda(A) \subseteq \left(\bigcup_{i=1}^{n} K_i \right) \cap \left(\bigcup_{i=1}^{n} K_i' \right).$$

Since similar matrices have the same spectrum, it is worthwhile trying to reduce the Gerschgorin radii using a similarity transformation $X^{-1}AX$. In the simplest case

$$X := D := \mathrm{diag}(d_1, d_2, \ldots, d_n) \qquad \text{with} \quad d_i > 0, \ i = 1, 2, \ldots, n$$

the similarity transformation is a row- and column-wise scaling

$$D^{-1}AD = (d_j a_{ij}/d_i).$$

Applying Theorem 15.1.14 on $D^{-1}AD$ and $(D^{-1}AD)^\top$, the Gerschgorin domains

$$\lambda(A) \subseteq \bigcup_{i=1}^{n} \{z \in \mathbb{C} : |z - a_{ii}| \le \frac{1}{d_i} \sum_{\substack{j=1 \\ j \ne i}}^{n} d_j |a_{ij}|\}$$

and

$$\lambda(A) \subseteq \bigcup_{j=1}^{n} \{z \in \mathbb{C} : |z - a_{jj}| \le \frac{1}{d_j} \sum_{\substack{i=1 \\ i \ne j}}^{n} d_i |a_{ij}|\}$$

are obtained.

15.2 Condition of Eigenvalue Problems

For diagonalizable matrices the following theorem supplies a condition estimate for all eigenvalues.

Theorem 15.2.1 (Bauer-Fike) *Let $A \in \mathbb{C}^{n \times n}$ and*

$$X^{-1}AX = \text{diag}(\lambda_1, \lambda_2, \ldots, \lambda_n).$$

If μ is an eigenvalue of the matrix $A + E$, i. e., A perturbed by $E \in \mathbb{C}^{n \times n}$, then

$$\min\{|\lambda_i - \mu| : \lambda_i \in \lambda(A)\} \leq \kappa_p(X)\|E\|_p, \tag{15.24}$$

where any p-norm can be chosen.

Proof: Bauer, Fike [105].

This theorem shows that the absolute condition of the eigenvalues of a diagonalizable matrix depends on the condition number

$$\kappa_p(X) = \|X\|_p \|X^{-1}\|_p$$

of the transformation matrix X and *not* on the condition number $\kappa_p(A)$ of the matrix of the original eigenproblem.

The columns of X are eigenvectors of A. In the case of simple eigenvalues, they are unique except for a factor. In the case of multiple eigenvalues, the column vectors of X can be chosen an invariant subspace. Thus,

$$\min\{\kappa_p(X) : X^{-1}AX = \text{diag}(\lambda_1, \ldots, \lambda_n)\}$$

is the condition number of the eigenvalue problem of a diagonalizable matrix.

Normal matrices can be transformed into diagonal form using unitary transformation matrices X satisfying

$$\|X\|_2 = \|X^{-1}\|_2 = 1 \qquad \text{and} \qquad \kappa_2(X) = 1.$$

It is for these matrices that the eigenvalue problem is always best conditioned. According to (15.24), for normal matrices—and, therefore, especially for symmetric matrices—the perturbations of the eigenvalues are of the same order as the perturbations of the matrix:

$$\min\{|\lambda_i - \mu| : \lambda_i \in \lambda(A)\} \leq \|E\|_2.$$

If A is not diagonalizable, then there exist no favorable condition estimates of the form (15.24). However, the non-normality of a matrix does not automatically imply that *all* eigenvalues are ill-conditioned. Often both well and ill conditioned eigenvalues appear. For this reason it is better to concentrate on the conditioning of a *simple* eigenvalue.

In order to derive condition estimates for simple eigenvalues the following analytical technique can be applied:

Let $\lambda_i \in \lambda(A)$ be an eigenvalue with algebraic multiplicity $n_i = 1$ and with right and left eigenvectors satisfying

$$Ax = \lambda_i x \quad \text{and} \quad y^H A = \lambda_i y^H \qquad \text{with} \quad \|x\|_2 = \|y\|_2 = 1.$$

Let

$$\tilde{A}(t) := A + tE, \qquad \|E\|_2 = 1$$

be a differentiable, parameterized perturbation of A. Then in a neighborhood of $t = 0$ there exist differentiable functions $x(t)$ and $\lambda_i(t)$ with

$$(A + tE)x(t) = \lambda_i(t)x(t) \qquad \|x(t)\|_2 \equiv 1,$$

where $\lambda_i(0) = \lambda_i$ and $x(0) = x$. Differentiation with respect to t leads to the following result for $t = 0$:

$$Ex + A\dot{x}(0) = \dot{\lambda}_i(0)x + \lambda_i \dot{x}(0).$$

Premultiplication by y^H leads to

$$y^H Ex = \dot{\lambda}_i(0)y^H x,$$

and furthermore to

$$|\dot{\lambda}_i(0)| = \left| \frac{y^H Ex}{y^H x} \right| \leq \frac{1}{|y^H x|}.$$

The first order condition number of a simple eigenvalue $\lambda_i \in \lambda(A)$ is therefore given by the reciprocal value of

$$s := |y^H x|.$$

This confirms the optimal conditioning of the eigenvalues of normal matrices, i.e., $s = 1$ because $y^H x = 1$. On the other hand, if $0 < s \ll 1$ then λ_i is ill-conditioned. In this case small perturbations may change the matrix A with the simple eigenvalue λ_i into a matrix $\tilde{A}$ with a *multiple* eigenvalue. Multiple eigenvalues of a matrix are generally very ill-conditioned.

All these considerations lead to a priori error estimates, which are not based on numerically obtained eigenvalue and/or eigenvector approximations. Knowledge of an approximate eigenvector $\tilde{x} \neq 0$ and its corresponding eigenvalue $\tilde{\lambda}$ enables much more accurate error estimates.

Theorem 15.2.2 *Let $A \in \mathbb{C}^{n \times n}$ be a diagonalizable matrix with*

$$X^{-1}AX = \operatorname{diag}(\lambda_1, \lambda_2, \ldots, \lambda_n).$$

For $\tilde{\lambda} \in \mathbb{C}$ and $\tilde{x} \in \mathbb{C}^n$ with $\tilde{x} \neq 0$, there exists an eigenvalue $\lambda_i \in \lambda(A)$ whose distance from $\tilde{\lambda}$ can be estimated with

$$|\tilde{\lambda} - \lambda_i| \leq \|X\| \, \|X^{-1}\| \frac{\|A\tilde{x} - \tilde{\lambda}\tilde{x}\|}{\|\tilde{x}\|} = \kappa(X) \frac{\|r\|}{\|\tilde{x}\|}.$$

If A is normal then there is an eigenvalue $\lambda_i \in \lambda(A)$ with

$$|\tilde{\lambda} - \lambda_i| \leq \frac{\|r\|_2}{\|\tilde{x}\|_2}. \tag{15.25}$$

Proof: Horn, Johnson [59].

The estimate (15.25) is remarkable because in the important case of Hermitian or symmetric matrices, a small residual r for numerically determined eigenpairs guarantees a small absolute error in the eigenvalue, regardless of the condition number $\kappa(X)$. This is extraordinary because with systems of linear equations a small residual does not automatically mean that the result vector has a small absolute error.

Condition of Eigenvectors

In contrast to the well conditioned *eigenvalues* of diagonalizable matrices, *eigenvectors* of such matrices often are ill-conditioned.

Example (Effect of Matrix Perturbations)   For

$$A = \begin{pmatrix} 1 & 0 \\ 0 & 1 \end{pmatrix} \quad \text{and} \quad E = \begin{pmatrix} e_{11} & e_{12} \\ 0 & 0 \end{pmatrix} \quad \text{with} \quad e_{11}, e_{12} \neq 0,$$

$\lambda(A + E) = \{1, 1 + e_{11}\}$ is valid, and the corresponding normed eigenvectors are

$$x_1 = \frac{1}{\sqrt{e_{11}^2 + e_{12}^2}} \begin{pmatrix} -e_{12} \\ e_{11} \end{pmatrix} \quad \text{and} \quad x_2 = \begin{pmatrix} 1 \\ 0 \end{pmatrix}.$$

By varying e_{11} and e_{12}, the eigenvector x_1 can be made to have any direction, no matter how small the absolute values of the perturbations e_{11} and e_{12} might be.

Setting $e_{11} = 0$, the eigenspace of the perturbed matrix $A + E$ is only one-dimensional, whereas A has two linearly independent eigenvectors.

Even a small residual of a numerically calculated eigenpair $(\tilde{\lambda}, \tilde{x})$ cannot guarantee that $\tilde{x}$ approximates the eigenvector x of A well.

Example (Small Residual)   The matrix

$$A = \begin{pmatrix} 1 & \varepsilon \\ \varepsilon & 1 \end{pmatrix}, \quad \varepsilon > 0$$

has the eigenvectors

$$x_1 = \begin{pmatrix} 1 \\ 1 \end{pmatrix} \quad \text{and} \quad x_2 = \begin{pmatrix} 1 \\ -1 \end{pmatrix}.$$

The perturbed eigenpair

$$\tilde{\lambda} = 1 \quad \text{and} \quad \tilde{x} = \begin{pmatrix} 1 \\ 0 \end{pmatrix}$$

has a residual

$$r = A\tilde{x} - \tilde{\lambda}\tilde{x} = \begin{pmatrix} 0 \\ \varepsilon \end{pmatrix}. \tag{15.26}$$

The smaller $\varepsilon > 0$ is chosen, the smaller the residual (15.26) can be made. However, the vector $\tilde{x}$ does not approximate either of the two eigenvectors x_1 or x_2 at all, however small ε might be.

The ill-conditioning of eigenvectors associated with multiple non-defective eigenvalues results from the fact that infinitely many different eigenvector bases exist for the corresponding invariant subspace.

15.3 The Power Method

The simplest method for determining the eigenvalue of largest absolute value and
its corresponding eigenvector of a matrix is the following iterative scheme:

Starting with an initial vector $x^{(0)} \in \mathbb{C}^n$, the sequence

$$x^{(k)} := Ax^{(k-1)}, \qquad k = 1, 2, 3, \ldots$$

is calculated. Since this is mathematically equivalent to

$$x^{(k)} = A^k x^{(0)}, \qquad k = 1, 2, 3, \ldots .$$

this technique is referred to as the *power method*. In order to simplify the convergence analysis of this iteration, it is assumed that the matrix A is diagonalizable,
i.e.,

$$X^{-1}AX = \operatorname{diag}(\lambda_1, \lambda_2, \ldots, \lambda_n)$$

and that it has a *dominant eigenvalue*, i.e., an eigenvalue whose magnitude is
larger than that of all other eigenvalues:

$$|\lambda_1| > |\lambda_2| \geq |\lambda_3| \geq \cdots \geq |\lambda_n| \geq 0. \tag{15.27}$$

The column vectors $x_1, \ldots, x_n$ of X are linearly independent eigenvectors of A
and thus form a basis for $\mathbb{C}^n$. $x^{(0)}$ can thus be represented as

$$x^{(0)} = \alpha_1 x_1 + \alpha_2 x_2 + \cdots + \alpha_n x_n.$$

It follows that

$$
\begin{aligned}
x^{(k)} &= A^k x^{(0)} \\
&= \alpha_1 A^k x_1 + \alpha_2 A^k x_2 + \cdots + \alpha_n A^k x_n \\
&= \alpha_1 \lambda_1^k x_1 + \alpha_2 \lambda_2^k x_2 + \cdots + \alpha_n \lambda_n^k x_n \\
&= \alpha_1 \lambda_1^k \left(x_1 + \frac{\alpha_2}{\alpha_1} \left(\frac{\lambda_2}{\lambda_1}\right)^k x_2 + \cdots + \frac{\alpha_n}{\alpha_1} \left(\frac{\lambda_n}{\lambda_1}\right)^k x_n \right),
\end{aligned}
\tag{15.28}
$$

provided $\alpha_1 \neq 0$. It follows from (15.27) that

$$\left|\frac{\lambda_i}{\lambda_1}\right| < 1 \qquad \text{for} \quad i = 2, 3, \ldots, n,$$

whence (15.28) can be qualitatively described by $\alpha_1 \lambda_1^k x_1$ as $k \to \infty$:

$$\lim_{k \to \infty} \left(\frac{A^k x^{(0)}}{\lambda_1^k} - \alpha_1 x_1 \right) = 0. \tag{15.29}$$

The sequence $\{x^{(k)}\}$ converges to zero if $|\lambda_1| < 1$ and diverges if $|\lambda_1| > 1$. The
relation (15.29) can be used to determine the eigenvector x_1 if the limit is finite
and non-zero. This can be achieved by appropriate scaling. For the sequences
$\{x^{(k)}\}$ and $\{\lambda^{(k)}\}$ defined by

do $k = 1, 2, 3, \ldots$
$\quad x^{(k)} := A x^{(k-1)}$
$\quad x^{(k)} := x^{(k)} / \|x^{(k)}\|$
$\quad \lambda^{(k)} := [x^{(k)}]^H A x^{(k)}$
end do

the relations

$$\text{dist}\left(\text{span}\{x^{(k)}\}, \text{span}\{x_1\}\right) = O\left(\left|\frac{\lambda_2}{\lambda_1}\right|^k\right),$$

$$\text{and} \qquad |\lambda^{(k)} - \lambda_1| = O\left(\left|\frac{\lambda_2}{\lambda_1}\right|^k\right)$$

hold.[2] The rate of convergence and, therefore, the efficiency of the power method crucially depends on the ratio $|\lambda_2|/|\lambda_1|$. If λ_1 and λ_2 are close in absolute value, then many iteration steps are necessary to achieve the desired accuracy. If, for instance, $|\lambda_2|/|\lambda_1| = 0.9$, then more than 22 iterations are required to improve the accuracy of the numerical approximations for x_1 and λ_1 by just one decimal place.

For problems with $|\lambda_2|/|\lambda_1| \ll 1$, the power method is an efficient algorithm to use for determining the dominant eigenvalue and its associated eigenvector. Particularly for very large, sparse matrices, this method has the advantage that the storage of A in an $n \times n$ array is not necessary: It suffices to evaluate the matrix-vector product Ax.

15.3.1 The Inverse Power Method

If all eigenvalues $\lambda(A)$ are non-zero, then there exists an inverse A^{-1} whose spectrum is

$$\lambda(A^{-1}) = \{1/\lambda_1, 1/\lambda_2, \ldots, 1/\lambda_n\}.$$

The eigenvectors of A and A^{-1} are identical. If $1/\lambda_n$ is a dominant eigenvalue of A^{-1}, i.e.,

$$1/|\lambda_n| > 1/|\lambda_{n-1}| \geq \cdots \geq 1/|\lambda_1| > 0,$$

then the power method can be applied to A^{-1}, and in this way the reciprocal of the eigenvalue of A with the *smallest* magnitude can be determined.

Spectral Shift

The inverse power method converges particularly quickly if $|\lambda_n|$ is significantly smaller than all other eigenvalues' absolute values. An example of this situation

---

[2]The distance function for subspaces has to be defined in an appropriate way (see e.g., Golub, Van Loan [50]).

occurs when an approximation μ for a simple real eigenvalue is already known and when this approximation is used for a *spectral shift*:

$$\lambda(A - \mu I) = \{\lambda_1 - \mu, \lambda_2 - \mu, \ldots, \lambda_n - \mu\}.$$

In this case the matrix $A - \mu I$ has an eigenvalue which is very close to zero, and the power method applied to $(A - \mu I)^{-1}$, i.e., the inverse power method, converges particularly quickly.

The practical implementation of the scheme

$$\begin{aligned}
&\textbf{do} \quad k = 1, 2, 3, \ldots \\
&\qquad \textbf{solve} \quad (A - \mu I)x^{(k)} = x^{(k-1)} \\
&\qquad x^{(k)} := x^{(k)}/\|x^{(k)}\| \\
&\qquad x^{(k)} := [x^{(k)}]^H A x^{(k)} \\
&\textbf{end do}
\end{aligned} \qquad\qquad (15.30)$$

requires modifications to ensure a reasonable level of efficiency. In particular, solving the system of linear equations (15.30), the computational effort of which is of order $O(n^3)$, should not be undertaken at every step. Instead, an LU decomposition outside the k-loop, or, in the symmetric case, a Cholesky factorization would be executed. The computational effort per step is thus reduced to the $O(n^2)$ operations of the back substitution in (15.30) and is, therefore, of the same order of magnitude as in the power method. The computational effort of the single LU factorization of $A - \mu I$ outside the loop has to be added only once.

If μ is a good approximation to one of the eigenvalues of A, then the matrix $A - \mu I$ has an eigenvalue $\lambda_i - \mu$ which is very close to zero, i.e., the iteration matrix is almost singular. Since each iteration requires the solution of a linear system with the matrix $A - \mu I$, serious doubts about the stability of the method arise. It can be shown, however, that the inverse power method produces acceptable results, despite the almost-singularity of $A - \mu I$ (Parlett [311]).

15.3.2 The Inverse Power Method with Spectral Shifts

The rate of convergence of the inverse power method can be increased greatly, if the translation parameter μ is set to the current eigenvalue approximation $\lambda^{(k)}$ after every step. However, with this approach the computational effort is increased because the LU decomposition has to be carried out at every step.

The convergence rate of the method obtained in this way—also called *Rayleigh quotient iteration*—is quadratic, whereas the inverse power method is only linear. If A is symmetric, the Rayleigh quotient iteration actually converges cubically.

15.4 The QR Algorithm

As similarity transformations do not change the spectrum of a matrix,

$$\lambda(X^{-1}AX) = \lambda(A),$$

it might be fruitful to transform the given matrix A into a similar matrix of simpler form whose eigenvalues can be obtained without significant effort.

For similarity transformations, orthogonal or, in the complex case, unitary matrices are very suitable, because their inverses are easily obtained:

$$Q^{-1} = Q^\top \quad \text{or} \quad Q^{-1} = Q^H.$$

In order to construct a unitarily similar matrix, A can firstly be decomposed into the product of a unitary matrix Q and a right (upper) triangular matrix R using QR factorization (see Section 13.14.2):

$$A = QR.$$

The factor matrices are then multiplied in the inverted order:

$$A_1 := RQ.$$

Now, because

$$A = QR = QRQQ^{-1} = QA_1Q^{-1},$$

A_1 is similar to A and therefore has the same eigenvalues. This process can be repeated iteratively:

$A_0 := A$
do $k = 1, 2, 3, \ldots$
    factorize $A_{k-1} = Q_k R_k$ (QR factorization)   (15.31)
    $A_k := R_k Q_k$
end do

By induction, the representation

$$A_k = R_k Q_k = Q_k^H (Q_k R_k) Q_k = Q_k^H A_{k-1} Q_k$$

leads to

$$A_k = (Q_1 Q_2 \cdots Q_k)^H A_0 (Q_1 Q_2 \cdots Q_k).$$

The product of the unitary matrices $Q_1, Q_2, \ldots, Q_k$ can be redefined as one unitary matrix

$$U_k := Q_1 Q_2 \cdots Q_k.$$

The relation

$$A_k = U_k^H A_0 U_k = U_k^H A U_k \qquad (15.32)$$

then shows that all the matrices $A_1, A_2, A_3, \ldots$ are unitarily similar to A. Less obvious are the convergence characteristics of this sequence of matrices: Under weak assumptions the sequence $\{A_k\}$ converges to a triangular matrix, i.e., the decomposition (15.32) gradually approximates the Schur factorization of A (Stoer, Bulirsch [75]).

15.4.1 The QR Algorithm with Spectral Shifts

The convergence of the subdiagonal elements of the matrices $A_1, A_2, A_3, \ldots$ to zero, can be characterized by

$$a_{ij}^{(k)} = O\left(\frac{|\lambda_i|^k}{|\lambda_j|^k}\right) \quad \text{as} \quad k \to \infty$$

provided the eigenvalues satisfy $|\lambda_1| > |\lambda_2| > \cdots > |\lambda_n|$ (Stoer, Bulirsch [75]). If two eigenvalues have nearly the same absolute values, $|\lambda_i| \approx |\lambda_j|$, the convergence is usually very slow. The rate of convergence can be increased, however, by carrying out a spectral shift at every step, similar to that which is done in the inverse power method (see Section 15.3.2): The QR factorization (15.31) is replaced by

$$\textbf{factorize } (\overline{A}_{k-1} - \mu_{k-1}I) = \overline{Q}_k \overline{R}_k. \tag{15.33}$$

The convergence rate of this modified method is characterized by

$$a_{ij}^{(k)} = O\left(\frac{|\lambda_i - \mu_k|^k}{|\lambda_j - \mu_k|^k}\right) \quad \text{as} \quad k \to \infty.$$

With appropriately chosen translation parameters μ_k the QR algorithm with spectral shifts converges cubically.

15.4.2 Efficiency Improvement of the QR Algorithm

The variants of the QR algorithm described above are much too inefficient for practical use because the factorizations (15.31) and (15.33) require a computational effort of order $O(n^3)$. However, if A_{k-1} or $\overline{A}_{k-1}$ is a Hessenberg matrix, then its QR decomposition requires only an effort of $O(n^2)$ and in the case of tridiagonal matrices the effort is only $O(n)$. Fortunately, the QR algorithm has the advantageous property that the matrices

$$A_1, A_2, A_3, \ldots \quad \text{or} \quad \overline{A}_1, \overline{A}_2, \overline{A}_3, \ldots$$

are all in Hessenberg or tridiagonal form, if A has a Hessenberg or tridiagonal sparsity structure (Golub, Van Loan [50]). Hence in the practical implementation of the QR algorithm, a preparatory step is carried out to reduce the given matrix to Hessenberg or tridiagonal form through a series of similarity transforms (see Section 15.6).

The entirety of all aspects that must be considered for an efficient implementation of the QR algorithm is beyond the scope of this book. Detailed discussions can be found in the books by Wilkinson [80], Stewart [74], Parlett [311] and Golub, Van Loan [50].

15.5 The Diagonal Reduction

The orthogonal similarity of a real symmetric matrix A to a diagonal matrix with the same eigenvalues

$$Q^T A Q = \text{diag}(\lambda_1, \lambda_2, \ldots, \lambda_n)$$

is the foundation of a whole class of algorithms. They all transform A into diagonal form by successively applying orthogonal similarity transformations. In this way all the eigenvalues of a symmetric matrix are determined simultaneously.

15.5.1 The Jacobi Method

The idea underlying the Jacobi method is to use plane rotations to make the non-diagonal entries of A disappear systematically:

$$\text{ND}(A) := \sum_{i=1}^{n} \sum_{\substack{j \neq i \\ j=1}}^{n} a_{ij}^2 \quad \rightarrow \quad 0.$$

Since the eigenvalues are roots of the characteristic polynomial of A, they cannot be determined using a finite process. Thus, the number of rotations needed in the Jacobi method is generally infinite. In practical implementations of the Jacobi method, the iteration is stopped as soon as the value of $\text{ND}(A)$ is below a given bound.

In the classical form of the Jacobi algorithm the off-diagonal matrix element with the largest absolute value is annihilated using a plane rotation. However, the subsequent rotations again destroy the zero-element just produced. Nevertheless, the Frobenius norm of the off-diagonal matrix elements is systematically reduced, until the off-diagonal entries are small enough to be considered zero.

Suppose that a_{pq}, with $1 \leq p < q \leq n$, is the off-diagonal element to be eliminated. The annihilation of a_{pq} is equivalent to the Schur factorization of a symmetric 2×2 matrix:

$$\begin{pmatrix} \cos\varphi & \sin\varphi \\ -\sin\varphi & \cos\varphi \end{pmatrix}^T \begin{pmatrix} a_{pp} & a_{pq} \\ a_{qp} & a_{qq} \end{pmatrix} \begin{pmatrix} \cos\varphi & \sin\varphi \\ -\sin\varphi & \cos\varphi \end{pmatrix} = \begin{pmatrix} b_{pp} & b_{pq} \\ b_{qp} & b_{qq} \end{pmatrix}.$$

The requirement $b_{qp} = b_{pq} = 0$ leads to the following trigonometric equation for determining the rotation angle φ

$$\cot 2\varphi = \frac{a_{qq} - a_{pp}}{2a_{pq}}.$$

Since the Frobenius norm is unitarily invariant, it follows that

$$\text{ND}(G^T(\varphi; p, q) \, A \, G(\varphi; p, q)) = \text{ND}(A) - 2a_{pq}^2.$$

Taking into account the fact that a_{pq} is (absolutely taken) the biggest off-diagonal entry, after k steps the following reduction of ND is obtained (Golub, Van Loan [50]):

$$\text{ND}(A^{(k)}) \leq \left(1 - \frac{2}{n(n-1)}\right)^k \text{ND}(A^{(0)}).$$

This estimate, which implies the convergence of the Jacobi method, holds for general symmetric matrices. For special types of matrices and after a sufficient number of iterations the Jacobi method converges *quadratically*.

15.6 The Hessenberg Reduction

While for an orthogonal transformation to diagonal form an infinite sequence of matrices $A, A^{(1)}, A^{(2)}, \ldots$ has to be terminated (because the eigenvalues cannot be determined exactly in a finite number of steps), the transformation of A to Hessenberg or, in the case of Hermitian/symmetric matrices, to tridiagonal form can be carried out with a *finite* number of unitary/orthogonal similarity transformations. After this preparatory step, all or some eigenvalues can be calculated numerically.

15.6.1 The Givens Algorithm

In the Givens algorithm similarity transformations are carried out using plane rotations (Givens rotations) in such a way that the off-diagonal entries are eliminated row by row. Using the rotation $G(\varphi; i, j)$, the matrix element at position $(i - 1, j)$ is eliminated. Unlike the Jacobi algorithm, off-diagonal entries which have been eliminated once are not changed back into non-zero elements. Therefore, any quadratic matrix can be transformed to Hessenberg form by performing $(n - 1)(n - 2)/2$ rotations.

15.6.2 The Householder Algorithm

Any quadratic matrix can be transformed into Hessenberg or tridiagonal form using Householder reflections as well as using plane rotations. In every step of the Householder algorithm whole rows and columns are transformed into the desired form.

15.7 LAPACK Programs

The program package LAPACK (see Section 13.15) is not only the de facto standard for the numerical solution of systems of linear equations, but also for solving eigenproblems.

As with the categories of LAPACK programs for linear systems there are two types of programs for eigenproblems:

Black box programs or **driver programs** are intended to solve problems simply and comfortably. They have relatively short parameter lists, which makes them easier to handle. However, more specially formulated problems cannot generally be solved using them.

Computational routines are implementations of (partial) algorithms. They allow for a greater influence on the algorithmic operations than do driver programs which are the *front ends* for the computational programs.

15.7.1 Symmetric Eigenproblems

LAPACK programs designed to solve real, symmetric eigenvalue problems

$$Ax = \lambda x \qquad \text{with} \quad A = A^\top \in \mathbb{R}^{n \times n} \tag{15.34}$$

make it possible to determine some or all eigenvalues $\lambda_i \in \lambda(A)$ and associated eigenvectors $x \neq 0$. For the complex Hermitian eigenproblem, this holds for

$$A \in \mathbb{C}^{n \times n} \qquad \text{with} \quad A = A^H. \tag{15.35}$$

Every symmetric or Hermitian matrix is orthogonally similar or unitarily similar to a real diagonal matrix. Accordingly, symmetric and Hermitian eigenproblems have only *real* eigenvalues. The matrix A in (15.35) has an orthonormal system of eigenvectors.

Simple driver programs for the symmetric (Hermitian) eigenproblems (15.34) and (15.35) are LAPACK/*s*ev and LAPACK/*h*ev. They calculate *all* eigenvalues and (as an option) the corresponding eigenvectors of symmetric and Hermitian matrices respectively.

Expert driver programs which calculate all or only certain eigenvalues and eigenvectors are LAPACK/*s*evx and LAPACK/*h*evx (see Table 15.1).

Table 15.1: LAPACK black box programs for symmetric eigenproblems.

matrix type	driver	REAL	COMPLEX
symmetric/Hermitian matrix	*simple*	ssyev	cheev
	expert	ssyevx	cheevx
symmetric/Hermitian matrix (packed storage)	*simple*	sspev	chpev
	expert	sspevx	chpevx
symmetric/Hermitian band matrix	*simple*	ssbev	chbev
	expert	ssbevx	chbevx
symmetric/Hermitian tridiagonal matrix	*simple*	sstev	
	expert	sstevx	

Algorithms and Computational Programs

The computational solution of eigenproblems is carried out in the following steps:

1. The real symmetric or complex Hermitian matrix A is reduced to a real symmetric tridiagonal matrix T in the corresponding LAPACK subroutines (see Table 15.2) by using orthogonal or unitary similarity transformations (see Section 15.6):

$$A = QTQ^\top \quad \text{or} \quad A = QTQ^H.$$

Table 15.2: LAPACK programs for tridiagonal reduction.

matrix type (storage)	REAL	COMPLEX
symmetric/Hermitian matrix	ssytrd	chetrd
symmetric/Hermitian matrix (packed storage)	ssptrd	chptrd
symmetric/Hermitian band matrix	ssbtrd	chbtrd

If the matrix A is real, then the transformation matrix Q can be computed explicitly using LAPACK/*qrgtr. For multiplication of Q with other matrices (without explicitly determining Q), the program LAPACK/*qrmtr is available. For complex matrices the programs LAPACK/*ungtr and LAPACK/*unmtr are provided.

2. The real, symmetric tridiagonal matrix T is then decomposed into

$$T = PDP^\top,$$

where P is orthogonal and D is a diagonal matrix. The elements in the diagonal of D are the eigenvalues of T, and the columns of P are the eigenvectors of T. The eigenvectors of A are the columns of QP.

LAPACK provides several programs for diagonalizing a symmetric, tridiagonal matrix $T = PDP^\top$. As to which program is used depends on whether all or just some eigenvalues and/or eigenvectors are required (see Table 15.3).

LAPACK/*steqr implements a special algorithm designed by Wilkinson, which alternates between the QR and the QL variant to process large matrices more efficiently than the simple QL variant alone would allow for (Greenbaum, Dongarra [213]).

LAPACK/*sterf implements a more efficient variant of the QR algorithm, which does not require square roots (Greenbaum, Dongarra [213]). It produces *all* eigenvalues of a symmetric tridiagonal matrix.

Table 15.3: LAPACK programs for the diagonalization of tridiagonal matrices.

matrix type	operation	REAL	COMPLEX
symmetric tridiagonal matrix	eigenvalues/-vectors	ssteqr	csteqr
	eigenvalues using a root-free QR	ssterf	
	eigenvalues using bisection	sstebz	
	eigenvectors using inverse power method	sstein	cstein
symmetric positive definite tridiagonal matrix	eigenvalues/-vectors	spteqr	cpteqr

LAPACK/*pteqr is only applicable to symmetric, *positive definite*, tridiagonal matrices. It uses a combination of Cholesky factorization and the bidiagonal QR iteration (see LAPACK/*bdsqr) and provides significantly better results than the other programs.

LAPACK/*stebz implements the bisection algorithm for the calculation of certain or all eigenvalues. All eigenvalues in a real interval or all eigenvalues between the ith and the jth are computed with very high accuracy. The speed of this program can be increased by reducing the demand for accuracy.

LAPACK/*stein uses the inverse power method to calculate certain or all eigenvectors from given eigenvalues.

LAPACK/*sterf is the fastest program for determining all of the eigenvalues of a small or medium-sized matrix. If, in addition, eigenvectors are required, then both LAPACK/*steqr and LAPACK/*pteqr are advantageous. The programs LAPACK/*stebz and LAPACK/*stein are preferable for the fast processing of large matrices. If, however, only a few eigenvalues and/or eigenvectors are required, then LAPACK/*stsbz and LAPACK/*stein are the most efficient programs.

15.7.2 Nonsymmetric Eigenproblems

LAPACK programs for nonsymmetric eigenvalue problems determine eigenvalues $\lambda_i \in \lambda(A)$ and the corresponding eigenvectors $x \neq 0$ for matrices with

$$A \neq A^\top \quad \text{or} \quad A \neq A^H. \tag{15.36}$$

These eigenproblems are solved using the Schur decomposition of A. In the case of real matrices this is the factorization

$$A = QTQ^\top,$$

where Q is an orthogonal matrix and T is a lower quasi-triangular matrix with blocks of order 1 or 2 in the main diagonal. The 2×2 blocks are associated with

the complex conjugate eigenpairs. In the case of complex matrices the factors of
the Schur decomposition

$$A = QTQ^H$$

are a lower triangular matrix T and a unitary matrix Q.

For every $k \in \{1, 2, \ldots, n\}$ the first k column vectors of the matrix Q, known as
Schur vectors, form an orthonormal basis of the subspace associated with the first
k eigenvalues (found in the main diagonal of T). Since this basis is orthonormal,
the Schur vectors are preferred to the eigenvectors in many applications.

Driver Programs

There are two types of driver programs for solving the problem (15.36): one for
the Schur decomposition, the other for the determination of the eigenvalues and
the eigenvectors of A (see Table 15.4).

LAPACK/*gees is a standard driver program for the complete or partial determi-
    nation of the Schur decomposition of a nonsymmetric matrix A. The order in
    which the eigenvalues occur is prescribed by the user.

LAPACK/*geesx is a special driver program which can calculate condition num-
    bers for eigenvalues and the right invariant subspaces associated with them.

LAPACK/*geev is a standard driver program for determining all eigenvalues and
    (as an option) left or right eigenvectors.

LAPACK/*geevx is a special driver program which controls the decomposition in
    such a way that the condition of the eigenvalues and the eigenvectors is as
    good as it can be.

Table 15.4: LAPACK black box programs for nonsymmetric eigenproblems.

matrix type	operation	*driver*	REAL	COMPLEX
general matrix	Schur decomposition	*simple*	sgees	cgees
		expert	sgeesx	cgeesx
	eigenvalues/-vectors	*simple*	sgeev	cgeev
		expert	sgeevx	cgeevx

Algorithms and Computational Programs

For the determination of eigenvalues, eigenvectors, Schur vectors, condition numbers etc. there are LAPACK programs for different matrix types (see Table 15.5).
The numerical solution of nonsymmetric eigenvalue problems is carried out in the
following steps:

Table 15.5: LAPACK programs for nonsymmetric eigenproblems.

type of matrix	operation	REAL	COMPLEX
general matrix	Hessenberg reduction	sgehrd	cgehrd
	balancing	sgebal	cgebal
	inverse transformation	sgebak	cgebak
Hessenberg matrix	Schur decomposition	shseqr	chseqr
	eigenvector computation	shsein	chsein
	using inverse power method		
(quasi-) triangular matrix	eigenvector computation	strevc	ctrevc
	rearrangement of	strexc	ctrexc
	the eigenvalues		
	Sylvester equation	strsyl	ctrsyl
	condition number of the	strsna	ctrsna
	eigenvalues/-vectors		
	condition number of the	strsen	ctrsen
	invariant subspaces		

1. To start with, the general matrix A is transformed into an upper Hessenberg matrix H in a preparatory step. If A is real this corresponds to the decomposition $A = QHQ^T$ with an orthogonal matrix Q. If A is complex, then $A = QHQ^H$ is obtained using a unitary matrix Q. This Hessenberg reduction is carried out using LAPACK/*gehrd, which produces Q in a decomposed form.

   If A is a real matrix, the orthogonal factor matrix Q can be determined either explicitly by using LAPACK/*orghr or it can be multiplied by another matrix without explicitly calculating Q by using LAPACK/*ormhr. If A is a complex matrix, then LAPACK/*unghr and LAPACK/*unmhr are used in a similar way.

2. The upper Hessenberg matrix H is reduced to Schur form $H = PTP^T$ (H real) or $H = PTP^H$ (H complex). The matrix P can be determined if desired. The program LAPACK/*hesqr retrieves the eigenvalues of A from the diagonal elements of T.

3. For given eigenvalues, corresponding eigenvectors can be calculated in various ways. LAPACK/*hsein determines the eigenvectors of H using the inverse power method, whereas LAPACK/*trevc determines the eigenvectors of T. Multiplication by Q (or QP) transforms the right eigenvectors of H (or of T) into the right eigenvectors of the initial matrix A. The same applies to the left eigenvectors.

LAPACK/*gebal is available for *balancing* matrices. It transforms the matrix A as close as possible into triangular form by using permutation matrices, and at the same time makes all vector norms, of both rows and columns of the matrix, nearly equal. These transformations increase the rate of convergence and, in some cases, the accuracy of subsequent calculations. LAPACK/*gebal carries out the balancing and LAPACK/*gebak the back transformation of the eigenvectors of the matrix obtained by LAPACK/*gebal.

In addition to these programs, there are four other programs available, which are designed to solve special problems:

LAPACK/*trexc serves to permute the positions of the eigenvalues in the main diagonal of the Schur form. In this way the order of the eigenvalues in the Schur canonical form can be prescribed.

LAPACK/*trsyl solves the *Sylvester equation* $BX + XC = D$ for the matrix X. The matrices B, C and D are input data, where B and C are (quasi-) triangular. This program is used by the following two programs although it may be used independently of both of them.

LAPACK/*trsna calculates the condition numbers of the eigenvalues and/or the right eigenvectors of a matrix T in Schur form. They are equal to the condition numbers of the eigenvalues and right eigenvectors of the original matrix A, from which T is determined. The condition numbers can be calculated for selected subsets or for all eigenpairs (Bai, Demmel, McKenney [97]).

LAPACK/*trsen places a selected subset of eigenvalues of the matrix T in Schur form onto the first positions of the diagonal of the matrix T and then calculates the condition numbers of the mean eigenvalue in the spectrum and of the right invariant subspace. These are equal to the condition numbers of the eigenvalue average and the right invariant subspace of the original matrix A, from which the matrix T is derived (Bai, Demmel,McKenney [97]).

15.7.3 Singular Value Decomposition (SVD)

In LAPACK there are programs for determining the *singular value decomposition* (SVD) of a real or complex $m \times n$ matrix A

$$A = USV^{\mathsf{T}} \quad \text{or} \quad A = USV^{H}. \tag{15.37}$$

U and V are orthogonal (or unitary) $m \times m$ or $n \times n$ matrices, and S is an $m \times n$ diagonal matrix with real diagonal elements

$$\sigma_1 \geq \sigma_2 \geq \ldots \sigma_{\min(m,n)} \geq 0.$$

The σ_i are the *singular values* of A, and the first $\min(m,n)$ column vectors of U and V are the *left* and *right singular vectors* of the matrix A. The singular values of A are the roots of the eigenvalues of $A^{\mathsf{T}}A$:

$$\sigma_i = \sqrt{\lambda_i}, \quad \lambda_i \in \lambda(A^{\mathsf{T}}A), \qquad i = 1, 2, \ldots, \min(m,n).$$

Accordingly, the singular values of A could be determined by calculating the eigenvalues of the symmetric matrix $A^{\mathsf{T}}A$. However, it is more efficient to use algorithms that calculate a singular value decomposition by using the matrix A only (and not $A^{\mathsf{T}}A$).

The driver programs LAPACK/*gesvd determine the singular value decomposition of general nonsymmetric matrices.

Algorithms and Computational Programs

The LAPACK programs listed in Table 15.6 serve to determine the singular value decomposition (15.37). The calculation is divided into two steps:

1. Firstly the matrix A is transformed into bidiagonal form:

$$A = U_1 B V_1^\mathsf{T} \quad \text{or} \quad A = U_1 B V_1^H$$

   using Householder transformations. U_1 and V_1 are orthogonal (unitary) matrices, and B is a real bidiagonal matrix. The program LAPACK/*gebrd carries out the calculations producing U_1 and V_1 in factorized form. For a real matrix A the matrices U_1 and V_1 can be explicitly determined with the program LAPACK/*orgbr, or they can be multiplied by other matrices, without explicitly determining them, by using LAPACK/*ormbr. The programs LAPACK/*ungbr and LAPACK/*unmbr are available for complex matrices A.

2. The singular value decomposition

$$B = U_2 S V_2^\mathsf{T}$$

   of the bidiagonal matrix B is calculated using the QR algorithm. U_2 and V_2 are orthogonal matrices, and S is a diagonal matrix. The singular vectors of A are $U = U_1 U_2$ and $V = V_1 V_2$. The calculations are carried out by the program LAPACK/*bdsqr. This offers the additional option of multiplying another matrix by the transposed right singular vectors. This possibility is useful when solving linear least squares problems.

Table 15.6: LAPACK programs for singular value decompositions.

matrix type	operation	REAL	COMPLEX
general matrix	bidiagonal reduction	sgebrd	cgebrd
orthogonal/ unitary matrix	generate matrix after bidiagonal reduction	sorgbr	cungbr
	multiply matrix after bidiagonal reduction	sormbr	cunmbr
bidiagonal matrix	singular values/vectors	sbdsqr	cbdsqr

15.7.4 Generalized, Symmetric Eigenproblems

There are programs for the numerical solution of generalized, symmetric definite eigenproblems

$$Ax = \lambda Bx, \qquad ABx = \lambda x, \qquad BAx = \lambda x \qquad (15.38)$$

to be found in LAPACK. A and B denote symmetric and Hermitian matrices respectively; and, furthermore, B is assumed to be positive definite. The driver programs for these problems in LAPACK (and as well those for *nonsymmetric* generalized eigenproblems) are listed in Table 15.7.

Table 15.7: LAPACK black box programs for generalized eigenvalue problems.

type of matrix (storage)	function	*driver*	REAL	COMPLEX
general matrices	Schur decomposition eigenvalues/vectors	*simple* *simple*	sgegs sgegv	cgegs cgegv
symmetric/ Hermitian matrices	Schur decomposition eigenvalues/vectors	*simple*	ssygv	chegv
symmetric/ Hermitian matrices (packed storage)	Schur decomposition eigenvalues/vectors	*simple*	sspgv	chpgv

Algorithms and Computational Programs

Each of the generalized eigenproblems (15.38) can be reduced to a symmetric eigenproblem after the decomposition of B into LL^T or $U^\mathsf{T}U$. Using $B = LL^\mathsf{T}$ leads to

$$Ax = \lambda Bx \quad \Rightarrow \quad (L^{-1}AL^{-\mathsf{T}})(L^\mathsf{T}x) = \lambda(L^\mathsf{T}x).$$

The eigenvalues of $Ax = \lambda Bx$ are thus also those of $Cy = \lambda y$, where C denotes the symmetric matrix $C := L^{-1}AL^{-\mathsf{T}}$ and $y := L^\mathsf{T}x$.

In a similar way,

$$ABx = \lambda x \quad \Rightarrow \quad (L^\mathsf{T}AL)(L^\mathsf{T}x) = \lambda(L^\mathsf{T}x)$$

and

$$BAx = \lambda x \quad \Rightarrow \quad (L^\mathsf{T}AL)(L^{-1}x) = \lambda(L^{-1}x)$$

are obtained. In exactly the same way, the use of $B = U^\mathsf{T}U$ leads to

$$Ax = \lambda Bx \quad \Rightarrow \quad (U^{-\mathsf{T}}AU^{-1})(Ux) = \lambda(Ux),$$

$$ABx = \lambda x \quad \Rightarrow \quad (UAU^\mathsf{T})(Ux) = \lambda(Ux)$$

and

$$BAx = \lambda x \quad \Rightarrow \quad (UAU^\mathsf{T})(U^{-\mathsf{T}}x) = \lambda(U^{-\mathsf{T}}x).$$

For a given matrix A and a given Cholesky decomposition of B, the program LAPACK/*gst overwrites the matrix A with the matrix C of the corresponding standard eigenproblem $Cy = \lambda y$. To compute the eigenvectors x of the generalized eigenproblem from the eigenvectors y of the standard eigenproblem, it is not necessary to use special programs because all calculations can be done with simple calls to BLAS-2 and BLAS-3 programs.

15.7.5 Generalized, Nonsymmetric Eigenproblems

There are programs available for the numerical calculation of the eigenvalues and the corresponding eigenvectors of the generalized, nonsymmetric eigenproblems

$$Ax = \lambda Bx \quad \text{or} \quad \mu Ay = By \tag{15.39}$$

where A and B are general (nonsymmetric) square matrices. The two eigenproblems in (15.39) are equivalent with $\mu = 1/\lambda$ and $x = y$, provided $\lambda \neq 0$ and $\mu \neq 0$.

If B is singular, then there exists $x \neq 0$ such that $Bx = 0$ and an *infinite eigenvalue* $\lambda = \infty$ may occur. On the other hand, the equivalent problem $\mu Ax = Bx$ has the corresponding eigenvalue $\mu = 0$ if A is regular. To handle these eigenvalues, the relevant LAPACK programs provide two values α and β with $\lambda = \alpha/\beta$ and $\mu = \beta/\alpha$ for every eigenvalue λ. The first main task of these programs is to determine all n pairs (α, β) and associated eigenvectors.

If the determinant of $A - zB$ vanishes for all $z \in \mathbb{C}$, then the generalized eigenproblem is said to be *singular*. Such a problem is characterized by $\alpha = \beta = 0$. Due to inevitable rounding errors the calculated values of α and β can also be very small. In this case the eigenproblem is very ill-conditioned (Stewart [359], Wilkinson [382], Demmel, Kågström [155]).

Driver Programs

For the nonsymmetric problem $Ax = \lambda Bx$ there are two standard driver programs available (see Table 15.7):

LAPACK/*gegs determines parts of or the complete generalized Schur decomposition of $A - \lambda B$;

LAPACK/*gegv calculates the generalized eigenvalues and (as an option) the left and right eigenvectors.

Algorithms and Computational Programs

The main task of the nonsymmetric, generalized eigenvalue problem is the determination of the generalized Schur decomposition of the matrices A and B. If A and B are both complex, then their generalized Schur decomposition is

$$A = QSZ^H \quad \text{and} \quad B = QPZ^H,$$

where Q and Z denote unitary and S and P upper triangular matrices. LAPACK programs normalize the matrix P so that all diagonal elements are non-negative. In the representation $\lambda_i = s_{ii}/p_{ii}$, the eigenvalues result directly from the diagonal elements of S and P. LAPACK programs output $\alpha_i = s_{ii}$ and $\beta_i = p_{ii}$.

When both A and B are real, then their generalized Schur decomposition is

$$A = QSZ^\top \quad \text{and} \quad B = QPZ^\top,$$

where Q and Z are orthogonal; P denotes an upper triangular matrix; and S denotes an upper quasi-triangular matrix.

The 1×1 diagonal blocks correspond to the real generalized eigenvalues, and the 2×2 diagonal blocks correspond to the conjugate complex eigenvalues. In this case P is normalized in such a way that the elements in the diagonal, corresponding to the blocks of order 1, are non-negative, whereas the diagonal blocks of P, corresponding to the blocks of order 2, are transformed into diagonal form. So in this case the eigenvalues can be determined without difficulty from the diagonal elements of S and P.

The column vectors of Q and Z are *generalized Schur vectors*, which each span pairs of *deflating subspaces* of A and B (Stewart [360]). Deflating subspaces are the generalization of invariant subspaces: The first k columns of Z span a right deflating subspace, which is mapped by both A and B onto the left deflating subspace spanned by the first k columns of Q. This pair of deflating subspaces corresponds to the first k eigenvalues of S and P.

Table 15.8: LAPACK computational programs for generalized nonsymmetric eigenproblems.

matrix type	operation	REAL	COMPLEX
general	Hessenberg reduction balancing inverse transformation	sgghrd sggbal sggbak	cgghrd cggbal cggbak
Hessenberg	Schur decomposition	shgeqz	chgeqz
(quasi) triangular matrix	eigenvectors	stgevc	ctgevc

The solution of a generalized nonsymmetric eigenproblem is carried out in the following steps (see Table 15.8):

1. The pair of matrices A, B is reduced to generalized upper Hessenberg form. If A and B are real, this decomposition is given by

$$A = UHV^\mathsf{T}, \quad B = UTV^\mathsf{T},$$

   where H is an upper Hessenberg matrix, T is an upper triangular matrix and both U and V are orthogonal matrices. For complex matrices A and B, the decomposition is given by

$$A = UHV^H, \quad B = UTV^H,$$

   where both U and V are unitary and H, T are as in the real case. This decomposition is carried out by the program LAPACK/*gghrd, which calculates H and T and, optionally, U and/or V. LAPACK/*gehrd does not compute the matrices U and V in decomposed form—unlike the corresponding program for standard nonsymmetric eigenvalue problems.

2. The pair H, T is reduced to generalized Schur form by LAPACK/*hgeqz:

$$H = QSZ^\mathsf{T}, \quad T = QPZ^\mathsf{T}$$

   (for real H and T) or

$$H = QSZ^H, \quad T = QPZ^H$$

   (for complex H and T). The values of α and β are calculated as well. The computation of matrices Z and Q is optional.

3. The left and/or right eigenvectors of S and P are calculated using the program LAPACK/*tgevc. The transformation of the right eigenvectors of S and P into the right eigenvectors of A and B (or of H and T) is optional.

Furthermore, there exist programs for balancing the matrices A and B before the reduction to generalized Hessenberg form. Firstly the matrices A and B are multiplied from the left and from the right by a permutation matrix to make them as similar as possible to triangular matrices. The subsequent scaling makes the norms of the rows and columns of A and B as close as possible to 1. In many cases these transformations increase the rate of convergence and the accuracy of subsequent calculations. Sometimes, however, the behavior of the problem deteriorates as a result of the balancing step.

LAPACK/*ggbal carries out the balancing, LAPACK/*ggbak serves for the back-transformation of eigenvectors of the balanced matrices. This is necessary because after the balancing a different Schur form is obtained.

Chapter 16

Large, Sparse Linear Systems

> *He who involves himself too much with the insignificant is incapable of doing greater things.*
>
> LA ROCHEFOUCAULD

Many mathematical models require a large number of variables to describe the state of the system adequately. Mostly, however, the interactions to be described are of a *local* nature. Therefore, only a few variables in a certain neighborhood are connected with each other.

This means that, although there are a large number of variables and equations, each equation involves only a small number of variables;[1] and only these variables have non-zero coefficients in the matrix of the corresponding system of linear (linearized) equations. When this is the case, the matrix is said to be *sparse*.

This matrix structure is found most distinctly in systems of linear equations which result from the discretization of ordinary or partial differential equations. In such linear systems the variables denote the values of one or several quantities at different states and/or at different times. For sufficiently fine discretization, a large number of mesh points is required. On the other hand, a differential equation describes only the *local* dependencies between the value and the derivatives of a variable at a certain point. If the method of finite differences (FDM) is used for discretization, i. e., the derivatives are approximated by difference quotients, then the resulting equation involves only variables belonging to adjoining points. Therefore the number of variables involved in one single equation is independent of the granularity of the discretization, i. e., the total number of variables.

Example (Discretization of Differential Equations) A second order, partial differential equation for a function $u(x, y)$ over a rectangular area can be solved as follows. First, new variables for the unknown values of u at the points of some suitably chosen mesh are introduced. Then, the discretization of the differential equation at some reference mesh point (denoted by a ∘ in Fig. 16.1) leads to a dependency between the value of u at the reference point and the values of u at neighboring mesh points (denoted by a •). For every interior mesh point, an equation involving 9 variables—provided that a 9-point discretization is used—is obtained.

Assuming that the values of u are given at the boundary points and numbering the variables denoting the unknown values at the inner mesh points row-wise results in a linear system whose coefficient matrix has the sparsity structure displayed in Fig. 16.2. Clearly, this matrix has a band structure (cf. Section 13.12), with many zero elements inside the band.

If the geometric form of the domain becomes more complicated, deviations from the band structure of the matrix occur. A very intricate sparsity structure is obtained if the finite element method (FEM) is used for discretization; but in this case also each equation involves only a small number of unknowns.

---

[1] It is assumed that the whole system *cannot* be broken up into independent subsystems, as these subsystems can be solved separately.

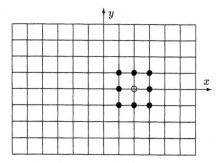

Figure 16.1: Discretization of a partial differential equation on a rectangular grid.

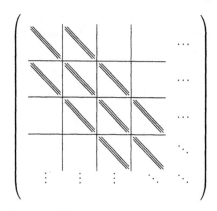

Figure 16.2: Block tridiagonal matrix associated with a discretized partial differential equation.

Typically, sparse systems consist of about 10^3 to 10^5 equations and variables, but only a few percent (or even less in the case of very large systems) of the total of n^2—that is 10^6 to 10^{10} (!)—matrix elements differ from zero.

Example (Automobile Chassis)   Matrix BCSSTK32 of the *Harwell Boeing Collection* [176] is a stiffness matrix derived from a static model of an automobile chassis. This matrix of order $n = 44\,609$ has about $2 \cdot 10^9$ matrix elements, but among them only $1\,029\,655$ non-zero elements. Therefore, the portion of non-zero elements is 0.05 %.

The unmodified Gaussian elimination requires storage for n^2 floating-point numbers and more than $2n^3/3$ floating-point operations. If this algorithm is applied to very large linear systems, the capacity of every computer is soon exceeded. Modern workstations are, at best, able to solve linear systems of equations with thousands of variables. A system with $n = 100\,000$ variables cannot be solved on even the most powerful supercomputers without exploiting the sparsity structure of the coefficient matrix.

It follows, therefore, that complex applications can be modeled and solved only if the respective linear systems are solved efficiently by special algorithms and software products (Dongarra, Van der Vorst [170]) which exploit the specific structure of the sparse coefficient matrices. Particularly in three-dimensional problems, perhaps with time as a fourth dimension, even a rather coarse description of the geometrical and physical dependencies results in an extremely large number of mesh points. But the numerical investigation of such problems which often *cannot* be reduced (e. g., by exploiting symmetry) into lower dimensional problems, is of particular interest.

Storage Schemes for Sparse Matrices

Obviously it is desirable to store only the non-zero coefficient values of a sparse matrix. At the same time, however, information about the positions of the stored coefficient values has to be recorded.

In the case of a *band matrix*, a compact storage can be implemented easily, e. g., by storing the diagonals as rows of a rectangular array and by maintaining the columns (cf. Section 13.19.3):

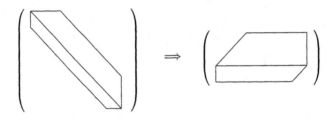

In the case of general sparse matrices (see, for instance, Fig. 16.3), the values of the non-zero elements have to be stored in a suitable order. The position of each element has to be stored separately. Which practical storage scheme is to be chosen for a specific procedure depends strongly, of course, on the selected algorithm.

16.1 Storage Schemes for Iterative Methods

Often iterative methods are used for solving sets of linear equations with large sparse coefficient matrices (Axelsson [93], Kelley [63]). The efficiency of each iterative method depends indirectly on the method chosen for storing the coefficient matrix and on the preconditioning matrix (cf. Section 16.10). Both choices affect the formation of matrix-vector products (cf. Section 16.11). The following sections provide an overview of some practical storage schemes. Often a certain format can be chosen in a natural way according to the distinctive features of the application problem.

In the description of the different storage schemes, $A \in \mathbb{R}^{n \times n}$ denotes the coefficient matrix, n denotes the number of rows and columns of the matrix, and the quantity non-zero$(A) \in \mathbb{N}$ denotes the number of non-zero elements of the matrix.

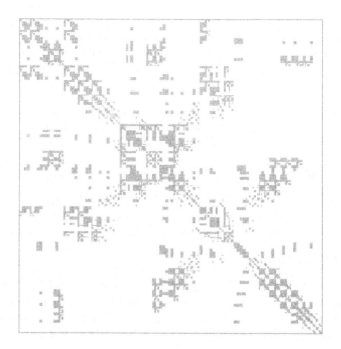

Figure 16.3: Sparsity structure of a symmetric sparse matrix.

Example (Matrix for Illustration) The different storage schemes are illustrated by the matrix

$$A := \begin{pmatrix} 11 & 12 & 0 & 14 & 0 & 0 \\ 0 & 22 & 23 & 0 & 0 & 0 \\ 31 & 0 & 33 & 34 & 0 & 0 \\ 0 & 42 & 0 & 44 & 45 & 46 \\ 0 & 0 & 0 & 0 & 55 & 56 \\ 0 & 0 & 0 & 0 & 65 & 66 \end{pmatrix}, \tag{16.1}$$

a small non-symmetric 6×6 matrix with 44 % non-zero elements:

$$\text{non-zero}(A) = 16.$$

This matrix with non-zero coefficient values $a_{ij} := 10i + j$ serves only to illustrate the different storage schemes. In practice, storage schemes for sparse matrices should never be used for matrices of such a small order or such a high percentage of non-zero elements.

16.1.1 Coordinate (COO) Format

The coordinate format is certainly the simplest storage scheme for sparse matrices. It consists of three arrays of size non-zero(A):

1. a REAL array **value** containing the floating-point values of the non-zero elements of A,

2. an INTEGER array **row_index** containing their row indices and

3. an INTEGER array **col_index** containing their column indices.

The order of the non-zero elements within the array is not important. Thus, this scheme is as general as the CRS format (cf. Section 16.1.2), but, in terms of memory requirements, it is not as efficient; the COO format requires storage for $3 \cdot$ non-zero(A) data elements. Because of its simplicity, this format is very commonly used.

Example (COO Format) Using COO format the matrix (16.1) can be specified by the three arrays given in the following table. The elements have been stored deliberately in a disordered manner to stress that no ordering of the elements is required.

value	65	56	46	34	44	45	22	66	11	12	31	55	23	33	14	42
row_index	6	5	4	3	4	4	2	6	1	1	3	5	2	3	1	4
col_index	5	6	6	4	4	5	2	6	1	2	1	5	3	3	4	2

Modified COO Format

The modified version of the COO format requires only *one* integer array containing the values $(i-1)n + j$ for the non-zero elements a_{ij}. Although this format is more efficient in terms of memory usage, there are two drawbacks to this scheme. Firstly, it requires additional computations to recover the original row and column indices; and secondly, for large matrices it may lead to integer overflow because the handling of the format involves integers which may be very large (of the order of n^2). For these reasons the modified COO format is seldom used in practice.

16.1.2 Compressed Row Storage (CRS) Format

The compressed row and column storage formats (cf. Section 16.1.4) unlike the COO format, do not make any assumptions about the structure of the sparse matrix nor do they store any redundant information. The CRS format is the basic format used by many software packages, e.g., SPARSKIT (Saad [336]).

Terminology (CRS = CSR) In literature on the subject and in some software documents (e.g., SPARSKIT documentation) the CRS format (*compressed row storage*) is often referred to as the CSR format (*compressed sparse row*). In the same way CSC format and CCS format are used synonymously.

The CRS format stores subsequent non-zero elements of the matrix rows in contiguous storage locations. The REAL array value of length non-zero(A) stores the floating-point values of the non-zero elements of the matrix A. The INTEGER array col_index contains the column indices of the elements a_{ij} as stored in the matrix A; that is, if value(k) $= a_{ij}$ then col_index(k) $= j$. The INTEGER array row_pointer of length $n+1$ contains the pointers to the beginning of each row in the arrays value and col_index:

$$\text{value(k)} = a_{ij} \quad \Rightarrow \quad \text{row\_pointer}(i) \le k < \text{row\_pointer}(i + 1).$$

By convention, the definition row_pointer$(n + 1) :=$ non-zero$(A) + 1$ is chosen.

The storage saving for this scheme is significant; only a total number of

$$2 \cdot \text{non-zero}(A) + n + 1$$

storage locations are required. In the case of a symmetric matrix, only the upper (or lower) triangular part of the matrix must be stored. The trade-off is a more complicated referencing.

Example (CRS Format) A possible representation of matrix (16.1) using CRS format:

value	11	12	14	22	23	31	33	34	42	44	45	46	55	56	65	66
col_index	1	2	4	2	3	1	3	4	2	4	5	6	5	6	5	6

row_pointer	1	4	6	9	13	15	17

16.1.3 Modified CRS (MRS) Format

The MRS format is a variation of the CRS format which differs from the CRS format in that the main diagonal of A is stored separately. The first n elements of the REAL array value contain the diagonal elements of A. The position $(n+1)$ of the array is not used. Starting from position $(n+2)$, the non-zero, off-diagonal elements of A are stored row-wise. The INTEGER array index contains the corresponding column indices, except for the first $n + 1$ positions, where the pointers to the beginning of each row are stored. Because in practice many matrices have a full main diagonal

$$(a_{11}, a_{22}, \ldots, a_{nn}) \quad \text{with} \quad a_{11} \neq 0, \ a_{22} \neq 0, \ \ldots, \ a_{nn} \neq 0,$$

this storage scheme is commonly used.

Example (MRS Format) Using MRS format the matrix (16.1) can be specified by the following two arrays:

value	11	22	33	44	55	66		12	14	23	31	34	42	45	46	56	65
index	8	10	11	13	16	17	18	2	4	3	1	4	2	5	6	6	5

16.1.4 Compressed Column Storage (CCS Format)

Compressed column storage, which is also called the *Harwell-Boeing* format [175], is analogous to compressed row storage. The CCS format is identical to the CRS format, except that the columns are traversed instead of the rows. In other words, the CCS format for A is the CRS format for $A^\top$.

16.1.5 Block Compressed Row Storage (BCRS) Format

If the sparse matrix A is comprised of dense, square blocks of non-zero elements in a regular pattern, this particular structure can be exploited by the block compressed row storage format (BCRS format) and the block compressed column storage format (BCCS format).

Block matrices typically arise from the discretization of partial differential equations in which there are several degrees of freedom associated with a grid point. The matrix is partitioned into small blocks with a size equal to the number of degrees of freedom, and each block is treated as a dense matrix, even though it has some zero elements.

If n_b is the dimension of each block, and non-zero-block(A) is the number of blocks containing non-zero elements, then the total storage required for the $n \times n$ matrix A is non-zero-block(A) $\cdot n_b^2$. The *block dimension* of A is then defined by $n_d := n/n_b$.

As with the CRS format, three numerical arrays are required for the BCRS format: a three-dimensional array for floating-point numbers `value(1:non-zero-block(A), 1:n_b, 1:n_b)` which stores the non-zero blocks in a (block) row-wise fashion, an integer array `col_index(1:non-zero-block(A))` which stores the actual column indices in the original matrix A of the $(1,1)$ elements of the non-zero blocks, and a pointer array `row_block(1 : n_d + 1)`, whose entries point to the beginning of each block row in `value(:,:,:)` and `col_index`. Compared to the CRS format, the saving of storage locations and the reduction of indirect addressing for the BCRS format can be significant for matrices with a large n_b.

16.1.6 Compressed Diagonal Storage (CDS) Format

If the matrix A has a band structure with a fairly constant bandwidth, then storing the sub-diagonals of the matrix A in consecutive locations may be advantageous. Not only can the vector `row_index` (resp. `col_index`) be omitted, but the non-zero elements can be arranged in a way which makes the computation of the matrix-vector product more efficient.

The band matrix A with left bandwidth p and right bandwidth q can be stored in a REAL array `value(1:n,-p:q)`. The declaration of the array with reversed dimensions `(-p:q,n)` originates from the LINPACK [11] band format. Usually, storage schemes for band matrices involve storing some of the zero elements; moreover, the format may even contain array elements that do not correspond to matrix elements at all.

Example (CDS Format)   As can be seen with matrix (16.1), using the CDS format on matrices with large bandwidth yields little or no storage benefits. Therefore the CDS format is only suitable for matrices with small bandwidth.

value(:,-2)	0	0	31	42	0	0
value(:,-1)	0	0	0	0	0	65
value(:,0)	11	22	33	44	55	66
value(:,1)	12	23	34	45	56	0
value(:,2)	0	0	0	46	0	0
value(:,3)	14	0	0	0	0	0

A generalization of the CDS format which is more suitable for matrices with varying bandwidth is discussed by Melhem [290]. This variant of the CDS format uses a *stripe* data structure to store the matrix A. This structure stores A more efficiently, but it makes the computation of the matrix-vector product slightly more expensive, as it involves an additional gather operation.

Example (Generalized CDS Format) Matrix (16.1) illustrates that the generalization of the CDS format (Melhem [290]) is also suitable for matrices with larger bandwidth:

row = 6

value(:,-1)	0	0	31	42	0	65
value(:,0)	11	22	33	44	55	66
value(:,1)	0	12	23	34	45	56
value(:,2)	0	14	0	0	46	0

16.1.7 LAPACK (BND) Format for Band Matrices

Band matrices represent the simplest form of sparse matrices, and with the CDS format a storage scheme which exploits the specific structure of band matrices has already been given. Another way of storing band matrices, the BND format (*band storage*), is used in the LAPACK package (cf. Section 13.19.3).

In the BND format the non-zero elements of the matrix A are stored in a rectangular REAL array value; the non-zero elements of the jth column of A are stored in the jth column of the array value. If p and q are the left and right bandwidth of A, then the array value has $p + q + 1$ rows. In addition an INTEGER parameter row is needed to indicate which row of value contains the lowest diagonal.

Example (BND Format) For band matrices with large bandwidths the storage benefits of the BND format are very poor. The 6×6 matrix (16.1) has 6 diagonals with non-zero elements. Their storage in BND format yields no advantages compared with the conventional storage scheme. Hence the BND format should only be used for matrices with small bandwidth.

row = 6

value(:,1)	0	0	0	14	0	0
value(:,2)	0	0	0	0	0	46
value(:,3)	0	12	23	34	45	56
value(:,4)	11	22	33	44	55	66
value(:,5)	0	0	0	0	65	0
value(:,6)	31	42	0	0	0	0

16.1.8 Jagged Diagonal Storage (JDS) Format

The JDS format can be useful for the implementation of algorithms from linear algebra on parallel and vector processors (Saad [335]). Like the CDS format, it gives a vector length essentially the size of the matrix. It saves more space than CDS, but it adds an additional gather operation.

A simplified variant of the JDS format is the ITPACK format, also called *Purdue storage*. Firstly, the non-zero elements are shifted left, and all rows are padded with zeros to give them equal length. The values of this matrix are then stored column-wise in the array value(:,:). Their column indices relating to the original matrix A are stored in the array col_index(:,:).

Example (ITPACK Format, Simplified JDS Format)   The non-zero elements of the 6×6 matrix A are shifted left, giving a 6×4 matrix:

$$\begin{pmatrix} 11 & 12 & 0 & 14 & 0 & 0 \\ 0 & 22 & 23 & 0 & 0 & 0 \\ 31 & 0 & 33 & 34 & 0 & 0 \\ 0 & 42 & 0 & 44 & 45 & 46 \\ 0 & 0 & 0 & 0 & 55 & 56 \\ 0 & 0 & 0 & 0 & 65 & 66 \end{pmatrix} \longrightarrow \begin{pmatrix} 11 & 12 & 14 & 0 \\ 22 & 23 & 0 & 0 \\ 31 & 33 & 34 & 0 \\ 42 & 44 & 45 & 46 \\ 55 & 56 & 0 & 0 \\ 65 & 66 & 0 & 0 \end{pmatrix} .$$

Storing the values and their indices column-wise gives:

value(:,1)	11	22	31	42	55	65
value(:,2)	12	23	33	44	56	66
value(:,3)	14	0	34	45	0	0
value(:,4)	0	0	0	46	0	0

col_index(:,1)	1	2	1	2	5	5
col_index(:,2)	2	3	3	4	6	6
col_index(:,3)	4	0	4	5	0	0
col_index(:,4)	0	0	0	6	0	0

It is clear that the padding zeros in this data structure may be disadvantageous, especially if the bandwidth of the matrix varies greatly. Therefore, in the JDS format, the rows of the matrix are reordered according to the number of non-zero elements per row. The compressed and permuted diagonals are then stored in a one-dimensional array. The new data structure is called a *jagged diagonal*.

Now the representation of the matrix A consists of a permutation array `perm(1:n)`, which describes the reordering of the rows; a floating-point array `jdiag(:)`, which contains the reordered rows, i.e., the jagged diagonal; an integer array `col_index(:)`, which contains the corresponding column indices and finally a pointer array `jd_pointer(:)`, whose elements indicate the beginning of each row in the jagged diagonal.

16.1.9 Skyline Storage (SKS) Format

This format for storing *skyline matrices* (also called *variable band matrices* or *profile matrices*; Duff, Erisman, Reid [174]) has been developed for direct methods, but it can also be used for storing the diagonal blocks in block matrix factorization methods.

The SKS format stores all rows into the floating-point array `value(:)`. The elements of the integer array `row_pointer(:)` point to the beginning of each row. The column indices of the non-zero elements stored in `value(:)` can easily be determined and are not stored.

If the matrix is symmetric, only the lower triangular part is stored. In the case of a non-symmetric matrix, the lower triangular part is stored in SKS format, and the upper triangular part is stored in a column-oriented SKS format. These two parts can be linked in a variety of ways (Saad [336]).

Example (SKS Format) This storage scheme is illustrated by the following *symmetric* 6×6 matrix:

$$A = \begin{pmatrix} 11 & 0 & 31 & 0 & 0 & 0 \\ 0 & 22 & 0 & 42 & 0 & 0 \\ 31 & 0 & 33 & 43 & 0 & 0 \\ 0 & 42 & 43 & 44 & 0 & 0 \\ 0 & 0 & 0 & 0 & 55 & 65 \\ 0 & 0 & 0 & 0 & 65 & 66 \end{pmatrix}. \tag{16.2}$$

Only the lower triangular elements (including the diagonal) are stored row-wise from the first non-zero element to the diagonal.

value	11	22	31	0	33	42	43	44	55	65	66

row_pointer	1	2	3	6	9	10	12

16.2 Storage Schemes for Symmetric Matrices

For symmetric matrices

$$\text{non-zero}(L_A + D_A) = \frac{\text{non-zero}(A) + \text{non-zero}(D_A)}{2} \qquad (16.3)$$

holds, where L_A is the strictly lower triangular part of A, and D_A is the diagonal of the matrix A. As the lower (or upper) triangular part (including the diagonal) is sufficient for a non-redundant storage, there is reason to suppose that the storage requirements of any format can be halved. However, not all storage schemes achieve that 50 % reduction.

The storage requirements of the COO format depend only on the number of non-zero elements of the matrix A. Therefore, using the COO format for symmetric matrices offers considerable storage benefits: the lengths of the three arrays decreases from non-zero(A) to non-zero($L_A + D_A$). The handling of the COO format is also simple because the two integer arrays storing the information about the positions of the matrix elements can be interpreted either as row or column indices. Therefore, the difference between the handling of non-symmetric and symmetric matrices is negligible.

The CRS format also has reduced storage requirements if it is applied to the triangular matrix $L_A + D_A$. The lengths of the arrays value and col_index are reduced to non-zero($L_A + D_A$); only the length of the array row_pointer remains unchanged. The whole matrix can easily be reconstructed by interpreting the data structures in accordance with the CRS format *and* the CCS format. This can be done by using the array col_index as row_index and the array row_pointer as col_pointer. Both the upper and the lower triangular part of the matrix can therefore be retrieved.

The CCS-, MRS-, and MCS-formats can be treated in the same way. If a given block matrix is symmetric with respect to the matrix blocks, the same method can be applied to the BCRS- and BCCS-format.

In the case of symmetric matrices, the CDS format becomes even more simple. As the right and the left side-diagonals are pairwise identical, only one of them has to be stored. Thus, since a matrix with bandwidth p requires only an array value(1:n,0:p), p rows of the array value can be saved. The array value(:,k) then contains both the kth and the $(-k)$th side-diagonal.

In the case of triangular matrices, the BND format does not differ from the CDS format. It must be taken into account, however, that the jth column of the array value corresponds to the jth column of the respective triangular matrix. The values must therefore be shifted in a straightforward manner when referencing a matrix element not contained in the respective triangular part.

The reduction of the storage requirements in the ITPACK format greatly depends on the structure of the matrix. The length of the arrays is determined by the row containing the most non-zero elements. If only a triangular matrix is stored instead of the whole (symmetric) matrix, the maximal number of non-zero elements in one row can remain almost unchanged or can dramatically decrease.

As a result, it is not possible to make general statements about storage reduction.

The SKS format for symmetric skyline matrices has already been discussed in Section 16.1.9.

16.3 Storage Schemes for Direct Methods

Storage schemes which make it possible to recover and manipulate elements of the same column as well as to efficiently insert fill-in elements generated by the elimination process are a prerequisite for elimination algorithms with partial (column) pivoting (cf. Section 16.5).

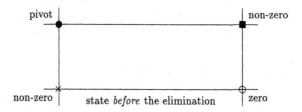

Figure 16.4: Fill-in at the position denoted ○, generated by a single elimination step.

Each time the situation depicted in Fig. 16.4 arises, the elimination of the element marked with an × based on the pivot element (denoted by ●) causes *fill-in* (i. e., the replacement of a zero element by a non-zero element ■) at position ○.

16.3.1 Band Format

If Gaussian elimination with partial pivoting is applied to a band matrix, then the positions generated by fill-in as a result of row interchanges are known a priori (cf. Section 16.5.2). These positions can thus be treated like non-zero matrix elements from the very beginning.

An $n \times n$ matrix is stored in a $(2p+q+1) \times n$ array, assuming that p and q are the left bandwidth and the right bandwidth respectively. The diagonals are stored row-wise one after the other in the array, starting at the bottom of the matrix. Elements located within the same column in the original matrix are always stored within the same column of the array. The first p rows are not used initially; they are reserved for fill-in elements generated by the elimination process.

Example (Band Format)   This and the following formats are illustrated by the 6×6 matrix:

$$A = \begin{pmatrix} 11 & 12 & 0 & 0 & 0 & 0 \\ 0 & 22 & 0 & 24 & 0 & 0 \\ 0 & 32 & 33 & 34 & 0 & 0 \\ 0 & 0 & 43 & 44 & 0 & 46 \\ 0 & 0 & 0 & 54 & 55 & 0 \\ 0 & 0 & 0 & 0 & 65 & 66 \end{pmatrix}. \tag{16.4}$$

This matrix is stored in the 5×6 array `value`.

value(1,:)	—	—	—	0	0	0
value(2,:)	—	—	0	24	0	46
value(3,:)	—	12	0	34	0	0
value(4,:)	11	22	33	44	55	66
value(5,:)	0	32	43	54	65	—

Positions denoted with — are not used by this format.

The rows of the original matrix can be found on the diagonals of the array. Thus, row interchanges and other operations arising during the course of the elimination process can be implemented easily. The implementation of the matrix-vector product is also straightforward. On the other hand, the number of storage locations can only be reduced if $2p+q+1$ is substantially smaller than n.

16.3.2 General Storage Schemes

Perhaps the most convenient way to specify a general sparse matrix is the COO format (cf. Section 16.1.1). Fill-in entries can be inserted easily by appending the new entries. The major difficulty with this data structure lies however in the inconvenience of accessing it by rows or columns. For example, it takes a long time to access all elements of a row or column, because in most cases the whole matrix has to be searched.

Therefore, it may be more suitable to use the CRS format (cf. Section 16.1.2), which allows immediate access to all elements of a row. If a partial (column) pivoting strategy is used, as in most cases, the whole array still has to be searched in order to get all column elements. For this reason the CRS format is not well suited to direct methods.

Linked Lists

The CRS format can be adapted for direct methods by supplying an additional INTEGER array row_list of length non-zero(A). Instead of storing the elements of a row one after the other, the elements in that array are stored using linked lists. For each element of the array value, the corresponding entry in row_list is the index of the next element in the same row. If no more elements are available in that row, the respective index is a null-pointer. Newly generated fill-in elements can only be appended to the end of the array. This attachment requires an update of row_list, which ensures that the new elements are inserted into the appropriate places.

Similarly, an additional array col_list can be used to speed up access to elements of the same row, too.

Example (CRS Format with Row- and Column-Linked Lists) To illustrate the use of this storage scheme, the matrix (16.4) is first stored without element $a_{46} = 46$. This element is then added and the two arrays row_list und col_list are updated:

	1	2	3	4	5	6	7	8	9	10	11	12	13	14
value	11	12	22	24	32	33	34	43	44	54	55	65	66	46
col_index	1	2	2	4	2	3	4	3	4	4	5	5	6	6
row_pointer	1	3	5	8	10	12								
row_list	2	0	4	0	6	7	0	9	14	11	0	13	0	0
col_list	0	3	5	7	0	8	9	0	10	0	12	0	14	0

If, for instance, one looks for element a_{45}, one proceeds as follows. Firstly, one looks for the first element of the fourth row; this element is given by row_pointer(4) = 8, but col_index(8) does not have the desired value, so the next entry has to be consulted. The position of the next element is given by row_list(8) = 9. As the element cannot be found there either, this procedure has to be repeated until the element is found or until the null-pointer indicates that the desired element is not stored in the array. In this case the corresponding entry in the original matrix must have been zero.

This modification of the CRS format requires a lot of additional storage as redundant information about the row indices is stored.

Cyclically Linked Lists

The following method uses *cyclically* linked lists, which do not store this redundant information. It is clear that a reduction of storage requirements results.

The non-zero elements of the matrix are stored in an one-dimensional array value of length $n +$ non-zero(A), skipping the first n positions. The ordering within the array elements is irrelevant. In addition two INTEGER arrays row and col of the same length are required.

The first n entries of the array row are pointers to an element of the respective row. The corresponding entry in the array row either contains a pointer to the next matrix element in that row, or if it was the last element, the index of the row itself is stored. Thus, cycles representing a row can be found in the array. If all values have been correctly assigned, the array contains a permutation of the numbers $\{1, 2, \ldots, n +$ non-zero$(A)\}$. The array col is constructed in exactly the same way except that columns are used instead of rows.

Example (Cyclically Linked Lists) Using cyclically linked lists the matrix (16.4) can be represented as:

	1	2	3	4	5	6	7	8	9	10
value	0	0	0	0	0	0	32	12	55	43
row	11	17	18	20	16	19	3	1	5	4
col	11	17	13	20	12	19	2	7	5	3

	11	12	13	14	15	16	17	18	19	20
value	11	65	33	46	24	54	22	34	66	44
row	8	6	7	10	2	9	15	13	12	14
col	1	9	10	6	4	15	8	16	14	18

This representation is not unique as the elements can be ordered within the array value without restriction.

This format offers quick access to whole rows and columns if only the values of the matrix elements and not their actual positions within the row or column are required. For really sparse matrices, determining the position of a matrix element within a row or column is not very expensive either. New elements can be appended at the end of the array value and the two arrays row and col can be updated without much effort.

16.4 Comparison of Storage Schemes

Four matrices of the *Harwell-Boeing Collection* (cf. Section 16.15.1) have been used for comparing different matrix storage schemes:

BCSSTM07: This symmetric 420×420 matrix originates from an eigenvalue problem in a dynamical analysis in structural engineering. This medium sized matrix has a relatively large number (4.11 %) of non-zero elements.

BCSSTK14: This symmetric matrix arises from static analysis in structural engineering (roof of the *Omni Coliseum*, Atlanta).

CIRPHYS is a non-symmetric matrix which occurs in the computer simulation of a physical model.

WATT1 is a non-symmetric band matrix derived from a problem in petroleum engineering.

For symmetric matrices, only the lower triangular part of the matrix is stored. For the non-symmetric matrices in skyline format, the lower triangular part is stored row-wise, and the upper one is stored column-wise. The storage requirements for each format are given in Tab. 16.1; both the REAL and INTEGER elements required by the method are listed. Obviously, CRS and the MRS are the most efficient formats with respect to storage requirements, but the COO format and the ITPACK format are also suitable.

Table 16.1: Comparison of storage requirements of four *Harwell-Boeing* matrices using different storage schemes.

		BCSSTM07	BCSSTK14	CIRPHYS	WATT1
order		420×420	$1\,806 \times 1\,806$	991×991	$1\,856 \times 1\,856$
number of matrix elements		176 400	3 261 636	982 081	3 444 736
non-zero elements		7 252	63 454	6 027	11 360
non-zeros to zeros ratio		4.11 %	1.95 %	0.61 %	0.33 %
symmetric		yes	yes	no	no
bandwidth		47	161	197/197	64/64
format		*storage requirements*			
COO	REAL	3 836	32 630	6 027	11 360
	INTEGER	7 672	65 260	12 054	22 720
CRS	REAL	3 836	32 630	6 027	11 360
	INTEGER	4 257	34 437	7 019	13 217
MRS	REAL	3 837	32 631	6 028	11 361
	INTEGER	3 837	32 631	6 028	11 361
BND	REAL	20 160	292 572	391 445	239 424
	INTEGER	1	1	1	1
ITPACK	REAL	6 720	54 180	15 856	12 992
	INTEGER	6 720	54 180	15 856	12 992
SKS	REAL	15 111	197 529	155 393	225 351
	INTEGER	421	1 807	1 984	3 714

16.5 Direct Methods

As to how many zero elements are lost due to *fill-in*, given that a certain element is chosen as the pivot element, can be analyzed with respect to the sparsity structure of the matrix. Clearly then, a pivot element providing a minimal amount of fill-in is selected. Therefore, in the elimination process all elements whose columns and rows have not yet supplied a pivot element are pivot candidates.

Strategies for determining "good" pivot elements fall into two categories: those which minimize the fill-in generated at each stage of the elimination process (*local* strategies) and those that confine the fill-in to a certain desirable form (by imposing certain restrictions, for example a band or a small number of columns). As this problem is known to be NP-complete (very hard to solve), only heuristic algorithms are computationally feasible for large problems. In practice, local strategies are used primarily as their implementation is not complicated.

Moreover, it must be noted that defining the best, or even a good sequence of pivot elements is crucial. A solution to the minimum fill-in problem may not be best for a number of reasons. One reason is that it can be very costly to compute, so that the overall computational costs may be less if the pivot sequence is computed in a way that allows more fill-in. The minimal fill-in solution may not permit as much exploitation of the available resources of vector and parallel computers as some pivot sequences generating more fill-in do, or the solution may require a very sophisticated data structure, the cost of which could override any saving of computational costs. These are just a few of the factors that demonstrate that "best" cannot be defined absolutely.

On the other hand the numerical stability of the algorithm has to be preserved. Numerical stability and minimization of fill-in are two conflicting properties. Balancing them is a prerequisite for a stable and efficient algorithm.

In practice, the element a_{ij}^*, which minimizes the fill-in is tested, to see if it is at least one of the "larger" elements (e. g., if it is larger than 25 % of the maximum value in that column). Only if this restriction is satisfied, is a_{ij}^* then selected as pivot element. Otherwise the second best element relating to fill-in minimization is tested. In this way the total amount of fill-in generally remains tolerable; but at the same time the numerical stability of the algorithm does not deteriorate noticeably.

The design of efficient algorithms for the factorization of large-scale sparse matrices seems to be much more costly than that for medium sized dense matrices; heuristic considerations must also be included in the design process.

16.5.1 Gaussian Elimination for Sparse Linear Systems

For an efficient processing of sparse systems, special storage schemes minimize operations on zero elements. Therefore, basically only the non-zero elements and information about their positions within the matrix are stored (cf. Section 16.3).

Using common Gaussian elimination for the solution of such systems may result in the generation of new non-zero elements at positions that previously had zero elements (*fill-in*). Depending on the matrix storage scheme used, the insertion of these new values into the data structure can be very expensive. Gaussian elimination should thus be adapted so as to minimize fill-in (Sherman [347]).

Matrices with Special Sparsity Structure

If information about the particular sparsity structure of a matrix is available, positions which are subject to fill-in may be determined in advance.

In many cases the reordering of the equations and variables can reduce the expected fill-in considerably. For instance, the *block arrowhead matrices*

$$
A_\searrow := \begin{pmatrix} A_1 & B_2 & \cdots & B_r \\ C_2 & A_2 & & \\ \vdots & & \ddots & \\ C_r & & & A_r \end{pmatrix} \qquad A_\searrow := \begin{pmatrix} \hat{A}_1 & & & \hat{B}_1 \\ & \ddots & & \vdots \\ & & \hat{A}_{r-1} & \hat{B}_{r-1} \\ \hat{C}_1 & \cdots & \hat{C}_{r-1} & \hat{A}_r \end{pmatrix}
$$

can be transformed into each other simply by inverting the order of rows and columns. If LU factorization is now applied to both matrices, then $A_\searrow$ is filled up almost completely, while $A_\searrow$ shows no fill-in outside the non-zero blocks.

16.5.2 Band Matrices

Band matrices show very favorable behavior (cf. Section 13.6.6) if Gaussian elimination *without pivoting* is applied. In this case no fill-in outside the original bandwidth is generated. Nevertheless the algorithm may become unstable, as there is no pivoting (cf. Section 13.11.3).

Example (LU Factorization of Band Matrices) Given a band matrix with left bandwidth p and right bandwidth q

$$
A = \begin{pmatrix}
* & * & * & & & 0 \\
* & * & * & * & & \\
& \ddots & \ddots & \ddots & \ddots & \\
& & * & * & * & * \\
& & & * & * & * \\
0 & & & & * & *
\end{pmatrix}, \tag{16.5}
$$

LU factorization without pivoting produces factor matrices of the form

$$
L = \begin{pmatrix}
1 & & & & 0 \\
* & 1 & & & \\
& \ddots & \ddots & & \\
& & * & 1 & \\
& & & * & 1 \\
0 & & & & * & 1
\end{pmatrix}
\qquad
U = \begin{pmatrix}
* & * & * & & & 0 \\
& * & * & * & & \\
& & \ddots & \ddots & \ddots & \\
& & & * & * & * \\
& & & & * & * \\
0 & & & & & *
\end{pmatrix}.
$$

The triangular matrix L has a left bandwidth p and U has a right bandwidth q.

To avoid possible algorithm instabilities, column pivoting can be applied. In this case new non-zero elements may be generated in the super-diagonals of the upper triangular matrix U, but only within a certain distance from the main diagonal. This distance is given by the sum of the left and the right bandwidth.

Example (Pivoting on a Band Matrix) Applying an LU factorization with column pivoting to the band matrix (16.5) leads to an upper triangular matrix U whose right bandwidth is at most $p+q$:

$$
U = \begin{pmatrix}
* & * & * & \diamond & & 0 \\
& * & * & * & \ddots & \\
& & \ddots & \ddots & \ddots & \diamond \\
& & & * & * & * \\
& & & & * & * \\
0 & & & & & *
\end{pmatrix}.
$$

Fill-in can take place in positions denoted $\diamond$. The structure of matrix L is preserved, but the values within the matrix are different from those obtained using LU factorization *without* pivoting.

This *a priori* information about band matrix fill-in can be exploited by providing enough storage locations for possible new super-diagonals at the very beginning of program execution.

16.5.3 Poisson Matrices

A particular category of sparse matrices comes from the discretization of the Poisson equation in two dimensions:

$$u_{xx} + u_{yy} = f(x, y). \tag{16.6}$$

In this equation $f : G \subset \mathbb{R}^2 \to \mathbb{R}$ is a given function, and the solution u satisfies the boundary condition

$$u(x, y) = g(x, y) \qquad \text{for} \quad (x, y) \in \text{boundary}(G),$$

where boundary(G) is the boundary of the region G, where u is sought. Selecting, for instance, the unit square $G := [0, 1] \times [0, 1]$ and superimposing a mesh of horizontal and vertical lines over the region G with a uniform spacing $h := 1/(N+1)$, the following mesh points in the interior of G are obtained:

$$(x_i, y_j) = \left(\frac{i}{N+1}, \frac{j}{N+1} \right), \qquad i, j \in \{1, \dots, N\}.$$

At these mesh points, the second partial derivatives of u are approximated by the second central differences

$$u_{xx}(x_i, y_j) \approx \frac{1}{h^2} \Big(u(x_{i-1}, y_j) - 2u(x_i, y_j) + u(x_{i+1}, y_j) \Big)$$

$$u_{yy}(x_i, y_j) \approx \frac{1}{h^2} \Big(u(x_i, y_{j-1}) - 2u(x_i, y_j) + u(x_i, y_{j+1}) \Big).$$

Using these approximations in the Poisson equation (16.6) (with the notation $u_{ij} := u(x_i, y_j)$ etc.) yields the system of algebraic equations

$$4u_{ij} - u_{i-1,j} - u_{i+1,j} - u_{i,j-1} - u_{i,j+1} = -h^2 f_{ij}, \qquad i, j \in \{1, \dots, N\}. \tag{16.7}$$

These equations together with the boundary conditions

$$u_{0,j} = g(0, y_j), \quad u_{N+1,j} = g(1, y_j), \quad u_{i,0} = g(x_i, 0), \quad u_{i,N+1} = g(x_i, 1).$$

define a discrete analog of the given problem. As can be seen in (16.7) a linear system with N^2 unknowns and the same number of equations can be obtained. Solving this system yields approximations u_{ij} of the solution $u(x_i, y_i)$ of the differential equation at the interior mesh points.

The difference equations (16.7) can be expressed as a system of linear equations whose matrix is

$$A := \begin{pmatrix} T & -I & & \mathbf{0} \\ -I & T & \ddots & \\ & \ddots & \ddots & -I \\ \mathbf{0} & & -I & T \end{pmatrix} \quad \text{with} \quad T := \begin{pmatrix} 4 & -1 & & \mathbf{0} \\ -1 & 4 & \ddots & \\ & \ddots & \ddots & -1 \\ \mathbf{0} & & -1 & 4 \end{pmatrix} \tag{16.8}$$

by row-wise numbering the mesh points $1, 2, \ldots, N^2$ and by ordering the equations and variables according to that arrangement.

This block-tridiagonal matrix (cf. Fig 16.2 on page 375) is called a *Poisson matrix* and has only five diagonals with non-zero elements. But the bandwidth of this matrix is very large—left and right bandwidths are both N. Generally, the interior of this band is almost filled up by the elimination process.

To avoid this fill-in, the structure of the sparse matrix can be changed by ordering the mesh points differently. A favorable structure for reducing fill-in can be obtained by dividing the mesh points into small areas, which are separated by layers of mesh points (*separating sets*). In the resulting subregions the mesh points are numbered in the usual way. The mesh points in the separating sets are numbered last, so that they have the highest numbers. In this way, each subdomain contains a small Poisson matrix with a small bandwidth. The elements of these subregions are only connected to each other and to the adjacent separating sets. Using this *domain decomposition*, a block arrowhead matrix $A_\searrow$ is generated. With regard to fill-in, this matrix has an advantageous form.

$$A = \begin{pmatrix} A_1 & & & & B_1^\mathsf{T} \\ & A_2 & & & B_2^\mathsf{T} \\ & & \ddots & & \vdots \\ & & & A_{r-1} & B_{r-1}^\mathsf{T} \\ B_1 & B_2 & \cdots & B_{r-1} & A_r \end{pmatrix}. \tag{16.9}$$

Example (Reordering of a Poisson Matrix) The consequences of domain decomposition are illustrated by a Poisson matrix, on which a rectangular mesh with 22 interior points is superimposed (Golub, Ortega [49]). This mesh is divided into the three areas D_1, D_2, and D_3, separated by the two separating sets S_1 and S_2. The mesh points are numbered in the way described above:

• 4	• 5	• 6	• 20	• 10	• 11	• 12	• 22	• 16	• 17	• 18
• 1	• 2	• 3	• 19	• 7	• 8	• 9	• 21	• 13	• 14	• 15
D_1			S_1	D_2			S_2	D_3		

By ordering the equations and variables according to their numbering, a coefficient matrix with the desired arrowhead structure can be constructed (see Fig. 16.5).

This figure illustrates that fill-in (denoted by ⋄) can occur only within the bandwidths of the respective blocks and in the last rows and columns. Gaussian elimination on the restructured Poisson matrix causes a fill-in of only 72 elements instead of 182 elements if the Poisson matrix generated by the natural ordering is used.

16.5.4 Matrices with General Sparsity Structure

In the case of general sparse matrices, fill-in is hard to predict. Therefore, it is necessary to minimize the *local* fill-in (caused by a single elimination step). Although the global minimum of fill-in cannot be reached, an essential reduction of fill-in can often be gained.

$$A = \left(\begin{array}{cccccccccccccccccccccc}
4 & -1 & & -1 & & & & & & & & & & & & & & & & & & \\
-1 & 4 & -1 & \diamond & -1 & & & & & & & & & & & & & & & & & \\
 & -1 & 4 & \diamond & \diamond & -1 & & & & & & & & & & & & & -1 & & & \\
-1 & \diamond & \diamond & 4 & -1 & \diamond & & & & & & & & & & & & & \diamond & & & \\
 & -1 & \diamond & -1 & 4 & -1 & & & & & & & & & & & & & \diamond & & & \\
 & & -1 & \diamond & -1 & 4 & & & & & & & & & & & & & \diamond & -1 & & \\
 & & & & & & 4 & -1 & & -1 & & & & & & & & & -1 & & & \\
 & & & & & & -1 & 4 & -1 & \diamond & -1 & & & & & & & & \diamond & & & \\
 & & & & & & & -1 & 4 & \diamond & \diamond & -1 & & & & & & & \diamond & & -1 & \\
 & & & & & & -1 & \diamond & \diamond & 4 & -1 & \diamond & & & & & & & \diamond & -1 & \diamond & \\
 & & & & & & & -1 & \diamond & -1 & 4 & -1 & & & & & & & \diamond & \diamond & \diamond & \\
 & & & & & & & & -1 & \diamond & -1 & 4 & & & & & & & \diamond & \diamond & \diamond & -1 \\
 & & & & & & & & & & & & 4 & -1 & & -1 & & & & & -1 & \\
 & & & & & & & & & & & & -1 & 4 & -1 & \diamond & -1 & & & & \diamond & \\
 & & & & & & & & & & & & & -1 & 4 & \diamond & \diamond & -1 & & & \diamond & \\
 & & & & & & & & & & & & -1 & \diamond & \diamond & 4 & -1 & \diamond & & & \diamond & -1 \\
 & & & & & & & & & & & & & -1 & \diamond & -1 & 4 & -1 & & & \diamond & \diamond \\
 & & & & & & & & & & & & & & -1 & \diamond & -1 & 4 & & & \diamond & \diamond \\
 & & -1 & \diamond & \diamond & \diamond & -1 & \diamond & \diamond & \diamond & \diamond & \diamond & & & & & & & 4 & -1 & \diamond & \diamond \\
 & & & & & -1 & & & & -1 & \diamond & \diamond & & & & & & & -1 & 4 & \diamond & \diamond \\
 & & & & & & & & -1 & \diamond & \diamond & \diamond & -1 & \diamond & \diamond & \diamond & \diamond & \diamond & \diamond & \diamond & 4 & -1 \\
 & & & & & & & & & & & -1 & & & & -1 & \diamond & \diamond & \diamond & \diamond & -1 & 4
\end{array}\right)$$

Figure 16.5: Poisson matrix restructured as a block arrowhead matrix and fill-in ($\diamond$) caused by the Gaussian elimination.

The local fill-in at the kth elimination step can be determined from the sparsity structure of the right lower $(n-k+1)\times(n-k+1)$ principle minor.

The required sparsity structure can be represented by the adjacency matrix B_k, which stores a one at each position where the original matrix has a non-zero element.

Example (Adjacency Matrix) For illustration purposes the 5×5 matrix

$$A := \begin{pmatrix} 3 & 4 & 0 & 0 & 2 \\ 0 & 6 & 3 & 1 & 0 \\ 1 & 0 & 1 & 0 & 0 \\ 0 & 2 & 0 & 0 & 1 \\ 0 & 0 & 2 & 3 & 0 \end{pmatrix} \tag{16.10}$$

is used. The adjacency matrix B_1 conveys the sparsity structure of the matrix A:

$$B_1 := \begin{pmatrix} 1 & 1 & 0 & 0 & 1 \\ 0 & 1 & 1 & 1 & 0 \\ 1 & 0 & 1 & 0 & 0 \\ 0 & 1 & 0 & 0 & 1 \\ 0 & 0 & 1 & 1 & 0 \end{pmatrix}.$$

Applying the first two steps of Gaussian elimination with column pivoting, the following adjacency matrices are obtained:

$$B_2 := \begin{pmatrix} 1 & 1 & 1 & 0 \\ \mathbf{1} & 1 & 0 & \mathbf{1} \\ 1 & 0 & 0 & 1 \\ 0 & 1 & 1 & 0 \end{pmatrix}, \qquad B_3 := \begin{pmatrix} 1 & \mathbf{1} & 1 \\ \mathbf{1} & \mathbf{1} & 1 \\ 1 & 1 & 0 \end{pmatrix}.$$

The positions of the new non-zero elements generated by the elimination steps are marked "**1**" (printed in bold type).

Theorem 16.5.1 (Local Fill-in) *Choosing the element* $a_{ij}^{(k-1)}$ *of the matrix* $A^{(k-1)}$ *as the pivot element at the kth elimination step, the local fill-in is given by the* $(i+1-k,\, j+1-k)$ *element of the matrix*

$$C_k = B_k(E - B_k)^\top B_k,$$

where B_k *is the adjacency matrix of the* $(n-k+1)\times(n-k+1)$ *principle minor of* $A^{(k-1)}$, *and* E *is a matrix of the same dimension with* $e_{ij} = 1$.

Using that theorem, the pivot element minimizing the local fill-in can be determined.

Example (Minimizing Local Fill-in) For the matrix (16.10) one computes:

$$C_1 = B_1(E - B_1)^\top B_1 = \begin{pmatrix} 2 & 3 & 7 & 5 & 1 \\ 4 & 4 & 3 & 1 & 4 \\ 1 & 4 & 2 & 2 & 3 \\ 2 & 1 & 5 & 3 & 0 \\ 3 & 4 & 1 & 0 & 4 \end{pmatrix}.$$

Both a_{54} and a_{45} can be selected as pivot elements, as they do not generate any fill-in. Choosing e. g., a_{54} leads to

$$B_2 = \begin{pmatrix} 1 & 1 & 0 & 0 \\ 0 & 1 & 1 & 0 \\ 1 & 0 & 0 & 1 \\ 1 & 0 & 1 & 1 \end{pmatrix}.$$

Thus, in the next elimination step, one obtains the matrix

$$C_2 = \begin{pmatrix} 2 & 1 & 2 & 2 \\ 4 & 1 & 1 & 3 \\ 1 & 3 & 2 & 0 \\ 3 & 4 & 2 & 1 \end{pmatrix}.$$

Therefore, choosing $a_{45}^{(1)}$ as pivot element will generate no fill-in in this elimination step.

In practice, this method is rarely used since the computations of the C_k involve many additional operations which make the method very expensive. The local fill-in is approximated most often by the so-called *Markowitz costs*:

Theorem 16.5.2 (Markowitz Costs) *If* $a_{ij}^{(k-1)}$ *is chosen as the pivot element in the kth step of the Gaussian elimination process, then the Markowitz costs* MK *of this element*

$$\mathrm{MK}(a_{ij}^{(k-1)}) = \left(\text{non-zero}(z_{i+1-k}^{(k-1)}) - 1\right)\left(\text{non} - \text{zero}(s_{j+1-k}^{(k-1)}) - 1\right)$$

are an upper bound for the local fill-in. $z_i^{(k-1)}$ *denotes the ith row and* $s_j^{(k-1)}$ *the jth column of the matrix* B_k.

Example (Markowitz Costs)   For the matrix (16.10) in the first elimination step, only the elements a_{54} and a_{45} have minimal Markowitz costs ($\mathrm{MK}(a_{45}) = \mathrm{MK}(a_{54}) = 1$). In the second elimination step, the Markowitz costs of $a_{34}^{(1)}$ are equal to those of $a_{45}^{(1)}$.

If the pivot element is chosen only with regard to Markowitz costs, the elimination algorithm may become numerically unstable. In practice the pivot element must thus satisfy additional conditions. For instance, the condition

$$|a_{ij}^{(k-1)}| \geq \rho \max \left\{ |a_{lj}^{(k-1)}| : l = k,\, k+1, \ldots, n \right\}$$

prevents the rounding error from growing too quickly, provided that a suitable $\rho \in (0, 1]$ has been chosen. A choice of a small value for ρ decreases the accuracy of the solution and simultaneously increases the number of possible pivot elements. There are therefore more possible ways of reducing the fill-in.

16.6 Iterative Methods

All iterative methods for solving the linear system $Ax = b$ are founded on the following basic idea: starting with one or several initial approximations

$$x^{(-s)}, x^{(-s+1)}, \ldots, x^{(0)} \in \mathbb{R}^n$$

a sequence $\{x^{(k)} : k = 1, 2, 3, \ldots\}$ is generated whose elements are intended to converge to the true solution x^*. In the most general case, the sequence $\{x^{(k)}\}$ is defined by

$$x^{(k+1)} := T_k(x^{(k)}, x^{(k-1)}, \ldots, x^{(k-s)}), \quad k = 0, 1, 2, \ldots . \tag{16.11}$$

This method is said to be non-stationary as the function T_k depends on k; the number of steps of this multistep method is s since $x^{(k+1)}$ depends on s subsequent iterates. Special cases of the iterative method (16.11) are the non-stationary and the stationary one-step methods

$$x^{(k+1)} := T_k(x^{(k)}) \quad \text{resp.} \quad x^{(k+1)} := T(x^{(k)}), \qquad k = 0, 1, 2, \ldots . \tag{16.12}$$

From (16.11) and (16.12) an *infinite* sequence is defined. To implement an algorithm for solving linear systems based on the function T_k or T, the sequence $\{x^{(k)}\}$ must be truncated after a finite number of iterations. Therefore, contrary to direct methods, a

$$\text{truncation error} \quad x^{(\text{stop})} - x^*,$$

which is inherent to the iterative method, always occurs. Generally, this inevitable error in the procedure is not a fundamental drawback to iterative methods since large sparse systems are often affected with data inaccuracies which greatly exceed the truncation error. If, for instance, the matrix is a result of the discretization of a partial differential equation, both the matrix A as well as the right-hand side b are perturbed by the inevitable discretization error. It makes no sense at all, therefore, to seek a solution to $Ax = b$ whose accuracy is substantially better than the discretization error of the differential equation.

Thus, the decision as to when to stop the method depends greatly on the given problem. A good stopping criterion should

1. stop the iteration as soon as the error $e^{(k)} := x^{(k)} - x^*$ is small enough,

2. stop if the error is no longer decreasing or is decreasing too slowly, and

3. limit the maximum amount of computational effort spent iterating.

Given the meta-algorithm

> **do** $k = 1, 2, 3, \dots$
> $\quad$ *Computation of $x^{(k)}$*
> $\quad$ *Computation of $r^{(k)} := Ax^{(k)} - b$*
> $\quad$ *Computation of $\|r^{(k)}\|$ and $\|x^{(k)}\|$*
> $\quad$ **if** $k \geq maxit$ $\quad$ **or**
> $\quad\quad$ $\|r^{(k)}\| \leq stop\_tol \, (norm\_A \, \|x^{(k)}\| + norm\_b)$ **then exit**
> **end do**

the user must supply the quantities *maxit*, *norm_A*, *norm_b*, and *stop_tol*:

- The integer parameter *maxit* is the maximum number of iterations the algorithm will be permitted to perform.

- The real parameter *norm_A* should be an approximation of $\|A\|$. In many cases the absolute value of the largest entry is adequate.

- The real parameter *norm_b* should be an approximation of $\|b\|$. Again a similar coarse approximation is adequate.

- The real parameter *stop_tol* specifies how small the the residual $r^{(k)} = Ax^{(k)} - b$ of the ultimate solution $x^{(k)}$ is required to be. One possible value for *stop_tol* is the relative data error of A and b. Considering that the entries of A have errors in the range $\pm 10^{-4} \|b\|$, choosing $stop\_tol \geq 10^{-4}$ is suitable. Specifying a smaller value for *stop_tol* is pointless as the algorithm will compute the solution no more accurately than its inherent uncertainty warrants. Generally, *stop_tol* should be chosen so that

$$eps < stop\_tol < 1.$$

It should be noted that if $x^{(k)}$ does not change much from step to step (in the proximity of the solution), then $\|x^{(k)}\|$ need not be recomputed. If $\|A\|$ is not available, the stopping criterion may be modified to $\|r^{(k)}\| \leq stop\_tol \|b\|$. In either case, the final error bound is $\|e^{(k)}\| \leq \|A^{-1}\| \|r^{(k)}\|$.

16.7 Minimization Methods

Many algorithms for the iterative solution of linear systems are based on the minimization of a quadratic function, which is a measure of the residual $Ax - b$. Let $A \in \mathbb{R}^{n \times n}$ be a symmetric positive definite matrix. The determination of the solution vector $x^* \in \mathbb{R}^n$ of the linear system

$$Ax = b \tag{16.13}$$

is equivalent to finding the minimum of the quadratic function

$$f : \mathbb{R}^n \to \mathbb{R} \quad \text{with} \quad f(x) := \tfrac{1}{2}\langle Ax, x \rangle - \langle b, x \rangle. \tag{16.14}$$

The quadratic form

$$\langle Ax, x \rangle = \sum_{i=1}^{n} \sum_{j=1}^{n} a_{ij} x_i x_j$$

is positive definite according to the analogous assumption on A. Thus, the quadratic function f has an unique minimum x^*. This minimum coincides with the solution of (16.13), as

$$\nabla f(x^*) = Ax^* - b = 0.$$

This coincidence is the basis of various iterative algorithms for the numerical solution of the linear system (16.13).

Starting with some initial approximation $x^{(0)} \in \mathbb{R}^n$, a non-vanishing direction vector $p \in \mathbb{R}^n$ is chosen and the step length $\alpha \in \mathbb{R}$ in

$$x^{(1)} := x^{(0)} + \alpha p$$

is determined so that $f(x^{(0)} + \alpha p)$ is minimized with respect to the parameter α. Because

$$
\begin{aligned}
f(x^{(0)} + \alpha p) &= \tfrac{1}{2}\langle A(x^{(0)} + \alpha p), (x^{(0)} + \alpha p) \rangle - \langle b, (x^{(0)} + \alpha p) \rangle \\
&= \tfrac{1}{2}\langle Ax^{(0)}, x^{(0)} \rangle + \alpha \langle Ax^{(0)}, p \rangle + \tfrac{1}{2}\alpha^2 \langle Ap, p \rangle - \langle b, x^{(0)} \rangle - \alpha \langle b, p \rangle
\end{aligned}
$$

the equation

$$\frac{\mathrm{d}f(x^{(0)} + \alpha p)}{\mathrm{d}\alpha} = \alpha \langle Ap, p \rangle + \langle Ax^{(0)} - b, p \rangle = 0 \tag{16.15}$$

is a necessary condition for the presence of the sought minimum. From (16.15) the *optimal* step length,

$$\alpha_{\min} := -\frac{\langle r^{(0)}, p \rangle}{\langle Ap, p \rangle} \quad \text{with} \quad r^{(0)} := Ax^{(0)} - b$$

can be obtained. That this particular choice of α in direction p really gives a minimum follows from

$$\frac{\mathrm{d}^2 f(x^{(0)} + \alpha p)}{\mathrm{d}\alpha^2} = \langle Ap, p \rangle > 0 \quad \text{for all} \quad p \neq 0.$$

The maximal decrement of the quadratic function f—the greatest possible reduction of f in direction p—achieved by the transition from $x^{(0)}$ to

$$x^{(1)} := x^{(0)} + \alpha_{\min} p$$

is given by

$$f(x^{(0)}) - f(x^{(1)}) = \frac{1}{2} \frac{\langle r^{(0)}, p \rangle^2}{\langle Ap, p \rangle} > 0 \qquad \text{for} \quad \langle r^{(0)}, p \rangle \neq 0. \qquad (16.16)$$

The direction vector p cannot be orthogonal to the residual vector $r^{(0)}$, as otherwise $x^{(1)} = x^{(0)}$ because $\alpha_{\min} = 0$.

In the case of two-dimensional functions, the minimization method can be illustrated geometrically (see Fig. 16.6).

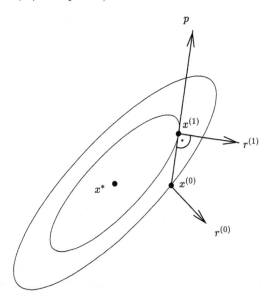

Figure 16.6: Minimization method for determining the solution x^* of a linear system with symmetric positive definite matrix.

The contour lines

$$\{x \in \mathbb{R}^2 \,:\, f(x) = \text{const.} > f(x^*)\}$$

are similar coaxial ellipses, whose common center represents the solution x^*. At the point $x^{(0)}$, the residual vector $r^{(0)}$, which can be expressed as gradient

$$\nabla f(x^{(0)}) = Ax^{(0)} - b = r^{(0)},$$

is orthogonal to the contour line through the initial approximation $x^{(0)}$.

A minimization step in direction p leads to $x^{(1)}$, where $f(x^{(0)} + \alpha p)$ is minimal. At that point $r^{(1)}$ is orthogonal to p. Therefore, $x^{(1)}$ is the tangential point where the minimization direction and the contour line meet.

Using the minimization principle, several iterative methods for the solution of linear systems with a symmetric positive definite matrix can be constructed. These methods differ in the choice of

$$p^{(1)}, p^{(2)}, p^{(3)}, \ldots \qquad \text{and} \qquad \alpha_1, \alpha_2, \alpha_3, \ldots .$$

16.7.1 The Gauss-Seidel Method

Choosing cyclically

$$p^{(1)} = e_1, \ p^{(2)} = e_2, \ \ldots, p^{(n)} = e_n, \ p^{(n+1)} = e_1, \ \ldots$$

for the directions used in the minimization process, one obtains

$$\alpha_{m+1} \ := \ -\frac{\langle r^{(m)}, p^{(m+1)} \rangle}{\langle Ap^{(m+1)}, p^{(m+1)} \rangle} = -\frac{\langle r^{(m)}, e_i \rangle}{\langle Ae_i, e_i \rangle} = -\frac{r_i^{(m)}}{a_{ii}}$$

$$x^{(m+1)} \ := \ x^{(m)} - \frac{r_i^{(m)}}{a_{ii}} e_i \qquad \text{where} \quad i := m \, (\text{mod} \, n) + 1. \qquad (16.17)$$

The minimization step (16.17) modifies only the ith component of the vector $x^{(m)}$:

$$x_i^{(m+1)} \ := \ x_i^{(m)} - (\sum_{j=1}^{n} a_{ij} x_j^{(m)} - b_i)/a_{ii}$$

$$= \ (b_i - \sum_{j=1}^{i-1} a_{ij} x_j^{(m)} - \sum_{j=i+1}^{n} a_{ij} x_j^{(m)})/a_{ii}.$$

The n components, which are the result of one cycle, are used to form the vector $x^{(k)}$. One obtains method (16.33), which is also known as the *method of successive displacements* or the *Gauss-Seidel iterative algorithm* (see Section 16.8.2).

The decrement in f pertinent to step (16.17) is given by (16.16):

$$f(x^{(m)}) - f(x^{(m+1)}) = \frac{1}{2} \frac{\left(r_i^{(m)}\right)^2}{a_{ii}}.$$

The sequence $\{f(x^{(m)})\}$ is therefore monotonically decreasing. Moreover, it is bounded below (i.e., f has a finite minimum), and hence convergent.

Extending the iteration

$$x^{(m+1)} := x^{(m)} + \omega \alpha_{m+1} p^{(m+1)}, \quad m = 0, 1, 2, \ldots$$

with a relaxation parameter ω yields the so-called *successive over-relaxation* or *SOR method* (see Section 16.8.3).

16.7.2 Gradient Methods

The general minimization principle (refer to the beginning of Section 16.7) does not determine the choice of the directions $p^{(1)}, p^{(2)}, p^{(3)}, \ldots$; only directions $p^{(k+1)}$ orthogonal to the residual $r^{(k)}$ must be avoided.

At the point $x^{(k)}$ the residual vector $r^{(k)}$ is the gradient of the function f being minimized. The direction of the gradient is the direction for which the function f has the greatest local rate of change. Thus, a natural choice of the minimization directions is given by

$$p^{(k+1)} := -r^{(k)}, \quad k = 0, 1, 2, \ldots,$$

i. e., the directions of maximal local decrease of f. The *method of steepest descent* can then be expressed as

$$x^{(k+1)} := x^{(k)} - \frac{\langle r^{(k)}, r^{(k)} \rangle}{\langle Ar^{(k)}, r^{(k)} \rangle} r^{(k)}, \quad k = 0, 1, 2, \ldots.$$

In practice, the convergence rate of this method can be very slow, particularly if the matrix is ill-conditioned. Although locally the best minimization directions are chosen, many iteration steps may be required in order to reach close proximity to the solution x^*.

16.7.3 The Jacobi Method

If the gradient $-r^{(k)}$ is chosen as the direction of the iteration, but the α_i are taken to be

$$\alpha_1 = \alpha_2 = \alpha_3 = \cdots = 1, \tag{16.18}$$

then the minimum value of f, to be found in the gradient direction, is *not* reached. The corresponding iterative method

$$x^{(k+1)} := x^{(k)} - r^{(k)}, \quad k = 0, 1, 2, \ldots$$

is known as the *Jacobi method* and is also called the *method of simultaneous displacements*. In this method each component $x_i^{(k)}$ is modified in such a way that the residual of the ith equation vanishes. The corrections $-r_1^{(k)}, -r_2^{(k)}, \ldots, -r_n^{(k)}$ are applied simultaneously; the name of the method is derived from this fact.

As the choice of the steplengths (16.18) does not depend on the problem at hand, the Jacobi method performs poorly compared to the Gauss-Seidel method.

16.7.4 The Conjugate Gradient Method

If the quadratic function (16.14) is diagonal viz.

$$f(x) = \tfrac{1}{2}\langle Dx, x \rangle - \langle b, x \rangle = \tfrac{1}{2} \sum_{i=1}^{n} \lambda_i x_i^2 - \sum_{i=1}^{n} b_i x_i \quad \text{where } D = \operatorname{diag}(\lambda_1, \lambda_2, \ldots, \lambda_n),$$

then n or less minimization steps along the coordinate directions $e_1, \ldots, e_n$ always lead to the minimum x^*.

Using the substitution $x := Py$, the quadratic function (16.14) can be transformed into diagonal form

$$
\begin{aligned}
f(x) &= \tfrac{1}{2}\langle Ax, x \rangle - \langle b, x \rangle \\
     &= \tfrac{1}{2}x^\top Ax - b^\top x \\
     &= \tfrac{1}{2}y^\top P^\top APy - b^\top Py \\
     &= \tfrac{1}{2}\langle Dy, y \rangle - \langle \bar{b}, y \rangle,
\end{aligned}
$$

provided that the column vectors $p_1, \ldots, p_n$ of P form a conjugate basis with respect to A.

Definition 16.7.1 (Conjugate Basis) *Two vectors $p, q \in \mathbb{R}^n$ are conjugate with respect to the symmetric positive definite matrix $A \in \mathbb{R}^{n \times n}$, if the orthogonality relation*

$$
\langle Ap, q \rangle = \langle p, Aq \rangle = \langle p, q \rangle_A = 0
$$

holds. Linearly independent vectors $p_1, \ldots, p_n$ with the property

$$
\langle p_i, Ap_j \rangle = \langle p_i, p_j \rangle_A = 0 \qquad \text{for all} \quad i \neq j
$$

are said to form a conjugate basis of $\mathbb{R}^n$ with respect to the matrix A.

If the vectors of a conjugate basis are chosen as minimization directions, then the solution x^* can be obtained in n or less iteration steps.

An obvious conjugate basis can be formed from the eigenvectors $x_1, \ldots, x_n$ of the matrix A, since

$$
\langle x_i, x_j \rangle_A = \langle x_i, Ax_j \rangle = x_i^\top Ax_j = \lambda_j x_i^\top x_j = 0
$$

holds for all $i \neq j$. The numerical solution of the complete eigenvalue problem requires, however, a much greater computational effort than the solution of the original linear system. Therefore, the determination of an eigenvector basis is out of the question. The same cost argument applies to orthogonalization methods. The crucial point in the application of this seemingly obvious method consists in finding a conjugate basis with a reasonable amount of work.

The *conjugate gradient* method—the *CG method*—begins in the same way as the gradient method, by choosing the negative gradient vector $p^{(1)} := -r^{(0)}$ as the first direction vector $p^{(1)}$ and seeking the minimum of f in that direction:

$$
x^{(1)} := x^{(0)} - \alpha_0 r^{(0)} \qquad \text{with} \quad \alpha_0 := -\frac{\langle r^{(0)}, p^{(1)} \rangle}{\langle Ap^{(1)}, p^{(1)} \rangle} = \frac{\langle r^{(0)}, r^{(0)} \rangle}{\langle Ar^{(0)}, r^{(0)} \rangle}.
$$

The second and all other direction vectors are linear combinations of $r^{(k-1)}$ and $p^{(k-1)}$:

$$
p^{(k)} := -r^{(k-1)} - \beta_{k-1} p^{(k-1)}, \qquad k = 2, 3, 4, \ldots . \tag{16.19}
$$

The coefficient β_{k-1} is chosen in such a way that $p^{(k)}$ and $p^{(k-1)}$ are conjugate vectors with respect to A.

$$\langle Ap^{(k)}, p^{(k-1)}\rangle = \langle p^{(k)}, Ap^{(k-1)}\rangle = 0. \tag{16.20}$$

From (16.19), (16.20), and from the orthogonality $\langle r^{(k-1)}, p^{(k-1)}\rangle = 0$ (cf. Golub, Van Loan [50]) it follows that

$$\beta_{k-1} = -\frac{\langle r^{(k-1)}, Ap^{(k-1)}\rangle}{\langle p^{(k-1)}, Ap^{(k-1)}\rangle} = \frac{\langle r^{(k-1)}, r^{(k-1)}\rangle}{\langle r^{(k-2)}, r^{(k-2)}\rangle}, \quad k = 2, 3, 4, \dots .$$

In the direction $p^{(k)}$ specified by the above definitions, the function f is minimized:

$$x^{(k+1)} := x^{(k-1)} + \alpha_k p^{(k)} \quad \text{with} \quad \alpha_k := -\frac{\langle r^{(k-1)}, p^{(k)}\rangle}{\langle Ap^{(k)}, p^{(k)}\rangle} = \frac{\langle r^{(k-1)}, r^{(k-1)}\rangle}{\langle p^{(k)}, Ap^{(k)}\rangle}.$$

The new residual vector $r^{(k)}$ can be calculated as

$$r^{(k)} = Ax^{(k)} - b = Ax^{(k-1)} + \alpha_k Ap^{(k)} - b = r^{(k-1)} + \alpha_k Ap^{(k)}.$$

This recursion formula reduces the computational effort, as the matrix-vector product $Ap^{(k)}$ has already been computed to determine the α_k.

Thus, the following **CG algorithm** is the result of the previous considerations:

> *choose an initial guess* $x^{(0)}$
> $r^{(0)} := Ax^{(0)} - b$
> **do** $k = 1, 2, 3, \dots$
>   **if** $k = 1$ **then**
> $p^{(1)} := -r^{(0)}$
>   **else**
> $\beta_{k-1} := \dfrac{\langle r^{(k-1)}, r^{(k-1)}\rangle}{\langle r^{(k-2)}, r^{(k-2)}\rangle}$
> $p^{(k)} := -r^{(k-1)} - \beta_{k-1}p^{(k-1)}$ (16.21)
>   **end if**
> $\alpha_k := \dfrac{\langle r^{(k-1)}, r^{(k-1)}\rangle}{\langle p^{(k)}, Ap^{(k)}\rangle}$
> $x^{(k)} := x^{(k-1)} + \alpha_k p^{(k)}$
> $r^{(k)} := r^{(k-1)} + \alpha_k Ap^{(k)}$ (16.22)
> **end do**

In the absence of rounding errors, the application of the CG method to linear systems with symmetric and positive definite coefficient matrices would produce the solution in n or less iterations. Thus, the CG method is—contrary to the other minimization methods—a *direct* method for the solution of linear systems. This property is lost however if the CG method is performed on a computer system as rounding errors cannot be neglected. It cannot be guaranteed that the direction vectors remain conjugate and that the residual vectors remain orthogonal. As a consequence, $r^{(n)}$ is different from the zero vector, and the computations can be continued in that iterative manner. This procedure can be justified by the fact that the CG method is a minimization method that also decreases the value of the quadratic function after the nth step.

16.7.5 The Krylov Method

In the CG method $x^{(k)}$ is constructed as a linear combination of the direction vectors and the initial guess:

$$x^{(k)} = \alpha_k p^{(k)} + \alpha_{k-1} p^{(k-1)} + \cdots + \alpha_1 p^{(1)} + x^{(0)}.$$

It follows that
$$x^{(k)} \in x^{(0)} + \text{span}\,\{p^{(1)}, p^{(2)}, \ldots, p^{(k)}\},$$

where the quadratic function (16.14) is minimized by the constructed vector $x^{(k)}$. From the equations (16.21) and (16.22) it follows that

$$p^{(k)} \in \text{span}\,\{r^{(0)}, Ar^{(0)}, A^2 r^{(0)}, \ldots, A^{k-1} r^{(0)}\}.$$

Definition 16.7.2 (Krylov Space)   *Given a vector* $r \in \mathbb{R}^n$ *and a matrix* $B \in \mathbb{R}^{n \times n}$, *then*

$$\mathcal{K}_i(r, B) := \text{span}\,\{r, Br, B^2 r, \ldots, B^{i-1} r\}$$

defines a subspace of $\mathbb{R}^n$. *This subspace is denoted as the ith Krylov space generated by* r *and* B.

The elements of the ith Krylov space are the values of *all* matrix polynomials

$$P_{i-1}(B) := c_0 + c_1 B + c_2 B^2 + \cdots + c_{i-1} B^{i-1}$$

of degree $i - 1$ applied to the vector r, whence

$$\mathcal{K}_i(r; B) = \{P_{i-1}(B)\, r : P_{i-1} \in \mathbb{P}_{i-1}\}.$$

Iterative methods with
$$x^{(i)} \in x^{(0)} + \mathcal{K}_i(r^{(0)}, A)$$

are therefore often called *Krylov*, or *polynomial methods*. A polynomial

$$R_i \in \mathbb{P}_i \qquad \text{with} \qquad R_i(0) = 1 \tag{16.23}$$

can be used to characterize the residual vector $r^{(i)}$:

$$r^{(i)} = Ax^{(i)} - b = R_i(A) r^{(0)}. \tag{16.24}$$

Therefore, every polynomial with property (16.23) is called a *residual polynomial*.

As can be seen from equation (16.24), the goal of an efficient Krylov method is to choose the residual polynomials $R_1, R_2, \ldots$ so that the residual vectors $r^{(1)}, r^{(2)}, \ldots$ are minimized in a certain sense. This minimization property can be derived e. g., from the minimization of the residual:

$$
\begin{aligned}
\|r^{(i)}\| &= \min\{\|Ax - b\| : x \in x^{(0)} + \mathcal{K}_i(r^{(0)}, A)\} \qquad (16.25)\\
&= \min\{\|R_i(A)r^{(0)}\| : R_i \in \mathbb{P}_i,\ R_i(0) = 1\},
\end{aligned}
$$

where $\|\ \|$ is a vector norm on $\mathbb{R}^n$, which can be chosen arbitrarily at each step of the iteration as it is done, for instance, in the QMR method (cf. Section 16.9.6). Choosing the A-norm defined by

$$
\|y\|_A := \sqrt{\langle y, y \rangle_A} = \sqrt{\langle y, Ay \rangle},
$$

one obtains the CG method (without preconditioning). All methods based on (16.25) are called *minimal residual methods* (*MR methods*).

A second approach to the required minimization property involves the orthogonality of the residual and leads to a subspace of $S_i \subseteq \mathbb{R}^n$ (cf. Fig. 16.7):

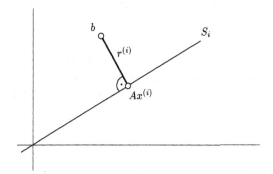

Figure 16.7: Minimization of the residual by choosing $r^{(i)}$ orthogonal to the space S_i.

$$
\langle r^{(i)}, s \rangle = 0 \qquad \text{forall} \quad s \in S_i. \qquad (16.26)
$$

In contrast to the sequence $x^{(1)}, x^{(2)}, \ldots$ defined by (16.25), the condition (16.26) cannot guarantee that for each i a new approximation $x^{(i)}$ is defined. Methods based on (16.26) are called *orthogonal residual methods* (*OR methods*).

In both the MR method and the OR method, the next iteration vector $p^{(i)}$ is determined solely from the basis of the Krylov space $\mathcal{K}_i(r^{(0)}, A)$ and from the subspace S_i respectively; no additional parameter of the iteration needs to be known *a priori*.

16.8 Stationary Iterative Methods

In Section 16.7 iterative methods for solving linear systems have been derived. All these methods are based on the minimization of a quadratic function. For linear systems with a symmetric and positive definite coefficient matrix, this approach arises naturally and intuitively. Iterative methods for linear systems with a *general* (non-symmetric and/or indefinite) matrix are derived more conveniently by specializing iterative methods for a system of nonlinear equations which are introduced in Section 14.1.

A system of nonlinear equations, given in *fixed-point form*

$$x = T(x)$$

can be solved by choosing an initial approximation $x^{(0)} \in \mathbb{R}^n$ and by successively calculating more accurate approximations using the iteration

$$x^{(k)} := T(x^{(k-1)}), \quad k = 1, 2, 3, \dots .$$

Definition 16.8.1 (Stationary Iteration) *An iteration $x^{(k+1)} := T(x^{(k)})$ is called stationary if the definition of the operator T does not depend on the iteration step.*

Stationary iterative methods for the solution of linear systems have the basic form

$$x^{(k)} := Bx^{(k-1)} + c, \quad k = 1, 2, 3, \dots \tag{16.27}$$

where $B \in \mathbb{R}^{n \times n}$ and $c \in \mathbb{R}^n$ do not depend on k. Initially, however, the linear system $Ax = b$ must be transformed into a form to which the iterative method is applicable. For instance, the coefficient matrix can be written as

$$A = L + D + U, \tag{16.28}$$

where $D = \text{diag}(a_{11}, \dots, a_{nn})$, L is a *strictly* lower triangular matrix and U is the respective upper triangular part:

$$L = \begin{pmatrix} 0 & & & & & \mathbf{0} \\ a_{21} & 0 & & & & \\ a_{31} & a_{32} & & & & \\ \vdots & \vdots & & & & \\ a_{n-1,1} & a_{n-1,2} & \cdots & 0 & & \\ a_{n1} & a_{n2} & \cdots & a_{n,n-1} & 0 \end{pmatrix}, \quad U = \begin{pmatrix} 0 & a_{12} & \cdots & & a_{1n} \\ & 0 & \cdots & & a_{2n} \\ & & & & \vdots \\ & & & & a_{n-2,n} \\ & & & & a_{n-1,n} \\ \mathbf{0} & & & & 0 \end{pmatrix}.$$

Using (16.28), the original linear system can be expressed as

$$Dx = b - Lx - Ux. \tag{16.29}$$

Provided that A is non-singular, it is possible to reorder D (by row and column interchanges if necessary) so that it is also non-singular. Therefore, all of the a_{ii} are non-zero, and (16.29) can be transformed into

$$x = D^{-1}b - D^{-1}(L + U)x.$$

Defining $B := -D^{-1}(L+U)$ and $c := D^{-1}b$, the fixed-point equation

$$x = T(x) := Bx + c$$

is obtained. The corresponding stationary iterative method (16.27) is the *Jacobi method* (cf. Section 16.7.3). Other iterative methods are obtained by decomposing A in other ways.

Among the stationary iterative methods, the following methods are best known:

The Jacobi method: A single iteration step of the Jacobi method corresponds to the local solution of the linear system for a single variable (cf. Section 16.7.3). The resulting method is easy to implement, but often converges very slowly.

The Gauss-Seidel method: This is like the Jacobi method, except that it uses updated approximations as soon as they are available (cf. Section 16.7.1). In general, it will converge faster than the Jacobi method, though still relatively slowly.

The SOR method: This can be derived from the Gauss-Seidel method by introducing an extrapolation parameter ω. For the (nearly) optimal choice of ω, the convergence speed can be increased substantially.

The SSOR method: The symmetric SOR method has no advantage over the SOR method. It is however useful as a preconditioning scheme for non-stationary methods (cf. Section 16.10.2).

16.8.1 The Jacobi Method

If $x_1, \ldots, x_{i-1}, x_{i+1}, \ldots, x_n$ are given, then solving the ith equation of the linear system $Ax = b$ with respect to x_i, the ith component of x, one obtains

$$x_i^{(k)} = (b_i - \sum_{j \neq i} a_{ij} x_j^{(k-1)})/a_{ii}$$

provided that $a_{ii} \neq 0$. This suggests the following algorithm—the *Jacobi method*:

do $k = 1, 2, 3, \ldots$
    do for $i \in \{1, 2, \ldots, n\}$
$$x_i^{(k)} := (b_i - \sum_{j=1}^{i-1} a_{ij} x_j^{(k-1)} - \sum_{j=i+1}^{n} a_{ij} x_j^{(k-1)})/a_{ii} \qquad (16.30)$$
    end do
    if *stopping criterion holds* **then exit**
end do

The order in which the equations are solved is irrelevant. In principal, the updates could be done simultaneously as indicated in the FORALL loop for i. Thus, the Jacobi method is also known as the *method of simultaneous displacements*. In matrix terms, (16.30) can be expressed as

$$x^{(k)} = -D^{-1}(L+U)x^{(k-1)} + D^{-1}b, \qquad k = 1,2,3,\ldots .$$

where D, L, and U represent the diagonal, the strictly lower triangular, and the strictly upper triangular parts of A.

Convergence Behavior

In Section 14.1 the convergence of the general stationary iteration

$$x^{(k)} := T(x^{(k-1)}), \quad k = 1,2,3,\ldots$$

to the solution x^* of the fixed-point equation $x = T(x)$ is characterized by the (contraction) Theorem 14.1.1 and hence by the contractivity of the operator T. In case of an affine mapping

$$T(x) := Bx + c \qquad\qquad\qquad (16.31)$$

the Lipschitz constant of T is given by $L = \|B\|$ because

$$\|T(x) - T(y)\| = \|Bx - By\| = \|B(x-y)\| \le \|B\| \, \|x - y\|.$$

The function (16.31) is thus contracting if and only if there exists a matrix norm $\|\ \|$, for which
$$\|B\| < 1.$$

If the matrix A is *strictly diagonally dominant*, i.e.,

$$|a_{ii}| > \sum_{j=1}^{i-1} |a_{ij}| + \sum_{j=i+1}^{n} |a_{ij}| \qquad \text{for all} \quad i = 1,2,\ldots,n,$$

then the row sum norm $\|\ \|_\infty$ of the iteration matrix $B_J = -D^{-1}(L+U)$ of the Jacobi method satisfies the convergence criterion

$$\|B_J\|_\infty < 1.$$

Hence, for every possible choice of the initial approximation $x^{(0)}$ the Jacobi method converges to the solution of that linear system.

The condition $\|B\| < 1$ is closely related to the spectral radius

$$\rho(B) := \max\{|\lambda_i| : \lambda_i \in \lambda(B), i = 1,2,\ldots,n\}$$

of the iteration matrix B.

Theorem 16.8.1 *For every $\varepsilon > 0$ there is a matrix norm with the property that*

$$\|B\| \le \rho(B) + \varepsilon \qquad \text{for all} \quad B \in \mathbb{R}^{n \times n}.$$

Proof: Ortega, Rheinboldt [67].

From this theorem it follows that the function (16.31) is contracting if the condition $\rho(B) < 1$ is satisfied.

If, on the other hand, an eigenvalue λ_i with $|\lambda_i| \ge 1$ exists, then for some eigenvector $x \ne 0$ corresponding to λ_i it follows that

$$B^k x = \lambda_i^k x, \quad k = 1, 2, 3, \dots,$$

and therefore (16.31) cannot be a contracting function. Hence, $\rho(B) < 1$ is a necessary and sufficient condition for the contractivity of (16.31).

The convergence rate of a stationary iterative method highly depends on the value of $\|B\|$ and $\rho(B)$. As in the nonlinear case (Chapter 14) the maximum number of steps k_{max} which must be executed to meet the accuracy requirement

$$\|x - x^{(k)}\| \le \varepsilon_{abs}$$

is given by

$$k_{max} = \left| \frac{\log\left(\varepsilon_{abs}\,(1 - \|B\|)/\|x^{(1)} - x^{(0)}\|\right)}{\log \|B\|} \right|.$$

Hence, the stronger the dominance of the diagonal elements, for instance, the smaller $\|B\|$ is, and the faster the convergence of the Jacobi method.

In all iterative methods the convergence rate is an important efficiency criterion. For dense matrices A the number of floating-point operations caused by the application of an iterative method is smaller than that of a direct method only if $k_{max} < n/3$ or $k_{max} < n/6$ in the case of symmetric matrices. Except when very modest accuracy is required and/or in other special cases, this condition is not satisfied by the problems typically occurring in practice.

Example (Discretization of Differential Equations) The convergence behavior of iterative methods can be investigated using ordinary or partial differential equations, to which these methods are often applied. For the two-point boundary value problem

$$\begin{aligned} -y''(x) &= f(x), & x \in [0,1] \\ y(0) &= \alpha, & \alpha \in \mathbb{R} \\ y(1) &= \beta, & \beta \in \mathbb{R} \end{aligned}$$

(compare with the example on page 196), by discretization (with a constant step size h) a system of linear equations with the coefficient matrix

$$A_h := \frac{1}{h^2} \begin{pmatrix} 2 & -1 & & 0 \\ -1 & 2 & \ddots & \\ & \ddots & \ddots & -1 \\ 0 & & -1 & 2 \end{pmatrix} \tag{16.32}$$

is obtained. The eigenvalues of A_h are

$$\lambda_{h,k} = \frac{1}{h^2}\left(2 - 2\cos\left(\frac{k\pi}{N+1}\right)\right) = \frac{4}{h^2}\sin^2\left(\frac{k\pi}{2N+2}\right), \quad k = 1,2,\dots,N,$$

the second part of this equation resulting from the identity $1 - \cos 2\varphi = 2\sin^2\varphi$ (Golub, Ortega [49]). The iteration matrix of the Jacobi method has the eigenvalues

$$\mu_{h,k} = \cos\left(\frac{k\pi}{N+1}\right), \quad k = 1,2,\dots,N,$$

and the same eigenvectors as (16.32), namely

$$v_{h,k} = \left(\sin\left(\frac{k\pi}{N+1}\right), \sin\left(\frac{2k\pi}{N+1}\right),\dots,\sin\left(\frac{Nk\pi}{N+1}\right)\right)^{\mathsf{T}}, \quad k = 1,2,\dots,N.$$

The larger k is the faster the function $\sin(k\pi x)$ oscillates; discrete values of $\sin(k\pi x)$ constitute the components of the kth eigenvector. Hence, for small values of k the eigenvectors are said to be "smooth", while for large values of k the respective eigenvectors are said to be "oscillating" (and there is no sharp boundary between the two types). Eigenvectors which correspond to the smallest eigenvalues are the ones which are most strongly damped when the Jacobi method is used. A modification of the Jacobi method that damps the most oscillating eigenvectors first (Golub, Ortega [49]) has been developed.

Multigrid Methods

Some iterative methods (such as the modified Jacobi method previously mentioned) tend to reduce the highly oscillating components of the error very quickly. This fact has initiated the development of methods based on several steps roughly outlined in the following:

1. the performance of some iterations of a basic method (e.g., of the Jacobi method) in order to smooth out the error;

2. the restriction of the discretization to a subset of the current grid points, the *coarse grid* and the numerical solution of the resulting projected problem;

3. the interpolation of the coarse grid solution leading back to the original (finer) grid and the performance of a number of iterations of the basic method again.

Steps 1 and 3 are called *pre-smoothing* and *post-smoothing* respectively; applying this method recursively results in a *multigrid method*. Usually the generation of increasingly coarser grids is stopped as soon as the efficient solution of the corresponding linear system with a direct method is feasible.

16.8.2 The Gauss-Seidel Method

The single equations of the linear system $Ax = b$ can be solved consecutively, whereby already computed values can be used immediately:

do $k = 1, 2, 3, \ldots$
    do $i = 1, 2, \ldots, n$

$$x_i^{(k)} := (b_i - \sum_{j=1}^{i-1} a_{ij} x_j^{(k)} - \sum_{j=i+1}^{n} a_{ij} x_j^{(k-1)})/a_{ii} \tag{16.33}$$

    end do
    if *stopping criterion holds* **then exit**
end do

In this case concurrent computations are not possible since each component of the new iterate depends on all previously computed components. Hence, the Gauss-Seidel method is sometimes called the *method of successive displacements*.

The vector $x^{(k)}$ depends on the order in which the equations are processed. If this order is changed, the components of the new iterate will also change.

If A is sparse, then the dependence of each component of the new iterate on previous components is not absolute. The presence of zeros in the matrix may remove the influence of some of the previous components. A judicious reordering of the equations may reduce the undesirable dependence, thus making concurrent updates to certain groups of components possible. However, reordering the equations can affect the rate at which the Gauss-Seidel method converges. A suitable order can enhance the rate of convergence; a poor order can degrade it.

In matrix terms, (same notation as for the Jacobi method) the Gauss-Seidel method can be defined by

$$\begin{aligned} x^{(k)} &:= (D+L)^{-1}(b - Ux^{(k-1)}) \\ &= x^{(k-1)} - (D+L)^{-1}(Ax^{(k-1)} - b), \qquad k = 1, 2, 3, \ldots . \end{aligned} \tag{16.34}$$

Convergence Behavior

As previously computed results are used in (16.33) as soon as possible, one may suppose that the Gauss-Seidel method converges faster then the Jacobi method. In some special cases this expectation can be proved:

Theorem 16.8.2 *If $A \in \mathbb{R}^{n \times n}$ is a regular, strictly diagonally dominant matrix, then the Gauss-Seidel method converges for every initial guess $x^{(0)}$. The rate of convergence is higher or at least not lower than for the Jacobi method.*

Proof: Hämmerlin, Hoffman [52].

Under even stronger restrictions to the class of matrices, it holds that the Gauss-Seidel method converges twice as fast as the Jacobi method (Ortega [305]). The faster convergence of the Gauss-Seidel method *cannot* be stated generally. In some case the Gauss-Seidel method diverges, although the Jacobi method converges. Convergence can be guaranteed for symmetric positive definite matrices, which can often be found in practice:

Theorem 16.8.3 *If $A \in \mathbb{R}^{n \times n}$ is a symmetric positive definite matrix, then the Gauss-Seidel method converges for every initial guess $x^{(0)}$.*

Proof: Golub, Van Loan [50].

Example (Discretization of Differential Equations) The tridiagonal matrix (16.32) arises from the solution of the boundary value problem of an ordinary differential equation if the second central difference quotient

$$(y_{i+1} - 2y_i + y_{i-1})/h^2 \approx y''(x_i)$$

is used for discretization. The matrix (16.32) is symmetric and positive definite. The Gauss-Seidel method can therefore be used for the solution of the resulting linear system.

16.8.3 The Successive Over-Relaxation (SOR) Method

The *successive over-relaxation method* is devised by adding extrapolation to the Gauss-Seidel method. The extrapolation is applied to each component separately. It takes the form of a weighted average between the previous iterate and the newly computed Gauss-Seidel iterate:

> **do** $k = 1, 2, 3, \ldots$
>> **do** $i = 1, 2, \ldots, n$
>>
>> $$x_i^{(k)} := \omega \left(b_i - \sum_{j=1}^{i-1} a_{ij} x_j^{(k)} - \sum_{j=i+1}^{n} a_{ij} x_j^{(k-1)} \right)/a_{ii} + (1 - \omega) x_i^{(k-1)}$$
>>
>> **end do**
>> **if** *stopping criterion holds* **then exit**
> **end do**

or in matrix terms

$$\begin{aligned}
x^{(k)} &= (D + \omega L)^{-1}[(1 - \omega)D - \omega U]x^{(k-1)} + \omega(D + \omega L)^{-1}b \\
&= x^{(k-1)} - \omega(D + \omega L)^{-1}(Ax^{(k-1)} - b), \qquad k = 1, 2, 3, \ldots .
\end{aligned} \qquad (16.35)$$

Obviously, for $\omega = 1$ the SOR method (16.35) is identical to the Gauss-Seidel method (16.34).

Because $\omega > 1$ in most cases, the term *over-relaxation* is used. The idea is to choose a value for the extrapolation factor ω that will accelerate the convergence rate of the algorithm.

Choosing the Relaxation Factor

In general, it is not possible to compute optimal values of ω in advance. All implementations of the SOR method therefore utilize some kind of approximation to determine a suitable value for ω. In some cases, heuristic estimates can be used.

Example (Discretization of Differential Equations) For matrices arising from the discretization of ordinary or partial differential equations, the heuristic estimate

$$\omega := 2 - O(h) \qquad \text{as} \quad h \to 0$$

is often used, where h characterizes the mesh spacing of the discretization.

The choice of ω is restricted by the following theorem:

Theorem 16.8.4 (Kahan) *The spectral radius of the iteration matrix B_ω of the SOR method satisfies*

$$\rho(B_\omega) \geq |\omega - 1| \qquad \text{for all} \quad \omega \in \mathbb{R}.$$

Proof: Hämmerlin, Hoffmann [52].

It follows that $\rho(B_\omega) < 1$ can hold only if $\omega \in (0, 2)$. Under certain additional assumptions, the convergence of the SOR method can be guaranteed.

Theorem 16.8.5 (Ostrowski, Reich) *If the coefficient matrix $A \in \mathbb{R}^{n \times n}$ is symmetric and positive definite, then the SOR method converges for any $\omega \in (0, 2)$ and for any initial guess $x^{(0)}$.*

Proof: Ortega [305].

Thus, the convergence is guaranteed for any $\omega \in (0, 2)$, although the choice of ω can significantly affect the rate at which the SOR method converges.

For coefficient matrices of a special class, e.g., for tridiagonal block matrices whose diagonal blocks are diagonal matrices, the spectral radius $\rho(B_J)$ of the Jacobi iteration matrix B_J can be used for determining the optimal value of ω:

$$\omega_{\text{opt}} = \frac{2}{1 + \sqrt{1 - \rho(B_J)^2}}.$$

This is seldom done since calculating the spectral radius of the Jacobi matrix requires an impractical amount of computation. However, relatively inexpensive rough approximations of $\rho(B_J)$ can yield reasonable estimates for ω_{opt}.

16.8.4 The Symmetric SOR (SSOR) Method

If the coefficient matrix A is assumed to be *symmetric*, then the symmetric overrelaxation method (SSOR method) combines two SOR sweeps in such a way that the resulting iteration matrix is similar to a symmetric matrix. To achieve this goal, the first matrix multiplication corresponds exactly to (16.35), while the second step updates the variables in reversed order. The similarity of the SSOR iteration matrix to a symmetric matrix permits the application of the SSOR matrix as a preconditioner for non-stationary iterative methods (see Section 16.10.2). Indeed, this is the primary motivation for the SSOR method since its convergence rate is usually slower than the convergence rate of an optimal SOR method.

In matrix terms, the SSOR method can be expressed as follows:

$$x^{(k)} := B_1 B_2 x^{(k-1)} + \omega(2 - \omega)(D + \omega U)^{-1} D(D + \omega L)^{-1} b$$

with

$$B_1 := (D + \omega U)^{-1}[(1 - \omega)D - \omega L],$$
$$B_2 := (D + \omega L)^{-1}[(1 - \omega)D - \omega U].$$

B_2 is simply the iteration matrix for the SOR method from (16.35), and matrix B_1 is the same, but the roles of L and U reversed.

16.9 Non-Stationary Iterative Methods

Non-stationary methods differ from stationary methods in that the computations involve information that changes at each iteration step. This continuously updated information consists primarily of inner products of residuals or other vectors arising from the method.

Definition 16.9.1 (Non-Stationary Iteration)   *An iteration*

$$x^{(k+1)} := T_k(x^{(k)}), \quad k = 0, 1, 2, \ldots$$

is said to be non-stationary, if the specification T_k of the iteration depends on the iteration step.

Some of the best known non-stationary iterative methods for solving linear systems are:

The **CG method** generates a sequence of conjugate vectors which are the residuals of the iterates (cf. Section 16.7.4). These vectors are also the gradients of a quadratic functional, the minimization of which is equivalent to the solution of the linear system. The CG method is extremely effective when the coefficient matrix is symmetric and positive definite.

The **MINRES method** and the **SYMMLQ method** are alternatives to the CG method which are used if the coefficient matrix is symmetric but possibly indefinite. The SYMMLQ method generates the same solutions as the CG method if the coefficient matrix is symmetric and positive definite.

The **CGNE method** and the **CGNR method** are specific CG methods for problems with non-symmetric and non-singular coefficient matrices. These methods are based on the fact that the matrices $AA^\top$ and $A^\top A$ are always symmetric and positive definite. The CGNE method solves the system $(AA^\top)y = b$ for y and then computes the solution $x = A^\top y$. The CGNR method solves $(A^\top A)x = \bar{b}$ for x where $\bar{b} = A^\top b$. The convergence of these methods may be slow since the spectrum of $AA^\top$ and $A^\top A$ will be less favorable than the spectrum of A.

The **GMRES method** computes a sequence of orthogonal vectors (as in the MINRES method), and combines them using a least-squares solve and update. However, unlike the MINRES method, it requires storing the whole sequence, so a large amount of storage is needed. This method is useful for general non-symmetric matrices.

The **BiCG method** generates two sequences of vectors: one based on a system with the original matrix A and one on $A^\top$, which are made mutually orthogonal, or *bi-orthogonal*. The BiCG method is useful when the matrix is non-symmetric and non-singular.

The **QMR method** applies a least-squares solve and update to the BiCG residuals, thereby smoothing out the irregular convergence behavior of the BiCG method. It also largely avoids the breakdown that can occur in the BiCG method.

The **CGS method** is a variant of the BiCG method that applies the updating operations of both sequences to the same vectors. An advantage is that this method does not need the multiplications by $A^\top$, but on the other hand convergence may be much more irregular than for the BiCG method.

The **BiCGSTAB method** is a variant of the BiCG method, like the CGS method. The difference is that the BiCGSTAB method uses different updates for the sequence corresponding to $A^\top$ in order to obtain smoother convergence than the CGS method.

Chebyshev iteration recursively determines polynomials with coefficients chosen to minimize the infinity norm of the residual. The coefficient matrix must be positive definite, and knowledge of the extremal eigenvalues is required. This method has the advantage that it requires no inner products.

16.9.1 The Conjugate Gradient Method (CG Method)

The conjugate gradient method (*CG method*) is an efficient algorithm for solving symmetric positive definite linear systems (cf. Section 16.7.4). The basis of the method is the use of search direction vectors to update the iterates and the residuals. Only a small number of vectors needs to be stored, however.

At each step of the method, update scalars are computed by evaluating two inner products in order for the iterates to satisfy certain orthogonality conditions. For a linear system with a symmetric positive definite matrix A, these conditions imply that the distance from the true solution to the iterates produced by the algorithm is minimized with respect to the underlying norm.

The iterates $x^{(i)}$ are updated at each step by a multiple α_i of the search direction vector $p^{(i)}$:

$$x^{(i)} := x^{(i-1)} + \alpha_i p^{(i)}.$$

Correspondingly, the residuals $r^{(i)} = Ax^{(i)} - b$ are updated according to

$$r^{(i)} := r^{(i-1)} + \alpha_i q^{(i)} \qquad \text{where} \qquad q^{(i)} := Ap^{(i)}. \tag{16.36}$$

The choice

$$\alpha = \alpha_i = \frac{r^{(i-1)^{\mathsf{T}}} r^{(i-1)}}{p^{(i)^{\mathsf{T}}} A p^{(i)}}$$

minimizes $r^{(i)^{\mathsf{T}}} A^{-1} r^{(i)}$ for all α_i in (16.36). The search directions are updated using the residuals:

$$p^{(i)} := -r^{(i-1)} - \beta_{i-1} p^{(i-1)}, \qquad (16.37)$$

where the choice

$$\beta_{i-1} = \frac{r^{(i-1)^{\mathsf{T}}} r^{(i-1)}}{r^{(i-2)^{\mathsf{T}}} r^{(i-2)}},$$

ensures that $p^{(i)}$ and $Ap^{(i-1)}$ and hence $r^{(i)}$ and $r^{(i-1)}$ are orthogonal. It can be shown that this choice of β_i makes $p^{(i)}$ and $r^{(j)}$ orthogonal to *all* previous $Ap^{(j)}$ and $r^{(j)}$ respectively.

Theory

The unpreconditioned CG method constructs $x^{(i)}$ as an element of the Krylov space $\mathcal{K}_i(r^{(0)}, A)$ shifted by the initial guess $x^{(0)}$,

$$x^{(i)} \in x^{(0)} + \operatorname{span}\{r^{(0)}, Ar^{(0)}, \ldots, A^{i-1} r^{(0)}\}, \qquad (16.38)$$

so that

$$f(x) = (x - x^*)^{\mathsf{T}} A (x - x^*)$$

is minimized by $x^{(i)}$, where x^* is the exact solution of $Ax = b$. The *existence* of the minimum

$$x^{(i)} \in x^{(0)} + \mathcal{K}_i(r^{(0)}, A), \quad i = 1, 2, 3, \ldots$$

is guaranteed only for symmetric positive definite matrices.

Convergence Behavior

As already mentioned in Section 16.7, the CG method needs at most n steps to obtain the exact solution x^* of a linear system with a symmetric positive definite matrix—as long as no rounding errors are considered. In practice rounding errors cannot be neglected, and hence, the CG method is not used as a direct method, but as an iterative method. The rate of convergence can be characterized by

$$\|x^{(i)} - x^*\|_A \leq 2\gamma^i \|x^{(0)} - x^*\|_A \qquad \text{and} \qquad (16.39)$$

$$\|x^{(i)} - x^*\|_2 \leq 2\sqrt{\kappa_2(A)}\, \gamma^i \|x^{(0)} - x^*\|_2,$$

where

$$\gamma := \frac{\sqrt{\kappa_2(A)} - 1}{\sqrt{\kappa_2(A)} + 1} \qquad \text{and} \qquad \|y\|_A := \sqrt{\langle y, Ay \rangle}$$

(Deuflhard, Hohmann [44], Golub, Ortega [49]).

Preconditioning

The convergence rate of the CG method depends on the condition number of the matrix A. For a matrix with the optimal condition number $\kappa_2(A) = 1$ it holds that $\gamma = 0$, and as $\kappa_2(A) \to \infty$ it follows that $\gamma \to 1$. The more badly the linear system is conditioned, the slower the convergence of the CG method becomes. The idea of *preconditioning* is to transform the original system—with the help of a regular matrix M—into one that is equivalent in the sense that it has the same solution, but has a more favorable condition number:

$$Ax = b \quad \Longleftrightarrow \quad M^{-1}Ax = M^{-1}b.$$

The better M approximates the matrix A, the smaller $\kappa_2(M^{-1}A)$ becomes, and the faster the CG method, applied to the transformed system, converges. For $M = A$ the optimal condition number $\kappa_2(M^{-1}A) = 1$ is obtained, but this, of course, is of no practical use as this choice requires the solution of a linear system as well. Since applying a preconditioner incurs extra effort, there is a trade-off between the cost of constructing and applying the preconditioner and the gain of improved convergence speed. A more detailed discussion of preconditioning is found in Section 16.10.

The **preconditioned CG method** has the following structure:

> *Choose an initial guess* $x^{(0)}$
> $r^{(0)} := Ax^{(0)} - b$
> **do** $k = 1, 2, 3, \ldots$
>   **solve** $Mz^{(k-1)} = r^{(k-1)}$
> $\varrho_{k-1} := \langle r^{(k-1)}, z^{(k-1)} \rangle$
>   **if** $k = 1$ **then**
> $p^{(1)} := -z^{(0)}$
>   **else**
> $\beta_{k-1} := \varrho_{k-1}/\varrho_{k-2}$
> $p^{(k)} := -z^{(k-1)} - \beta_{k-1}p^{(k-1)}$
>   **end if**
> $q^{(k)} := Ap^{(k)}$
> $\alpha_k := \dfrac{\varrho_{k-1}}{\langle p^{(k)}, q^{(k)} \rangle}$
> $x^{(k)} := x^{(k-1)} + \alpha_k p^{(k)}$
> $r^{(k)} := r^{(k-1)} + \alpha_k q^{(k)}$
> **end do**

Relation (16.39) indicates that the number of iterations necessary for reaching a relative reduction of ε in the error is proportional to $\sqrt{\kappa_2}$, which implies *linear* convergence for the CG method.

    If the extremal eigenvalues of the matrix $M^{-1}A$ are well separated, then even *superlinear* convergence is often observed; in this case, the convergence rate increases at each iteration. This phenomenon is explained by the fact that CG methods first eliminate error components in the direction of eigenvectors associated with extremal eigenvalues. After these error components have been eliminated, the method proceeds as if the respective eigenvalues did not exist in the system; at this point the convergence rate depends on a reduced system with a (much) smaller condition number. For an analysis of this acceleration phenomenon, see Van der Sluis, Van der Vorst [369].

16.9.2 The CG Method on the Normal Equations

The CG method is based on the principle of minimizing the quadratic function

$$f(x) := \tfrac{1}{2}x^\top Ax - b^\top x, \tag{16.40}$$

but it can be applied only to linear systems with a symmetric positive definite matrix A. If this requirement is not satisfied, the principle of minimization can be applied to the residual

$$r(x) := Ax - b.$$

Minimizing the function

$$f_R(x) := \|r(x)\|_2^2 = (Ax - b)^\top (Ax - b) = x^\top A^\top Ax - 2b^\top Ax + b^\top b$$

by applying a CG method is the same as solving the *normal equations*

$$A^\top Ax = A^\top b$$

using this CG method. This algorithm for solving general linear systems is called the **CGNR method**. The letter N refers to normal equations, and the letter R stands for the minimization of the residual.

If the error

$$e(x) := x - A^{-1}b$$

or the function

$$f_E(x) := \|e(x)\|_2^2 = x^\top x - 2b^\top A^{-\top} x + b^\top A^{-\top} A^{-1} b, \tag{16.41}$$

is minimized instead of the residual, then another iterative method can be obtained. The substitution $x := A^\top y$ transforms (16.41) into

$$f_E(y) = y^\top A A^\top y - 2b^\top y + b^\top A^{-\top} A^{-1} b.$$

The application of a CG method for minimizing f_E is therefore equivalent to the solution of the system

$$A A^\top y = b$$

using a CG method. The resulting algorithm is called the **CGNE method**; the letter E stands for the error function, and the letter N refers to normal equations.

The CGNE and CGNR methods are the simplest methods for solving nonsymmetric or indefinite systems. Since other methods for such systems are generally more complicated than the CG method, transforming the system into a symmetric positive definite system and then applying the CG method is attractive.

While this approach is easy to understand and to code, the convergence rate of the CGNE method and the CGNR method now depends on $(\kappa(A))^2$, the square of the condition number of the original coefficient matrix. The convergence rate may therefore be unacceptably slow (Nachtigal, Reddy, Trefethen [295]). In addition to there being a large number of iterations, the computational costs are doubled at each iteration; the matrices $A^{\top}A$ or $AA^{\top}$ are not computed explicitly, so each iteration requires *two matrix-vector* multiplications.

16.9.3 The MINRES and the SYMMLQ Method

The minimal residual method and the symmetric LQ method are methods that can be applied to symmetric *indefinite* systems. If the CG method is applied to such a system, a breakdown of the algorithm can occur.

The MINRES method is a Krylov method based on the principle of minimizing the residual; it also uses the Euclidean norm for quantifying the residual (16.25). The SYMMLQ method, on the other hand, is an orthogonal residual method that does not minimize anything explicitly (it simply assures that all residuals remain orthogonal). Both methods are variants of the CG method that avoid the LU-factorization.

16.9.4 The Generalized Minimal Residual (GMRES) Method

The generalized minimal residual method—the GMRES method—is an extension of the MINRES method to *non-symmetric* systems (Saad, Schultz [337]). In the GMRES method, an orthogonal basis for the Krylov space

$$\text{span}\{r^{(0)}, Ar^{(0)}, A^2r^{(0)}, \ldots\}$$

is formed explicitly:

$$w^{(i)} := Av^{(i)}$$
$$\textbf{do } k = 1, 2, \ldots, i$$
$$\quad w^{(i)} := w^{(i)} - \langle w^{(i)}, v^{(k)}\rangle v^{(k)}$$
$$\textbf{end do}$$
$$v^{(i+1)} := w^{(i)}/\|w^{(i)}\|$$

Obviously, this is a modified Gram-Schmidt orthogonalization procedure. Applied to the Krylov sequence $\{A^k r^{(0)}\}$ this orthogonalization is called the *Arnoldi method* (Arnoldi [91]).

The GMRES method has the advantage that the norm of the residuals can be computed without the iterates having been formed. Thus, the expensive computation of the iterate can be postponed until the residual norm is deemed small enough.

The iterates $x^{(i)}$ are constructed as

$$x^{(i)} = x^{(0)} + y_1 v^{(1)} + \ldots + y_i v^{(i)},$$

where the coefficients y_k have been chosen to minimize $\|b - Ax^{(i)}\|$.

The most popular form of the GMRES method is based on the orthogonalization method mentioned above, and uses restarts to reduce storage requirements. If no restarts are used, the GMRES method will converge in no more than n (order of A) steps. Of course, this is of no practical value when n is large. Moreover, the storage and computational requirements are prohibitive. The usual way to overcome these limitations is by restarting the iteration. After a chosen number m of iterations, the accumulated data is cleared and the intermediate results are used as initial data for the next m iterations. The difficulty is in choosing an appropriate value for m. Unfortunately, there are examples for which the GMRES method stagnates and for which convergence takes place only at the nth step. For such systems, any choice of m less than n fails to converge.

Implementation

The major drawback of the GMRES method is that the amount of work and storage required per iteration rises linearly with the iteration count. In this regard the restarted GMRES method is also not satisfactory, as there are no definite rules governing the choice of the restart parameter m.

A pseudocode for the restarted GMRES(m) algorithm can be found in the TEMPLATES [5].

16.9.5 The Bi-Conjugate Gradient (BiCG) Method

The CG method is not suitable for non-symmetric systems because the residual vectors cannot be made orthogonal using short recurrences. The GMRES method retains the orthogonality of the residuals by using long recurrences, at the cost of a larger storage requirement (cf. Section 16.9.4). The BiCG method takes another approach, replacing the orthogonal sequence of residuals with two mutually orthogonal sequences, at the price of no longer providing a minimization of the residuals.

The update relations for the residuals in the CG methods are extended in the BiCG method with similar relations based on $A^\top$ instead of A. The update therefore involves two sequences of residuals

$$r^{(i)} := r^{(i-1)} + \alpha_i A p^{(i)}, \qquad \tilde{r}^{(i)} := \tilde{r}^{(i-1)} + \alpha_i A^\top \tilde{p}^{(i)}.$$

and two sequences of search directions

$$p^{(i)} := -r^{(i-1)} - \beta_{i-1} p^{(i-1)}, \qquad \tilde{p}^{(i)} := -\tilde{r}^{(i-1)} - \beta_{i-1} \tilde{p}^{(i-1)}.$$

The choices

$$\alpha_i := \frac{\tilde{r}^{(i-1)^\top} r^{(i-1)}}{\tilde{p}^{(i)^\top} A p^{(i)}}, \qquad \beta_i := \frac{\tilde{r}^{(i)^\top} r^{(i)}}{\tilde{r}^{(i-1)^\top} r^{(i-1)}}$$

ensure the *bi-orthogonality*

$$\tilde{r}^{(i)^\top} r^{(j)} = \tilde{p}^{(i)^\top} A p^{(j)} = 0 \qquad \text{if} \quad i \neq j.$$

Convergence Behavior

Few theoretical results are known about the convergence of the BiCG method. For symmetric positive definite systems the method produces the same results as the CG method, but at twice the cost per iteration (the matrix-vector multiplication must be done for A *and* $A^\top$). For non-symmetric matrices it has been shown that in phases of the iterative process where a significant reduction of the norm of the residuals takes place, the method is more or less comparable to the GMRES method without restarts. In practice it is often observed that the convergence behavior may be quite irregular, and the BiCG method may even break down.

Implementation

The BiCG method requires the computation of *two* matrix-vector products: $A p^{(k)}$ and $A^\top \tilde{p}^{(k)}$. In some applications the latter may be impossible to perform, for instance, if the matrix A is not given explicitly, and the matrix-vector product is only given as a subprogram.

It is difficult to make a fair comparison between the GMRES method and the BiCG method, considering both reliability and efficiency. The GMRES method, for instance, really minimizes the residuals, but at the cost of strongly increased storage demands. Although the BiCG method does not minimize the residuals, its accuracy is often comparable to that of the GMRES method. It does, however, involve a cost of twice the number of matrix-vector products per iteration step. Some variants of the BiCG method (CGS method, BiCGSTAB method) have been proposed to obtain increased efficiency in certain circumstances. These variants are discussed in Sections 16.9.7 and 16.9.8.

16.9.6 The Quasi-Minimal Residual (QMR) Method

The QMR method of Freund and Nachtigal [200], [201] attempts to overcome irregular convergence behavior and the possibility of a break down of the BiCG method. Like the GMRES method, the QMR method is based on the least squares principle. Since the constructed basis for the Krylov subspace is *bi-orthogonal*, rather than orthogonal as in the GMRES method, the solution obtained is viewed as a *quasi-minimal* residual solution. Additionally, the QMR method uses *look-ahead* techniques to avoid breakdowns, which makes it more robust than the BiCG method. Also, the convergence behavior of the QMR method is typically much smoother than that of the BiCG method. The QMR method converges about as rapidly as the GMRES method.

16.9.7 The Squared CG (CGS) Method

In the BiCG method, the residual vector $r^{(i)}$ can be regarded as the product of $r^{(0)}$ and an ith degree polynomial in A, that is

$$r^{(i)} = P_i(A)r^{(0)}.$$

This same polynomial can be used to obtain $\tilde{r}^{(i)} = P_i(A^\top)\tilde{r}^{(0)}$, so that

$$\rho_i := \langle \tilde{r}^{(i)}, r^{(i)} \rangle = \langle P_i(A^\top)\tilde{r}^{(0)}, P_i(A)r^{(0)} \rangle = \langle \tilde{r}^{(0)}, P_i^2(A)r^{(0)} \rangle. \qquad (16.42)$$

This suggests that if $P_i(A)$ reduces the initial residual $r^{(0)}$ to a smaller vector $r^{(i)}$, then it might be advantageous to apply this contraction operator again and compute $P_i^2(A)r^{(0)}$. This approach leads to the squared CG method (the CGS method), proposed by Sonneveld [356].

The CGS method requires about the same number of operations per iteration as the BiCG method, but does not involve computations with $A^\top$. Hence, in circumstances where computation with $A^\top$ is impractical, for instance, if A is not formed explicitly, the CGS method may be attractive.

Convergence Behavior

Often one observes a speed of convergence in the CGS method that is about twice that of the BiCG method, which is in agreement with the observation that the same contraction operator is applied twice. On the other hand the convergence behavior of the CGS method is often highly irregular. Moreover, the CGS method may diverge even if the starting vector is close to the solution.

16.9.8 The Stabilized BiCG (BiCGSTAB) Method

The stabilized BiCG (BiCGSTAB) method was developed to solve non-symmetric linear systems while avoiding the often irregular convergence patterns found in the squared CG method (Van der Vorst [370]). Instead of computing the CGS sequence $\{P_i^2(A)r^{(0)}\}$, the BiCGSTAB method computes the sequence $\{Q_i(A)P_i(A)r^{(0)}\}$, where Q_i is an ith degree polynomial describing a steepest descent update.

The BiCGSTAB method requires two matrix-vector products and four inner products, i.e., two inner products more than the BiCG method and the CGS method.

Convergence Behavior

The BiCGSTAB method often converges about as rapidly as the CGS method. BiCGSTAB can be interpreted as the combination of BiCG and a repeatedly applied GMRES(1). At least locally, a residual vector is minimized, so that considerably smoother convergence behavior is obtained. On the other hand, if the local GMRES(1) step stagnates, then the Krylov subspace is not expanded, and the BiCGSTAB method will break down. This type of breakdown may be avoided by combining the BiCG method with other methods. One such alternative is the BiCGSTAB2 method suggested by Gutknecht [215].

16.9.9 Chebyshev Iteration

The Chebyshev iteration is another non-stationary iterative method for solving non-symmetric linear systems. Contrary to the other non-stationary methods, this method avoids the computation of inner products (Golub, Van Loan [50]). For some special memory hierarchies these inner products are an efficiency bottleneck. The price one pays for avoiding inner products is that the method requires knowledge of the spectrum of the coefficient matrix A. This difficulty can be overcome by using a variant of the algorithm developed by Manteuffel [282].

Comparison with other Methods

Comparing the pseudocode of the Chebyshev iteration (TEMPLATES [5]) with the pseudocode of the CG method reveals a high degree of similarity, except that inner products are computed in the CG method and not in the Chebyshev iteration.

The Chebyshev iteration has one advantage over the GMRES method: only short recurrences are used. Therefore only a small additional amount of storage is required. On the other hand, the GMRES method is guaranteed to generate the smallest residual over the current search space. Finally, the GMRES and the BiCG methods may be more effective in practice because of superlinear convergence behavior which cannot be expected of the Chebyshev iteration.

In circumstances where the computation of inner products is a bottleneck, it may be advantageous to start with a CG method and to compute estimates of the extremal eigenvalues obtained from the CG coefficients. After sufficient convergence of these approximations one can switch over to Chebyshev iteration. A similar strategy may be adopted for a switch from the GMRES or the BiCG method to the Chebyshev iteration.

Convergence Behavior

For a symmetric coefficient matrix, the Chebyshev iteration has the same error bound as the CG method, provided the parameters for the Chebyshev iteration are based on the (exact) extremal eigenvalues. There is a severe penalty for overestimating or underestimating the eigenvalues. For example, if for a symmetric coefficient matrix $\lambda_{\max}$ is underestimated, then the method may diverge. If it is overestimated, then very slow convergence may occur. Similar statements can be made for non-symmetric coefficient matrices. This implies that fairly accurate bounds on the spectrum of A are needed for the method to be effective (in contrast to the CG method or the GMRES method).

16.10 Preconditioning

A matrix M is called a *preconditioner* if the matrix M enables a transformation of the linear system into an equivalent system (with the same solution) which has more favorable spectral properties:

$$Ax = b \quad \Longleftrightarrow \quad M^{-1}Ax = M^{-1}b.$$

There is a choice between finding a matrix that approximates A, for which solving a system is easier than solving one with A, or finding a matrix M that approximates A^{-1}. The majority of preconditioners fall in the first category; an example of the second category is discussed in Section 16.10.6.

Since applying a preconditioner involves extra cost, there is a trade-off between the cost of constructing and applying the preconditioner and the gain of increased convergence speed. Most preconditioners involve an amount of computational work proportional to the number n of variables in their application. They thus multiply the amount of computational work per iteration by a constant factor. On the other hand, the number of iterations is usually only improved by a constant, i.e., the saving in computational work is independent of the matrix size.

In practice, the matrix M is factorized as $M = M_1 M_2$, and the linear system is transformed into

$$M_1^{-1}AM_2^{-1}(M_2 x) = M_1^{-1}b. \tag{16.43}$$

The matrices M_1 und M_2 are called the *left*- and *right* preconditioners, respectively. An iterative method can be preconditioned according to the following scheme:

1. The transformation of the right-hand side: $b \longmapsto M_1^{-1}b$,

2. The application of the original (unpreconditioned) iterative method to the linear system with the modified coefficient matrix $M_1^{-1}AM_2^{-1}$,

3. The computation of $x = M_2^{-1}y$, where y is the solution resulting from the previous step.

Since symmetry and definiteness are crucial to the success of some iterative methods, transformation (16.43) is preferable to the simple approach $M^{-1}A$, because $M_1^{-1}AM_2^{-1}$ preserves these properties provided $M_1 = M_2^{\mathsf{T}}$.

16.10.1 Jacobi Preconditioning

The *Jacobi* or *point-preconditioner* consists of just the diagonal of the matrix A:

$$M := \operatorname{diag}(a_{11}, a_{22}, \ldots, a_{nn}).$$

Theoretically there is no need for any extra storage, but in practice storage is allocated for the reciprocals of the diagonal elements in order to avoid repeated (unnecessary) division operations.

If the index set $J := \{1, \ldots, n\}$ is partitioned into disjoint subsets J_i, a block version can be derived:

$$m_{ij} := \begin{cases} a_{ij}, & \text{if } i, j \in J_k \\ 0 & \text{otherwise.} \end{cases}$$

The preconditioner is now a block-diagonal matrix; this is the *Jacobi block-preconditioning*.

Jacobi preconditioners need very little storage, and they are easy to implement, even on parallel computers. On the other hand, more sophisticated preconditioners usually yield much better improvements in the rate of convergence.

16.10.2 SSOR Preconditioning

If the symmetric matrix A is decomposed as $A = D + L + L^\top$, then the *SSOR matrix* is defined as

$$M := (D + L)D^{-1}(D + L)^\top,$$

or, parameterized by ω,

$$M_\omega := \frac{1}{2 - \omega}(\tfrac{1}{\omega}D + L)(\tfrac{1}{\omega}D)^{-1}(\tfrac{1}{\omega}D + L)^\top.$$

Using the optimal value of the relaxation factor ω reduces the number of iterations, but in practice, this optimal value is prohibitively expensive to compute. As M is given in factored form, this method shares many of the properties of preconditioning methods based on incomplete factorizations (see Section 16.10.3). Since the factorization of the SSOR preconditioner is given *a priori*, there is no danger of a breakdown in the construction phase.

16.10.3 Incomplete Factorization

A factorization is said to be *incomplete* if certain fill-in elements (zero elements that would be non-zero if they were in an exact factorization) are ignored during the factorization process. The matrix M is then given by $M := \tilde{L}\tilde{U}$, and the efficiency of the preconditioner depends on how well $\tilde{L}\tilde{U}$ approximates A.

Computing an Incomplete Factorization

The construction of an incomplete factorization may break down (due to a division by a zero pivot element) or result in indefinite matrices (indicated by a negative pivot element). The existence of an incomplete factorization is guaranteed for many factorization strategies, if the original matrix A has certain properties (Meijerink, Van der Vorst [289]).

An important consideration for incomplete factorization preconditioners is the cost of the factorization process. These costs bear fruit only if the iterative method without preconditioning converges very slowly, or if the same matrix M can be used for solving several linear systems, as can be done, for instance, in Newton's method for large nonlinear systems with a sparse Jacobian matrix.

16.10.4 Incomplete Block Factorization

The starting point for an incomplete block factorization is a partitioning of
the matrix A. An incomplete factorization is then performed using the matrix blocks as basic entities. The most important difference from the methods of
Section 16.10.3 arises in the inversion of the pivot blocks. Normally, inverting a
scalar is easily accomplished. Inverting pivot blocks is more difficult for two reasons. Firstly, inverting the pivot block is likely to be a costly operation. Secondly,
initially all diagonal blocks of the matrix may be sparse, and the maintenance of
this type of structure is desirable. Hence the need arises to approximate inverses.

Approximation of Inverses

In any case, the demand for easy and efficient computation excludes the computation of the full inverse, and the use of the respective band matrix. The simplest
approximation to A^{-1} is the diagonal matrix D^{-1} of the reciprocals of the diagonal elements of A. Other possibilities were considered by Axelsson, Eijkhout [94]
and Axelsson, Polman [95].

Banded approximations to the inverse of band matrices are theoretically jus-
tifiable. For instance, in the context of partial differential equations, where the
diagonal blocks of the coefficient matrix are usually strongly diagonally dominant.

Block Tridiagonal Matrices

In many applications, a block tridiagonal structure can be found in the coefficient
matrix. For such matrices all properties follow from the treatment of the pivot
blocks since no fill-in can occur outside the diagonal blocks A_{ii}. The recursion
which generates the pivot blocks also takes a simple form. Let A be the block
indexed, coefficient matrix, and let $\{X_i : i = 1, 2, \ldots, n\}$ be the sequence of
pivots, then the sequence $\{Y_i : i = 1, 2, \ldots, n\}$ of approximations to the inverse
of the pivots can be derived as outlined in the following scheme:

$$X_1 := A_{11}$$
$$\textbf{do} \quad i = 1, 2, 3, \ldots$$
$$\quad compute \quad Y_i :\approx X_i^{-1}$$
$$\quad X_{i+1} := A_{i+1,i+1} - A_{i+1,i} Y_i A_{i,i+1}$$
$$\textbf{end do}$$

Parallel Incomplete Block Factorization

One reason that block factorizations are of interest is that they are more suitable
for vector computers and multiprocessor machines. Consider the block factorization

$$A = (D + L)D^{-1}(D + U) = (D + L)(I + D^{-1}U),$$

where D is the block diagonal matrix of pivot blocks. Then an incomplete block
factorization can be obtained in two alternative ways:

1. replacing D with $X := \text{diag}(X_i)$ which leads to

$$C := (X + L)(I + X^{-1}U).$$

2. replacing D with $Y := \text{diag}(Y_i)$ which leads to

$$C := (Y^{-1} + L)(I + YU).$$

It is clear that for factorizations of the first type, solving a system means solving smaller systems with X_i matrices. For factorizations of the second type, Y_i blocks are very often used in multiplications. Therefore these factorizations are advantageous for vector computers.

16.10.5 Incomplete LQ Factorization

Instead of an incomplete LU factorization, an incomplete LQ factorization can also be constructed (Saad [334]). The idea is to orthogonalize the rows of the matrix using a Gram-Schmidt process. It turns out that the resulting incomplete factor L can be viewed as an incomplete Cholesky factor of the matrix $AA^\top$. Experiments show that using L in a CG process for $L^{-1}AA^\top L^{-\top}y = b$ is effective for some relevant problems.

16.10.6 Polynomial Preconditioning

It has already been mentioned that there is a class of preconditioning matrices which approximate A^{-1}; *polynomial* preconditioners are part of this class. Suppose that A can be represented in the form $A = I - B$, where the spectral radius of B is less than 1. Then, using the Neumann series, the inverse of A can be written as

$$A^{-1} = I + B + B^2 + B^3 + \cdots.$$

By truncating this infinite series, an approximation for A^{-1} can be derived.

Dubois, Greenbaum and Rodrigue [173] investigated the relationship between a simple preconditioning based on a splitting $A = M - N$ and a polynomially preconditioned method with

$$M_p^{-1} = \left(\sum_{i=0}^{p-1}(I - M^{-1}A)^i\right)M^{-1}.$$

Their main result is that for conventional iterative methods, k steps of the polynomially preconditioned method are exactly equivalent to kp steps of the original method. For accelerated methods, specifically Chebyshev iteration, the preconditioned algorithm can improve the number of iterations by, at most, a factor of p. Although there is no gain in the number of times a matrix-vector product is evaluated, polynomial preconditioning does eliminate a large fraction of the inner products and update operations, so there may be an overall increase in efficiency.

More abstractly, a polynomial preconditioner can be defined as any polynomial $M = P_n(A)$ normalized so that $P(0) = 1$. The polynomial minimizing the deviation $\|I - M^{-1}A\|$ will be the best choice for the polynomial preconditioner. Using the infinity norm, Chebyshev polynomials are obtained. The required estimates on the spectrum of A can be derived from the CG algorithm itself.

16.11 Matrix-Vector Products

Efficient iterative methods for solving systems of linear equations are based on the matrix-vector products

$$y = Ax \quad \text{und} \quad y = A^{\mathsf{T}}x.$$

Special algorithms adapted for the CRS format (cf. Section 16.1.2) and the CDS format (cf. Section 16.1.6) are presented in the following two sections.

16.11.1 Matrix-Vector Product Using the CRS Format

The implementation of the matrix-vector product $y = Ax$ using the CRS format is based on the usual algorithm

$$y_i := y_i + a_{ij}x_j, \qquad i = 1, 2, \ldots, n, \quad j = 1, 2, \ldots, n.$$

For a matrix stored in CRS format, the matrix-vector algorithm is then given by

> **do** $i = 1, 2, \ldots, n$
> $y(i) := 0$
>   **do** $j = \texttt{row\_pointer}(i), \ldots, \texttt{row\_pointer}(i+1) - 1$
> $y(i) := y(i) + \texttt{value}(j) \cdot x(\texttt{col\_index}(j))$
>   **end do**
> **end do**

Since this algorithm only multiplies non-zero matrix entries, the operation count is two times the number of non-zero elements in A. The number of operations is reduced significantly as compared with the 2^n operations needed for dense matrices.

Traversing columns of the matrix is an extremely inefficient operation for matrices stored in CRS format. Hence, for the transposed product $y = A^{\mathsf{T}}x$, the order of the loops is changed:

$$y_i := y_i + a_{ij}x_j \qquad j = 1, 2, \ldots, n, \quad i = 1, 2, \ldots, n. \tag{16.44}$$

The algorithm is then given by

> **do** $i = 1, 2, \ldots, n$
> $y(i) := 0$
> **end do**

$$\textbf{do } j = 1, 2, \ldots, n$$
$$\quad \textbf{do } i = \texttt{row\_pointer}(j), \ldots, \texttt{row\_pointer}(j+1) - 1$$
$$\quad\quad y(\texttt{col\_index}(i)) := y(\texttt{col\_index}(i)) + \texttt{value}(i) \cdot x(j)$$
$$\quad \textbf{end do}$$
$$\textbf{end do}$$

16.11.2 Matrix-Vector Product Using the CDS Format

To take advantage of the CDS format, the index of the outer loop is shifted. Replacing j by $i + j$ transforms (16.44) into

$$y_i := y_i + a_{i,i+j} x_{i+j}, \qquad i = 1, 2, \ldots, n, \quad i+j = 1, 2, \ldots, n.$$

With the index i in the inner loop, it can be seen that $a_{i,i+j}$ accesses the jth diagonal (where the main diagonal has the number 0). The algorithm then has a doubly nested loop with the outer loop running through the diagonals, from the left most to the right most diagonal. The bounds for the inner loop follow from the requirements that $1 \le i \le n$ and $1 \le i+j \le n$.

$$\textbf{do } i = 1, 2, \ldots, n$$
$$\quad y(i) := 0$$
$$\textbf{end do}$$
$$\textbf{do diag} = -\text{diag\_left}, \ldots, \text{diag\_right}$$
$$\quad \textbf{do loc} = \max(1, 1 - \text{diag}), \ldots, \min(n, n - \text{diag})$$
$$\quad\quad y(\text{loc}) := y(\text{loc}) + \texttt{value}(\text{loc}, \text{diag}) \cdot x(\text{loc} + \text{diag})$$
$$\quad \textbf{end do}$$
$$\textbf{end do}$$

Using the update formula

$$y_i := y_i + a_{i+j,i+j-j} x_{i+j} \qquad i = 1, 2, \ldots, n, \quad i+j = 1, 2, \ldots, n,$$

the following algorithm for $y = A^{\top} x$ can be obtained:

$$\textbf{do } i = 1, 2, \ldots, n$$
$$\quad y(i) := 0$$
$$\textbf{end do}$$
$$\textbf{do diag} = -\text{diag\_left}, \ldots, \text{diag\_right}$$
$$\quad \textbf{do loc} = \max(1, 1 - \text{diag}), \ldots, \min(n, n - \text{diag})$$
$$\quad\quad y(\text{loc}) := y(\text{loc}) + \texttt{value}(\text{loc} + \text{diag}, -\text{diag}) \cdot x(\text{loc}+\text{diag})$$
$$\quad \textbf{end do}$$
$$\textbf{end do}$$

Thereby no indirect addressing is used; so the algorithm is vectorizable with vector lengths of essentially the matrix order n.

16.12 Parallelism

Accelerating iterative methods by parallelizing the most time consuming operations as efficiently as possible is often a basic requirement for the applicability of an iterative method on multiprocessor machines.

Inner products of two vectors can easily be parallelized; each processor computes the inner product of the corresponding segments of each vector (*local inner products*).

Matrix-vector products are often parallelized on shared-memory machines by splitting the matrix into strips corresponding to the vector segments. Each processor then computes the matrix-vector product of a strip.

For distributed-memory machines, there may be a problem if each processor has only a segment of the vector in its local memory. This may lead to serious communication bottlenecks.

Preconditioning is often the most problematic part of parallelizing an iterative method for solving systems of linear equations.

A more comprehensive overview of the problems arising from the parallelization of iterative methods for solving linear systems can be found in TEMPLATES [5].

16.13 Selecting an Iterative Method

Generally, the solutions obtained using iterative methods are less accurate than results computed using direct methods. This is not a fundamental drawback for iterative methods: stopping the iterative process, if the desired accuracy is reached, is an effective way to limit the computational effort. On the other hand, the sparsity of the matrix impedes the reuse of data, and the indirect addressing needed for compressed storage schemes reduces performance, resulting in increased runtime.

16.13.1 Properties of Iterative Methods

This section gives a short summary of the most important iterative methods for solving systems of linear equations. For each method the following characteristics are summarized:

Field of application: Not every method will work on every problem type, so knowledge of the relevant problem properties is the main criterion for selecting an iterative method.

Efficiency: Methods differ in the operations that they perform. However, the fact that one method uses more operations than another is not necessarily a reason for rejecting it.

The Jacobi Method

- Extremely easy to use, but slow convergence. Unless the matrix A is strongly diagonally dominant, this method is recommended only as a preconditioner for non-stationary methods.

- Parallelization is easily accomplished.

The Gauss-Seidel Method

- Typically faster convergence than the Jacobi method, but in general not competitive with non-stationary methods.

- Applicable to strictly diagonally dominant, or symmetric positive definite matrices.

- Parallelization properties depend on the structure of the matrix A. Different orderings of the unknowns result in different degrees of parallelism.

Successive Over-Relaxation (the SOR Method)

- Accelerates convergence of the Gauss-Seidel method with $\omega > 1$ (*over*-relaxation); may yield convergence when the Gauss-Seidel method fails ($0 < \omega < 1$, *under*-relaxation).

- The rate of convergence critically depends on the choice of ω; the optimal value ω_{opt} may be estimated from the spectral radius of the Jacobi iteration matrix under certain conditions.

- Parallelization properties are the same as those of the Gauss-Seidel method.

The Conjugate Gradient (CG) Method

- Applicable to symmetric positive definite systems.

- Rate of convergence depends on the condition number $\kappa(A)$; if the extremal eigenvalues are well separated, superlinear convergence behavior can result.

- Parallelization properties are largely independent of the coefficient matrix A, but greatly depend on the preconditioner.

The Generalized Minimal Residual (GMRES) Method

- Applicable to non-symmetric matrices.

- The smallest possible residual is reached in a fixed number of iterations, but these steps become increasingly expensive.

- Restarting is necessary in order to limit the increasing storage requirements and work per iteration step. When to restart depends on A and the preconditioner; this method requires skill and experience.

The Bi-Conjugate Gradient (BiCG) Method

- Applicable to non-symmetric matrices.

- Requires matrix-vector products with A and A^T. This excludes the method in cases where the matrix is only implicitly given as an operator.

- Parallelization properties are similar to those in the CG method; the two matrix-vector products are independent, so they can be done in parallel.

The Quasi-Minimal Residual (QMR) Method

- Applicable to non-symmetric matrices.

- Smoother and faster convergence than with the BiCG method.

- Computational costs per iteration are slightly higher than for the BiCG method.

- Parallelization properties are the same as in the BiCG method.

The Conjugate Gradient Squared (CGS) Method

- Applicable to non-symmetric matrices.

- Converges (diverges) typically about twice as fast as BiCG. Convergence behavior is often quite irregular, which may lead to a loss of accuracy. Tends to diverge if the starting iterate is close to the solution.

- Computational costs per iteration are similar to BiCG, but A^T is not required.

The Bi-Conjugate Gradient Stabilized (BiCGSTAB) Method

- Applicable to non-symmetric matrices.

- Computational costs per iteration are similar to those in the BiCG and CGS method, but A^T is not required.

- An alternative to the CGS method that avoids the irregular convergence pattern of the CGS method while maintaining about the same rate of convergence.

Chebyshev Iteration

- Applicable to non-symmetric matrices.

- Explicit knowledge of the spectrum of the matrix is required. In the case of a symmetric coefficient matrix, the iteration parameters can easily be obtained from the two extremal eigenvalues, which can be estimated either directly from the matrix, or by applying a few iterations of the CG method.

- The adaptive Chebyshev method can be used in combination e. g., with the CG method or the GMRES method to continue the iteration once suitable bounds on the spectrum have been obtained from these methods.

Selecting the *best* method for a given class of linear systems is largely a matter of trial and error. Of course some theoretical background, as given in Freund, Golub, Nachtigal [199], is a must. The selection also depends on how much storage is available (GMRES method) on the availability of A^T (BiCG and QMR method) and on how expensive the matrix-vector products are in comparison to inner products.

16.13.2 Case Study: Comparison of Iterative Methods

The TEMPLATES programs (cf. Section 16.16.2) have been used for experimentally comparing several iterative methods. They have also been modified to work with large matrices stored in the CRS format. For matrix operations the SPARSKIT programs (cf. Section 16.15.3) have been used.

The test data to which the methods have been applied are the matrices BC-SSTK14 and WATT1 from the *Harwell-Boeing collection* (cf. Section 16.15.1). These matrices have already been used in the comparison of the different storage schemes in Section 16.4 where a description of the matrices is given.

Performing the Tests

Firstly, the matrix was multiplied by the vector $u = (1\ 1\ 1\ \ldots\ 1)^T$. The result of this multiplication was used as the right-hand side b of the linear system.

The numerical solutions produced by several programs were compared to the vector u. The absolute value of the maximum difference is displayed in the column *error* in the tables on page 434.

The performance measurements were executed on an HP workstation with a peak performance of 50 Mflop/s. The timing results given in Tables 16.2 and 16.3 are not the optimum values obtainable on that workstation. They only serve the purpose of comparing the different methods.

Generally, the use of the Jacobi preconditioner leads to an increased efficiency; in some cases divergent methods even converged.

16.14 Software for Sparse Systems

As for linear systems with dense coefficient matrices, there are also commercial software products and public-domain software for large, sparse systems. Both offer facilities for storing and manipulating sparse matrices as well as methods for solving mathematical problems (linear systems, eigenvalue problems, etc.).

A major difference between software for sparse linear algebra and software for other problems dealt with in this book is that the standardization of program interfaces and functional properties of the software modules does not yet exist. This is due to the fact that the data structures for storing matrices and efficient methods for solving problems greatly depend on the problem itself.

Table 16.2: Results for the symmetric positive definite matrix **BCSSTK14**.

method	preconditioner	number of iterations	error	total runtime	runtime per iteration
Gauss-Seidel		3 000	$6.30 \cdot 10^{-2}$	88.6 s	26 ms
SOR $\omega = 1.3$		3 000	$1.09 \cdot 10^{-2}$	88.5 s	26 ms
$\omega = 0.7$		3 000	$2.11 \cdot 10^{-1}$	88.5 s	26 ms
CG	—	1 000	> 1	23.8 s	24 ms
	Jacobi	625	$3.58 \cdot 10^{-11}$	14.9 s	24 ms
CGS	—	1 000	> 1	47.7 s	48 ms
	Jacobi	463	$1.16 \cdot 10^{-9}$	22.1 s	48 ms
BiCG	—	1 000	> 1	46.7 s	47 ms
	Jacobi	625	$3.58 \cdot 10^{-11}$	29.2 s	47 ms
BiCGSTAB	—	1 000	> 1	48.1 s	48 ms
	Jacobi	455	$4.16 \cdot 10^{-10}$	21.9 s	48 ms
GMRES	—	1 000	> 1	28.2 s	28 ms
	Jacobi	457	$9.16 \cdot 10^{-2}$	12.9 s	28 ms

Table 16.3: Results for the non-symmetric matrix **WATT1**.

method	preconditioner	number of iterations	error	total runtime	runtime per iteration
Gauss-Seidel		3 000	$5.19 \cdot 10^{-4}$	25.2 s	8.4 ms
SOR $\omega = 1.3$		3 000	$6.24 \cdot 10^{-7}$	25.2 s	8.4 ms
$\omega = 0.7$		3 000	$1.92 \cdot 10^{-2}$	25.2 s	8.4 ms
CG	—	495	$1.80 \cdot 10^{-7}$	3.6 s	7.2 ms
	Jacobi	146	$3.55 \cdot 10^{-8}$	1.1 s	7.3 ms
CGS	—	500	$5.32 \cdot 10^{-4}$	6.8 s	13.5 ms
	Jacobi	91	$1.95 \cdot 10^{-7}$	1.3 s	13.8 ms
BiCG	—	377	$1.37 \cdot 10^{-7}$	4.9 s	12.9 ms
	Jacobi	148	$3.01 \cdot 10^{-9}$	1.9 s	13.1 ms
BiCGSTAB	—	500	$7.06 \cdot 10^{-6}$	7.5 s	14.9 ms
	Jacobi	103	$2.09 \cdot 10^{-7}$	1.6 s	15.0 ms
GMRES	—	1 000	$2.47 \cdot 10^{-1}$	8.7 s	8.7 ms
	Jacobi	823	$2.93 \cdot 10^{-8}$	7.1 s	8.7 ms

Systems of Linear Equations

For linear systems with a large sparse coefficient matrix, many software products are available.

When selecting a suitable program, the user has to decide whether to use *direct* or *iterative* methods for solving the system. The following survey indicates which methods are better for certain applications:

	direct methods	iterative methods
accuracy	not susceptible	can be chosen
computational costs	predictable	mostly unpredictable, but often low
new right-hand sides	fast	no saving of time
storage requirements	more	less
initial guess	not required	mostly advantageous
parameters	not necessary	essential
black box usage	possible	often not feasible
robustness	yes	no

For some classes of problems, especially for very large linear systems (e. g., three-dimensional problems arising from applications in hydrodynamics), the increase of speed gained through the use of iterative methods is often so high (factor $10-100$) that a direct method is not suitable at all.

In the following sections no stand-alone programs are listed; only whole program packages and main parts of important software libraries have been entered. Stand-alone programs written in Fortran 77 can be found, for instance, in NETLIB/LINALG, and C programs in NETLIB/C.

Software products which are only suitable for a specific class of matrices (e. g., Toeplitz matrices) have been omitted deliberately.

Eigenvalue Problems, Least Squares Problems, etc.

A discussion of the theory dealing with the solution of eigenvalue and singular value problems as well as least squares problems for sparse matrices exceeds the scope of this book. Software products designed to solve such problems are not covered either.

The references given in Section 7 should however enable the reader to get an idea of the software available (e. g., by using GAMS, cf. Section 7.3.9).

16.15 Basic Software

16.15.1 The Harwell-Boeing Collection

The *Harwell-Boeing Collection* is a compilation of large sparse matrices arising from different fields of application (mainly physics and engineering). A specific storage scheme for storing these matrices, the *Harwell-Boeing* format, is available. Essentially, it corresponds to an extended CRS format with additional information (for example, the order of the matrix, the number of non-zero elements, etc.). In the *Harwell-Boeing* format, it is possible to store only the sparsity structure of the matrix (i. e., the positions of the non-zero elements are stored, not the values themselves). It is also possible to store dense matrices together with one or more ·right-hand sides (including vectors representing the solution or starting vectors for an iterative method).

Information about the Harwell-Boeing collection and software can be obtained via

```
http://www.rl.ac.uk/departments/ccd/numerical
       /harwellboeing/harwellboeing.html
```

The matrices of the *Harwell-Boeing Collection* can be obtained via anonymous FTP from `orion.cerfacs.fr`. The data can be found in directory `pub/harwell_boeing` in compressed format. Binary mode must therefore be selected before transferring the data. The names of the data files are `*.data.z`, where ∗ symbolizes the respective matrix collections.

A comprehensive user's guide in POSTSCRIPT format includes a detailed description of the *Harwell-Boeing* format, and of all the matrices in the collection.

16.15.2 SPARSE-BLAS

Analogous to the definition of BLAS (for dense vectors and matrices), a lot of effort has been made to standardize the interfaces of basic operations in sparse linear algebra. One of the first proposals of a definition, regarding only vector operations (not regarding matrix-vector or matrix-matrix operations) was published in 1991 (Dodson, Grimes, Lewis [159]).

Storage Schemes

Sparse vectors can be stored either in compressed or in uncompressed form in the SPARSE-BLAS. In the case of uncompressed storage mode, a vector v of order n corresponds to a one-dimensional array **v** of the same length n, which stores *all* components (including zeros) of the vector as well as an integer array index of length nn, where nn is the number of non-zero elements of the vector. The array index stores the indices of the non-zero elements of **v**. Thus, for $i = 1, 2, \ldots, nn$, the quantity **v(index(i))** runs through the non-zero elements of **v**.

In the compressed storage mode, a sparse vector v now corresponds to a one-dimensional array **v** of length nn which contains only the *non-zero elements* of the vector, as well as the integer array index of same length. The array index contains the same information as in the uncompressed storage scheme.

Subprograms

Of course, all BLAS-1 routines can be applied to sparse vectors stored in the uncompressed storage scheme. Certainly, this raises computational costs to $O(n)$, whereas, if only the non-zero elements are considered in the computations then merely $O(\mathbf{nn})$ operations are required.

A number of BLAS-1 routines (e. g., those for computing the Euclidean norm of a vector) can, however, also be applied directly to vectors stored in compressed form by simply passing the array **v** as the vector argument and **nn** as the length argument to the corresponding procedure. Of course, the resulting computational costs are also $O(\mathbf{nn})$.

SPARSE-BLAS subprograms must therefore be regarded as an extension of the existing BLAS-1 programs for operations which cannot be applied directly to sparse vectors in compressed form. In this case, the routines BLAS/*dot, BLAS/*axpy, and BLAS/*rot (for the Givens rotation of a vector) are replaced in the SPARSE-BLAS package by the programs *doti, *axpyi, and *roti. They perform their respective operations on sparse vectors. In the case of *doti and *axpyi, it must be noted that one of the two vector operands must be stored in the compressed form, while the other is specified in uncompressed form; otherwise the operations cannot be implemented with $O(\mathbf{nn})$ computational costs.

SPARSE-BLAS offers also subroutines for converting a storage scheme into another. The routine *sctr (*scatter*) transforms the compressed format into the uncompressed one; the routines *gthr (*gather*) and *gthrz (*gather and zero*) perform the inverse operation. In case of *gthrz, all non-zero elements of the uncompressed representation are explicitly set to zero during the conversion.

It is questionable whether the SPARSE-BLAS package will gain the same importance for algorithms for sparse matrices as the BLAS package already has for dense matrices. Specifically, it is questionable whether SPARSE-BLAS programs can actually utilize advanced computer systems efficiently. This is the more so as BLAS-1 routines which have the same level of complexity as SPARSE-BLAS routines achieve only a *poor* floating-point performance on such computers. Additionally, the success and the wide spread use of BLAS is based on the fact that BLAS was developed simultaneously with LINPACK and LAPACK, both packages of great importance. Important software packages using SPARSE-BLAS have not yet been developed.

A portable implementation of SPARSE-BLAS written in Fortran 77 can be obtained from the NETLIB in the directory sparse-blas.

16.15.3 SPARSKIT

SPARSKIT is a basic tool kit for sparse matrix computations written in Fortran 77. The CRS format is the basic format used in this package. SPARSKIT includes various procedures:

Input/Output routines allow files containing matrices stored in the Harwell-Boeing format (cf. Section 16.15.1) to be read and written. Additionally there are some utilities for visualizing the sparsity structure of the matrix.

Format conversion routines: Most of the routines in SPARSKIT use the CRS format internally, but other commonly used storage schemes are also supported (COO, CCS, MRS, BCRS, BND, CDS, JDS, SKS; see Section 16.1). There are numerous format conversion routines which convert one storage scheme into another.

Unary matrix operations: There is a module consisting of a number of routines for transposing the matrix, extracting a specified part of the matrix, and retrieving information about the sparsity structure of the matrix.

Binary matrix operations: There is a module comprising routines for computing products and sums of sparse matrices in different ways.

Matrix-vector operations: This module offers (in its current form) certain triangular matrix solution routines in addition to the matrix-vector product which is especially important for iterative methods.

Matrix generation routines set up certain types of sparse matrices (e. g., matrices arising from 5-point or 7-point discretization of differential equations).

Statistics and information routines provide information on the sparsity structure of matrices stored in the *Harwell-Boeing* format.

A brochure of about 30 pages offers information about the SPARSKIT package and its use. Each program is well described and should be easy to use.

The SPARSKIT package is public domain and has been made accessible through the Internet. It is available through anonymous FTP from `ftp.cs.umn.edu` as the file `SPARSKIT2.tar.Z` in the directory `/dept/sparse`.

16.16 Dedicated Software Packages

16.16.1 ITPACK

The ITPACK software packages have mainly been developed at the University of Texas in Austin. It forms the initial basis of the software modules for the iterative solution of linear systems contained in the ELLPACK package (cf. Section 7.6.2). ITPACK consists (currently) of five independent software packages; three of them—ITPACK 2C, ITPACKV 2D, and NSPCG—are available in NETLIB (in directory `itpack`). More details about the files stored in this directory and how to obtain the other software packages—ITPACK 3A and ITPACK 3B—can be obtained from the documentation file `info.tex`.

Information concerning the algorithms used in the ITPACK packages along with full details of the package can be found in Hageman, Young [53].

ITPACK 2C

ITPACK 2C consists of seven modules for the iterative solution of linear systems with a symmetric positive definite coefficient matrix. The coefficient matrix must be supplied in the CRS format. In addition to the Jacobi, the SOR and the SSOR method, the so-called *Richardson iteration* is also used as a basic iterative method. The convergence rate of these basic methods can be increased by using acceleration methods. To approximate the critical parameters optimally (e. g., the relaxation factor ω), adaptive methods that adjust the parameter dynamically to the observed convergence behavior have been implemented.

ITPACKV 2D

ITPACKV 2D is a vectorized version of the ITPACK 2C package. The algorithms used are exactly the same as those in ITPACK 2C. The sparse matrix storage scheme has been changed to a variant of the JDS format (which is therefore often called the ITPACK format) to improve vectorization. This package is intended for vector computers like the Cray Y-MP.

ITPACK 3A

ITPACK 3A is an extension of ITPACK 2C to general, possibly *non-symmetric* systems. Some acceleration routines have been implemented.

ITPACK 3B

ITPACK 3B contains basically the same algorithms as ITPACK 3A. It provides an ELLPACK-style preprocessor that gives the user more flexibility in specifying the linear system and the iterative solution method.

NSPCG

NSPCG is an extension to ITPACK 2C and ITPACK 3A. It contains basic iterative methods and acceleration schemes intended for both symmetric and non-symmetric systems. The matrix can be represented in any of several sparse matrix storage formats. In addition, it is a possible for the user to supply a routine for computing the matrix-vector product that avoids storing the matrix explicitly.

As the NSPCG package is the most flexible one, it should be given preference over the other ITPACK packages.

16.16.2 TEMPLATES

In TEMPLATES [5], a book which focuses on the use of iterative methods for solving large, sparse systems of linear equations, the use of *templates* is introduced. A template is a description of a general algorithm rather than the executable object code or the source code, which is more commonly found in a conventional software library. The templates have been presented in the book [5] in an Algol style, which is readily translatable into any target language commonly used.

Implementations in Fortran 77 and C have been made freely available from NETLIB (cf. Section 7.3.7). They can be found in the `linalg` directory as well as in the `templates` directory; they contain routines for the Jacobi, SOR, CG, CGS, GMRES, BiCG, BiCGSTAB, QMR, and Chebyshev methods. The versions of the templates available are listed in the table below:

file name	contents
`sctemplates.shar`	C routines, single precision
`dctemplates.shar`	C routines, double precision
`sftemplates.shar`	Fortran 77 routines, single precision
`dftemplates.shar`	Fortran 77 routines, double precision
`mltemplates.shar`	MATLAB routines

16.16.3 SLAP

SLAP is a package of iterative methods for solving systems of linear equations developed at Lawrence Livermore National Laboratory (LLNL).

The SLAP package contains non-stationary methods such as the CG, CGNE, BiCG, CGS and GMRES methods as well as the Jacobi and Gauss-Seidel methods. The coefficient matrix can be specified in the COO format or in the modified CCS format (the MCS format), or indirectly by supplying a subroutine that performs the matrix-vector product Ax. If the iterative method also requires the product with the transposed coefficient matrix A^T, a corresponding subroutine must be supplied as well.

If the system matrix has been specified explicitly, several basic iterative methods can be combined with preconditioning schemes. Jacobi preconditioning and preconditioning based on incomplete LU factorizations (and in the case of symmetric positive definite matrices) incomplete Cholesky factorizations are available. If, however, the coefficient matrix is specified indirectly by a corresponding matrix-vector multiplication routine, preconditioning must be implemented by the user himself by supplying a subroutine that solves the system $Mz = r$ (and $M^T z = r$ if necessary).

The SLAP package is available from NETLIB (directory `slap`) and is also a part of the SLATEC library.

16.16.4 Y12M

Y12M solves sparse systems of linear equations using a specially adapted Gaussian elimination. The subroutine `y12ma` is an easy to use *black box program* based on three subroutines:

`y12mb` converts the matrix supplied by the user (in the COO format) into the internal storage scheme.

y12mc performs an LU factorization of the matrix. To a great extent the pivoting strategy can be controlled to preserve both the numerical stability of the computations and the sparsity structure of the coefficient matrix. Moreover, a pivot sequence derived from a previous factorization of a matrix of the same sparsity structure can be used to reduce computational costs.

y12md solves the sparse system of linear equations using the LU factorization computed by y12mc.

The main task of y12ma is to call these three subroutines using suitable parameter settings. y12mf does the same except that it can perform an additional iterative refinement.

Y12M can be obtained from the y12m directory of NETLIB.

16.16.5 UMFPACK

UMFPACK is a program package for solving sparse, non-symmetric systems of linear equations using LU factorization. It contains subroutines for generating suitable pivot sequences, for constructing symbolic and numerical LU factorizations, and for computing the solution to a given LU factorization. The coefficient matrix must be specified in the COO format.

The implementation of the algorithms in UMFPACK—in contrast to many other software products—offers extensive opportunities for vectorization and parallelization. Previous[2] implementations of direct methods have been disappointing in this respect. A detailed description of the algorithms used in UMFPACK is given in Davis, Duff [149].

UMFPACK can be obtained from the NETLIB; the directory linalg contains the file umfpack.shar. UMFPACK is also part of the Harwell library.

16.16.6 PIM

The PIM (*Parallel Iterative Methods*) package contains a number of non-stationary iterative methods (e. g., CG, BiCG, CGS; BiCGSTAB, GMRES, CGNR, CGNE), which have been implemented in such a way as to be executable on uniprocessor computers as well as on multiprocessor systems. According to the SPMD (*Single Program Multiple Data*) paradigm, several copies of one and the same subroutine can be executed concurrently on several processors. If the subroutine is executed on a conventional (uniprocessor) computer, all computations are done by that single subroutine.

The PIM package itself does not contain any parallel elements; the user himself must supply subroutines implementing the parallel computations of matrix-vector products and global reduction operations according to the SPMD paradigm. If preconditioning is desired, an additional subroutine solving $Mz = r$ must be

---

[2]Version 1.1 of UMFPACK was released in January 1995.

supplied by the user. Moreover, the distribution of the data to their respective processors must be managed by the user.

PIM is available via anonymous FTP from `unix.hensa.ac.uk`. The directory `misc/netlib` contains the file `pim/pim.tar.Z`.

16.17 Routines from Software Libraries

This section covers those parts of the large numerical program libraries which are dedicated to the solution of linear systems with a sparse coefficient matrix.

16.17.1 IMSL Software Libraries

General Matrices

The IMSL Fortran library contains three subroutines for the solution of systems of linear equations with a general sparse coefficient matrix. The coefficient matrix must be supplied by the user in the COO format.

`IMSL/MATH-LIBRARY/lftxg` performs an LU factorization of the matrix A. The subroutine `IMSL/MATH-LIBRARY/lfsxg` then solves the linear system by using the two factor matrices L and U. Given a matrix A and a right-hand side b, `IMSL/MATH-LIBRARY/lslxg` solves the respective linear system by calling the two former subroutines.

In `IMSL/MATH-LIBRARY/lftxg`, a direct method which estimates the local fill-in by the Markowitz costs has been implemented. By specifying certain parameters the user can control both the numerical stability of the method and the fill-in generated by the method. If the accuracy of the solution needs to be improved, this program offers iterative refinement as well (cf. Section 13.13.3).

For sparse, *complex* matrices an analogous set of subroutines with identical parameter lists is available. The names of those subprograms are identical to those for real matrices, except that the letter x (indicating real type) has been changed to z (indicating complex type).

Symmetric Positive Definite Matrices

For the solution of linear systems with symmetric positive definite matrices, the IMSL library offers a direct method based on Cholesky factorization.

Given a matrix A, whose lower triangular part is specified in the COO format, `IMSL/MATH-LIBRARY/lscxd` performs a *symbolic* Cholesky factorization. This program computes a minimum degree ordering or uses a user-supplied ordering to set up the sparse data structure for the Cholesky factor L. The routine `IMSL/MATH-LIBRARY/lnfxd` then computes the *numerical* entries in L very efficiently as all fill-in arising in the Cholesky factorization has already been considered in the sparse data structure for L. The numerical computations can be carried out in one of two ways. The first method performs a factorization using a multi-frontal technique. This option requires more storage but in certain cases is faster. The second method is just the standard factorization method based on the sparse storage scheme.

Given the Cholesky factors of the coefficient matrix A and the right-hand side b, IMSL/MATH-LIBRARY/lfsxd solves the system of linear equations. The subroutine IMSL/MATH-LIBRARY/lslxd performs all computations and subroutine calls necessary for the solution of a given linear system.

An analogous set of subroutines is provided for the direct solution (using Cholesky factorization) of linear systems with a sparse *Hermitian* positive definite coefficient matrix.

In addition, the subroutine IMSL/MATH-LIBRARY/pcgrc is part of the IMSL Fortran library; it implements a preconditioned CG method. As the interface of this routine has been designed using the *reverse communication* technique, program control is returned to the calling routine after each step of the iteration, requesting the solution of the linear system $Mz^{(k-1)} = r^{(k-1)}$. The subroutine IMSL/MATH-LIBRARY/jcgrc is a special case of the above routine, with the diagonal of the matrix A used as the preconditioning matrix.

16.17.2 NAG Software Libraries

General Matrices

The NAG Fortran library, like the IMSL Fortran library, contains subprograms for the solution of sparse linear systems based on direct methods. NAG/f01brf performs an LU factorization of a matrix specified in the COO format. The numerical stability of the method can be controlled by several user definable parameters.

The pivot sequence obtained by the application of NAG/f01brf can be reused by the subprogram NAG/f01bsf if a matrix of the same sparsity structure has to be factorized. NAG/f04axf solves the linear system for a single right-hand side b provided that an LU factorization of the matrix A has already been computed.

For the iterative solution of linear systems, the program NAG/f04qaf is available, which is based on an implementation (NETLIB/TOMS/583—done by Paige and Saunders [307]) of the *Lanczos algorithm* (Golub, Van Loan [50]). This program is also suitable for the solution of linear least squares problems. Instead of specifying the coefficient matrix explicitly, the user must supply a subprogram that implements the matrix-vector products Ax and $A^{\mathsf{T}}x$.

Symmetric Matrices

For the iterative solution of sparse symmetric linear systems, the NAG Fortran library provides a suite of routines NAG/f11gaf, NAG/f11gbf and NAG/f11gcf, which implement either the conjugate gradient method, or a Lanczos method based on SYMMLQ. If the coefficient matrix is known to be positive definite the CG method should be chosen; the Lanczos method is more robust but less efficient for positive definite matrices. These routines allow a choice of termination criteria and the norms used in them, permit monitoring of the approximate solution, and can return norm and singular value estimates.

The library also provides various preconditioning routines. NAG/f11jaf computes a preconditioning matrix based on incomplete Cholesky factorization. The amount of fill-in occurring in the incomplete factorization can be controlled by specifying either the level of fill, or the drop tolerance. Diagonal Markowitz pivoting may optionally be employed, and the factorization can be modified to preserve row sums. Another routine NAG/f11jdf can be used to provide an SSOR preconditioner.

These basic routines are also combined to provide black box routines for the iterative solution of sparse symmetric linear systems using CG or SYMMLQ, preconditioned by incomplete Cholesky (NAG/f11jcf) or by Jacobi or SSOR (NAG/f11jef).

16.17.3 Harwell Library

The Harwell library contains very efficient subprograms for the direct solution of linear systems with sparse coefficient matrices. The algorithms and subprograms developed for the Harwell library (mainly by I. S. Duff) often set the standards for other libraries. If the direct methods offered by other libraries are not sufficient in some way, acquisition of (parts of) the Harwell Library should be considered.

The Harwell library offers subroutines for solving linear systems with positive definite or indefinite, symmetric or non-symmetric coefficient matrices and for the solution of linear least squares problems. Multiple right-hand sides can be handled as well as complex matrices. The methods use a pivoting strategy that ensures both the minimization of fill-in and numerical stability of the algorithm. Iterative refinement and block partitioning is provided too.

The mathematical and algorithmic basis of the methods implemented in the Harwell library are described in detail by Duff, Erisman und Reid [174]. The Harwell library catalog comprises a full listing of all relevant subprograms together with a short description of these programs. A more comprehensive description of the programs can be found in the accompanying documentation. Inquiries and orders concerning the Harwell library can be made via

```
http://www.rl.ac.uk/departments/ccd/numerical/hsl/hsl.html
```

Some subroutines from older versions of the Harwell library are freely available from NETLIB in the harwell directory. It must be noted, however, that these have been replaced by improved algorithms in the latest versions of the Harwell library.

Chapter 17

Random Numbers

Ist das Unscharfe nicht oft gerade das, was wir brauchen ? [1]

LUDWIG WITTGENSTEIN

To implement stochastic algorithms in computer programs, for example, to use the Monte Carlo method, sequences of numbers that behave exactly like samples of independent random variables are required. The construction of such sequences of numbers is called the *generation of random numbers*.

What is a random number? For example, as to whether the number 6, generated by throwing a six on a dice, is a random number, is impossible to say. However, this problem is not relevant. What is important is whether the k-tuple

$$(4, 1, 4, 3, 3, 6, 3, 1, 1, 5, 3, 5, 4, 2, 3, 4, 3)$$

represents random numbers. The origin of a single number is not important but the characteristics of *sequences* of numbers are.

Definition 17.0.1 (Random Numbers) *A k-tuple of numbers that is compatible with the statistical hypothesis to be a realization of a k-variate random vector with independently and identically distributed components from a distribution function F is called a k-tuple of random numbers with the distribution function F.*

17.1 Random Number Generators

The Definition 17.0.1 provides the basis for the generation of random numbers on a computer: Sequences of random numbers can be the result of deterministic algorithms if they show the same behavior as samples of a genuine random variable. Accordingly, a wealth of algorithms that generate number sequences with many desirable stochastic properties have been developed. Such algorithms are called *random number generators*.

Since random number generators are based on deterministic, i.e., on completely *predictable* computations, their output sequence cannot be of a stochastic nature. For this reason numbers generated in this way are referred to as *pseudo-random* numbers.

There are two kinds of random numbers: *Uniform* and *non-uniform* random numbers. Uniform random numbers have the properties of random samples

---

[1] Isn't the imprecise often exactly what we need?

from the (continuous) uniform distribution on $[0, 1]^n$, whereas non-uniformly distributed random numbers have either a non-constant density function and/or a domain B, which is different from the hypercube $[0, 1]^n$.

17.2 The Generation of Uniform Random Numbers

In this section the basic techniques for generating uniform random numbers are described. The generation of non-uniform random numbers follows in Section 17.3. Detailed discussions of this topic can be found in Dagpunar [39], Kalos, Whitlock [62] and Knuth [254].

The starting point for all generators of uniform random numbers is a function $f : Z \to Z$, whose domain is a *finite* set of numbers Z (usually the INTEGER numbers of the computer). With this function a sequence $\{z_i\}$ is defined recursively:

$$z_i := f(z_{i-1}, \ldots, z_{i-k}), \quad i = k+1, \, k+2, \, k+3, \, \ldots \, .$$

The initial values $z_1, \ldots, z_k$ of this recursion must be specified appropriately.

17.2.1 Congruential Generators

The most important method for generating uniform random numbers is the *mixed-congruential method*. Every random number generator that uses linear congruential sequences is characterized by three natural numbers, m, a and c. m is a large positive integer and $a \in \{1, 2, \ldots, m-1\}$ is a natural number which is relatively prime to m. m is called the *modulus* and a the *multiplier* of the random number generator. c, an arbitrary element of the set $\{0, 1, \ldots, m-1\}$ is an additive constant. If $c = 0$ the method is referred to as *multiplicative congruential*.

For any $z_1 \in \{0, 1, \ldots, m-1\}$ (or $z_1 \in \{1, 2, \ldots, m-1\}$ for $c = 0$) a sequence of numbers $\{z_i\} \subset \mathbb{Z}$ is defined by

$$z_{i+1} := az_i + c \mod m, \quad i = 1, 2, 3, \ldots \, . \tag{17.1}$$

The initial value z_1 is called the *seed* of the random number generator. After the normalization transformation

$$x_i = z_i/m, \quad i = 1, 2, 3, \ldots \, .$$

the sequence $\{x_i\}$ approximates a random sample taken from the uniform distribution over $[0, 1]$. Clearly every sequence $\{z_i\}$ defined by (17.1)—and therefore the corresponding sequence $\{x_i\}$—is *periodic* with period $T \leq m$.

Example (Period of Congruential Generators) The two random number generators

$$z_{i+1} := 69\,069\,z_i \qquad \mod 2^{32}$$
$$z_{i+1} := 69\,069\,z_i + 1 \qquad \mod 2^{32}$$

have periods $T = 2^{31} = 2.15 \cdot 10^9$ and $T = 2^{32} = 4.29 \cdot 10^9$, provided the seeds are chosen from the following sets:

$$z_1 \in \{1, 3, 5, 7, \dots, 2^{32} - 1\} \quad \text{and}$$

$$z_1 \in \{0, 1, 2, 3, \dots, 2^{32} - 1\}$$

respectively.

For some Monte Carlo studies (e. g., multi-dimensional problems) the seemingly adequate period of $4.29 \cdot 10^9$ is unacceptably short. To increase this period, an extension from

$$z_{i+1} := f(z_i) \qquad \text{to} \qquad z_{i+1} := \overline{f}(z_i, z_{i-1}) = a_1 z_i + a_2 z_{i-1} + c \bmod m,$$

i. e., a two-step generator, can be put into practice. The maximally attainable period increases in this way from m to m^2.

Example (Two-Step Congruential Generator) With

$$z_{i+1} := 1999 z_i + 4444 z_{i-1} \bmod 2^{31} - 1$$

the—in most cases acceptable—period $T = 2^{62} = 4.61 \cdot 10^{18}$ is obtained.

The periodicity could give reason to doubt the applicability of the linear congruential method for the generation of practically useful random numbers. However, despite periodicity, finite subsequences $\{x_i, x_{i+1}, \dots, x_j\}$ can show many characteristics of a genuine random sequence, provided the parameters m, a and c are chosen suitably. Appropriate values for the parameters can be determined by theoretical means complemented by extensive testing.

The linear congruential method is in widespread practical use on account of its simplicity, despite its periodicity and other disadvantageous properties (Niederreiter [299]).

Fibonacci Generators

The *Fibonacci recursion*

$$z_{i+1} := z_i + z_{i-1} \mod m, \qquad i = 1, 2, 3, \dots$$

yields a sequence of numbers with unacceptable stochastic properties. These numbers, however, can be transformed into a sequence of useful random numbers. Relevant techniques are discussed later (cf. Section 17.2.3).

The *modified (delayed) Fibonacci generators* (with parameters $r, s \in \mathbb{N}$, $r > s$)

$$z_{i+1} := z_{i-r} - z_{i-s} \mod m, \qquad i = 1, 2, 3, \dots, \tag{17.2}$$

which require $r + 1$ initial values $z_1, z_0, z_{-1}, \dots, z_{-r+1}$, provide satisfactory sequences of random numbers, even without supplementary improvement techniques.

Example (Modified Fibonacci Generators)   The random number generators

$$z_{i+1} := z_{i-16} - z_{i-4} \qquad \mathrm{mod}\ 2^{32}$$
$$z_{i+1} := z_{i-54} - z_{i-23} \qquad \mathrm{mod}\ 2^{32}$$
$$z_{i+1} := z_{i-606} - z_{i-273} \qquad \mathrm{mod}\ 2^{32}$$

have periods of $2.8 \cdot 10^{14}$, $7.7 \cdot 10^{25}$ and 10^{192} respectively.

Subtraction Generators with Carry

Even longer periods than those of the modified Fibonacci generators (17.2) can be obtained by including a carry:

$$z_{i+1} := z_{i-s} - z_{i-r} - c \quad \mathrm{mod}\ m, \qquad i = 1, 2, 3, \dots .$$

The carry bit $c \in \{0, 1\}$ is defined by the following algorithm:

> **do** $i = 1, 2, 3, \dots$
>> $t := z_{i-r} - z_{i-s} - c \quad \mathrm{mod}\ m$
>> **if** $t \geq 0$ **then**
>>> $z_i := t; \quad c := 0$
>> **else**
>>> $z_i := t + m; \quad c := 1$
>> **end if**
> **end do**

Example (Subtraction Generator with Carry)   The generator

$$z_{i+1} := z_{i-23} - z_{i-36} - c \quad \mathrm{mod}\ 2^{32}$$

has a period of $4.1 \cdot 10^{354}$.

In the relevant literature a multitude of further methods for the generation of uniform random numbers can be found (see e.g., Niederreiter [299]).

17.2.2 Uniform Random Vectors

Uniform random vectors can be generated easily by combining n successively generated one-dimensional uniform random numbers into n-vectors: For a one-dimensional random sequence $\{x_i\}$ the corresponding sequence of n-dimensional random vectors $\{y_i\}$ is defined by

$$y_i := (x_{(i-1)n+1}, \dots, x_{in})^{\top}, \qquad i = 1, 2, 3, \dots . \tag{17.3}$$

By using the *matrix congruential method*

$$y_{i+1} := A y_i \bmod m, \qquad A \in \mathbb{R}^{n \times n}, \quad y_i \in \mathbb{R}^n, \quad i = 1, 2, 3, \dots,$$

which is an immediate generalization of the linear congruential method, uniformly distributed random vectors can be generated directly.

17.2.3 Improving Random Number Generators

In general, sequences of n-dimensional vectors that are produced by a one-dimensional congruential (random number) generator are not uniformly distributed over the hypercube $[0,1]^n$, as Fig. 17.1 illustrates: The n-tuples $y_1, y_2, y_3, \ldots$ are situated on a few parallel hyperplanes of $\mathbb{R}^n$.

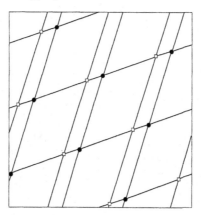

Figure 17.1: The results of the generator $x_{i+1} = 11x_i \bmod 64$ with $x_0 = 1$ for the generation of 2D-random numbers $(x_0, x_1), (x_2, x_3), \ldots$ (symbol ●) or $(x_1, x_2), (x_3, x_4), \ldots$ (symbol □).

Various methods have been developed to enable a supplementary improvement of a sequence $\{x_i\}$ of random numbers. One of the simplest methods, requiring little additional work, is based on the *self-permutation* of the sequence $\{x_i\}$. At first an array $F(1), F(2), \ldots, F(k)$, where $k \approx 100$, is initialized with the first k values of a sequence of random numbers. Then, for each element x_i of the sequence $\{x_{k+1}, x_{k+2}, \ldots\}$ an index

$$j := \lfloor kx_i \rfloor + 1$$

is calculated, and the number found in $F(j)$ is used as the new random number. Afterwards the value x_i is stored in $F(j)$.

Example (Turbo C) The built-in random number generator of Turbo C does not produce sequences that approximate random samples of the standard uniform distribution very well (see Fig. 17.2). Only a supplementary improvement, using, for instance, the permutation method introduced above (the method of McLaren and Marsaglia), leads to an acceptable sequence of uniform random numbers (see Fig. 17.3).

17.3 The Generation of Non-uniform Random Numbers

Non-uniform random numbers represent a random sample from a non-uniform probability distribution P on a domain $B \subseteq \mathbb{R}^n$. Usually non-uniform random numbers from the distribution P are obtained by appropriately transforming uniform random numbers. In this section the basic techniques for doing

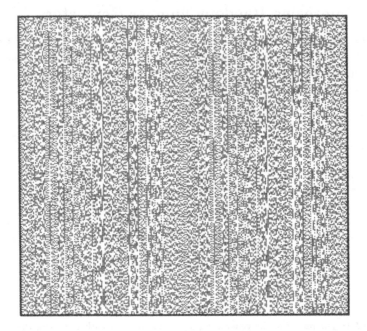

Figure 17.2: A bit sequence generated with the random number generator of Turbo C.

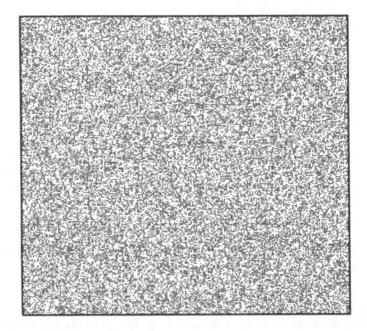

Figure 17.3: A bit sequence of the Turbo C random number generator after a supplementary improvement using the permutation method of McLaren and Marsaglia.

this are explained. Devroye [158] provides a detailed description of the topic. For many statistical standard distributions special methods for random number generation (and their implementations in computer programs) are available (cf. Sections 17.5.2 and 17.5.3).

17.3.1 The Inversion Method for Univariate Distributions

If a one-dimensional random variable has a continuous, strictly monotonic distribution function F, then the inverse function F^{-1}, defined on $[0, 1]$, exists. This is also true for distribution functions F that are not continuous and/or not strictly monotonic if the inverse function F^{-1} is defined in a more general way. If for example $F(x) = \alpha$ for all $x \in (a, b)$ then F^{-1} at α can be assigned to any $x \in (a, b)$. It can be shown that with this definition F^{-1} is well defined. A more detailed discussion of this topic is beyond the scope of this book.

For a random sample $X_1, X_2, \ldots$ from the standard uniform distribution, the transformed sequence of random variables $F^{-1}(X_1)$, $F^{-1}(X_2), \ldots$ is a random sample from a univariate distribution with the distribution function F.

Example (Cauchy Distribution) A Cauchy distributed random variable has a probability density function

$$v_C(x) := \frac{C}{\pi(C^2 + x^2)} \tag{17.4}$$

with the parameter $C \in \mathbb{R}_+$. Integration of v_C leads to the distribution function

$$F_C(x) = \frac{1}{\pi} \arctan\left(\frac{x}{C}\right) - \frac{1}{2},$$

whose inverse is given by

$$F_C^{-1}(y) = C \tan\left(\pi\left(y + \frac{1}{2}\right)\right).$$

For a sequence of uniform random numbers $y_1, y_2, \ldots$ on $[0, 1]$ the transformed values

$$x_1 := F_C^{-1}(y_1), \quad x_2 := F_C^{-1}(y_2), \ldots$$

form a sequence of random numbers with Cauchy distribution.

Theoretically, for any given univariate distribution, corresponding random numbers can be generated by using the inversion method. In practice however, difficulties appear since the inverse function F^{-1} is often not available in analytic form. In this case the determination of the transformed sequence $F^{-1}(X_i)$, $i = 1, 2, 3, \ldots$, is more difficult.

Example (Normal Distribution) A closed form expression exists neither for the normal ($N(0, 1)$) distribution function

$$\Phi(x) = \frac{1}{\sqrt{2\pi}} \int\limits_{-\infty}^{x} e^{-t^2/2}\, dt \quad \textit{(Gaussian distribution function)},$$

nor for its inverse Φ^{-1}. There are however approximations for Φ^{-1} using polynomials or rational functions, which can be utilized to generate $N(0, 1)$ distributed random numbers.

To generate normally distributed random numbers on high performance computers, specially developed approximations for Φ^{-1} are available (Marsaglia [284]).

17.3.2 The Rejection Method

In its simplest form, the rejection method can generate random numbers uniformly distributed on a set B, assuming that random numbers with (continuous) uniform distribution are available on a superset $B' \supset B$.

Theorem 17.3.1 *Let $X_1, X_2, \ldots$ be a sequence of independent, uniformly distributed random variables on $B' \subseteq \mathbb{R}^n$. For $B \subset B'$ let $i_{\min}$ be defined by*

$$i_{\min} := \min\{ i \ : \ X_i \in B \}.$$

Then the random variable $Y := X_{i_{\min}}$ is uniformly distributed on B.

Proof: Devroye [158].

If $x_1, x_2, \ldots$ is a random sample obtained from the uniform distribution on B', then—according to Theorem 17.3.1—the values $x_{i_1}, x_{i_2}, \ldots$; $i_1 < i_2 < \cdots$; which are obtained from $x_1, x_2, \ldots$ by *rejecting* all samples $x_i \notin B$ (i.e., by removing them from the sequence), represent a random sample obtained from the uniform distribution on B. In practice the domain B is transformed into $\overline{B} \subset [0,1]^n$ such that B' can be chosen as $B' = [0,1]^n$. The sequence $x_1, x_2, \ldots$ is generated by one of the uniform random number generators described in Section 17.2.

The following theorem is the basis of an extended rejection method that is useful for the generation of random numbers for arbitrary probability densities.

Theorem 17.3.2 *Let X be a random variable with a density v on $B \subseteq \mathbb{R}^n$ and U be a random variable uniformly distributed on $[0,1]$. Furthermore, U is assumed to be independent of X. Then for any constant $c > 0$ the random variable $(X, cUv(X))$ is uniformly distributed on*

$$H := \{ (x, u) \in \mathbb{R}^{n+1} \ : \ x \in B, \ u \in [0, c\,v(x)] \}.$$

If, conversely, a random variable (X, V) is uniformly distributed on H, then X is distributed on B with probability density v.

Proof: Devroye [158].

Now let w be a probability density on B with

$$w(x) \leq c\,v(x) \qquad \text{for all} \quad x \in B. \tag{17.5}$$

Furthermore, let $X_1, X_2, \ldots$ be a sequence of independent, identically distributed random variables with density v and let $U_1, U_2, \ldots$ be a sequence of independent random variables uniformly distributed on $[0,1]$.

According to Theorem 17.3.2 the random variables

$$(X_i, c\,U_i\,v(X_i)), \quad i = 1, 2, \ldots,$$

are uniformly distributed on

$$H = \{ (x, u) \in \mathbb{R}^{N+1} \ : \ x \in B, \ u \in [0, c\,v(x)] \}.$$

Let $(X_{i_j}, c\,U_{i_j}\,v(X_{i_j}))$, $j = 1, 2, \ldots$, be the subsequence of random variables obtained from $(X_i, c\,U_i\,v(X_i))$, $i = 1, 2, \ldots$, by rejecting those elements for which

$$(X_i, c\,U_i\,v(X_i)) \notin H' = \{\,(x, u) \in \mathbb{R}^{n+1} \;:\; x \in B,\; u \in [0, w(x)]\,\},$$

i.e.,

$$c\,U_i\,v(X_i) > w(X_i) \quad \Longleftrightarrow \quad c\,U_i\frac{v(X_i)}{w(X_i)} > 1.$$

Then, according to Theorem 17.3.1, the sequence $(X_{i_j}, c\,U_{i_j}\,v(X_{i_j}))$, $j = 1, 2, \ldots$, is uniformly distributed on H'. Finally, Theorem 17.3.2 guarantees that the X_{i_j}, $j = 1, 2, \ldots$, are distributed on B with density w.

This procedure allows for the generation of random numbers with density w on B assuming that the random numbers with density v that satisfy the condition (17.5) with a suitable constant $c > 0$ are available. In practice it is very difficult to find a suitable v.

Example (The Rejection Method) The generation of normally distributed random numbers can be carried out by using Cauchy distributed random numbers. The quotient of the density

$$w(x) = \frac{1}{\sqrt{2\pi}}\exp\left(-\frac{x^2}{2}\right)$$

of the standard normal distribution and the density $v = v_1$ (17.4) of the Cauchy distribution with parameter $C = 1$ has the absolute maximum $\sqrt{2\pi/e} \approx 1.520$. Therefore, for the constant c in (17.5), any value $c \geq \sqrt{2\pi/e}$ can be chosen.

From a sequence $u_1, u_2, u_3 \ldots$ of uniformly distributed random numbers on $[0, 1]$ and an independent sequence $x_1, x_2, x_3 \ldots$ of Cauchy distributed random numbers (with $C = 1$) a subsequence $x_{i_1}, x_{i_2}, x_{i_3}, \ldots$ of $x_1, x_2, x_3 \ldots$ is constructed by rejecting those x_i with

$$c\,u_i\frac{v(x_i)}{w(x_i)} > 1.$$

The resulting sequence has a standardized normal distribution $N(0, 1)$.

17.3.3 The Composition Method

A special case arises when the density w of a probability distribution on B can be written as a linear combination of probability densities $w_1, \ldots, w_J$ on B:

$$w(x) = \sum_{j=1}^{J} p_j w_j(x) \qquad \text{for all} \quad x \in B,$$

and the weights p_j satisfy

$$\sum_{j=1}^{J} p_j = 1 \quad \text{and} \quad p_j \geq 0, \quad j = 1, 2, \ldots, J.$$

If $Z_1, Z_2, \ldots$ denotes a sequence of independent, discrete random variables, which assume values from $\{1, 2, \ldots, J\}$ with a probability

$$P(\{j\}) = p_j, \quad j = 1, 2, \ldots, J,$$

and $X_1, X_2, X_3, \ldots$ is a sequence of independent random variables on B with densities $w_{Z_1}, w_{Z_2}, w_{Z_3}, \ldots$, then the random variables $\{X_i\}$ are distributed on B with density w.

Practically, this method is used in situations in which a multi-dimensional domain B is too complicated to be dealt with. In this case B is split up into simpler disjoint subdomains $B_1, \ldots, B_J$:

$$B = B_1 \cup B_2 \cup \cdots \cup B_J, \qquad B_j \cap B_k = \emptyset \quad \text{for} \quad j \neq k.$$

The densities $w_1, \ldots, w_J$ are then defined by

$$w_j := \frac{c_{B_j} w}{\int_{B_j} w(x)\, dx}, \quad j = 1, 2, \ldots, J.$$

Example (The Composition Method)   For the generation of random numbers that are uniformly distributed on a plane polygonal domain B, the domain B is triangulated, i.e., reduced to a finite set of disjoint triangles $B_1, B_2, \ldots, B_J$. The densities $w_1, w_2, \ldots, w_J$ and the discrete probability distribution $(p_1, p_2, \ldots, p_J)$ are then given by

$$w_j := \mathrm{vol}(B_j)^{-1} c_{B_j}, \quad p_j := \frac{\mathrm{vol}(B_j)}{\mathrm{vol}(B)}, \quad j = 1, 2, \ldots, J.$$

Therefore, with a method for generating random numbers uniformly distributed on a triangular domain, uniform random numbers in any polygonal domain can be obtained by using the composition method.

17.4 Testing Random Number Generators

As to whether a random number generator is useful or not depends on whether the produced values are distributed independently with respect to the intended distribution function F. Some extreme forms of deviation from the ideal behavior of a generator for uniform random numbers are shown in Fig. 17.4.

Distribution tests are empirical tests used to investigate whether a sample of random numbers $X_1, \ldots, X_n$ is actually distributed according to a given distribution function F. The two most important tests of this type are the chi-squared test and the Kolmogorov-Smirnov test, both of which are found in the IMSL and the NAG libraries.

Tests for stochastic independence are based on quantities whose distribution can be determined by presupposing the stochastic independence of the investigated random numbers. Programs for these tests can be found in the IMSL and the NAG libraries.

17.5 Software for Generating Random Numbers

17.5.1 Programming Languages

Many programming languages provide intrinsic procedures for generating pseudo-random numbers: In Fortran 90 the predefined subroutine RANDOM_NUMBER generates random numbers that are uniformly distributed on the interval $[0, 1)$.

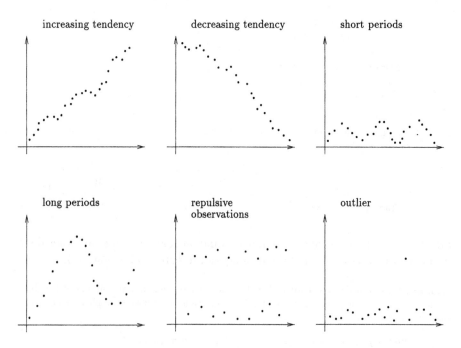

Figure 17.4: Various forms of non-independency and non-uniform distribution.

Example (Volume of the Unit Sphere) To approximate the volume of the n-dimensional unit sphere $\{x \in \mathbb{R}^n : \|x\|_2 \leq 1\}$, a Monte Carlo method can be used: Random points are generated within the enclosing hypercube $\{x \in \mathbb{R}^n : \|x\|_\infty \leq 1\}$. The proportion of the *hits* (points within the sphere) among the total number of points provides an approximate value for the ratio of the volume of the sphere to the volume 2^n of the cube.

```
PROGRAM volume_of_sphere
   INTEGER, PARAMETER  ::  number_tests   = 10000
   INTEGER, PARAMETER  ::  max_dimension  =    11
   INTEGER             ::  dimension, test
   REAL                ::  hit, frequency, volume

   DO dimension = 2, max_dimension
      hit = 0.
      DO test = 1, number_tests
         IF (in_the_sphere(dimension)) hit = hit + 1.
      END DO
      frequency = hit/number_tests
      volume    = frequency*(2.**dimension)
      ...
   END DO
END PROGRAM volume_of_sphere

FUNCTION in_the_sphere (dimension) RESULT (sphere_inside)
   LOGICAL                    ::  sphere_inside
   INTEGER, INTENT (IN)       ::  dimension
   REAL, DIMENSION (dimension)  ::  vector

   CALL RANDOM_NUMBER (vector)
```

```
    IF (SUM (vector**2) < 1.) THEN
        sphere_inside = .TRUE.
    ELSE
        sphere_inside = .FALSE.
    END IF
END FUNCTION in_the_sphere
```

From this program the following results (illustrating the behavior *volume_of_sphere* $\rightarrow 0$ as $n \rightarrow \infty$) can be obtained:

n	2	3	4	5	6	7	8	9	10	11
volume	3.15	4.18	4.97	5.24	4.98	4.77	4.04	3.33	2.25	1.67

Starting with uniform random numbers, random samples from any other distribution can be obtained using the techniques presented in Section 17.3.

Example (Normal Distribution)   For the generation of approximately normally distributed random numbers with given mean and standard deviation the following simple Fortran 90 function procedure can be used:

```
FUNCTION random_normal (mean, deviation) RESULT (normal)
    REAL, INTENT (IN)      ::  mean, deviation
    REAL                   ::  normal
    REAL, DIMENSION (12)   ::  random_uniform

    CALL RANDOM_NUMBER (random_uniform)
    normal = mean + deviation*(SUM (random_uniform) - 6.)
END FUNCTION random_normal
```

Initial Values for Random Number Generators

The choice of the initial values $x_1, \ldots, x_k$ of the intrinsic procedures implementing random number generators in scientific programming languages can either be left up to the system or deliberately made by the user. The latter possibility is advantageous e. g., when reproducible sequences of random numbers are to be generated while testing a simulation program. Otherwise, the effects of changes in the tested program would not be distinguishable from the effects of using a new sequence of random numbers.

Example (Fortran 90)   With the aid of the intrinsic procedure RANDOM_SEED, information about the number k of the initial values of the Fortran 90 random number generator can be obtained:

```
CALL RANDOM_SEED (SIZE = k)
```

The call

```
INTEGER, DIMENSION (k)   ::   initial_values
...

CALL RANDOM_SEED (PUT = initial_values)
```

makes sure that the INTEGER numbers

`initial_values(1),...,initial_values(k)`

are used as initial values. With

`CALL RANDOM_SEED (GET = initial_values),`

the current initial values (set by the user's program or by the system) are stored in the array `initial_values`. If a reinitialization of the sequence $\{x_i\}$ is desired, then this can be done using

`CALL RANDOM_SEED`

(without a parameter list).

17.5.2 IMSL Library

In the *statistical analysis* part of the IMSL library, there are about 50 subprograms that are, along with other subprograms, dedicated to the following functions:

Univariate Continuous Distributions: There are random number generators designed for continuous uniform distribution, normal distribution, logarithmic normal distribution, exponential distribution, Cauchy distribution, beta distribution, gamma distribution, χ^2 distribution, t distribution, Weibull distribution etc.

Univariate Discrete Distributions: There are random number generators for discrete uniform distribution, binomial distribution, geometric and hypergeometric distribution, Poisson distribution etc.

Multivariate Distributions: There are random number generators for the multivariate normal distribution, continuous uniform distribution in and on the unit sphere etc.

Stochastic Processes (ARMA- and non-homogeneous Poisson processes) are simulated.

Random Permutations and Samples are facilitated.

Moreover, the user can prescribe arbitrary continuous or discrete distribution functions. This allows the generation of random numbers with arbitrary distribution.

17.5.3 NAG Library

The NAG library contains around 40 subprograms to support statistical functions:

Univariate Continuous Distributions: There are random number generators designed for the continuous uniform distribution, normal distribution, logarithmic normal distribution, exponential distribution, Cauchy distribution, beta distribution, gamma distribution, χ^2 distribution, t distribution, F distribution, Weibull distribution etc.

Univariate Discrete Distributions: There are random number generators for the discrete uniform distribution, binomial distribution, hypergeometric distribution, Poisson distribution etc.

Multivariate Distributions: Normally distributed random vectors can be generated.

Stochastic Processes (ARMA- and VARMA-processes) are simulated.

Random Permutations and Samples are facilitated.

Glossary of Notation

$M_1 \backslash M_2$	Difference of two sets (M_1 minus M_2)
$M_1 \times M_2$	Cartesian product of M_1 and M_2
M^n	n-fold Cartesian product of the set M
$M^\perp$	Orthogonal complement
$\{x, y, z, \dots\}$	Set whose members are x, y, z, $\dots$
$\{x_i\}$	Sequence whose ith term is x_i, $i \in I \subseteq \mathbb{Z}$
(a, b)	Interval excluding the endpoints
$[a, b]$	Interval including the endpoints
$f : M_1 \to M_2$	Function (mapping) from M_1 to M_2
$f^{(m)}$	mth derivative of the function f
$S := A$	Definition: S is defined by A
$\approx$	Approximate equality
$\sim$	Asymptotic equivalence
$\equiv$	Identity
$\doteq$	Correspondence
$\lvert \cdot \rvert$	Absolute value
$\lVert \cdot \rVert_F$	Frobenius norm
$\lVert \cdot \rVert_2$	Euclidean norm
$\lVert \cdot \rVert_\infty$	Maximum norm (uniform norm)
$\lVert \cdot \rVert_p$	l_p- or L_p-norm
$\lVert \cdot \rVert_{p,w}$	l_p- or L_p-norm weighted with w
$\langle u, v \rangle$	Inner product (scalar product) of the vectors u and v
$\oplus$	Direct sum of two spaces
$\circ : \mathbb{R} \times \mathbb{R} \to \mathbb{R}$	Binary (arithmetic) operation
$\square : \mathbb{R} \to \mathbb{F}$	Rounding function
$\boxplus\ \boxminus\ \boxdot\ \boxplus$	Floating-point addition, subtraction, multiplication, division
$\boxdot$	Floating-point operation, general
$\lceil x \rceil$	Smallest integer greater than or equal to x
$\lfloor x \rfloor$	Greatest integer less than or equal to x
$\Delta_k g$	Function g discretized at k sampling points
$\kappa_p(A)$	Condition number $\lVert A \rVert_p \lVert A^+ \rVert_p$ of the matrix A
$\lambda_i(x; K)$	Lebesgue function w. r. t. the ith row of node matrix K
$\lambda(A)$	Spectrum (set of eigenvalues) of the matrix A
$\rho(A)$	Spectral radius of the matrix A
$\rho_A,\ \rho_b$	Relative error of the matrix A, of the vector b
μs	Microsecond (10^{-6} s)
∇	Gradient
$\varphi_{d,i},\ \sigma_{d,i}$	Basis function (basis spline)
Φ	Gaussian integral (standard normal distribution function)
τ	Error tolerance of a numerical problem
$\displaystyle\int_a^b f(x)\,dx$	Definite integral of f over $[a, b]$
$\displaystyle\sum_{j=m}^k f(j)$	Sum $f(m) + f(m+1) + \cdots + f(k)$
$\sum{}'$	Dashed sum: Halving of the first term
$\sum{}''$	Double dashed sum: Halving of the first term and the last term
ω_c	Nyquist frequency

$A^\top$, $v^\top$	Transpose of the matrix A, the vector v
A^H	Conjugate transpose of the matrix A
A^{-1}	Inverse of the matrix A
$A^{-\top}$	Inverse of the matrix $A^\top$
A^+	Generalized inverse (pseudo-inverse) of the matrix A
$b_{d,i}$	Bernstein polynomial
$\mathbb{C}$	Set of all complex numbers
$\mathbb{C}^{m \times n}$	Set of all complex $m \times n$ matrices
$C(\Omega)$, C	Set of all functions continuous on Ω or $\mathbb{R}^n$ respectively
$C^m(\Omega)$, C^m	Set of all functions having m continuous derivatives on Ω or $\mathbb{R}^n$ respectively
c^*	Parameter vector of a best approximating function
$\mathrm{cond}_p(A)$	Condition number $\|A\|_p \|A^+\|_p$ of the matrix A
$D(\cdot, \cdot)$	Distance function
$D_{p,w}(\cdot, \cdot)$	Distance function w.r.t. $\| \cdot \|_{p,w}$
$\mathcal{D}$	Data set
$\tilde{\mathcal{D}}$	Perturbed data set
d	Degree
$\det(A)$	Determinant of the matrix A
$\mathrm{diag}(a_{11}, \ldots, a_{nn})$	Diagonal matrix
$\mathrm{diam}(M)$	Diameter of the set M
$\dim$	Dimension of a space
$\mathrm{dist}(\cdot, \cdot)$	Distance function
div	Divergence of a vector field
$E_d(t; a_0, \ldots, a_d, \alpha_1, \ldots, \alpha_d)$	Exponential sum $a_0 + a_1 e^{\alpha_1 t} + \cdots + a_d e^{\alpha_d t}$
eps	Relative machine accuracy
$\mathcal{F}$	Set of functions
$\mathbb{F}$	Set of all floating-point numbers
$\mathbb{F}_D$	Set of all *denormalized* floating-point numbers
$\mathbb{F}_N$	Set of all *normalized* floating-point numbers
flop	Floating-point operation
$\mathcal{G}$, $\mathcal{G}_k$	Set of functions (with k parameters)
g^*	Best approximating function
grad	Gradient
H_f	Hessian matrix $\nabla^2 f = (\mathrm{grad} f)'$
Hz	Hertz (cycles per second)
$I(f; a, b)$	Definite integral of $f : \mathbb{R} \to \mathbb{R}$ over $[a, b]$
$I(f; B)$	Definite integral of $f : \mathbb{R}^n \to \mathbb{R}$ over $B \subseteq \mathbb{R}^n$
KB	Kilobyte (10^3 bytes, also 2^{10} bytes)
K_c	Condition number w.r.t. coefficients
K_e	Node matrix, equidistant
K_f	Condition number w.r.t. function values
$k_{F \leftarrow x}$	Absolute (first order) condition number
$K_{F \leftarrow x}$	Relative (first order) condition number
K_T	Node matrix, Chebyshev zeros
K_U	Node matrix, Chebyshev extrema
$\mathcal{K}_i(r, B)$	Krylov space
$l_i : \mathcal{F} \to \mathbb{R}$	Linear functional
l_p	Set of all real sequences $\{a_i\}$ with convergent $\sum_{i=1}^{\infty} |a_i|^p$
$L^p[a, b]$, L^p	Set of all functions f, for which $I(|f|^p; a, b)$ or $I(|f|^p; -\infty, \infty)$ exists
$\mathrm{Lip}(F, B)$	Lipschitz norm of F on the set B
ln	Natural logarithm (base e)
$\log$, $\log_b$	Logarithm (base b)

$\max M$	Maximum value of the elements of the set M
MB	Megabyte (10^6 bytes, also 2^{20} bytes)
$\min M$	Minimum value of the elements of the set M
MK	Markowitz cost function
mod	Modulo function
ms	Millisecond (10^{-3} s)
$\mathcal{N}(F)$	Null space of the mapping F
$\mathbb{N}$	Set of all positive integers
$\mathbb{N}_0$	Set of all non-negative integers
$ND(A)$	Sum of the squares of all non-diagonal elements of the matrix A
non-zero(A)	Number of non-zero elements of the matrix A
non-zero-block(B)	Number of non-zero elements of the block matrix B
ns	Nanosecond (10^{-9} s)
$O(\cdot)$	Order of convergence (Landau's asymptotic growth notation)
n	Dimension
$n_{1/2}$	Vector length which achieves $r_\infty/2$
$N^{\mathrm{a}}(\cdot)$	Adaptive information
$N^{\mathrm{na}}(\cdot)$	Non-adaptive information
$N_{d,i}$	B-spline function
$\mathbb{P}, \mathbb{P}_d$	Set of all univariate polynomials (of degree not exceeding d)
$\mathbb{P}^n, \mathbb{P}^n_d$	Set of all n-variate polynomials (of degree not exceeding d)
$P_d(x; c_0, \ldots, c_d)$	Polynomial $c_0 + c_1 x + c_2 x^2 + \cdots + c_d x^d$
P_d^*	Best approximating polynomial of degree not exceeding d
P_F	Floating-point performance [flop/s]
P_I	Instruction performance
r_∞	Asymptotic throughput
$\mathbb{R}$	Set of all real numbers
$\mathbb{R}_+$	Set of all positive real numbers
$\mathbb{R}_+^0$	Set of all non-negative real numbers
$\mathbb{R}^{m \times n}$	Set of all real $m \times n$ matrices
$\mathbb{R}_D$	Set of all real numbers covered by $\mathbb{F}_D$
$\mathbb{R}_N$	Set of all real numbers covered by $\mathbb{F}_N$
$\mathbb{R}_{\mathrm{overflow}}$	Set of all real numbers *not* covered by $\mathbb{F}$
$\mathcal{R}(F)$	Range space of the mapping F
rank(A)	Rank of the matrix A
$S_d(x; a_0, \ldots, a_d, b_1, \ldots, b_d)$	Trigonometric polynomial $a_0/2 + \sum_{j=1}^{d}(a_j \cos jx + b_j \sin jx)$
span	Linear hull
$T_d \in \mathbb{P}_d$	Chebyshev polynomial of degree not exceeding d
$w(\cdot)$	Weight function
w_i	ith integration weight
$\mathbb{Z}$	Set of all integers
$zz \cdots z_2$, $zz \cdots z_{10}$, $zz \cdots z_{16}$	Number in binary, decimal- or hexadecimal representation

Bibliography

Reference Books

[1] M. Abramowitz, I. A. Stegun: *Handbook of Mathematical Functions*, 10th ed. National Bureau of Standards, Appl. Math. Ser. No. 55, U. S. Government Printing Office, 1972.

[2] J. C. Adams, W. S. Brainerd, J. T. Martin, B. T. Smith, J. L. Wagener: *Fortran 90 Handbook—Complete ANSI/ISO Reference*. McGraw-Hill, New York 1992.

[3] E. Anderson, Z. Bai, C. Bischof, J. Demmel, J. J. Dongarra, J. Du Croz, A. Greenbaum, S. Hammarling, A. McKenney, S. Ostrouchov, D. C. Sorensen: LAPACK *User's Guide*, 2nd ed. SIAM Press, Philadelphia 1995.

[4] R. E. Bank: PLTMG—*A Software Package for Solving Elliptic Partial Differential Equations—User's Guide 7.0*. SIAM Press, Philadelphia 1994.

[5] R. Barrett, M. Berry, T. Chan, J. Demmel, J. Donato, J. J. Dongarra, V. Eijkhout, R. Pozo, C. Romine, H. A. Van der Vorst: TEMPLATES *for the Solution of Linear Systems— Building Blocks for Iterative Methods*. SIAM Press, Philadelphia 1993.

[6] J. A. Brytschkow, O. I. Maritschew, A. P. Prudnikow: *Tables of Indefinite Integrals*. Gordon & Breach Science Publishers, New York 1989.

[7] J. Choi, J. J. Dongarra, D. W. Walker, SCALAPACK *Reference Manual—Parallel Factorization Routines (LU, QR, and Cholesky), and Parallel Reduction Routines (HRD, TRD, and BRD)*. Technical Report TM-12471, Mathematical Sciences Section, Oak Ridge National Laboratory, 1994.

[8] W. J. Cody, W. Waite: *Software Manual for the Elementary Functions*. Prentice-Hall, Englewood Cliffs 1981.

[9] T. F. Coleman, C. Van Loan: *Handbook for Matrix Computations*. SIAM Press, Philadelphia 1988.

[10] W. R. Cowell (Ed.): *Sources and Development of Mathematical Software*. Prentice-Hall, Englewood Cliffs 1984.

[11] J. J. Dongarra, J. R. Bunch, C. B. Moler, G. W. Stewart: LINPACK *User's Guide*. SIAM Press, Philadelphia 1979.

[12] J. J. Dongarra, R. A. Van de Geijn, R. C. Whaley: *A User's Guide to the* BLACS, Version 1.0, Technical Report, University of Tennessee, 1995.

[13] G. Engeln-Müllges, F. Uhlig: *Numerical Algorithms with C*. Springer-Verlag, Berlin Heidelberg New York Tokyo 1996.

[14] G. Engeln-Müllges, F. Uhlig: *Numerical Methods with Fortran*. Springer-Verlag, Berlin Heidelberg New York Tokyo 1996.

[15] H. Engesser, V. Claus, A. Schwill: *Encyclopedia of Information Technology*. Ellis Horwood, Chichester 1992.

[16] B. S. Garbow, J. M. Boyle, J. J. Dongarra, C. B. Moler: *Matrix Eigensystem Routines— EISPACK Guide Extension*. Lecture Notes in Computer Science Vol. 51, Springer-Verlag, Berlin Heidelberg New York Tokyo 1977.

[17] J. F. Hart et al.: *Computer Approximations*. Wiley, New York 1968.

[18] High Performance Fortran Forum (HPFF): *High Performance Fortran Language Specification*. Version 1.1, 1994.

[19] E. Krol: *The Whole Internet—User's Guide and Catalog*. O'Reilly, Sebastopol 1992.

[20] J. J. Moré, S. J. Wright: *Optimization Software Guide*. SIAM Press, Philadelphia 1993.

[21] NAG Ltd.: *NAG Fortran Library Manual—Mark 17*. Oxford 1995.

[22] R. Piessens, E. de Doncker, C. W. Ueberhuber, D. K. Kahaner: QUADPACK—*A Subroutine Package for Automatic Integration*. Springer-Verlag, Berlin Heidelberg New York Tokyo 1983.

[23] S. Pittner, J. Schneid, C. W. Ueberhuber: *Wavelet Literature Survey*. Technical University Vienna, Wien 1993.

[24] W. H. Press, B. P. Flannery, S. A. Teukolsky, W. T. Vetterling: *Numerical Recipes in Fortran—The Art of Scientific Computing*, 2nd ed. Cambridge University Press, Cambridge 1992.

[25] W. H. Press, B. P. Flannery, S. A. Teukolsky, W. T. Vetterling: *Numerical Recipes in Fortran 90—The Art of Parallel Scientific Computing*. Cambridge University Press, Cambridge 1996.

[26] W. H. Press, B. P. Flannery, S. A. Teukolsky, W. T. Vetterling: *Numerical Recipes in C—The Art of Scientific Computing*, 2nd ed. Cambridge University Press, Cambridge 1992.

[27] W. H. Press, B. P. Flannery, S. A. Teukolsky, W. T. Vetterling: *Numerical Recipes in Fortran—Example Book*, 2nd ed. Cambridge University Press, Cambridge 1992.

[28] W. H. Press, B. P. Flannery, S. A. Teukolsky, W. T. Vetterling: *Numerical Recipes in C—Example Book*, 2nd ed. Cambridge University Press, Cambridge 1992.

[29] J. R. Rice, R. F. Boisvert: *Solving Elliptic Problems Using* ELLPACK. Springer-Verlag, Berlin Heidelberg New York Tokyo 1985.

[30] B. T. Smith, J. M. Boyle, J. J. Dongarra, B. S. Garbow, Y. Ikebe, V. C. Klema, C. B. Moler: *Matrix Eigensystem Routines—*EISPACK *Guide*. Lecture Notes in Computer Science, Vol. 6, 2nd ed. Springer-Verlag, Berlin Heidelberg New York Tokyo 1990.

[31] Visual Numerics Inc.: IMSL MATH/LIBRARY—*User's Manual*, Version 3.0, Houston 1994.

[32] Visual Numerics Inc.: IMSL STAT/LIBRARY—*User's Manual*, Version 3.0, Houston 1994.

[33] D. Zwillinger (Ed. in Chief): *CRC Standard Mathematical Tables and Formulae*, 30th ed. CRC Press, Boca Raton New York London Tokyo 1996.

Textbooks

[34] D. P. Bertsekas, J. N. Tsitsiklis: *Parallel and Distributed Computation—Numerical Methods*. Prentice-Hall, Englewood Cliffs 1989.

[35] Å. Björck: *Numerical Methods for Least Squares Problems*. SIAM Press, Philadelphia 1996.

[36] H. D. Brunk: *An Introduction to Mathematical Statistics*, 2nd ed. Blaisdell, New York 1965.

[37] E. W. Cheney: *Introduction to Approximation Theory*. McGraw-Hill, New York 1966.

[38] J. B. Conway: *A Course in Functional Analysis*. Springer-Verlag, Berlin Heidelberg New York Tokyo 1984.

[39] J. Dagpunar: *Principles of Random Variate Generation*. Clarendon Press, Oxford 1988.

[40] P. J. Davis: *Interpolation and Approximation*. Blaisdell, New York 1963 (reprinted 1975 by Dover Publications).

[41] P. J. Davis, P. Rabinowitz: *Methods of Numerical Integration*, 2nd ed. Academic Press, New York 1984.

[42] C. de Boor: *A Practical Guide to Splines*. Springer-Verlag, Berlin Heidelberg New York Tokyo 1978.

[43] J. E. Dennis, R. B. Schnabel: *Numerical Methods for Unconstrained Optimization and Nonlinear Equations*. Prentice-Hall, Englewood Cliffs 1983 (reprinted 1996, SIAM Press).

[44] P. Deuflhard, A. Hohmann: *Numerical Analysis—A First Course in Scientific Computation.* de Gruyter, Berlin New York 1995.

[45] K. Dowd: *High Performance Computing.* O'Reilly & Associates, Sebastopol 1993.

[46] H. Engels: *Numerical Quadrature and Cubature.* Academic Press, New York 1980.

[47] G. Evans: *Practical Numerical Integration.* Wiley, Chichester 1993.

[48] G. H. Golub, J. M. Ortega: *Scientific Computing and Differential Equations—An Introduction to Numerical Methods.* Academic Press, New York 1991.

[49] G. H. Golub, J. M. Ortega: *Scientific Computing—An Introduction with Parallel Computing.* Academic Press, New York 1993.

[50] G. H. Golub, C. F. Van Loan: *Matrix Computations*, 2nd ed. Johns Hopkins University Press, Baltimore 1989.

[51] W. Hackbusch: *Elliptic Differential Equations—Theory and Numerical Treatment.* Springer-Verlag, Berlin Heidelberg New York Tokyo 1993.

[52] G. Hämmerlin, K. H. Hoffmann: *Numerical Mathematics.* Springer-Verlag, Berlin Heidelberg New York Tokyo 1991.

[53] L. A. Hageman, D. M. Young: *Applied Iterative Methods.* Academic Press, New York London 1981.

[54] R. Hamming: *Numerical Methods for Scientists and Engineers.* McGraw-Hill, New York 1962 (reprinted 1987).

[55] J. L. Hennessy, D. A. Patterson: *Computer Architecture—A Quantitative Approach*, 2nd ed. Morgan Kaufmann, San Mateo 1995.

[56] P. Henrici: *Elements of Numerical Analysis.* Wiley, New York 1964.

[57] N. Higham: *Accuracy and Stability of Numerical Algorithms.* SIAM Press, Philadelphia 1996.

[58] F. B. Hildebrand: *Introduction to Numerical Analysis.* McGraw-Hill, New York 1974 (reprinted 1987).

[59] R. A. Horn, C. R. Johnson: *Matrix Analysis.* Cambridge University Press, Cambridge 1985 (reprinted 1990).

[60] R. A. Horn, C. R. Johnson: *Topics in Matrix Analysis*, paperback ed. Cambridge University Press, Cambridge 1994.

[61] E. Isaacson, H. B. Keller: *Analysis of Numerical Methods.* Wiley, New York 1966 (reprinted 1994).

[62] M. H. Kalos, P. A. Whitlock: *Monte Carlo Methods.* Wiley, New York 1986.

[63] C. T. Kelley: *Iterative Methods for Linear and Nonlinear Equations.* SIAM Press, Philadelphia 1995.

[64] A. R. Krommer, C. W. Ueberhuber: *Computational Integration.* SIAM Press, Philadelphia, 1996.

[65] C. L. Lawson, R. J. Hanson: *Solving Least Squares Problems.* Prentice-Hall, Englewood Cliffs 1974 (reprinted 1995, SIAM Press).

[66] P. Linz: *Theoretical Numerical Analysis.* Wiley, New York 1979.

[67] J. M. Ortega, W. C. Rheinboldt: *Iterative Solution of Nonlinear Equations in Several Variables.* Academic Press, New York London 1970.

[68] D. A. Patterson, J. L. Hennessy: *Computer Organization and Design—The Hardware / Software Interface.* Morgan Kaufmann, San Mateo 1994.

[69] C. S. Rees, S. M. Shah, C. V. Stanojevic: *Theory and Applications of Fourier Analysis.* Marcel Dekker, New York Basel 1981.

[70] J. R. Rice: *Matrix Computations and Mathematical Software.* McGraw-Hill, New York 1981.

[71] J. R. Rice: *Numerical Methods, Software, and Analysis.* McGraw-Hill, New York 1983.

[72] H. R. Schwarz, (J. Waldvogel): *Numerical Analysis—A Comprehensive Introduction.* Wiley, Chichester 1989.

[73] H. Schwetlick: *Numerische Lösung nichtlinearer Gleichungen.* Oldenbourg, München Wien 1979.

[74] G. W. Stewart: *Introduction to Matrix Computations.* Academic Press, New York 1974.

[75] J. Stoer, R. Bulirsch: *Introduction to Numerical Analysis,* 2nd ed. Springer-Verlag, Berlin Heidelberg New York Tokyo 1993.

[76] G. Strang: *Linear Algebra and its Applications,* 3rd ed. Academic Press, New York 1988.

[77] A. H. Stroud: *Numerical Quadrature and Solution of Ordinary Differential Equations.* Springer-Verlag, Berlin Heidelberg New York Tokyo 1974.

[78] C. W. Ueberhuber, P. Meditz: *Software-Entwicklung in Fortran 90.* Springer-Verlag, Wien New York 1993.

[79] H. Werner, R. Schaback: *Praktische Mathematik II.* Springer-Verlag, Berlin Heidelberg New York Tokyo 1979.

[80] J. H. Wilkinson: *The Algebraic Eigenvalue Problem.* Oxford University Press, London 1965 (reprinted 1988).

Technical Literature

[81] C. A. Addison, J. Allwright, N. Binsted, N. Bishop, B. Carpenter, P. Dalloz, J. D. Gee, V. Getov, T. Hey, R. W. Hockney, M. Lemke, J. Merlin, M. Pinches, C. Scott, I. Wolton: *The Genesis Distributed-Memory Benchmarks. Part 1—Methodology and General Relativity Benchmark with Results for the SUPRENUM Computer.* Concurrency—Practice and Experience 5-1 (1993), pp. 1–22.

[82] A. V. Aho, J. E. Hopcroft, J. D. Ullman: *Data Structures and Algorithms.* Addison-Wesley, Reading 1983.

[83] H. Akima: *A New Method of Interpolation and Smooth Curve Fitting Based on Local Procedures.* J. ACM 17 (1970), pp. 589–602.

[84] H. Akima: *A Method of Bivariate Interpolation and Smooth Surface Fitting for Irregularly Distributed Data Points.* ACM Trans. Math. Softw. 4 (1978), pp. 148–159.

[85] E. L. Allgower, K. Georg: *Numerical Continuation—An Introduction.* Springer-Verlag, Berlin Heidelberg New York Tokyo 1990.

[86] G. S. Almasi, A. Gottlieb: *Highly Parallel Computing.* Benjamin/Cummings, Redwood City 1989.

[87] L. Ammann, J. Van Ness: *A Routine for Converting Regression Algorithms into Corresponding Orthogonal Regression Algorithms.* ACM Trans. Math. Softw. 14 (1988), pp. 76–87.

[88] L.-E. Andersson, T. Elfving: *An Algorithm for Constrained Interpolation.* SIAM J. Sci. Stat. Comp. 8 (1987), pp. 1012–1025.

[89] M. A. Arbib, J. A. Robinson (Eds.): *Natural and Artificial Parallel Computation.* MIT Press, Cambridge 1990.

[90] M. Arioli, J. Demmel, I. S. Duff: *Solving Sparse Linear Systems with Sparse Backward Error.* SIAM J. Matrix Anal. Appl. 10 (1989), pp. 165–190.

[91] W. Arnoldi: *The Principle of Minimized Iterations in the Solution of the Matrix Eigenvalue Problem.* Quart. Appl. Math. 9 (1951), pp. 165–190.

[92] K. Atkinson: *The Numerical Solution of Laplace's Equation in Three Dimensions.* SIAM J. Num. Anal. 19 (1982), pp. 263-274.

[93] O. Axelsson: *Iterative Solution Methods.* Cambridge University Press, Cambridge 1996.

[94] O. Axelsson, V. Eijkhout: *Vectorizable Preconditioners for Elliptic Difference Equations in Three Space Dimensions.* J. Comput. Appl. Math. 27 (1989), pp. 299–321.

[95] O. Axelsson, B. Polman: *On Approximate Factorization Methods for Block Matrices Suitable for Vector and Parallel Processors.* Linear Algebra Appl. 77 (1986), pp. 3–26.

[96] L. Bacchelli-Montefusco, G. Casciola: C^1 *Surface Interpolation.* ACM Trans. Math. Softw. 15 (1989), pp. 365–374.

[97] Z. Bai, J. Demmel, A. McKenney: *On the Conditioning of the Nonsymmetric Eigenproblem.* Technical Report CS-89-86, Computer Science Dept., University of Tennessee, 1989.

[98] D. H. Bailey: *Extra High Speed Matrix Multiplication on the Cray-2.* SIAM J. Sci. Stat. Comput. 9 (1988), pp. 603–607.

[99] D. H. Bailey: MPFUN—*A Portable High Performance Multiprecision Package.* NASA Ames Tech. Report RNR-90-022, 1990.

[100] D. H. Bailey: *Automatic Translation of Fortran Programs to Multiprecision.* NASA Ames Tech. Report RNR-91-025, 1991.

[101] D. H. Bailey: *A Fortran-90 Based Multiprecision System.* NASA Ames Tech. Report RNR-94-013, 1994.

[102] D. H. Bailey, H. D. Simon, J. T. Barton, M. J. Fouts: *Floating-Point Arithmetic in Future Supercomputers.* Int. J. Supercomput. Appl. 3-3 (1989), pp. 86–90.

[103] C. T. H. Baker: *On the Nature of Certain Quadrature Formulas and their Errors.* SIAM J. Numer. Anal. 5 (1968), pp. 783–804.

[104] B. A. Barsky: *Exponential and Polynomial Methods for Applying Tension to an Interpolating Spline Curve.* Comput. Vision Graph. Image Process. 1 (1984), pp. 1–18.

[105] F. L. Bauer, C. F. Fike: *Norms and Exclusion Theorems.* Numer. Math. 2 (1960), pp. 123–144.

[106] F. L. Bauer, H. Rutishauser, E. Stiefel: *New Aspects in Numerical Quadrature.* Proceedings of Symposia in Applied Mathematics, Amer. Math. Soc. 15 (1963), pp. 199–219.

[107] R. K. Beatson: *On the Convergence of Some Cubic Spline Interpolation Schemes.* SIAM J. Numer. Anal. 23 (1986), pp. 903–912.

[108] M. Beckers, R. Cools: *A Relation between Cubature Formulae of Trigonometric Degree and Lattice Rules.* Report TW 181, Department of Computer Science, Katholieke Universiteit Leuven, 1992.

[109] M. Beckers, A. Haegemans: *Transformation of Integrands for Lattice Rules,* in "Numerical Integration—Recent Developments, Software and Applications" (T. O. Espelid, A. Genz, Eds.). Kluwer, Dordrecht 1992, pp. 329–340.

[110] J. Berntsen, T. O. Espelid: DCUTRI—*An Algorithm for Adaptive Cubature over a Collection of Triangles.* ACM Trans. Math. Softw. 18 (1992), pp. 329–342.

[111] J. Berntsen, T. O. Espelid, A. Genz: *An Adaptive Algorithm for the Approximate Calculation of Multiple Integrals.* ACM Trans. Math. Softw. 17 (1991), pp. 437–451.

[112] S. Bershader, T. Kraay, J. Holland: *The Giant Fourier Transform,* in "Scientific Applications of the Connection Machine" (H. D. Simon, Ed.). World Scientific, Singapore New Jersey London Hong Kong 1989.

[113] C. Bischof: LAPACK—*Portable lineare Algebra-Software für Supercomputer.* Informationstechnik 34 (1992), pp. 44–49.

[114] C. Bischof, P. T. P. Tang: *Generalized Incremental Condition Estimation.* Technical Report CS-91-132, Computer Science Dept., University of Tennessee, 1991.

[115] C. Bischof, P. T. P. Tang: *Robust Incremental Condition Estimation.* Technical Report CS-91-133, Computer Science Dept., University of Tennessee, 1991.

[116] G. E. Blelloch: *Vector Models for Data-Parallel Computing.* MIT Press, Cambridge London 1990.

[117] J. L. Blue: *A Portable Fortran Program to Find the Euclidean Norm.* ACM Trans. Math. Softw. 4 (1978), pp. 15–23.

[118] P. T. Boggs, R. H. Byrd and R. B. Schnabel: *A Stable and Efficient Algorithm for Non-linear Orthogonal Distance Regression.* SIAM J. Sci. Stat. Comput. 8 (1987), pp. 1052–1078.

[119] R. F. Boisvert: *A Fourth-Order-Accurate Fourier Method for the Helmholtz Equation in Three Dimensions.* ACM Trans. Math. Softw. 13 (1987), pp. 221–234.

[120] R. F. Boisvert, S. E. Howe, D. K. Kahaner: GAMS—*A Framework for the Management of Scientific Software.* ACM Trans. Math. Softw. 11 (1985), pp. 313–356.

[121] P. Bolzern, G. Fronza, E. Runca, C. W. Ueberhuber: *Statistical Analysis of Winter Sulphur Dioxide Concentration Data in Vienna.* Atmospheric Environment 16 (1982), pp. 1899–1906.

[122] M. Bourdeau, A. Pitre: *Tables of Good Lattices in Four and Five Dimensions.* Numer. Math. 47 (1985), pp. 39–43.

[123] H. Braß: *Quadraturverfahren.* Vandenhoeck und Ruprecht, Göttingen 1977.

[124] R. P. Brent: *An Algorithm with Guaranteed Convergence for Finding a Zero of a Function.* Computer J. 14 (1971), pp. 422–425.

[125] R. P. Brent: *A Fortran Multiple-Precision Arithmetic Package.* ACM Trans. Math. Softw. 4 (1978), pp. 57–70.

[126] R. P. Brent: *Algorithm 524—A Fortran Multiple-Precision Arithmetic Package.* ACM Trans. Math. Softw. 4 (1978), pp. 71–81.

[127] K. W. Brodlie: *Methods for Drawing Curves,* in "Fundamental Algorithms for Computer Graphics" (R. A. Earnshaw, Ed.). Springer-Verlag, Berlin Heidelberg New York Tokyo 1985, pp. 303–323.

[128] M. Bronstein: *Integration of Elementary Functions.* J. Symbolic Computation 9 (1990), pp. 117–173.

[129] C. G. Broyden: *A Class of Methods for Solving Nonlinear Simultaneous Equations.* Math. Comp. 19 (1965), pp. 577–593.

[130] J. C. P. Bus, T. J. Dekker: *Two Efficient Algorithms with Guaranteed Convergence for Finding a Zero of a Function.* ACM Trans. Math. Softw. 1 (1975), pp. 330–345.

[131] K. R. Butterfield: *The Computation of all Derivatives of a B-Spline Basis.* J. Inst. Math. Appl. 17 (1976), pp. 15–25.

[132] P. L. Butzer, R. L. Stens: *Sampling Theory for Not Necessarily Band-Limited Functions—A Historical Overview.* SIAM Review 34 (1992), pp. 40–53.

[133] G. D. Byrne, C. A. Hall (Eds.): *Numerical Solution of Systems of Nonlinear Algebraic Equations.* Academic Press, New York London 1973.

[134] S. Cambanis, E. Masry: *Trapezoidal Stratified Monte Carlo Integration.* SIAM J. Numer. Anal. 29 (1992), pp. 284–301.

[135] R. Carter: *Y-MP Floating-Point and Cholesky Factorization.* International Journal of High Speed Computing 3 (1991), pp. 215–222.

[136] J. Choi, J. J. Dongarra, D. W. Walker: *A Set of Parallel Block Basic Linear Algebra Subprograms.* Technical Report TM-12468, Mathematical Sciences Section, Oak Ridge National Laboratory, 1994.

[137] J. Choi, J. J. Dongarra, D. W. Walker: SCALAPACK I—*Parallel Factorization Routines (LU, QR, and Cholesky).* Technical Report TM-12470, Oak Ridge National Laboratory, Mathematical Sciences Section, 1994.

[138] W. J. Cody: *The* FUNPACK *Package of Special Function Subroutines.* ACM Trans. Math. Softw. 1 (1975), pp. 13–25.

[139] J. W. Cooley, J. W. Tukey: *An Algorithm for the Machine Calculation of Complex Fourier Series.* Math. Comp. 19 (1965), pp. 297–301.

[140] R. Cools: *A Survey of Methods for Constructing Cubature Formulae,* in "Numerical Integration—Recent Developments, Software and Applications" (T. O. Espelid, A. Genz, Eds.). Kluwer, Dordrecht 1992, pp. 1–24.

[141] R. Cools, P. Rabinowitz: *Monomial Cubature Rules Since "Stroud"—A Compilation.* Report TW 161, Department of Computer Science, Katholieke Universiteit Leuven, 1991.

[142] W. A. Coppel: *Stability and Asymptotic Behavior of Differential Equations.* Heath, Boston, 1965.

[143] P. Costantini: *Co-monotone Interpolating Splines of Arbitrary Degree—a Local Approach.* SIAM J. Sci. Stat. Comp. 8 (1987), pp. 1026–1034.

[144] W. R. Cowell (Ed.): *Portability of Mathematical Software.* Lecture Notes in Computer Science, Vol. 57, Springer-Verlag, New York 1977.

[145] M. G. Cox: *The Numerical Evaluation of B-Splines.* J. Inst. Math. Appl. 10 (1972), pp. 134–149.

[146] J. H. Davenport: *On the Integration of Algebraic Functions.* Lecture Notes in Computer Science, Vol. 102, Springer-Verlag, Berlin Heidelberg New York Tokyo 1981.

[147] J. H. Davenport: *Integration—Formal and Numeric Approaches,* in "Tools, Methods and Languages for Scientific and Engineering Computation" (B. Ford, J. C. Rault, F. Thomasset, Eds.). North-Holland, Amsterdam New York Oxford 1984, pp. 417–426.

[148] J. H. Davenport, Y. Siret, E. Tournier: *Computer Algebra—Systems and Algorithms for Algebraic Computation,* 2nd ed. Academic Press, New York 1993.

[149] T. A. Davis, I. S. Duff: *An Unsymmetric-Pattern Multifrontal Method for Sparse LU Factorization.* Technical Report TR-94-038, Computer and Information Science Dept., University of Florida, 1994.

[150] C. de Boor: CADRE—*An Algorithm for Numerical Quadrature,* in "Mathematical Software" (J. R. Rice, Ed.). Academic Press, New York 1971, pp. 417–449.

[151] C. de Boor: *On Calculating with B-Splines.* J. Approx. Theory 6 (1972), pp. 50–62.

[152] C. de Boor, A. Pinkus: *Proof of the Conjecture of Bernstein and Erdös concerning the Optimal Nodes for Polynomial Interpolation.* J. Approx. Theory 24 (1978), pp. 289–303.

[153] E. de Doncker: *Asymptotic Expansions and Their Application in Numerical Integration,* in "Numerical Integration—Recent Developments, Software and Applications" (P. Keast, G. Fairweather, Eds.). Reidel, Dordrecht 1987, pp. 141–151.

[154] T. J. Dekker: *A Floating-point Technique for Extending the Available Precision.* Numer. Math. 18 (1971), pp. 224–242.

[155] J. Demmel, B. Kågström: *Computing Stable Eigendecompositions of Matrix Pencils.* Lin. Alg. Appl. 88/89-4 (1987), pp. 139–186.

[156] J. E. Dennis Jr., J. J. Moré: *Quasi-Newton Methods, Motivation and Theory.* SIAM Review 19 (1977), pp. 46–89.

[157] R. A. De Vore, G. G. Lorentz: *Constructive Approximation.* Springer-Verlag, Berlin Heidelberg, New York Tokyo 1993.

[158] L. Devroye: *Non-Uniform Random Variate Generation.* Springer-Verlag, Berlin Heidelberg New York Tokyo 1986.

[159] D. S. Dodson, R. G. Grimes, J. G. Lewis: *Sparse Extensions to the Fortran Basic Linear Algebra Subprograms.* ACM Trans. Math. Softw. 17 (1991), pp. 253–263, 264–272.

[160] J. J. Dongarra: *The* LINPACK *Benchmark—An Explanation,* in "Evaluating Supercomputers" (A. J. Van der Steen, Ed.). Chapman and Hall, London 1990, pp. 1–21.

[161] J. J. Dongarra, J. Du Croz, I. S. Duff, S. Hammarling: *A Set of Level 3 Basic Linear Algebra Subprograms.* ACM Trans. Math. Softw. 16 (1990), pp. 1–17, 18–28.

[162] J. J. Dongarra: *Performance of Various Computers Using Standard Linear Equations Software.* Technical Report CS-89-85, Computer Science Dept., University of Tennessee, 1994.

[163] J. J. Dongarra, J. Du Croz, S. Hammarling, R. J. Hanson: *An Extended Set of Fortran Basic Linear Algebra Subprograms.* ACM Trans. Math. Softw. 14 (1988), pp. 1–17, 18–32.

[164] J. J. Dongarra, I. S. Duff, D. C. Sorensen, H. A. Van der Vorst: *Solving Linear Systems on Vector and Shared Memory Computers.* SIAM Press, Philadelphia 1991.

[165] J. J. Dongarra, E. Grosse: *Distribution of Mathematical Software via Electronic Mail.* Comm. ACM 30 (1987), pp. 403–407.

[166] J. J. Dongarra, F. G. Gustavson, A. Karp: *Implementing Linear Algebra Algorithms for Dense Matrices on a Vector Pipeline Machine.* SIAM Review 26 (1984), pp. 91–112.

[167] J. J. Dongarra, P. Mayes, G. Radicati: *The IBM RISC System/6000 and Linear Algebra Operations.* Technical Report CS-90-12, Computer Science Dept., University of Tennessee, 1990.

[168] J. J. Dongarra, R. Pozo, D. W. Walker: LAPACK++ *V. 1.0—Users' Guide.* University of Tennessee, Knoxville, 1994.

[169] J. J. Dongarra, R. Pozo, D. W. Walker: LAPACK++—*A Design Overview of Object-Oriented Extensions for High Performance Linear Algebra.* Computer Science Report, University of Tennessee, 1993.

[170] J. J. Dongarra, H. A. Van der Vorst: *Performance of Various Computers Using Standard Sparse Linear Equations Solving Techniques,* in "Computer Benchmarks" (J. J. Dongarra, W. Gentzsch, Eds.). Elsevier, New York 1993, pp. 177–188.

[171] C. C. Douglas, M. Heroux, G. Slishman, R. M. Smith: GEMMW—*A Portable Level 3 BLAS Winograd Variant of Strassen's Matrix-Matrix Multiply Algorithm.* J. Computational Physics 110 (1994), pp. 1–10.

[172] Z. Drezner: *Computation of the Multivariate Normal Integral.* ACM Trans. Math. Softw. 18 (1992), pp. 470–480.

[173] D. Dubois, A. Greenbaum, G. Rodrigue: *Approximating the Inverse of a Matrix for Use in Iterative Algorithms on Vector Processors.* Computing 22 (1979), pp. 257–268.

[174] I. S. Duff, A. Erisman, J. Reid: *Direct Methods for Sparse Matrices,* paperback ed. Oxford University Press, Oxford 1989.

[175] I. S. Duff, R. G. Grimes, J. G. Lewis: *Sparse Matrix Test Problems.* ACM Trans. Math. Softw. 15 (1989), pp. 1–14.

[176] I. S. Duff, R. G. Grimes, J. G. Lewis: *User's Guide for the Harwell-Boeing Sparse Matrix Collection* (Release I). CERFACS-Report TR/PA/92/86, Toulouse, 1992. Available via anonymous-FTP: orion.cerfacs.fr.

[177] R. A. Earnshaw (Ed.): *Fundamental Algorithms for Computer Graphics.* Springer-Verlag, Berlin Heidelberg New York Tokyo 1985 (reprinted 1991).

[178] H. Ekblom: L_p-*Methods for Robust Regression.* BIT 14 (1974), pp. 22–32.

[179] D. F. Elliot, K. R. Rao: *Fast Transforms:—Algorithms, Analyses, Applications.* Academic Press, New York 1982.

[180] T. M. R. Ellis, D. H. McLain: *Algorithm 514—A New Method of Cubic Curve Fitting Using Local Data.* ACM Trans. Math. Softw. 3 (1977), pp. 175–178.

[181] M. P. Epstein: *On the Influence of Parameterization in Parametric Interpolation.* SIAM J. Numer. Anal. 13 (1976), pp. 261–268.

[182] P. Erdős: *Problems and Results on the Theory of Interpolation.* Acta Math. Acad. Sci. Hungar., 12 (1961), pp. 235–244.

[183] P. Erdős, P. Vértesi: *On the Almost Everywhere Divergence of Lagrange Interpolatory Polynomials for Arbitrary Systems of Nodes.* Acta Math. Acad. Sci. Hungar. 36 (1980), pp. 71–89.

[184] T. O. Espelid: DQAINT—*An Algorithm for Adaptive Quadrature (of a Vector Function) over a Collection of Finite Intervals,* in "Numerical Integration—Recent Developments, Software and Applications" (T. O. Espelid, A. Genz, Eds.). Kluwer, Dordrecht 1992, pp. 341–342.

[185] G. Farin: *Splines in CAD/CAM.* Surveys on Mathematics for Industry 1 (1991), pp. 39–73.

[186] H. Faure: *Discrépances de suites associées à un système de numération (en dimension s)*. Acta Arith. 41 (1982), pp. 337–351.

[187] L. Fejér: *Mechanische Quadraturen mit positiven Cotes'schen Zahlen*. Math. Z. 37 (1933), pp. 287–310.

[188] S. I. Feldman, D. M. Gay, M. W. Maimone, N. L. Schryer: *A Fortran-to-C Converter*. Technical Report No. 149, AT&T Bell Laboratories, 1993.

[189] A. Ferscha: *Modellierung und Leistungsanalyse paralleler Systeme mit dem PRM-Netz Modell*. Oldenburg Verlag, München Wien 1995.

[190] R. Fletcher, J. A. Grant, M. D. Hebden: *The Calculation of Linear Best L_p-Approximations*. Computer J. 14 (1971), pp. 276–279.

[191] R. Fletcher, C. Reeves: *Function Minimization by Conjugate Gradients*. Computer Journal 7 (1964), pp. 149–154.

[192] T. A. Foley: *Interpolation with Interval and Point Tension Controls Using Cubic Weighted ν-Splines*. ACM Trans. Math. Softw. 13 (1987), pp. 68–96.

[193] B. Ford, F. Chatelin (Eds.): *Problem Solving Environments for Scientific Computing*. North-Holland, Amsterdam 1987.

[194] L. Fox, I. B. Parker: *Chebyshev Polynomials in Numerical Analysis*. Oxford University Press, London 1968.

[195] R. Frank, J. Schneid, C. W. Ueberhuber: *The Concept of B-Convergence*. SIAM J. Numer. Anal. 18 (1981), pp. 753–780.

[196] R. Frank, J. Schneid, C. W. Ueberhuber: *Stability Properties of Implicit Runge-Kutta Methods*. SIAM J. Numer. Anal. 22 (1985), pp. 497–515.

[197] R. Frank, J. Schneid, C. W. Ueberhuber: *Order Results for Implicit Runge-Kutta Methods Applied to Stiff Systems*. SIAM J. Numer. Anal. 22 (1985), pp. 515–534.

[198] R. Franke, G. Nielson: *Smooth Interpolation of Large Sets of Scattered Data*. Int. J. Numer. Methods Eng. 15 (1980), pp. 1691–1704.

[199] R. Freund, G. H. Golub, N. Nachtigal: *Iterative Solution of Linear Systems*. Acta Numerica 1, 1992, pp. 57–100.

[200] R. Freund, N. Nachtigal: *QMR—A Quasi-Minimal Residual Method for Non-Hermitian Linear Systems*. Numer. Math. 60 (1991), pp. 315–339.

[201] R. Freund, N. Nachtigal: *An Implementation of the QMR Method Based on Two Coupled Two-Term Recurrences*. Tech. Report 92.15, RIACS, NASA Ames, 1992.

[202] F. N. Fritsch, J. Butland: *A Method for Constructing Local Monotone Piecewise Cubic Interpolants*. SIAM J. Sci. Stat. Comp. 5 (1984), pp. 300–304.

[203] F. N. Fritsch, R. E. Carlson: *Monotone Piecewise Cubic Interpolation*. SIAM J. Numer. Anal. 17 (1980), pp. 238–246.

[204] F. N. Fritsch, D. K. Kahaner, J. N. Lyness: *Double Integration Using One-Dimensional Adaptive Quadrature Routines—a Software Interface Problem*. ACM Trans. Math. Softw. 7 (1981), pp. 46–75.

[205] P. W. Gaffney, J. W. Wooten, K. A. Kessel, W. R. McKinney: NITPACK—*An Interactive Tree Package*. ACM Trans. Math. Softw. 9 (1983), pp. 395–417.

[206] E. Gallopoulos, E. N. Houstis, J. R. Rice: *Problem Solving Environments for Computational Science*. Computational Science and Engineering Nr. 2 Vol. 1 (1994), pp. 11–23.

[207] K. O. Geddes: *Algorithms for Computer Algebra*. Kluwer, Dordrecht 1992.

[208] J. D. Gee, M. D. Hill, D. Pnevmatikatos, A. J. Smith: *Cache Performance of the SPEC92 Benchmark Suite*. IEEE Micro 13 (1993), pp. 17–27.

[209] W. M. Gentleman: *Implementing Clenshaw-Curtis Quadrature*. Comm. ACM 15 (1972), pp. 337–342, 343–346.

[210] A. Genz: *Statistics Applications of Subregion Adaptive Multiple Numerical Integration*, in "Numerical Integration—Recent Developments, Software and Applications" (T. O. Espelid, A. Genz, Eds.). Kluwer, Dordrecht 1992, pp. 267–280.

[211] D. Goldberg: *What Every Computer Scientist Should Know About Floating-Point Arithmetic.* ACM Computing Surveys 23 (1991), pp. 5–48.

[212] G. H. Golub, V. Pereyra: *Differentiation of Pseudo-Inverses and Nonlinear Least Squares Problems Whose Variables Separate.* SIAM J. Numer. Anal. 10 (1973), pp. 413–432.

[213] A. Greenbaum, J. J. Dongarra: *Experiments with QL/QR Methods for the Symmetric Tridiagonal Eigenproblem.* Technical Report CS-89-92, Computer Science Dept., University of Tennessee, 1989.

[214] E. Grosse: *A Catalogue of Algorithms for Approximation,* in "Algorithms for Approximation II" (J. C. Mason, M. G. Cox, Eds.). Chapman and Hall, London New York 1990, pp. 479–514.

[215] M. H. Gutknecht: *Variants of Bi-CGSTAB for Matrices with Complex Spectrum.* Tech. Report 91-14, IPS ETH, Zürich 1991.

[216] S. Haber: *A Modified Monte Carlo Quadrature.* Math. Comp. 20 (1966), pp. 361–368.

[217] S. Haber: *A Modified Monte Carlo Quadrature II.* Math. Comp. 21 (1967), pp. 388–397.

[218] H. Hancock: *Elliptic Integrals.* Dover Publication, New York 1917.

[219] J. Handy: *The Cache Memory Book.* Academic Press, San Diego 1993.

[220] J. G. Hayes: *The Optimal Hull Form Parameters.* Proc. NATO Seminar on Numerical Methods Applied to Ship Building, Oslo 1964.

[221] J. G. Hayes: *Numerical Approximation to Functions and Data.* Athlone Press, London 1970.

[222] N. Higham: *Efficient Algorithms for Computing the Condition Number of a Tridiagonal Matrix.* SIAM J. Sci. Stat. Comput. 7 (1986), pp. 82–109.

[223] N. Higham: *A Survey of Condition Number Estimates for Triangular Matrices.* SIAM Review 29 (1987), pp. 575–596.

[224] N. Higham: *Fortran 77 Codes for Estimating the One-Norm of a Real or Complex Matrix, with Applications to Condition Estimation.* ACM Trans. Math. Softw. 14 (1988), pp. 381–396.

[225] N. Higham: *The Accuracy of Floating-Point Summation.* SIAM J. Sci. Comput. 14 (1993), pp. 783-799.

[226] D. R. Hill, C. B. Moler: *Experiments in Computational Matrix Algebra.* Birkhäuser, Basel 1988.

[227] E. Hlawka: *Funktionen von beschränkter Variation in der Theorie der Gleichverteilung.* Ann. Math. Pur. Appl. 54 (1961), pp. 325–333.

[228] R. W. Hockney, C. R. Jesshope: *Parallel Computers 2.* Adam Hilger, Bristol 1988.

[229] A. S. Householder: *The Numerical Treatment of a Single Nonlinear Equation.* McGraw-Hill, New York 1970.

[230] E. N. Houstis, J. R. Rice, T. Papatheodorou: PARALLEL ELLPACK—*An Expert System for Parallel Processing of Partial Differential Equations.* Purdue University, Report CSD-TR-831, 1988.

[231] E. N. Houstis, J. R. Rice, R. Vichnevetsky (Eds.): *Intelligent Mathematical Software Systems.* North-Holland, Amsterdam 1990.

[232] L. K. Hua, Y. Wang: *Applications of Number Theory to Numerical Analysis.* Springer-Verlag, Berlin Heidelberg New York Tokyo 1981.

[233] P. J. Huber: *Robust Regression—Asymptotics, Conjectures and Monte Carlo.* Annals of Statistics 1 (1973), pp. 799–821.

[234] P. J. Huber: *Robust Statistics.* Wiley, New York 1981.

[235] J. M. Hyman: *Accurate Monotonicity Preserving Cubic Interpolation.* SIAM J. on Scientific and Statistical Computation 4 (1983), pp. 645–654.

[236] J. P. Imhof: *On the Method for Numerical Integration of Clenshaw and Curtis.* Numer. Math. 5 (1963), pp. 138–141.

[237] M. Iri, S. Moriguti, Y. Takasawa: *On a Certain Quadrature Formula* (japan.), Kokyuroku of the Research Institute for Mathematical Sciences, Kyoto University, 91 (1970), pp. 82–118.

[238] L. D. Irvine, S. P. Marin, P. W. Smith: *Constrained Interpolation and Smoothing*. Constructive Approximation 2 (1986) pp. 129–151.

[239] ISO/IEC DIS 10967-1 : 1994: *International Standard—Information Technology—Language Independent Arithmetic—Part 1—Integer and Floating Point Arithmetic*. 1994.

[240] R. Jain: *The Art of Computer Systems Performance Analysis—Techniques for Experimental Design, Measurement and Simulation*. Wiley, New York 1991.

[241] M. A. Jenkins: *Algorithm 493—Zeroes of a Real Polynomial*. ACM Trans. Math. Softw. 1 (1975), pp. 178–189.

[242] M. A. Jenkins, J. F. Traub: *A Three-Stage Algorithm for Real Polynomials Using Quadratic Iteration*. SIAM J. Numer. Anal. 7 (1970), pp. 545–566.

[243] A. J. Jerri: *The Shannon Sampling—its Various Extensions and Applications—a Tutorial Review*. Proc. IEEE 65 (1977), pp. 1565–1596.

[244] S. Joe, I. H. Sloan: *Imbedded Lattice Rules for Multidimensional Integration*. SIAM J. Numer. Anal. 29 (1992), pp. 1119–1135.

[245] D. S. Johnson, M. R. Garey: *A 71/60 Theorem for Bin Packing*. J. Complexity 1 (1985), pp. 65–106.

[246] D. W. Juedes: *A Taxonomy of Automatic Differentiation Tools*, in "Automatic Differentiation of Algorithms—Theory, Implementation and Application" (A. Griewank, F. Corliss, Eds.). SIAM Press, Philadelphia 1991, pp. 315–329.

[247] D. K. Kahaner: *Numerical Quadrature by the ε-Algorithm*. Math. Comp. 26 (1972), pp. 689–693.

[248] N. Karmarkar, R. M. Karp: *An Efficient Approximation Scheme for the One Dimensional Bin Packing Problem*. 23rd Annu. Symp. Found. Comput. Sci., IEEE Computer Society, 1982, pp. 312–320.

[249] L. Kaufmann: *A Variable Projection Method for Solving Separable Nonlinear Least Squares Problems*. BIT 15 (1975), pp. 49–57.

[250] G. Kedem, S. K. Zaremba: *A Table of Good Lattice Points in Three Dimensions*. Numer. Math. 23 (1974), pp. 175–180.

[251] H. L. Keng, W. Yuan: *Applications of Number Theory to Numerical Analysis*. Springer-Verlag, Berlin Heidelberg New York Tokyo 1981.

[252] T. King: *Dynamic Data Structures—Theory and Application*. Academic Press, San Diego 1992.

[253] M. Klerer, F. Grossman: *Error Rates in Tables of Indefinite Intergrals*. Indust. Math. 18 (1968), pp. 31–62.

[254] D. E. Knuth: *The Art of Computer Programming*. Vol. 2—*Seminumerical Algorithms*, 2nd ed. Addison-Wesley, Reading 1981.

[255] P. Kogge: *The Architecture of Pipelined Computers*. McGraw-Hill, New York 1981.

[256] A. R. Krommer, C. W. Ueberhuber: *Architecture Adaptive Algorithms*. Parallel Computing 19 (1993), pp. 409–435.

[257] A. R. Krommer, C. W. Ueberhuber: *Lattice Rules for High-Dimensional Integration*. Technical Report SciPaC/TR 93-3, Scientific Parallel Computation Group, Technical University Vienna, Wien 1993.

[258] A. R. Krommer, C. W. Ueberhuber: *Numerical Integration on Advanced Computer Systems*. Lecture Notes in Computer Science, Vol. 848, Springer-Verlag, Berlin Heidelberg New York Tokyo 1994.

[259] A. S. Kronrod: *Nodes and Weights of Quadrature Formulas*. Consultants Bureau, New York 1965.

[260] V. I. Krylov: *Approximate Calculation of Integrals*. Macmillan, New York London 1962.

[261] U. W. Kulisch, W. L. Miranker: *The Arithmetic of the Digital Computer—A New Approach.* SIAM Review 28 (1986), pp. 1–40.

[262] U. W. Kulisch, W. L. Miranker: *Computer Arithmetic in Theory and Practice.* Academic Press, New York 1981.

[263] J. Laderman, V. Pan, X.-H. Sha: *On Practical Acceleration of Matrix Multiplication.* Linear Algebra Appl. 162–164 (1992), pp. 557–588.

[264] M. S. Lam, E. E. Rothberg, M. E. Wolf: *The Cache Performance and Optimizations of Blocked Algorithms.* Computer Architecture News 21 (1993), pp. 63–74.

[265] C. Lanczos: *Discourse on Fourier Series.* Oliver and Boyd, Edinburgh London 1966.

[266] C. L. Lawson, R. J. Hanson, D. Kincaid, F. T. Krogh: *Basic Linear Algebra Subprograms for Fortran Usage.* ACM Trans. Math. Softw. 5 (1979), pp. 308–323.

[267] A. R. Lebeck, D. A. Wood: *Cache Profiling and the SPEC Benchmarks—A Case Study.* IEEE Computer, October 1994, pp. 15–26.

[268] P. Ling: *A Set of High Performance Level 3 BLAS Structured and Tuned for the IBM 3090 VF and Implemented in Fortran 77.* Journal of Supercomputing 7 (1993), pp. 323–355.

[269] P. R. Lipow, F. Stenger: *How Slowly Can Quadrature Formulas Converge.* Math. Comp. 26 (1972), pp. 917–922.

[270] D. B. Loveman: *High Performance Fortran.* IEEE Parallel and Distributed Technology 2 (1993), pp. 25–42.

[271] J. Lund, K. L. Bowers: *Sinc Methods for Quadrature and Differential Equations.* SIAM Press, Philadelphia 1992.

[272] T. Lyche: *Discrete Cubic Spline Interpolation.* BIT 16 (1976), pp. 281–290.

[273] J. N. Lyness: *An Introduction to Lattice Rules and their Generator Matrices.* IMA J. Numer. Anal. 9 (1989), pp. 405–419.

[274] J. N. Lyness, J. J. Kaganove: *Comments on the Nature of Automatic Quadrature Routines.* ACM Trans. Math. Softw. 2 (1976), pp. 65–81.

[275] J. N. Lyness, B. W. Ninham: *Numerical Quadrature and Asymptotic Expansions.* Math. Comp. 21 (1967), pp. 162–178.

[276] J. N. Lyness, I. H. Sloan: *Some Properties of Rank-2 Lattice Rules.* Math. Comp. 53 (1989), pp. 627–637.

[277] J. N. Lyness, T. Soerevik: *A Search Program for Finding Optimal Integration Lattices.* Computing 47 (1991), pp. 103–120.

[278] J. N. Lyness, T. Soerevik: *An Algorithm for Finding Optimal Integration Lattices of Composite Order.* BIT 32 (1992), pp. 665–675.

[279] T. Macdonald: *C for Numerical Computing.* J. Supercomput. 5 (1991), pp. 31–48.

[280] D. Maisonneuve: *Recherche et utilisation des "bons treillis",* in "Applications of Number Theory to Numerical Analysis" (S. K. Zaremba, Ed.). Academic Press, New York 1972, pp. 121–201.

[281] M. Malcolm, R. Simpson: *Local Versus Global Strategies for Adaptive Quadrature.* ACM Trans. Math. Softw. 1 (1975), pp. 129–146.

[282] T. Manteuffel: *The Tchebychev Iteration for Nonsymmetric Linear Systems.* Numer. Math. 28 (1977), pp. 307–327.

[283] D. W. Marquardt: *An Algorithm for Least Squares Estimation of Nonlinear Parameters.* J. SIAM 11 (1963), pp. 431–441.

[284] G. Marsaglia: *Normal (Gaussian) Random Variables for Supercomputers.* J. Supercomput. 5 (1991), pp. 49–55.

[285] J. C. Mason, M. G. Cox: *Scientific Software Systems.* Chapman and Hall, London New York 1990.

[286] E. Masry, S. Cambanis: *Trapezoidal Monte Carlo Integration.* SIAM J. Numer. Anal. 27 (1990), pp. 225–246.

[287] E. W. Mayr: *Theoretical Aspects of Parallel Computation*, in "VLSI and Parallel Computation" (R. Suaya, G. Birtwistle, Eds.). Morgan Kaufmann, San Mateo 1990, pp. 85–139.

[288] G. P. McKeown: *Iterated Interpolation Using a Systolic Array*. ACM Trans. Math. Softw. 12 (1986), pp. 162–170.

[289] J. Meijerink, H. A. Van der Vorst: *An Iterative Solution Method for Linear Systems of Which the Coefficient Matrix is a Symmetric M-matrix*. Math. Comp. 31 (1977), pp. 148–162.

[290] R. Melhem: *Toward Efficient Implementation of Preconditioned Conjugate Gradient Methods on Vector Supercomputers*. Internat. J. Supercomp. Appl. 1 (1987), pp. 77–98.

[291] J. P. Mesirov (Ed.): *Very Large Scale Computation in the 21st Century*. SIAM Press, Philadelphia 1991.

[292] W. F. Mitchell: *Optimal Multilevel Iterative Methods for Adaptive Grids*. SIAM J. Sci. Statist. Comput. 13 (1992), pp. 146–167.

[293] J. J. Moré, M. Y. Cosnard: *Numerical Solution of Nonlinear Equations*. ACM Trans. Math. Softw. 5 (1979), pp. 64–85.

[294] D. E. Müller: *A Method for Solving Algebraic Equations Using an Automatic Computer*. Math. Tables Aids Comput. 10 (1956), pp. 208–215.

[295] N. Nachtigal, S. Reddy, L. Trefethen: *How Fast are Nonsymmetric Matrix Iterations?* SIAM J. Mat. Anal. Appl. 13 (1992), pp. 778–795.

[296] P. Naur: *Machine Dependent Programming in Common Languages*. BIT 7 (1967), pp. 123–131.

[297] J. A. Nelder, R. Mead: *A Simplex Method for Function Minimization*. Computer Journal 7 (1965), pp. 308–313.

[298] H. Niederreiter: *Quasi-Monte Carlo Methods and Pseudorandom Numbers*. Bull. Amer. Math. Soc. 84 (1978), pp. 957–1041.

[299] H. Niederreiter: *Random Number Generation and Quasi-Monte Carlo Methods*. SIAM Press, Philadelphia 1992.

[300] G. Nielson: *Some Piecewise Polynomial Alternatives to Splines Under Tension*, in "Computer Aided Geometric Design" (R. E. Barnhill, R. F. Riesenfeld, Eds.). Academic Press, New York San Francisco London 1974.

[301] G. Nielson, B. D. Shriver: *Visualization in Scientific Computing*. IEEE Press, Los Alamitos 1990.

[302] H. J. Nussbaumer: *Fast Fourier Transform and Convolution Algorithms*. Springer-Verlag, Berlin Heidelberg New York Tokyo 1981 (reprinted 1990).

[303] D. P. O'Leary, O. Widlund: *Capacitance Matrix Methods for the Helmholtz Equation on General 3-Dimensional Regions*. Math. Comp. 33 (1979), pp. 849–880.

[304] T. I. Ören: *Concepts for Advanced Computer Assisted Modeling*, in "Methodology in Systems Modeling and Simulation" (B. P. Zeigler, M. S. Elzas, G. J. Klir, T. I. Ören, Eds.). North-Holland, Amsterdam New York Oxford 1979.

[305] J. M. Ortega: *Numerical Analysis—A Second Course*. SIAM Press, Philadelphia 1990.

[306] A. M. Ostrowski: *On Two Problems in Abstract Algebra Connected with Horner's Rule*. Studies in Math. and Mech. presented to Richard von Mises, Academic Press, New York 1954, pp. 40–68.

[307] C. C. Page, M. A. Saunders: *LSQR: An Algorithm for Sparse Linear Equations and Sparse Least-Squares*. ACM Trans. Math. Software 8 (1982), pp. 43–71.

[308] V. Pan: *Methods of Computing Values of Polynomials*. Russian Math. Surveys 21 (1966), pp. 105–136.

[309] V. Pan: *How Can We Speed Up Matrix Multiplication?* SIAM Rev. 26 (1984), pp. 393–415.

[310] V. Pan: *Complexity of Computations with Matrices and Polynomials*. SIAM Rev. 34 (1992), pp. 225–262.

[311] B. N. Parlett: *The Symmetric Eigenvalue Problem*. Prentice Hall, Englewood Cliffs 1980.

[312] T. N. L. Patterson: *The Optimum Addition of Points to Quadrature Formulae*. Math. Comp. 22 (1968), pp. 847–856.

[313] J. L. Peterson: *Petri Net Theory and the Modeling of Systems*. Prentice Hall, Englewood Cliffs 1981.

[314] R. Piessens: *Modified Clenshaw-Curtis Integration and Applications to Numerical Computation of Integral Transforms*, in "Numerical Integration—Recent Developments, Software and Applications" (P. Keast, G. Fairweather, Eds.). Reidel, Dordrecht 1987, pp. 35–41.

[315] R. Piessens, M. Branders: *A Note on the Optimal Addition of Abscissas to Quadrature Formulas of Gauss and Lobatto Type*. Math. Comp. 28 (1974), pp. 135–140, 344–347.

[316] D. R. Powell, J. R. Macdonald: *A Rapidly Converging Iterative Method for the Solution of the Generalized Nonlinear Least Squares Problem*. Computer J. 15 (1972), pp. 148–155.

[317] M. J. D. Powell: *A Hybrid Method for Nonlinear Equations*, in "Numerical Methods for Nonlinear Algebraic Equations" (P. Rabinowitz, Ed.). Gordon and Breach, London 1970.

[318] M. J. D. Powell, P. L. Toint: *On the Estimation of Sparse Hessian Matrices*. SIAM J. Numer. Anal. 16 (1979), pp. 1060–1074.

[319] J. G. Proakis, D. G. Manolakis: *Digital Signal Processing*, 3rd ed. Macmillan, New York 1995.

[320] J. S. Quarterman, S. Carl-Mitchell: *The Internet Connection—System Connectivity and Configuration*. Addison-Wesley, Reading 1994.

[321] R. J. Renka: *Multivariate Interpolation of Large Sets of Scattered Data*. ACM Trans. Math. Softw. 14 (1988), pp. 139–148.

[322] R. J. Renka, A. K. Cline: *A Triangle-Based C^1 Interpolation Method*. Rocky Mt. J. Math. 14 (1984), pp. 223–237.

[323] R. F. Reisenfeld: *Homogeneous Coordinates and Projective Planes in Computer Graphics*. IEEE Computer Graphics and Applications 1 (1981), pp. 50–56.

[324] W. C. Rheinboldt: *Numerical Analysis of Parametrized Nonlinear Equations*. Wiley, New York 1986.

[325] J. R. Rice: *Parallel Algorithms for Adaptive Quadrature II—Metalgorithm Correctness*. Acta Informat. 5 (1975), pp. 273–285.

[326] J. R. Rice (Ed.): *Mathematical Aspects of Scientific Software*. Springer-Verlag, Berlin Heidelberg New York Tokyo 1988.

[327] A. Riddle: *Mathematical Power Tools*. IEEE Spectrum Nov. 1994, pp. 35–47.

[328] R. Rivest: *Cryptography* in "Handbook of Theoretical Computer Science" (J. van Leeuwen, Ed.). North Holland, Amsterdam, 1990.

[329] T. J. Rivlin: *The Chebyshev Polynomials*, 2nd ed. Wiley, New York 1990.

[330] Y. Robert: *The Impact of Vector and Parallel Architectures on the Gaussian Elimination Algorithm*. Manchester University Press, New York Brisbane Toronto 1991.

[331] M. Rosenlicht: *Integration in Finite Terms*. Amer. Math. Monthly 79 (1972), pp. 963–972.

[332] A. Ruhe: *Fitting Empirical Data by Positive Sums of Exponentials*. SIAM J. Sci. Stat. Comp. 1 (1980), pp. 481–498.

[333] C. Runge: *Über empirische Funktionen und die Interpolation zwischen äquidistanten Ordinaten*. Z. Math. u. Physik 46 (1901), pp. 224–243.

[334] Y. Saad: *Preconditioning Techniques for Indefinite and Nonsymmetric Linear Systems*. J. Comput. Appl. Math. 24 (1988), pp. 89–105.

[335] Y. Saad: *Krylov Subspace Methods on Supercomputers.* SIAM J. Sci. Statist. Comput. 10 (1989), pp. 1200–1232.

[336] Y. Saad: SPARSKIT—*A Basic Tool Kit for Sparse Matrix Computation.* Tech. Report CSRD TR 1029, CSRD, University of Illinois, Urbana 1990.

[337] Y. Saad, M. Schultz: GMRES—*A Generalized Minimal Residual Algorithm for Solving Nonsymmetric Linear Systems.* SIAM J. Sci. Statist. Comput. 7 (1986), pp. 856–869.

[338] T. W. Sag, G. Szekeres: *Numerical Evaluation of High-Dimensional Integrals.* Math. Comp. 18 (1964), pp. 245–253.

[339] K. Salkauskas, C^1 *Splines for Interpolation of Rapidly Varying Data.* Rocky Mt. J. Math. 14 (1984), pp. 239–250.

[340] R. Salmon, M. Slater: *Computer Graphics—Systems and Concepts.* Addison-Wesley, Wokingham 1987.

[341] W. M. Schmidt: *Irregularities of Distribution.* Acta Arith. 21 (1972), pp. 45–50.

[342] D. G. Schweikert: *An Interpolation Curve Using a Spline in Tension.* J. Math. & Physics 45 (1966), pp. 312–317.

[343] T. I. Seidman, R. J. Korsan: *Endpoint Formulas for Interpolatory Cubic Splines.* Math. Comp. 26 (1972), pp. 897–900.

[344] Z. Sekera: *Vectorization and Parallelization on High Performance Computers.* Computer Physics Communications 73 (1992), pp. 113–138.

[345] S. Selberherr: *Analysis and Simulation of Semiconductor Devices.* Springer-Verlag, Berlin Heidelberg New York Tokyo 1984.

[346] D. Shanks: *Non-linear Transformation of Divergent and Slowly Convergent Sequences.* J. Math. Phys. 34 (1955), pp. 1–42.

[347] A. H. Sherman: *Algorithms for Sparse Gauss Elimination with Partial Pivoting.* ACM Trans. Math. Softw. 4 (1978), pp. 330–338.

[348] L. L. Shumaker: *On Shape Preserving Quadratic Spline Interpolation.* SIAM J. Numer. Anal. 20 (1983), pp. 854–864.

[349] K. Sikorski: *Bisection is Optimal.* Numer. Math. 40 (1982), pp. 111–117.

[350] I. H. Sloan: *Numerical Integration in High Dimensions—The Lattice Rule Approach,* in "Numerical Integration—Recent Developments, Software and Applications" (T. O. Espelid, A. Genz, Eds.). Kluwer, Dordrecht 1992, pp. 55–69.

[351] I. H. Sloan, S. Joe: *Lattice Methods for Multiple Integration.* Clarendon Press, Oxford 1994.

[352] I. H. Sloan, P. J. Kachoyan: *Lattice Methods for Multiple Integration—Theory, Error Analysis and Examples.* SIAM J. Numer. Anal. 24 (1987), pp. 116–128.

[353] D. M. Smith: *A Fortran Package for Floating-Point Multiple-Precision Arithmetic.* ACM Trans. Math. Softw. 17 (1991), pp. 273–283.

[354] B. T. Smith, J. M. Boyle, J. J. Dongarra, B. S. Garbow, Y. Ikebe, V. C. Klema, C. B. Moler: *Matrix Eigensystem Routines—*EISPACK *Guide,* 2nd ed. Springer-Verlag, Berlin Heidelberg New York Tokyo 1976 (reprinted 1988).

[355] I. M. Sobol: *The Distribution of Points in a Cube and the Approximate Evaluation of Integrals.* Zh. Vychisl. Mat. i Math. Fiz. 7 (1967), pp. 784–802.

[356] P. Sonneveld: *CGS, a Fast Lanczos-type Solver for Nonsymmetric Linear Systems.* SIAM J. Sci. Statist. Comput. 10 (1989), pp. 36–52.

[357] D. C. Sorensen: *Newton's Method with a Model Trust Region Modification.* SIAM J. Numer. Anal. 19 (1982), pp. 409–426.

[358] W. Stegmüller: *Unvollständigkeit und Unbeweisbarkeit* (2. Aufl.). Springer Verlag, Berlin Heidelberg New York, Tokyo, 1970.

[359] G. W. Stewart: *On the Sensitivity of the Eigenvalue Problem* $Ax = \lambda Bx$. SIAM J. Num. Anal. 9-4 (1972), pp. 669–686.

[360] G. W. Stewart: *Error and Perturbation Bounds for Subspaces Associated with Certain Eigenvalue Problems.* SIAM Review 15-10 (1973), pp. 727–764.

[361] V. Strassen: *Gaussian Elimination Is not Optimal.* Numer. Math. 13 (1969), pp. 354–356.

[362] A. H. Stroud: *Approximate Calculation of Multiple Integrals.* Prentice-Hall, Englewood Cliffs 1971.

[363] E. E. Swartzlander (Ed.): *Computer Arithmetic—I, II.* IEEE Computer Society Press, Los Alamitos 1990.

[364] G. Tomas, C. W. Ueberhuber: *Visualization of Scientific Parallel Programs.* Lecture Notes in Computer Science, Vol. 771, Springer-Verlag, Berlin Heidelberg New York Tokyo 1994.

[365] J. F. Traub: *Complexity of Approximately Solved Problems.* J. Complexity 1 (1985), pp. 3–10.

[366] J. F. Traub, H. Wozniakowski: *A General Theory of Optimal Algorithms.* Academic Press, New York 1980.

[367] J. F. Traub, H. Wozniakowski: *Information and Computation*, in "Advances in Computers, Vol. 23" (M. C. Yovits, Ed.). Academic Press, New York London 1984, pp. 35–92.

[368] J. F. Traub, H. Wozniakowski: *On the Optimal Solution of Large Linear Systems.* J. Assoc. Comput. Mach. 31 (1984), pp. 545–559.

[369] A. Van der Sluis, H. A. Van der Vorst: *The Rate of Convergence of Conjugate Gradients.* Numer. Math. 48 (1986) pp. 543–560.

[370] H. A. Van der Vorst: *Bi-CGSTAB—A Fast and Smoothly Converging Variant of Bi-CG for the Solution of Nonsymmetric Linear Systems.* SIAM J. Sci. Statist. Comput. 13 (1992), pp. 631–644.

[371] S. Van Huffel, J. Vandewalle: *The Total Least Square Problem—Computational Aspects and Analysis.* SIAM Press, Philadelphia 1991.

[372] C. Van Loan: *Computational Framework for the Fast Fourier Transform.* SIAM Press, Philadelphia 1992.

[373] G. W. Wasilkowski: *Average Case Optimality.* J. Complexity 1 (1985), pp. 107–117.

[374] G. W. Wasilkowski, F. Gao: *On the Power of Adaptive Information for Functions with Singularities.* Math. Comp. 58 (1992), pp. 285–304.

[375] A. B. Watson: *Image Compression Using the Discrete Cosine Transform.* Mathematica Journal 4 (1994), Issue 1, pp. 81–88.

[376] L. T. Watson, S. C. Billups, A. P. Morgan: HOMPACK—*A Suite of Codes for Globally Convergent Homotopy Algorithms.* ACM Trans. Math. Softw. 13 (1987), pp. 281–310.

[377] P.-Å. Wedin: *Perturbation Theory for Pseudo-Inverses.* BIT 13 (1973), pp. 217–232.

[378] R. P. Weicker: *Dhrystone—A Synthetic Systems Programming Benchmark.* Commun. ACM 27-10 (1984), pp. 1013–1030.

[379] S. Weiss, J. E. Smith: *POWER and PowerPC.* Morgan Kaufmann, San Francisco 1994.

[380] R. C. Whaley: *Basic Linear Algebra Communication Subprograms—Analysis and Implementation Across Multiple Parallel Architectures.* LAPACK Working Note 73, Technical Report, University of Tennessee, 1994.

[381] J. H. Wilkinson: *Rounding Errors in Algebraic Processes.* Prentice-Hall, Englewood Cliffs 1963 (reprinted 1994 by Dover Publications).

[382] J. H. Wilkinson: *Kronecker's Canonical Form and the QZ Algorithm.* Lin. Alg. Appl. 28 (1979), pp. 285–303.

[383] H. Wozniakowski: *A Survey of Information-Based Complexity.* J. Complexity 1 (1985), pp. 11–44.

[384] P. Wynn: *On a Device for Computing the $e_m(S_n)$ Transformation.* Mathematical Tables and Aids to Computing 10 (1956), pp. 91–96.

[385] P. Wynn: *On the Convergence and Stability of the Epsilon Algorithm.* SIAM J. Numer. Anal. 3 (1966), pp. 91–122.

[386] F. Ziegler: *Mechanics of Solids and Fluids*, 2nd ed. Springer-Verlag, Berlin Heidelberg New York Tokyo 1994.

[387] A. Zygmund: *Trigonometric Series—I, II*, 2nd ed. Cambridge University Press, Cambridge 1988.

Author Index

Subject Index

Springer
and the
environment

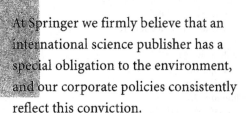

At Springer we firmly believe that an international science publisher has a special obligation to the environment, and our corporate policies consistently reflect this conviction.

We also expect our business partners – paper mills, printers, packaging manufacturers, etc. – to commit themselves to using materials and production processes that do not harm the environment. The paper in this book is made from low- or no-chlorine pulp and is acid free, in conformance with international standards for paper permanency.